Precipitation: Advances in Measurement, Estimation and Prediction

Silas Michaelides

Precipitation: Advances in Measurement, Estimation and Prediction

 Springer

Dr. Silas Michaelides
Meteorological Service
CY-1418 Nicosia
Cyprus

ISBN: 978-3-540-77654-3   e-ISBN: 978-3-540-77655-0

Library of Congress Control Number: 2007943078

© 2008 Springer-Verlag Berlin Heidelberg

This work is subject to copyright. All rights are reserved, whether the whole or part of the material is concerned, specifically the rights of translation, reprinting, reuse of illustrations, recitation, broadcasting, reproduction on microfilm or in any other way, and storage in data banks. Duplication of this publication or parts thereof is permitted only under the provisions of the German Copyright Law of September 9, 1965, in its current version, and permission for use must always be obtained from Springer. Violations are liable to prosecution under the German Copyright Law.

The use of general descriptive names, registered names, trademarks, etc. in this publication does not imply, even in the absence of a specific statement, that such names are exempt from the relevant protective laws and regulations and therefore free for general use.

Cover design: deblik, Berlin

Printed on acid-free paper

9 8 7 6 5 4 3 2 1

springer.com

This book is dedicated to my wife Fyllitsa

Silas Michaelides
Editor

Editorial

This book is the outcome of contributions from scientists who were invited to expose their latest findings on precipitation research and in particular, on the measurement, estimation and prediction of precipitation. In this respect, the book comprises a state-of-the-art coverage of the most modern views and approaches in the study of precipitation. In addition, the 20 Chapters that this book consists of provide an insight into the evolutionary aspects of their respective disciplines; also, many of the authors attempt to project into the future by providing an outlook of the planned and expected developments in their respective areas of research.

The Chapters presented in this book are mostly written by selected scientists who presented their advances in precipitation research during activities at the 2006 and 2007 General Assemblies of the European Geophysical Union (EGU) that I convened, at the kind invitation of its Atmospheric Sciences Division. However, in order to give a more complete picture of the subject, other invited experts were asked to supplement with additional Chapters.

The readers of this volume are presented with a blend of theoretical, mathematical and technical treatise of precipitation science. Large parts of many Chapters are devoted to authentic applications of technological and theoretical advances: from local field experiments to country-scale campaigns and, beyond these, to multinational space endeavors. Also, the book reveals the high level of scientific ingenuity, the systematic exploitation of modern technological knowledge and the extent of scientific collaboration and networking that were employed by the scientific community in tackling a very complex issue.

Bearing in mind the above, the book is addressed to those who are involved in precipitation research, but also to those researchers from the wider area of atmospheric sciences whose interests touch on this extremely important weather phenomenon. Moreover, the book aims at introducing newcomers in the field of precipitation science to the various up-to-date scientific facets of the subject, by exposing the full dimensions of the measurement, estimation and prediction of precipitation. I trust that this volume will become a valuable source of inspiration for the scientific endeavors of all scientists working on the multifaceted physical phenomenon of precipitation.

This book is the result of an intense collaborative effort and close interaction between the Authors and the Editor, on the one hand, and the Editor and the Publishers, on the other hand. In this respect, I wish to express my deepest appreciation to all and each one of the 51 esteemed colleagues, scientists and researchers who contributed to this book for their valuable writings but also for their patience during the compilation of this volume. I also wish to thank the Publishers for their kind invitation to lead and coordinate this effort which turned out to be a great experience for me. Finally, I am grateful to my wife Fyllitsa for her valuable support during the writing and compilation of this book.

Dr. Silas Michaelides
Editor

Prologue

As mankind faces up to the various pressing environmental and climatic problems of the twenty-first century, protecting freshwater resources is to be found at the forefront. The availability of freshwater for human consumption, agriculture and industry is of concern to all nations, particularly those in the arid zones where prolonged droughts have already created immense human suffering, population displacement and erosion of arable resources. Ultimately, precipitation is the foremost source of freshwater. With the exception of ancient artesian deposits and deep aquifers, which themselves can only be recharged by precipitation once depleted, mankind largely depends upon precipitation to supply inland lakes, rivers, wetlands and reservoirs of all types for its freshwater stores – including the buildup of snow in mountains for the eventual Spring runoff. Understanding the physical processes which control and produce precipitation and the development of models to predict precipitation, are responsibilities left to scientists, especially those who are specialists in precipitation physics, measurement, remote sensing estimation, model formulation, and verification.

It is notable than some 60% of the world's population that is impacted by shortages of freshwater live within the 21 countries that surround the Mediterranean Sea – the centerpiece of a basin whose water budget is of central concern to the European Union and even more so to its neighbors to the East and South where current and pending water shortages are extreme. Preserving a fresh water supply to residents of these nations and elsewhere has become a prime responsibility of national and local governments, as well as of individuals – and more recently of international organizations whose well conceived policies are able to help assist governments and individuals to protect, preserve, conserve, and utilize water in the best possible fashion. By the same token, scientists are left with the responsibility of finding the optimal means to measure precipitation, to understand how its production is influenced by climate change, aerosol effects, and land use change – and ultimately to predict its distribution and those additional elements of regional and global water cycles that affect man's life and health. It is easy enough to overlook these issues when conducting research, seeking funding for research, teaching and supervising students concerning precipitation science and running models associated with precipitation and water resources; without scientific commitment in helping solve mankind's central problems with water and water conservation, scientists would not be exercising their very best skills. That is why books such as this up-to-date compendium are so very important.

The book's Chapters are organized into four thematic Parts, entitled: I. Measurement, II. Estimation (via space, ground and underwater remote sensing), III. Prediction, and IV. Integration – with each Part covering a selection of distinct views.

The first Part addresses measurement techniques and quality control based on new technology instruments, including the 2D-Video-Distrometer and the Droplet Spectrometer which obtain measurements of accumulated rainfall by actually counting and integrating the water volumes of individual droplets. Such technology enables diversified quantization and segmentation of rainfall because it resolves the process down to its fundamental unit metric.

The second Part addresses the remote sensing of rainfall- which has traditionally been a problem in transforming backscatter observations (i.e., reflectivity factor measurements) from non-coherent, non-polarimetric, single frequency ground radar systems into estimates of rain rate – but in recent decades has undergone a technology revolution into the use of Doppler, polarimetric, and multi-frequency radar systems operated on the ground, on ships and on aircraft, plus the use of passive microwave radiometers and high frequency radars operating on Earth-pointing spacecraft. New remote sensing technology has even been used in the ocean to estimate rain rates by measurement of under water acoustic waves produced by rainfall noise on the ocean surface. The Chapters in this Part provide a selection of new ideas concerning the remote sensing of precipitation using the newer technologies, including a view to the future.

The third Part then moves to the prediction of precipitation through the use of different types of prognostic modeling systems: (a) the ensemble numerical weather prediction (E-NWP) model, (b) the Limited Area Model (LAM), and (c) the advection-based Nowcasting Model (NM) which can be used with either time-lapse ground radar images or optical-infrared satellite images.

Finally, the fourth Part addresses the integration of precipitation research. Amongst other issues, this Part addresses the research that took place within the *Voltaire Project*, a European-wide project that addressed many of the same issues addressed within this book's compilation of Chapters, but closely focused on verification and validation of precipitation observations.

It is left as a challenge to the reader to help devise and guide future research programs concerning precipitation physics, measurement, estimation, prediction and validation. These remain as imperative research topics for the experimental, applications and operational agencies along with the academic research departments charged with understanding, monitoring and predicting precipitation and the stores of freshwater resources that mankind depends upon for its livelihood, health, food production and commerce. The publication of findings of these research programs is crucial in moving the science forward and in creating understanding of all aspects of precipitation – knowledge that is sublimely important to the world community. Therefore, I offer my gratitude as a scientist and as a friend to Dr. Silas Michaelides for the very fine effort he has put forth to deliver this book and its contents – provided by an international body of scientists – into the open literature.

Professor Eric A. Smith

Contributors

Accadia, C., MSc
EUMETSAT
Am Kavalleriesand 3, D-64295 Darmstadt
Germany

Ahrens, B., Prof
Institute for Atmosphere and Environment
Goethe-University Frankfurt
Altenhoeferallee 1, D-60438 Frankfurt/Main
Germany

Amitai, E., PhD
(a) College of Science, George Mason University,
Fairfax, VA, 22030
USA
(b) NASA Goddard Space Flight Center 613.1, Greenbelt, MD 20771
USA

Anagnostou, E.N., Prof
(a) Hellenic Center for Marine Research, Institute of Inland Waters
Anavissos-Attikis, Greece
(b) Civil and Environmental Engineering, University of Connecticut
Storrs, CT
USA

Anagnostou, M.N., PhD
Hellenic Center for Marine Research, Institute of Inland Waters
Anavissos-Attikis
Greece

Anselm, M.
Technische Universität München
St.-Kilians-Platz 1, 83670 Bad Heilbrunn
Germany

Bechini, R., MPhys
Arpa Piemonte, Area Previsione e Monitoraggio Ambientale
Corso U.Sovietica 216, 10134 Torino
Italy

Bendix, J., Prof. Dr
Laboratory of Climatology and Remote Sensing, University of Marburg
Deutschhausstr. 10, 35032 Marburg
Germany

Bringi, V.N., Prof
Dept. of Electrical and Computer Engineering
Colorado State University
Fort Collins, CO 80523-1373
USA

Campana, V., MMath
Arpa Piemonte, Area Previsione e Monitoraggio Ambientale
Corso U.Sovietica 216, 10134 Torino
Italy

Casaioli, M., MSc
Agency for Environmental Protection and Technical Services (APAT)
Via Curtatone 3, I- 00185 Rome
Italy

Casale, R., PhD
Research Directorate General
European Commission
B1049 Brussels
Belgium

Chandrasekhar, V., Prof
Electrical and Computer Engineering, Colorado State University
CO 80523, Fort Collins, CO
USA

Cremonini, R., MPhys
Arpa Piemonte, Area Previsione e Monitoraggio Ambientale
Corso U. Sovietica 216, 10134 Torino
Italy

Duchon, C., Professor Emeritus
School of Meteorology, University of Oklahoma
120 David L. Boren Blvd.,
Norman, OK 73072-7307
USA

Dufournet, Y., MSc
IRCTR, Delft University of Technology
Mekelweg 4, 2628 CD
Delft
The Netherlands

Ebert, E.E., Dr
Centre for Australian Weather and Climate Research
Bureau of Meteorology
GPO Box 1289, Melbourne, Victoria 3001
Australia

Einfalt, T., Dr
Hydro & meteo GmbH & Co. KG
Breite Str. 6-8
D-23552 Lübeck
Germany

Gabella, M., PhD
Politecnico di Torino
Electronics Department
Corso Duca degli Abruzzi 24
10129 Torino, Turin
Italy

Glasl, S.
Technische Universität München
St.-Kilians-Platz 1, 83670 Bad Heilbrunn
Germany

Grecu, M., PhD
(a) Goddard Earth Sciences and Technology Center, University of Maryland
Baltimore County, Baltimore
USA
(b) NASA Goddard Space Flight Center
8800 Greenbelt Road, 20771, Greenbelt, MD
USA

Gultepe, I., PhD
(a) Cloud Physics and Severe Weather Research Section Science and Technology Branch, Environment Canada
4905 Dufferin Street, Downsview, Ontario,
Canada, M3H 5T4
(b) Département des Sciences de la Terre et de l'Atmosphère
Université du Québec à Montréal
Montréal, Québec,
Canada, H3C 3P8

Haiden, T., Dr
Central Institute for Meteorology and Geodynamics (ZAMG)
Hohe Warte 38, A-1190 Vienna
Austria

Hou, A., Dr
Earth Science Division, Code 610.1
NASA Goddard Space Flight Center
Greenbelt, MD 20771
USA

Jaun, S., Dipl. ETH
Institute for Atmospheric and Climate Science, ETH Zurich
Universitätsstrasse 16, CH-8092 Zürich
Switzerland

Kokhanovsky, A.A., Dr
Institute of Remote Sensing, University of Bremen
Otto Hahn Allee 1, 28334 Bremen
Germany

Kummerow, C., Prof
Dept. of Atmospheric Science
Colorado State University
Fort Collins, CO 80523
USA

Lammer, G., Dipl.-Ing.(FH)
Institute of Applied Systems Technology
JOANNEUM RESEARCH
Inffeldgasse 12
A-8010 Graz
Austria

Lensky, I.M., PhD
Department of Geography and Environment, Bar-Ilan University
Ramat-Gan 52900
Israel

Levizzani, V., Dr
Institute of Atmospheric Sciences and Climate, National Research Council
via Gobetti 101, I-40129 Bologna
Italy

Mariani, S., MSc
(a) Agency for Environmental Protection and Technical Services (APAT)
 Via Curtatone 3, I-00185 Rome
 Italy
(b) Department of Mathematics, University of Ferrara
 Via Machiavelli 35, I-44100 Ferrara
 Italy

Michaelides, S., Dr
Meteorological Service
CY-1418 Nicosia
Cyprus

Moisseev, D., Dr
Electrical and Computer Engineering, Colorado State University
CO 80523, Fort Collins, CO
USA

Nauss, T., Dr
Laboratory of Climatology and Remote Sensing, University of Marburg
Deutschhausstr. 10, 35032 Marburg
Germany

Nurmi, P., PhLic
Meteorological Research
Finnish Meteorological Institute
P.O. Box 503
FI-00101 Helsinki
Finland

Nystuen, J.A., PhD
Applied Physics Laboratory, University of Washington
1013 NE 40th Street, Seattle, WA 98105
USA

Perona, G., Prof
Politecnico di Torino
Electronics Department
Corso Duca degli Abruzzi 24
10129 Torino
Italy

Randeu, W.L., Prof
Graz University of Technology
Inffeldgasse 12, A-8010 Graz
Austria

Rossa, A.M., Dr
Centro Meteorologico di Teolo, ARPA Veneto
Via Marconi 55, I-35037 Teolo (PD)
Italy

Russchenberg, H., Dr
IRCTR, Delft University of Technology
Mekelweg 4, 2628 CD, Delft
The Netherlands

Schönhuber, M., Dr
Institute of Applied Systems Technology
JOANNEUM RESEARCH
Inffeldgasse 12
A-8010 Graz
Austria

Shepherd, J.M., Associate Professor
University of Georgia, Atmospheric Sciences Program, Geography Department
GG Building
Athens, GA 30602
USA

Skofronick-Jackson, G., Dr
Hydrospheric and Biospheric Sciences Laboratory
NASA Goddard Space Flight Center
Greenbelt, MD 20771
USA

Smith, E.A., Prof
NASA/Goddard Space Flight Center
Laboratory for Atmospheres/Code 613.1
Greenbelt, MD 20771
USA

Spek, L., MSc
IRCTR, Delft University of Technology
Mekelweg 4, 2628 CD, Delft
The Netherlands

Steinheimer, M., Mag
Central Institute for Meteorology and Geodynamics Geophysics (ZAMG)
Althanstrasse 14, A-1090 Vienna
Austria

Tartaglione, N., MSc
Department of Physics, University of Camerino
Via Madonna delle Carceri 9, I-62032 Camerino
Italy

Thies, B.
Laboratory of Climatology and Remote Sensing, University of Marburg
Deutschhausstr. 10, 35032 Marburg
Germany

Thurai, M., PhD
Dept. of Electrical and Computer Engineering
Colorado State University
Fort Collins, CO 80523-1373
USA

Tomassone, L., MEng
Arpa Piemonte, Area Previsione e Monitoraggio Ambientale
Corso U.Sovietica 216, 10134 Torino
Italy

Turek, A.
Laboratory of Climatology and Remote Sensing, University of Marburg
Deutschhausstr. 10, 35032 Marburg
Germany

Unal, C., MSc
IRCTR, Delft University of Technology
Mekelweg 4, 2628 CD Delft
The Netherlands

Contents

Part I. Measurement of precipitation 1

Chapter 1 – The 2D-Video-Distrometer 3
Michael Schönhuber, Günter Lammer, Walter L. Randeu

1.1 Introduction 3
1.2 About distrometer types 4
1.3 Principle of measurement by 2D-Video-Distrometer 6
  1.3.1 Design of the instrument 8
  1.3.2 Measurable and derived quantities 12
1.4 Current implementation 20
  1.4.1 Specifications 21
  1.4.2 Maintenance procedures 21
1.5 Experiences 23
1.6 Scientific merits 24
1.7 Outlook 28
References 29

Chapter 2 – Using vibrating-wire technology for precipitation measurements 33
Claude E. Duchon

2.1 Introduction 33
2.2 Principles of operation 35
2.3 Description of field site and data acquisition 36
2.4 Advantages of using three vibrating wires 41
2.5 Calibration-verification 44
2.6 Temperature sensitivity 47
2.7 Rain rate estimation 50
2.8 Very low precipitation events 53
2.9 Summary 56
References 58

Chapter 3 – Measurements of light rain, drizzle and heavy fog 59
Ismail Gultepe

3.1 Introduction 59
3.2 FRAM field projects and observations 62
    3.2.1 FD12P measurements 63
    3.2.2 VRG101 measurements 64
    3.2.3 POSS measurements 65
    3.2.4 Total Precipitation Sensor (TPS) measurements ... 67
    3.2.5 FMD and CIP measurements 68
3.3 Analysis 68
3.4 Results 69
    3.4.1 Case studies 69
    3.4.2 Overall comparisons 70
3.5 Discussion 73
    3.5.1 Light precipitation and drizzle measurements 75
    3.5.2 Visibility calculations 75
    3.5.3 Uncertainties 78
3.6 Conclusions 79
References 80

Chapter 4 – The Droplet Spectrometer – a measuring concept for detailed precipitation characterization 83
Sebastian Glasl, Magnus Anselm

4.1 Introduction 83
4.2 Physical basis 84
    4.2.1 Drop size calculation 84
    4.2.2 Calibration 86
4.3 The measuring concept 87
    4.3.1 The droplet sensor 87
    4.3.2 The software 'Rainalyser' 89
4.4 Discussion and applications 93
    4.4.1 Measuring range 93
    4.4.2 Influence of wind 94
    4.4.3 Drop shapes and drag coefficient 94
    4.4.4 Significance of the impulse of the drops 94
    4.4.5 Application possibilities 95
4.5 Future plans and improvements 96
4.6 Appendix 97
References 99

Chapter 5 – Quality control of precipitation data 101
Thomas Einfalt, Silas Michaelides

5.1 Introduction 101
5.2 Quality Control of rain gauge data 102
    5.2.1 Gaps in the data 103
    5.2.2 Physically impossible values 103
    5.2.3 Constant values 103
    5.2.4 Values above set thresholds 103
    5.2.5 Improbable zero values 104
    5.2.6 Unusually low daily values 104
    5.2.7 Unusually high daily values 104
    5.2.8 Data check time series 104
    5.2.9 Station data quality 105
    5.2.10 Generalization and future work 106
    5.2.11 Conclusion: what can we do automatically? 106
5.3 Quality Control of radar data 106
    5.3.1 Data Quality report of COST 717 107
    5.3.2 Error sources 108
    5.3.3 Data Quality Index 111
    5.3.4 Correction methods 114
5.4 Future developments 123
References 123

Part II. Estimation of precipitation 127

i. Space estimation 129

Chapter 6 – Global precipitation measurement 131
Arthur Y. Hou, Gail Skofronick-Jackson, Christian D. Kummerow, James Marshall Shepherd

6.1 Introduction 131
6.2 Microwave precipitation sensors 135
6.3 Rainfall measurement with combined use of active
    and passive techniques 140
6.4 The Global Precipitation Measurement (GPM) mission 143
    6.4.1 GPM mission concept and status 145
    6.4.2 GPM core sensor instrumentation 148
    6.4.3 Ground validation plans 151
6.5 Precipitation retrieval algorithm methodologies 153
    6.5.1 Active retrieval methods 155
    6.5.2 Combined retrieval methods for GPM 157

6.5.3	Passive retrieval methods	159
6.5.4	Merged microwave/infrared methods	160
6.6 Summary	162
References	164

Chapter 7 – Operational discrimination of raining from non-raining clouds in mid-latitudes using multispectral satellite data 171

Thomas Nauss, Boris Thies, Andreas Turek, Jörg Bendix, Alexander Kokhanovsky

7.1	Introduction	171
7.2	Conceptual model for the discrimination of raining from non-raining mid-latitude cloud systems	172
7.3	Retrieval of the cloud properties using multispectral satellite data	173
7.4	Application of the conceptual model to Meteosat Second Generation SEVIRI data	175
	7.4.1 The daytime approach	175
	7.4.2 The night-time approach	178
7.5	Evaluation of the new rain area delineation scheme	183
	7.5.1 Evaluation study using daytime scenes	184
	7.5.2 Evaluation study using night-time scenes	186
7.6	Conclusions	188
References		190

Chapter 8 – Estimation of precipitation from space-based platforms 195

Itamar M. Lensky, Vincenzo Levizzani

8.1	Introduction	195
8.2	Estimating rainfall from space	196
	8.2.1 VIS/IR	197
	8.2.2 Passive microwave	198
	8.2.3 Active sensors	200
	8.2.4 Blended techniques	202
8.3	Retrieval of precipitation formation processes using microphysical data	205
	8.3.1 Rain estimates using microphysical considerations	205
	8.3.2 Retrieval of precipitation formation processes	207
	8.3.3 Future developments	212
8.4	Abbreviation	212
References		213

Chapter 9 – Combined radar–radiometer retrievals from satellite observations 219
Mircea Grecu, Emmanouil N. Anagnostou

9.1 Introduction 219
9.2 Background 220
9.3 General formulation 223
9.4 Concluding remarks 228
References 228

Part II. Estimation of precipitation

ii. Ground estimation 231

Chapter 10 – Rain microstructure from polarimetric radar and advanced disdrometers 233
Merhala Thurai, V. N. Bringi

10.1 Introduction 234
    10.1.1 Background 234
    10.1.2 Rain microstructure: relevance 235
    10.1.3 Relating rain microstructure to polarimetric radar measurements 238
10.2 Drop size distributions 242
    10.2.1 Variability 242
    10.2.2 DSD models 243
    10.2.3 DSD estimation from polarimetric radar measurements 248
    10.2.4 DSD estimation from advanced disdrometers 254
    10.2.5 Global DSD characteristics 257
    10.2.6 Seasonal variation 259
10.3 Drop shapes 263
    10.3.1 Axis ratio measurements from an artificial rain experiment 263
    10.3.2 Drop contours 265
    10.3.3 Consistency with polarimetric radar measurements 268
10.4 Drop orientation angles 269
10.5 Fall velocities 274
10.6 Summary 276
References 279

Chapter 11 – On the use of spectral polarimetry to observe ice cloud microphysics with radar ... 285
Herman Russchenberg, Lennert Spek, Dmitri Moisseev, Christine Unal, Yann Dufournet, Chandrasekhar Venkatachalam

11.1 Introduction ... 286
11.2 The concept of spectral polarimetry ... 287
11.3 Microphysical model of ice particles ... 288
    11.3.1 The shape of ice crystals ... 289
    11.3.2 Canting angles of ice crystals ... 290
    11.3.3 Mass density of ice crystals ... 290
    11.3.4 Velocity of ice crystals ... 291
    11.3.5 Bulk parameters ... 292
11.4 Radar observables of ice particles ... 293
11.5 Retrieval of microphysical parameters ... 296
    11.5.1 Dependence on DSD parameters of plates and aggregates ... 296
    11.5.2 The curve fitting procedure ... 297
    11.5.3 Quality of retrieval technique ... 302
11.6 Application to radar data ... 303
    11.6.1 Retrieval algorithm results ... 304
    11.6.2 Comparison of IWC with LWC ... 304
    11.6.3 Relation between IWC and reflectivity ... 307
    11.6.4 Influence of the shape parameter of the DSD ... 308
11.7 Summary and conclusions ... 310
References ... 311

Chapter 12 – Performance of algorithms for rainfall retrieval from dual-polarization X-band radar measurements ... 313
Marios N. Anagnostou, Emmanouil N. Anagnostou

12.1 Introduction ... 313
12.2 X-band dual-polarization systems ... 316
12.3 Attenuation correction schemes for X-band dual-polarization radar observations ... 318
12.4 Rainfall estimation algorithms ... 319
    12.4.1 Review of microphysical retrieval algorithms ... 319
    12.4.2 Rainfall retrieval algorithms ... 324
    12.4.3 Data ... 325

12.5   Algorithm evaluation 328
       12.5.1   Evaluation of the DSD retrieval techniques 329
       12.5.2   Evaluation of rainfall retrieval techniques 333
12.6   Closing remarks 337
References 337

Part II. Estimation of precipitation

iii. Underwater estimation 341

Chapter 13 – Underwater acoustic measurements of rainfall 343
Eyal Amitai, Jeffrey A. Nystuen

13.1   Introduction 343
       13.1.1   Why measure rainfall at sea? 343
       13.1.2   Why listen to rainfall underwater? 344
       13.1.3   What instrumentation is used to measure rainfall at sea? 344
       13.1.4   Using sound to measure drop size distribution and rain rate 345
13.2   Listening to rainfall in a shallow water pond 348
13.3   Oceanic field studies of the acoustic measurement of rainfall 349
13.4   Listening to rainfall 2000 meters underwater – the Ionian Sea Rainfall Experiment 350
       13.4.1   Rain type classification and wind speed estimates 358
13.5   Conclusions and outlook 360
References 361

Part III. Prediction of precipitation 365

Chapter 14 – Probabilistic evaluation of ensemble precipitation forecasts 367
Bodo Ahrens, Simon Jaun

14.1   Introduction 367
14.2   Rain station precipitation data 370
14.3   Forecast data by the limited-area prediction system COSMO-LEPS 371
14.4   Observational references 373
14.5   Skill scores 376

14.6   Results and discussion 379
14.7   Conclusions 384
References 386

Chapter 15 – Improved nowcasting of precipitation based on convective analysis fields 389
Thomas Haiden, Martin Steinheimer

15.1   Introduction 389
15.2   The INCA system 393
15.3   Advection forecast 397
15.4   Convective analysis fields 401
15.5   Cell evolution algorithm 403
15.6   Verification and parameter sensitivity 407
15.7   Orographic effects in convective initiation 412
15.8   Conclusions 415
References 416

Chapter 16 – Overview of methods for the verification of quantitative precipitation forecasts 419
Andrea Rossa, Pertti Nurmi, Elizabeth Ebert

16.1   Introduction 419
16.2   Traditional verification of QPF and limitations for high resolution verification 423
         16.2.1 Common scores 424
         16.2.2 The double penalty issue 429
16.3   Scale-dependent techniques 433
         16.3.1 Neighborhood methods 433
         16.3.2 Spatial decomposition methods 437
16.4   Object and entity-based techniques 438
16.5   Stratification 440
         16.5.1 Seasonal, geographical and temporal stratification 441
         16.5.2 Weather-type dependent stratification 442
16.6   Which verification approach should I use? 448
References 449

Chapter 17 – Objective verification of spatial precipitation forecasts 453
Nazario Tartaglione, Stefano Mariani, Christophe Accadia, Silas Michaelides, Marco Casaioli

17.1 Introduction 453
17.2 The problem of observations in objective verification 456
17.3 Use of rainfall adjusted field for verifying precipitation 458
17.4 Statistical interpretation of position errors as derived by object-oriented methods 461
17.5 Assessing the difference between CMS indices from two different forecast systems 466
17.6 Conclusions 467
References 469

Part IV. Integration of measurement, estimation and prediction of precipitation 473

Chapter 18 – Combined use of weather radar and limited area model for wintertime precipitation type discrimination 475
Roberto Cremonini, Renzo Bechini, Valentina Campana, Luca Tomassone

18.1 Introduction 475
18.2 Data source and precipitation type discriminating algorithms 478
    18.2.1 Data sources 478
    18.2.2 Precipitation type discriminating algorithms 480
18.3 Algorithm's validation 482
18.4 Results 485
    18.4.1 Ground network 2 m air temperature 485
    18.4.2 LAMI freezing level 486
    18.4.3 LAMI wet-bulb temperature 488
18.5 Summary and conclusions 489
References 490

Chapter 19 – Adjusting ground radar using space TRMM Precipitation Radar 493
Marco Gabella, Silas Michaelides

19.1 Introduction 494
    19.1.1 Monitoring hardware stability and measurements' reproducibility 494
    19.1.2 Calibration versus absolute calibration 494

19.1.3 Adjustment 495
19.1.4 Why to adjust Ground-based Radar (GR) data? 496
19.2 Radar/Gauge factor: range-dependence as seen by gauges 497
19.2.1 Adjustment not directly related to physical variables 498
19.2.2 Adjustment factor related to some physical variables 499
19.3 Comparing ground-based and spaceborne radar 500
19.3.1 Range-dependence as seen by the TPR 501
19.4 Instrumentation and data description 503
19.4.1 The TRMM Precipitation Radar (TPR) 503
19.4.2 The Ground-based Radar (GR) in Cyprus 504
19.5 Results 505
19.5.1 Bias and range-dependence derived from single overpasses 505
19.5.2 A robust range-adjustment equation: integrating more overpasses 507
19.5.3 Comparing TPR and GR echoes 508
19.6 Summary and lessons learned 510
References 512

Chapter 20 – Implementing a multiplatform precipitation experiment 515

Giovanni Perona, Marco Gabella, Riccardo Casale

20.1 Introduction 515
20.1.1 Scientific/technological objectives of the VOLTAIRE project 517
20.1.2 Project organization 518
20.2 VOLTAIRE project summary and recommendations 520
20.2.1 Summary 520
20.2.2 Main lessons learned 523
20.2.3 Recommendations 524
20.3 VOLTAIRE technical conclusions 526
20.4 Outlook for QPE using radar 527
20.4.1 Where we stand today 527
20.4.2 Proposed solution: use of many inexpensive, redundant, short-range radars 528
20.5 General conclusions 529
20.6 Appendix 530
References 530

Author index 533

Subject index 535

Part I. Measurement of precipitation

1 The 2D-Video-Distrometer

Michael Schönhuber[1], Günter Lammer[1], Walter L. Randeu[2]

[1]Institute of Applied Systems Technology, JOANNEUM RESEARCH, Graz, Austria
[2] Graz University of Technology, Graz, Austria

Table of contents

1.1 Introduction 3
1.2 About distrometer types.. 4
1.3 Principle of measurement by 2D-Video-Distrometer 6
    1.3.1 Design of the instrument 8
    1.3.2 Measurable and derived quantities..................................... 12
1.4 Current implementation 20
    1.4.1 Specifications 21
    1.4.2 Maintenance procedures... 21
1.5 Experiences... 23
1.6 Scientific merits 24
1.7 Outlook 28
References 29

1.1 Introduction

Detailed knowledge on tropospheric precipitation microstructure is one of the bases in various fields of sciences and applications, like terrestrial and satellite radio transmission, remote sensing of precipitation, generally tropospheric wave propagation and atmospheric sciences.

    In the field of telecommunications, precipitation causes several unwanted effects on Earth-satellite as well as on terrestrial links. System design has to consider that and has to take care for appropriate countermeasures. Statistical approaches allow quantitative answers on questions about precipitation's impact on wave propagation. Thus probabilities are given, that e.g., rain induced attenuation or phase delay exceeds a certain threshold for a given set of parameters (location, frequency etc.). Increasingly demanding applications (higher frequencies, frequency re-use and multiple satellite links) require answers on increasingly complex questions.

Remote sensing technologies aim at measuring precipitation parameters at far distances, using either spaceborne or ground-based radars and radiometers. Such observations permit better climatological characterization, on a global as well as on a regional scale. Global keyword terms like greenhouse effect, global atmospheric warming, tropical rainfall, the Earth's energy and water cycle, etc. immediately indicate the urgent need for such observations. Speaking in local scale, short-term climatological considerations play a more important role. Weather fore- and nowcasting help in many various ways in everyday life. To mention only but a few examples of the numerous points of interest: flood and storm warnings, control of air and road traffic, control of hydroelectric power plants, water resources management, etc.

1.2 About distrometer types

Drop size distribution meters are called disdrometers, often with this very spelling being used. Within this Chapter, however, the spelling distrometer is preferred, indicating a device for measuring *distributions*, not limited to raindrops only, but also suited for other particular matter (amongst which are snow flakes and hail stones). The 2D-Video-Distrometer (2DVD) has been produced and marketed under this very name since more than a decade.

Based on different technologies, distrometers have been developed to get detailed information on precipitation microstructure in point monitoring observations. Most distrometer types rely either on measurement of precipitation particles' mechanical impact onto some sensor, or on optical methods. Whereas measurements by electro-mechanical distrometers are restricted to drop size distribution information, optical, especially imaging distrometers, provide more comprehensive information.

A well known electromechanical instrument is the RD69 distrometer (Joss and Waldvogel 1967). The measurement principle of this system is based on the automatic compensation of the force produced by a rain drop falling upon the sensor. This automatic force compensation together with raindrop fall velocities taken from literature models allows obtaining a value for the drop size. On this basis, rain rate and drop size information are provided by electro-mechanical distrometers.

A number of authors (Hauser et al. 1984; Delahaye et al. 1995; Löffler-Mang 1997) report on distrometers using measurement of light extinction. This distrometer type uses a light source, a collimating system (to produce a flat beam of light, which constitutes the sensitive measurement area), followed by a focusing system and a receiver (optical detector). Whenever a rain drop falls through the measuring area, at the receiver a negative pulse is detected in the light intensity, due to the extinction arising from the drop's shadow. The shape (amplitude and duration) of this pulse allows giving a value of the drop's size and speed. Light extinction distrometers provide rain rate and drop sizes, whereby the drops' fall velocity is obtained from actual measurement.

The optical array probe distrometer scans the measuring area not by one individual signal only, but by an array of optical receiver elements, illuminated by a suitable light-source. Whenever a raindrop passes, a number of array elements is shadowed, depending on the width of the particle. Further information is gained from the number of scan-lines affected by such drop. Knollenberg (1970) reports on the development of an optical array system. A one-dimensional system with a measurement area of 46 cm^2 (23 cm by 2 cm) has been developed.

Line-scan camera distrometers work on a similar basis as the optical array probe distrometer. In comparison to the system reported by Knollenberg (1970), a lens is used which allows a focused measurement in a certain distance from the camera. A distrometer of this type is the Snow Spectrograph developed by Atmospheric Science, Swiss Federal Institute of Technology, Switzerland. A report on comparative measurements with the Snow Spectrograph and other precipitation gauges is given by Steiner (1988).

Finally, matrix camera distrometers take a full picture of particles. As with any camera, the measurement volume is slightly pyramidal and not a total regular cuboid. Frank et al. (1994) describe a distrometer of that type. The measurement volume depends on the size of the particles, since the depth of the visibility range depends on the camera's focus and the size of the objects.

The development of an imaging 2D-Video-Distrometer (2DVD) by JOANNEUM RESEARCH, Graz, Austria, in cooperation with ESA/ESTEC began in 1991, when polarimetric weather radar data revealed unexpectedly high differential reflectivity (ZDR) values. The motivation for the development was to obtain continuous measurements not only of the drop size distribution (DSD) but also of the shapes and the fall velocities of all precipitation particles and types. The 2DVD is based on two high speed line scan cameras. The specific alignment enables the 2DVD to measure the velocity, the front and side view of each particle falling through the measurement area of approximately 100 cm² (10 cm by 10 cm), moreover a precise time stamp is recorded and the location where within the measuring area such particle has arrived. The imaging grid resolution for particles is finer than 0.2 mm, in both horizontal and vertical direction, when fall velocity is less than 10 m/s which is typically the case for raindrops, melting snowflakes and snowflakes. 2DVD data do not use literature-derived estimates for rain drop shapes or fall velocities, instead these parameters are measured. 2DVD provided quantities like rain rate, rain drop size distributions and furthermore, 2DVD-derived weather radar reflectivities for rain thus are based on measured precipitation characteristics without the need to rely on literature models. 2DVD data furthermore allow separating contributions from individual fractions in mixed phased events. So the 2DVD is deemed to be the sensor system delivering a maximum of information on precipitation micro structure and derived bulk parameters, for rain, hail, snow and mixed-phase ensembles.

1.3 Principle of measurement by 2D-Video-Distrometer

The 2D-Video-Distrometer principle of measurement shall be discussed in more detail: as shown in Fig. 1A two optical systems are used. Each of the optical systems consists of a line scan camera (with the scan line aligned horizontally) and a background illumination device. The two optical systems are orthogonally aligned against each other with a height offset between 6 and 7 mm. The background illumination device focuses the light of a standard halogen bulb onto the camera. Intense light is seen by the camera. Any obstacle (i.e. hydrometeors) between the background illumination device and the camera results in a blockage of light over a certain number of camera pixels. The number of shadowed pixels corresponds to the width of the particle. The time the particle shadows the scan line depends on its height and fall velocity. The cameras recognize each particle line by line, disassembled into slices. Caused by the orthogonal alignment of the two optical systems, the front and side view of each particle falling through the measurement area are measured.

The small vertical offset of 6–7 mm between the two optical systems allows measuring the fall velocity. Once the fall velocity and number of affected lines are known, the height to width ratio of the particles may truly be calculated from the cameras' measurements. The inlet in the housing of the Sensor Unit (SU) is a square with a size of 25 cm by 25 cm.

To minimize the unwanted effect of splashing from the rims of the housing, the inlet is wider (25 cm) than the cameras' field of view (approximately 10 cm in the middle of the inlet). Thus, an active area of approximately 25 cm (depth of field of view underneath the housing's inlet) by 10 cm (width of field of view) is obtained for each camera. Based on the alignment of the two optical systems, a common area is seen by both cameras in the middle of the inlet, a square called intersection area, which is shown in Fig. 1A.

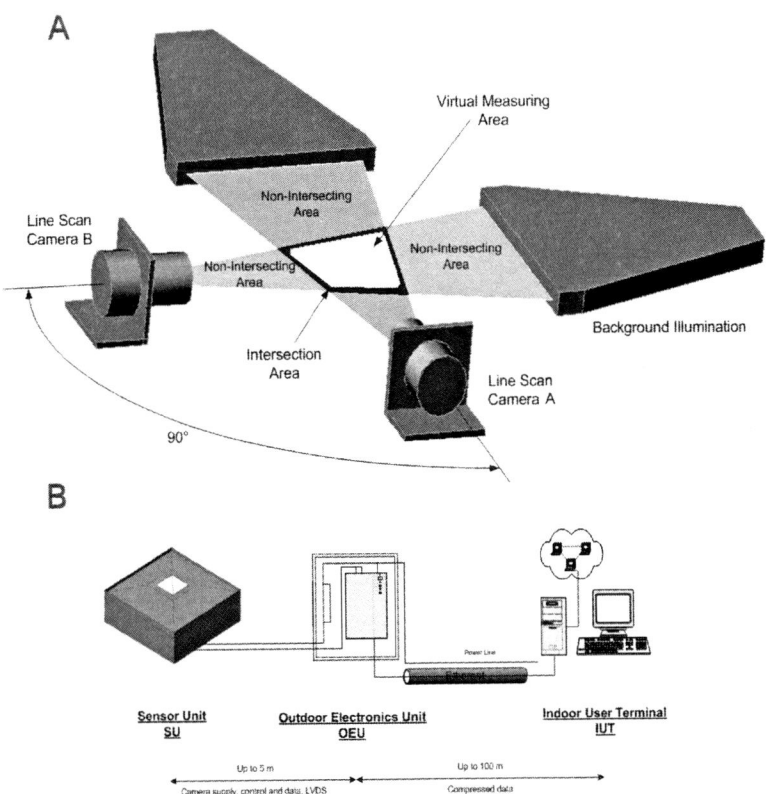

Fig. 1. Measurement principle and components of the 2D-Video-Distrometer

The intersection area has a size of approximately 10 cm by 10 cm and it represents the virtual measuring area. Here, the location can be identified where a particle passes. The 2DVD preprocessing considers only particles fully visible in both systems, i.e. those falling in their whole extent through the virtual measurement area are counted. Particles falling through the non-intersecting parts of the inlet or hitting an 'edge' of the virtual measurement area are stored in the raw data as well, but are not considered by the preprocessing module. This results in a decrease of the measuring area as a function of particle size and the resulting area is called the effective measuring area. For example, a ball of nearly 10 cm in diameter could only be measured passing exactly through the middle of the area, its effective measuring area is nearly zero. For particles of 10 mm in diameter (which is bigger than natural raindrops) a measuring area of around 90 mm by 90 mm is available.

In case of hail, which is not as dense as rain, data recorded in the non-intersecting area may also be used and help to increase the sample size and its statistical significance. In the non intersecting area, however, exists the limitation, that only one view (front or side view) is measured; therefore no velocity and consequently no true height to width ratio is available.

1.3.1 Design of the instrument

To withstand the requirements of precipitation measurement even in adverse climatic conditions, the 2D-Video-Distrometer has been carefully designed and its manufacturing includes thorough quality control. The selection of materials, coating for housings and electronic components, as well as the software implementation for a continuous data acquisition, pre-processing and recording were demanding tasks during development. Detailed descriptions of the structure of the instrument are given by Urban et al. (1994) and by Schönhuber et al. (1993, 1996a), a field test report is given by Schönhuber et al. (1996b). Figure 1B presents a block diagram of the system components.

The three main units are:

- The Sensor Unit (SU) with two optical systems mounted inside, each system consisting of camera, lenses, mirrors and the illumination device.
- The Outdoor Electronics Unit (OEU) has to be located close to the SU. The OEU unit supplies and controls the two line scan cameras,

acquires and preprocesses the raw data. The raw data is transported to the Indoor User Terminal via Ethernet.
- The Indoor User Terminal (IUT) stores and preprocesses the received data and simultaneously offers various analyses in online mode to the user, via a GUI.
- Optionally a three-dimensional ultrasonic type wind sensor and a single tipping bucket rain gauge may be connected to the system.

Mechanical structure of the sensor unit

Two generations of the 2D-Video-Distrometer have been developed: the classic tall and the low-profile version. Turbulences around the classic tall instrument setup would distort measurements in windy conditions, as this was indicated by simulations (Nešpor et al. 2000). Consequently, the next generation's design resulted in the low-profile 2D-Video-Distrometer with the instrument's height decreased by a factor 3 down to 35 cm. It is assumed that the flat and smooth shape of the low-profile 2D-Video-Distrometer makes turbulence effects negligible.

The alignment of the optical components, such as camera mirrors and illumination device, is designed to be insensitive to temperature variations, what guarantees continued measurement precision. This has been achieved by a selection of construction materials with suitable temperature coefficients, special shape and structure. To minimize the mechanical influences onto the calibration status, the mechanical structure of the Sensor Unit is split up into two parts. The inner part, the optics carrier, holding all optical parts is connected to the outer part, the housing, by shock absorbers. This guarantees that the adjustment of the optical parts, such as camera, mirrors and illumination device, is not impaired by mechanical influences like heavy wind or hail, by manual effects like opening the housing and other. The outer part is the housing of the inner one and shields it against wind and water.

Special care is taken to avoid splashing onto the optical surfaces (mirrors and lenses), since such splashes are seen by the cameras as permanently shadowed section(s). The mirrors and lenses of the optical paths have been protected by slit plates against splashes. With the low-profile version 2DVD, sufficiently large distances between path slits and optical elements have been achieved, totally preventing splashes from reaching the optical surfaces, as it was the case under extreme weather conditions with the first 2DVD generation, the classic tall version.

The instrument is designed in such way that it may be opened for maintenance purposes quickly and that all components are easily accessible.

The electronics

The design of the electronics is a modular one. Individual parts can be replaced if necessary, off-the-shelf products are used as far as possible. On the market the ability of electronics hardware is subject to constant improvements and upgrades. The modular design of the 2DVD allows taking advantage of such improvements whenever relevant. So it was possible to double the measurement speed (and resolution), when faster cameras became available.

The Sensor Unit (SU) contains two high speed line scan cameras and two illumination units.

The Outdoor Electronics Unit (OEU) cabinet consists of:

- Power supply unit;
- Industrial computer chassis equipped with:

1. Slot CPU card
2. Camera interface cards
3. Network interface card
4. Timer card
5. Video card.

The Indoor User Terminal (IUT) is a standard desktop PC or a laptop, suitably configured.

The software

This Section lists the tasks carried out by the software. It is arranged according to software executing on the OEU, on the IUT in online mode only and on IUT in online or offline mode. IUT offline mode refers to a situation where a user displays or analyses data of past days, or at some PC working with archived data. In more detail, the tasks are:

- OEU:

1. Control of the two line scan cameras;
2. Acquisition, compression and storage of the cameras raw (shadow) data;
3. Calibration Procedure for the Sensor Unit.

- IUT online:

1. Automatic documentation of data acquisition process in a log file;
2. Storage of two cameras' raw data streams;
3. Combination of the two cameras' data streams and object identification;
4. Compression and storage of preprocessed (hydrometeor) data.

- IUT online or offline:

1. Display of the raw data in synchronous and asynchronous mode (Fig. 2A, screendump of VIEW_AB raw data viewer);
2. Display of the video signal of both optical systems after automatic hourly calibration;
3. Display and analysis of measured and derived quantities (Fig. 2B, screendump of VIEW_HYD hydrometeor data viewer) such as:
- rain rate, a comparison of the 2DVD-measured data to an optionally connected rain gauge data may be done;
- rain drop size distribution, a comparison of the measured data to the models by Marshall and Palmer (1948) or by Joss et al. (1970) may be done;
- vertical velocity versus diameter, a comparison of the measured data to the models by Atlas et al. (1973) and Gunn and Kinzer (1949) may be done;
- estimate of rain drops' horizontal velocity;
- oblateness (height/width ratio) versus diameter, a comparison of the measured data to literature models by Pruppacher and Beard (1970) and by Poiares Baptista (1992) may be done;
- front and side view of each particle;
- indication of location where within the measuring area the particle has arrived;
- a numerical table with values of time stamp, equivolumetric sphere diameter, vertical velocity, oblateness (height/width ratio) and effective measurement area for each particle. The data of an optionally connected three-dimensional ultrasonic type wind sensor such as wind speed and direction is displayed.

The software allows the user to choose the integration time between 15 s and 24 h. Filters with parameters such as observation time, drop diameter, oblateness of the particle or the area where the particle passes through the measurement area can be configured by the user.

Figure 2A shows the visualization of the raw data (viewer program: VIEW_AB) and Fig. 2B that of the processed data (viewer program: VIEW_HYD), as they appear on the IUT screen. In the first column of the left hand side the shadowed pixels of the camera A and on the next column the shadowed pixels of the camera B are displayed. The x-axis indicates the number of pixels of the line scan camera, the y-axis corresponds to the time, given as continuously counted number of lines since midnight. On the upper left side of each column, numeric information on line number and time is displayed. Figure 2B shows the

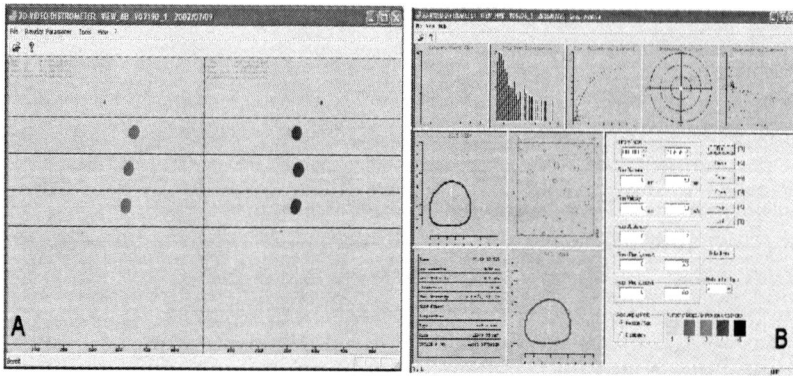

Fig. 2. VIEW_AB raw data display and VIEW_HYD main menu at the IUT

main menu of the IUT VIEW_HYD software. VIEW_HYD displays and analyzes the measurable and the derived quantities.

1.3.2 Measurable and derived quantities

Raw data

The primary measurable quantity is the shadow data of the two line scan cameras. Every approximately 18 µs, each of the two cameras scans the measurement area for any obstacles blocking the background illumination.

The intensity of the light received by the line scan camera is represented by an 8 bit AD-conversion from 0 to 255 for each pixel of the line scan camera.

The background illumination intensity significantly varies throughout the length of the scan line and the slopes of the blockages are not of infinite steepness. That causes the results of the edge detection algorithm being dependent not only on the size of the object but also on the location of the obstacle. To overcome this unwanted effect, each camera pixel signal is multiplied to a level higher than the measured intensity, obtaining a flat reference signal. Thus, observing any obstacles and applying the edge detection algorithm yields a relative threshold, i.e. thresholds are related to the background illumination intensity levels and not to a threshold given in absolute numbers of the AD-converter output.

The raw shadow data are compressed. Together with the information indicating any blocked areas within the 630 pixel line, the precise time stamp is stored, given in multiples of the line scan period (which lasts about 18 µs). This line counter is reset every midnight. Data of optionally connected instruments, like wind sensor or rain gauge, are marked with a synchronous time stamp. The raw data is transferred via Ethernet every 3 s from the OEU to the IUT and stored on the IUT disk drive.

In the line data stream of each camera, time information is given by the line number, furthermore position and size information of the blocked areas are included. The first preprocessing step on the IUT side is the identification of coherent elements, in case of raindrops this may be done easily. Since raindrops do not reveal branched structures it is sufficient to check if any blockages in subsequent lines overlap.

In case of snowflakes, identification of coherent elements in each camera's data stream is a somewhat more complex task: branches, holes in the shadow areas and several different portions recorded in one scan line at the same time may occur.

The next processing step is the synchronization and combination of the two data streams. The time scales of the two camera streams have to be synchronized and the elements measured by each of the two cameras have to be matched, meaning an identification has to be accomplished as to which of the images seen by camera A and by camera B stem from the very same particle. Considering the number density of drops in natural rain, such matching algorithm may easily be carried out. With splashes from the housing reaching the measuring planes, which may happen in intense wind and heavy precipitation conditions, the matching algorithm is not such easy task any longer; nevertheless it succeeds to disregard most of the unwanted splashes. Camera A is the geometrically upper plane, therefore, the element has first to pass the camera A plane and then passes the camera B plane. The difference of the line numbers between an element passing camera A and passing camera B is inversely related to the fall velocity. So the algorithm for each particle recorded by camera A sets a plausibility time window, within which the same particle has to appear in camera B if it hit the intersecting part of the measuring area. Further plausibility checks, like differences in height between pictures of cameras A and B, are applied. If a picture pair recorded by camera A and camera B is found as plausible to stem from the same particle, it is entered into the (preprocessed) hydrometeor data file as front and side view.

Fall velocity and scaling of the views

The fall velocity is measured via the time a particle takes for proceeding from the upper camera A to the lower camera B. The distance between the two planes is between 6 and 7 mm.

Four significant time stamps are to be considered for such process. At time t_0 a particle enters the measurement plane of camera A, at time t_2 the particle enters the measurement plane B. Times t_1 and t_3 are for leaving accordingly. To minimize quantization effects, the average of (t_2-t_0) and (t_3-t_1) is used.

The fall velocity (v) is calculated as

$$v = \frac{2d}{(t_2 - t_0) + (t_3 - t_1)} \left[\frac{m}{s}\right], \tag{1}$$

where:
d   vertical distance between Camera A and B planes
t_0 time when particle enters measuring plane of camera A
t_2 time when particle enters measuring plane of camera B
t_1 time when particle has left measuring plane of camera A
t_3 time when particle has left measuring plane of camera B

The fall velocity of calibration spheres may be calculated from each camera's data individually, considering that the oblateness of the spheres is 1. The distance of the measuring planes may then be determined, even taking into account slight lack of perfect parallelism between measurement planes A and B. The effect of the two optical planes slightly being slanted against each other is then corrected for during data preprocessing.

The lens system used with the cameras requires a fan beam correction. Since the location of the particle within the virtual measurement area is known, the distances of the particle from the cameras is known as well and therefore the pixel width may truly be calculated according to the rules of geometrical optics.

In addition to such scaling of height and width by use of fall velocity and distance from the lens, the linearity of the relationship between size of object and number of shadowed scan lines and pixel numbers has to be checked. Generally, there is no linear relationship between number of shadowed pixels and object width. Therefore, data preprocessing has to restore the correct scaling. After hardware calibration, i.e. after having ensured that all optical elements (cameras, lenses, mirrors and illumination devices) are properly aligned and adjusted, a statistically significant number of calibration spheres with known discrete diameters is recorded and reproduced. The oblateness is calculated as the geometrical mean of the ratios height divided by width for front and side view. The mean of the deviations from the nominals for each calibration sphere diameter is analyzed and the result stored in a lookup table (Table 1) of adjustment factors for application during data preprocessing. Non-integer diameters require appropriate interpolations. It is visible in Table 1, that for bigger particles from approximately 3 mm onwards a practically linear representation is given, whereas smaller particles at this 50% threshold are being underestimated.

The equivolumetric sphere diameter

Once the correct scaling for the height of one scan line and for the pixel widths in the two cameras is found, the equivolumetric sphere diameter of raindrops and hailstones can be calculated. The method used is the following: the scan lines divide a particle into several slices. In each of these slices the particle is assumed to form an elliptical cylinder. The two main axes of such an ellipse are defined by the lengths of the blockages in the two cameras for this scan line. The height of the slice is calculated from the line frequency and the particle fall velocity. Summing up all the slices of one particle, the water or ice volume of a

Table 1. Example for the adjustment factor lookup table, generated from throwing a large number of spheres with known diameters (as shown in line 1) into the measuring area of the 2DVD

	Factors in percent for relative threshold of 50%									
diameter [mm]	0.50	1.00	1.50	2.00	3.00	4.00	5.00	6.00	8.00	10.00
spheres found	362	293	336	300	511	196	261	197	121	144
height A[%]	−35.59	−8.48	−2.95	−1.39	−0.42	−0.12	0.05	0.20	0.10	0.16
height B[%]	−34.86	−9.38	−3.44	−1.74	−0.55	−0.19	−0.06	0.08	0.10	0.10
width A[%]	−27.81	−6.14	−1.83	−0.88	−0.15	−0.01	−0.20	−0.10	0.01	0.08
width B[%]	−26.51	−6.10	−2.35	−1.28	0.02	0.22	0.61	0.41	0.37	0.39

raindrop or a hailstone is obtained, which can straightforward be translated into its equivolumetric sphere diameter.

In case of snowflakes, this method fails. The internal three-dimensional structure cannot be recorded by the cameras. There are considerations to combine front- and side view and fall velocity to produce an estimate for the water content of snowflakes. Front- and side view represents the size of the flake and may perhaps allow an estimate of its drag parameter in air. Knowing size, drag parameter and fall velocity, the weight (being proportional to the water content) might be estimated.

Summarizing, it should be pointed out that the 2D-Video-Distrometer provides calibrated size and shape measurements of raindrops and hailstones down to 0.5 mm equivolumetric sphere diameter. Calibration using even smaller spheres would be possible. There is practically no upper limit for raindrops and hailstones to be measured given by the optical paths, since the width of natural hydrometeors never will exceed the width of the optical planes, i.e. approximately 10 cm.

Table 2 presents an evaluation of calibration spheres, their diameter, velocity and their oblateness. Whereas the algorithm enforces that the mean values in diameter and oblateness fit the nominals, their standard deviation describes the effect of quantization. To allow such data being applicable for rain drops, it was necessary to dispense the calibration spheres from a height above the 2DVD resulting in the natural terminal fall velocity of raindrops of same size.

In practical measurements, it is found that variations of drop oblateness are much bigger than to be expected due to quantization effects, indicating the presence and influence of drop oscillations (Thurai and Bringi 2005).

Rain rate

The calculation of rain rate is based on the following information: equivolumetric sphere diameter (water volume) of raindrops, their time stamps, size of their effective measuring area and the definition of an integration time interval. Herein, the area relevant for weighting a drop's contribution to the rain rate (or drop size distribution) is called effective measuring area.

The effective measuring area is calculated for each particle individually considering the following rules:

Table 2. Evaluation of calibration spheres with fall velocities corresponding to terminal velocity of equally sized rain drops

Nominal sphere diameter	Mean diameter	Std. dev. diameter	Mean velocity	Std. dev. velocity	Mean oblateness	Std. dev. oblateness
mm	mm	mm	m/s	m/s	1	1
1.5	1.5034	0.0584	5.6892	0.2644	1.0004	0.0416
2.0	2.0046	0.0225	6.5134	0.4661	1.0008	0.0291
3.0	3.0052	0.0157	8.0518	0.0813	1.0005	0.0226
4.0	4.0027	0.0140	8.7378	0.3536	1.0015	0.0218
5.0	5.0037	0.0170	9.4125	0.2771	0.9994	0.0167
6.0	5.9989	0.0270	9.6309	0.1402	0.9998	0.0159
7.0	7.0085	0.0204	9.6824	0.1015	1.0024	0.0161
8.0	8.0054	0.0277	9.7506	0.0811	0.9993	0.0122
10.0	10.0009	0.0249	9.7075	0.0929	0.9995	0.0142

- The maximum number of pixels (width of the cameras measuring planes) available for data acquisition process is known.
- It is checked if the raindrop is fully visible. In both cameras the left- and rightmost pixels must not be blocked. If not, such drop is presently not relevant for analysis. The left- and rightmost pixels are not considered for area calculation.
- The user may define a subset of the virtual measuring area as the user-defined active area. It is checked if the center of the hydrometeor is contained in the user-defined active area. If not, this raindrop is presently not relevant for analysis.
- From the number of measuring pixels at both edges half of the drop's maximum horizontal dimension (in pixel numbers) seen by such camera is subtracted, plus two further pixels being subtracted (left and rightmost). This is done for both cameras. Thus the biggest possible area for such a raindrop is found.
- The effective measuring area for the present drop is determined by the common part of the biggest possible and the user defined active area.
- The pixel widths in the center of the present drop's effective measuring area are found by the rules of geometrical optics.

Though the effective measuring area is always an irregular four sided figure its size is calculated as a rectangle. As the side lengths of such a rectangle the effective measuring area's widths at its center are used. The validity of that approximation has been investigated by comparing the precise with the approximate results, the error is always less than 0.0016%.

The integration interval for rain rate calculation may either be determined directly as a time period or via a rain amount, when simulating a tipping-bucket rain gauge. The data acquisition and display system allows time intervals of 15 s or longer. However, this limit is an arbitrary choice.

Each drop is tested if it has fallen within the integration interval limits. The contribution of a single drop to the rain amount is given by the quotient of the drop's volume and its effective measuring area. The contributions of all drops to the rain amount in the integration time interval are summed up. The rainfall rate (R) is yielded when dividing the rain amount by the corresponding time interval:

$$R = 3600 \; \frac{1}{\Delta t} \sum_{i=1}^{n} \frac{V_i}{A_i} \left[\frac{mm}{h}\right], \tag{2}$$

where:
Δt time interval [s]
i drop number
n total number of fully visible drops measured in time interval Δt
V_i volume of drop i [mm³]
A_i effective measuring area for drop i [mm²]

Drop size distribution (DSD)

Calculation of DSD needs the following information: equivolumetric sphere diameter of raindrops, their time stamps, the sizes of their effective measuring areas, their fall velocities, definition of an integration interval and of a size class width. Since the DSD is the number of drops per unit volume and per unit size, a particular size class is composed as follows:

$$N(D_i) = \frac{1}{\Delta t \; \Delta D} \sum_{j=1}^{m_i} \frac{1}{A_j \; v_j} \left[\frac{1}{m^3 mm}\right], \tag{3}$$

where:
Δt time interval [s]
i denotes particular drop size class
j denotes particular drop within size class i and the time interval Δt
m_i number of drops within size class i and time interval Δt.
D_i mean diameter of class i [mm]

ΔD   width of drop size class [mm]
A_j effective measuring area for drop j [m²]
v_j fall velocity of drop j [m/s]

Raindrop orientation angles and horizontal velocity

As already mentioned in the introduction, precise knowledge on the shape of raindrops is of utmost importance when studying their interaction with electromagnetic waves, especially when interpreting polarimetric radar data or frequency reuse in signal transmission. Well defined models exist describing the shape of raindrops. They are always assumed to be bodies of revolution with a vertical rotation axis, the models give either their oblateness defined as the ratio of the maximum vertical to the maximum horizontal dimension (e.g., Poiares Baptista 1992), or their shape by means of $r = f(\theta)$ functions in polar coordinates (e.g., Pruppacher and Pitter 1971). In a first approximation raindrops are modeled as ellipsoids. The polarization status is not affected by non-canted ellipsoids. Canted ellipsoids will generate differential phase shift and differential attenuation and affect the polarization status. Not only rotation of polarization planes occurs, orthogonality is lost as well.

In telecommunications, that unwanted behavior causes effects like crosstalk, one polarization's channel talks into the other. The degree of crosstalk is measured as cross-polar discrimination (XPD). Considering backscatter, non-spherical canted raindrops cause cross-polar reflectivity. To the best of the authors' knowledge the canting angles of raindrops could not be measured up to now on a continuous basis under field conditions, the 2D-Video-Distrometer is the first instrument producing such data.

Since the 2D-Video-Distrometer applies line scan cameras, distortions are introduced in the views whenever objects are moving not only vertically but also horizontally. Explicitly, it is stated here that the distortions caused by horizontal motion do not affect the measurement of equivolumetric sphere diameter, fall velocity and the ratio of a hydrometeor's maximum vertical to its maximum horizontal extension.

Nevertheless, an approximation and a method, precise for bodies of revolution, have been developed to recover from the distortions by line scanning.

Distortions may approximately be recovered, assuming that a drops' top has to reside exactly above its bottom, which is perfectly true for spheres and holds quite well in case of small rain drops, too. The principle of this correction algorithm is to consider a line from the contour's bottom to its top and to put it fully upright, this in both views.

The amount such line has to be put up allows an approximate determination of the drop's horizontal velocity.

However, since this approximation method does not consider axes of symmetry in the drops' views, possibly existing non-zero orientation angles are ignored. The ratio of the drop's maximum vertical to its maximum horizontal extension, therefore, is only an approximate measure for its oblateness. The precise calculation would result in the oblateness as the geometrical means of the ratios of the drop's axes of revolution to their biggest extension orthogonal to those axes. In analogy, the same holds for determination of the horizontal velocity. For a precise reading the drop's orientation angle should be taken into account, introducing a natural displacement of the drop's top against its bottom.

For bodies of revolution a method has been developed to precisely recover distortions by line scanning by Schönhuber (1998). The orthogonality of a drop's axis of revolution to its true orthogonal set of lines may be recovered. For the set of lines orthogonal to an axis of symmetry this axis is a bisection line in each view. Through distortion by line scanning and horizontal movement orthogonality is lost. The axis of symmetry, however, still stays the bisection line to this set of originally orthogonal lines. This characteristic may be used and through vector analysis orthogonality and thus the true shape and orientation may be recovered.

Application of the precise method to front and side view allows composing the orientation angles of the drop. In windy conditions some drops are observed with rather irregular shapes. They hardly allow recognition of an axis of symmetry. The use of this method is, therefore, recommended at moderate wind speed only.

1.4 Current implementation

The 2D-Video-Distromter is manufactured in a small scale series production, with first deliveries having done in 1996. Twenty-two units have been manufactured up to now. The original classic tall 2D-Video-Distrometer version in a second generation has been upgraded to the low-profile 2DVD version (Fig. 3). This was done in reaction to publications by Nešpor et al. (2000), indicating wind induced measurement errors caused by the shape of the classic tall Sensor Unit housing. Consideration of experiences gained, technical progress in hardware (faster line scan cameras) and user feedback in combination

Fig. 3. Low-profile 2D-Video-Distrometer, Outdoor Electronics Unit (OEU) and Sensor Unit (SU)

with continuous small improvements have formed an outstanding instrument.

1.4.1 Specifications

Table 3 lists performance specifications for the low-profile version, being nearly identical to those of the earlier classic tall unit and Table 4 gives dimensions and weight.

1.4.2 Maintenance procedures

Maintenance includes check of calibration status, cleaning of optical surfaces, replacing bulbs of the illumination units and maintaining the data archive. Checking calibration status is rather easy. The user drops a few high precision stainless steel spheres through the measuring area and checks if these are correctly recorded in size and height/width ratio. Intervals for cleaning of optical surfaces depend on environmental conditions, how much dust settles on mirrors and lenses. Nominal life

Table 3. Performance specifications (low-profile 2DVD)

resolution (horizontal)	better than 0.19 mm
resolution (vertical)	better than 0.19 mm for fall velocities less than 10 m/s
vertical velocity accuracy	better than 4% for velocities less than 10 m/s
sampling area	approximately 100 × 100 mm²
rain rate compared to tipping bucket	differences typically less than 10%
mains voltage	100–240 V at 50/60 Hz
power consumption (SU, OEU, IUT)	approximately 500 W

Table 4. Dimensions and weight (low-profile 2DVD)

	Sensor Unit (SU)	Outdoor Electronic Unit (OEU)
approx. length [mm]	1100	750
approx. width [mm]	1100	600
approx. height [mm]	350	300
approx. weight [kg]	85	45

time of bulbs is 2000 h, a period of not even full 3 months. Though the bulbs life cycle usually exceeds the nominal life time by far, it is nevertheless an issue repeatedly requiring the user's attention. Maintaining the data archive is straightforward: the 2DVD for each day creates a separate data set and the wealth of information is stored in highly compressed binary files.

A critical issue with 2DVD operation is the precise alignment of optical components. Caused by the short exposure time of 18 µs and by the simultaneously required high f-stop values the necessary background illumination intensity is obtained at exact optical path alignment only. A well aligned and well calibrated 2DVD typically keeps its calibration status for several months. With necessary replacement of bulbs or other similar interventions necessity for a recalibration might arise.

1.5 Experiences

The 2D-Video-Distrometer has been used in various experiments, collecting a vast amount of data, in various climatic conditions, from Alpine mountain conditions to tropical rain measurements. A few employments under specific conditions shall be mentioned explicitly, in the following.

Under Contract with ESA/ESTEC, a tropical rain measurement campaign was carried out in Lae, Papua New Guinea from April to September 1995. The zero degree isotherm in the tropical regions is generally higher than in moderate climate. This is one of the reasons that evolution of rain events, especially of the drop size distribution, may differ significantly from rain in moderate climate. The measurement campaign was a good success confirming that drop size distributions in tropical rain are characterized by a much bigger number of small drops than in moderate climate (Schönhuber 1998).

In cooperative research efforts of the Institute of Applied Systems Technology, JOANNEUM RESEARCH, Graz/Austria and the Department of Electrical Engineering, Colorado State University Ft. Collins CO/USA a 2D-Video-Distrometer participated in storm chasing field campaigns 1995 and 1996 in Colorado. Mounted in one of the vans of Colorado State University the 2D-Video-Distrometer was sent right into storm cores. An excess of big drops compared with Marshall-Palmer (1948) drop size distribution often was observed, only polarimetric radar observations could give reliable predictions of the rain rate (Hubbert et al. 1997).

In winter 1996/97, the Institute of Applied System Technology, JOANNEUM RESEARCH, Graz, Austria and Atmospheric Science, Swiss Federal Institute of Technology, Zürich Switzerland cooperated in a snowfall and melting particles measurement campaign. A number of sensors collected a comprehensive data set, the experiment was focusing on the microstructure of the transition region from ice to water (melting layer). Data from three radar systems were acquired, three types of distrometers, a number of tipping-bucket and several other instruments were involved (Barthazy 1998).

In winter 2000 and 2001, the 2D-Video-Distrometer contributed to the Snow Avalanche Monitoring and Prognosis by Laser Equipment (SAMPLE) Project at Mt. Erzberg/Austria. In cooperation with the Technical University Graz/Austria and the Department of Avalanche and Torrent Research, Innsbruck, Austria, precipitation measurements with this instrument have been carried out to obtain information on the layering and mechanical stability of the snow cover and resulting avalanche risk (Moser et al. 2001).

Drop oscillation and canting angle experiments (Thurai et al. 2005) were done in cooperation with the Department of Electrical Engineering, Colorado State University, USA in 2004 from an approximately 80 m high railroad bridge (Jauntal, Carinthia, Austria). Using fire brigade water hoses artificial heavy showers were produced including a high number of big drops.

The measurement principle and the specifications of the 2D-Video-Distrometer make it an ideal instrument also for all kinds of industrial measurements wherever drops or particles have to be measured in real-time. Examples for these applications are measurement of nozzle properties or the efficiency of fire retardants dispensed from aircrafts.

1.6 Scientific merits

The usefulness of data collected by the 2D-Video-Distromter is documented in a large number of publications. In the following, only some recent ones shall be mentioned.

Attenuation and cross-polar discrimination were studied by Thurai and Bringi (2005), with a discussion on drop oscillations given in Thurai et al. (2005). Intercomparisons regarding drop size distributions, rainfall rates and derived radar reflectivities, involving data from different instruments, were carried out by Bringi et al. (2005) and Kanofsky et al. (2005). A climatological analysis of drop size distributions based on 2DVD data is presented by Schuur et al. (2005). Winter storm observations by radar and 2DVD are discussed by Ikeda et al. (2005). Distrometer derived Z-S relations are presented by Tokay et al. (2007). These publications confirm that the special suitability of the 2DVD data type for investigations of precipitation microstructure is due to its ability to measure precipitation particles with no upper limit in size, to give reliable information also on shapes of rain drops and to record contours and fall velocities of melting and solid particles.

Figure 4A shows a 6.6 mm drop with a canting angle of 4° in its front view. Based on such data, predictions of transmission signal attenuation and polarization crosstalk are possible.

Drop size distribution and drop shapes are of fundamental importance when interpreting polarimetric radar data. Frequently, radar measurements present higher values for the differential reflectivity Z_{DR}

than can ever be expected due to the widely used Marshall and Palmer (1948) raindrop size distribution together with one of the standard shape models. The 2DVD measurements show that the standard literature models for drop size distributions may considerably underestimate the number of big drops, especially in moderate climate convective events, an example is shown in Fig. 4B.

The use of higher frequencies for radio transmission services calls for better knowledge of scattering and absorption by solid and mixed phase precipitation. Figure 4C shows the front view of a snowflake and Fig. 4D a vertical velocity versus diameter graph for a mixed phase event. Whereas smaller particles are completely melted as indicated by their fall speed following the rain drop fall speed models, bigger particles are not melted completely. Their fall speed is significantly smaller than those of equal size raindrops.

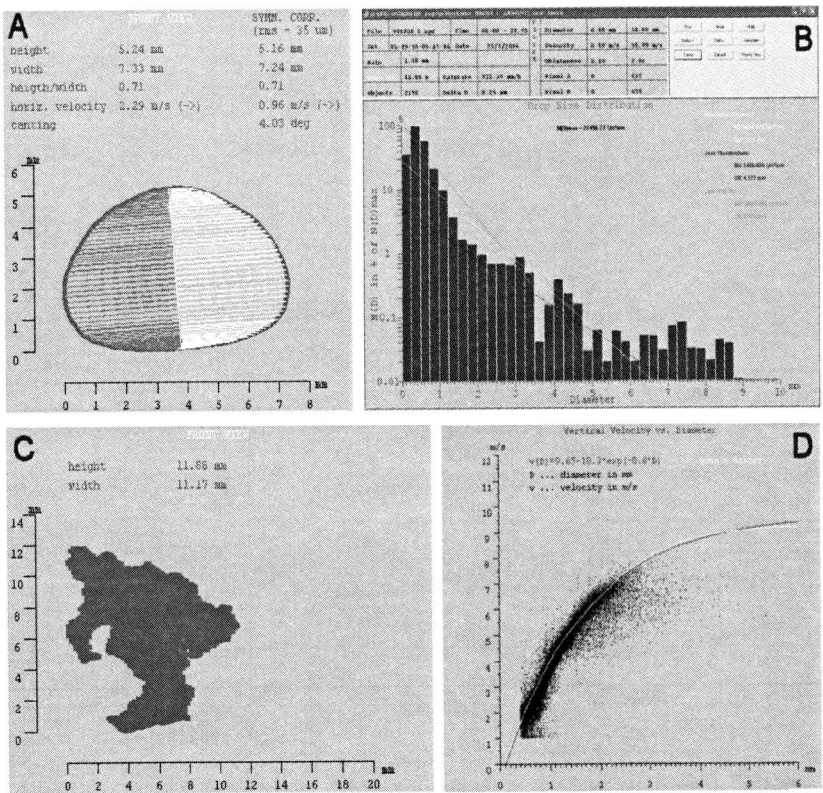

Fig. 4. Large raindrop, DSD with many big drops, snowflake, fall velocity of mixed phase event

Unique system performance analysis

To the authors' knowledge, no other distrometer provides such detailed system performance analysis and thus validation of results. Any potential calibration problems are immediately detected since particle contours would appear as distorted. In case of inhomogeneous filling of the measuring area for whatever reason, the virtual top view does indicate that.

Measurement of big precipitation particles

With the 2D-Video-Distrometer there is practically no upper limit in measuring precipitation particle sizes. In natural rain the authors have found drops with equivolumetric sphere diameters of up to 7.9 mm. Especially in mid-latitude convective events, even for rather moderate rain rates drops in the size range of about 7 mm were recorded. These observations answer the question of unexpectedly high weather radar differential reflectivity readings. Snowflakes or hailstones exceed this size.

Independent measurement of fall velocity

2DVD fall velocity readings are independent; they are not based on any literature model or other assumptions. Such information is of special value when analyzing mixed phase events. Figure 4D lower right hand side gives fall velocities versus diameter for a mixed phase event, clearly indicating that small particles already are fully melted to raindrops. Bigger particles still fall slower than raindrops of same size, indicating that they consist of a water-ice mixture. For comparison, two rain drops' fall velocity literature models are indicated, i.e. those by Gunn and Kinzer (1949) and Atlas et al. (1973). Moreover fall velocity information may be used for estimating the water equivalent of snowflakes and melting particles. Though the three-dimensional microphysical structure is not fully represented by contour data, Schönhuber et al. (2000) describe a method to estimate the water equivalent on basis of a detailed contour inhomogeneity and fall velocity analysis.

Measurement of drops' axis ratio and orientation angles

From the 2DVD contour data rain drops' axis ratio values may be obtained as well as their orientation angles. Such knowledge is especially important when predicting crosstalk for frequency reuse radio transmission and for interpretation of polarimetric weather radar reflectivities.

Figure 4A upper left hand side shows an example of a drop's front view, with axis ratio and canting angle indicated.

Oscillation of raindrops

Raindrops may oscillate in shape. Their oblateness may oscillate around its mean value, i.e. the oblateness of equilibrium state. As may be shown theoretically such oscillations have a non negligible impact on the interpretation of polarimetric weather radar reflectivity. The degree of natural drop oscillations needs, therefore, to be determined experimentally. The 2D-Video-Distrometer allows studying drop oscillations.

Measurement of solid and mixed phase particles

Though knowledge on solid and mixed phase particles is of special importance for a thorough understanding of atmospheric situations and processes, relevant measurements are less available than rain recordings. The 2DVD is able to provide front and side contour information also of melting and solid particles. Such information may be used for enhanced comparisons to weather radar reflectivities, measured close above the distrometer. With mixed phase events the portions of e.g., rain and hail may be separated in the 2DVD-recorded data set and the wave propagation parameters separately be derived, summed up and then in total predicted for such precipitation type. Figure 4D shows a vertical velocity versus diameter diagram of a mixed phase particle event. The different vertical velocities for identical diameters may be used to distinguish between solid and mixed phase particles.

Measurement of snowflakes

Solid precipitation particles like snowflakes or ice needles are not rotation-symmetric. Therefore, some methods applicable for rain are not valid any longer. As described by Teschl et al. (2006), 2D-Video-Distrometer front and side view data from a snowflake can be used to approximate a three-dimensional object. This object is one possible shape of the measured particle. A CAD tool may be used for geometrically modeling the object by its finite elements. As a first approach, such particle is assumed to consist fully of ice, improved modeling may consider air inclusions. Such data may be taken as a basis for the calculation of scattering amplitudes of individual snowflakes. Figure 4C shows a front view of a snowflake. Snowflakes present totally irregular shapes including holes in the view, therefore, the graphics representation for front and side view has been changed from drawing the contours to fully coloring the blocked area.

1.7 Outlook

The 2DVD has proved to be a useful instrument, providing a unique set of data which adds substantial value to various fields of precipitation related research activities. Nevertheless, the need for further improvement is a big concern of the authors, thus plans and laboratory prototypes for a third-generation 2DVD have been set up. Existing 2DVD units are built in a way that the optical paths are three-dimensionally arranged, this requires a voluminous and sturdy construction to mount the optical elements onto. Alternatively, a concept for a purely horizontal arrangement of the optical paths has been developed; basic experiments have successfully confirmed its feasibility and usefulness. Onto a rigid supporting structure of some 80 × 80 cm², the optical elements are mounted within a horizontal plane, in a way that the geometry for measurements finally results in the same data type and precision as provided by the existing bigger Sensor Units. The weight of the intended Sensor Unit thus may be decreased down to about 40 kg, a reduction by more than 50%. The compact design will allow omission of the alignment mechanics; instead the optical components will firmly be mounted during manufacture with no need for realignment by the user. Furthermore, tests already gave confidence that for this configuration the presently used standard halogen bulbs possibly may be replaced by LED illumination systems, increasing life time drastically, while simultaneously decreasing power consumption. The 2D-Video-Distrometer is based on high speed line scan cameras, generating a data rate of 40 MB/sec each. Such data rate still requires special handling, which presently is implemented by special camera interface slot-cards integrated into a Personal Computer (PC) located close to the Sensor Unit. This PC is housed within the Outdoor Electronics Unit. Recently, intelligent line scan cameras are available on the market, offering sufficient embedded computing power for online data compression. Data may be compressed directly by the intelligent line scan camera applying a shading algorithm, then a threshold detection algorithm and finally by composing a run length code which after further compression is sent to the Indoor User Terminal. A proof of this concept has successfully been carried out. Such setup allows to omit the Outdoor Electronics Unit and to reduce the whole setup to Sensor Unit plus Indoor User Terminal. A notebook PC can serve as indoor User Terminal. With the classic tall and also the low-profile 2DVD versions the distance of the optical slits to the rim of the housing is about 7 cm, in case of extreme wind speeds potentially hindering drops with slanted trajectories from reaching the measuring area. Though unwanted inhomogeneous filling of the measuring area is recognized in the virtual top view representation of the measuring area, with a new mechanical setup the distance of the optical slits to the rim of the housing shall be reduced and this unwanted effect then be made negligible. Finally, a next generation 2D-Video-Distrometer shall be available, offering the proven 2D-Video-Distrometer data type at costs and operation conditions acceptable not only for research organizations and universities, but for a wider user community.

References

Atlas D, Shrivastava RC, Sekhon RS (1973) Doppler radar characteristics of precipitation at vertical incidence. Rev Geophys Space GE 2:1–35

Barthazy E (1998) Microphysical properties of the melting layer. Dissertation submitted to Swiss Federal Institute of Technology (ETH) Zürich. http://e-collection.ethbib.ethz.ch/ecol-pool/diss/fulltext/eth12687.pdf

Bringi VN, Thurai M, Nakagawa K, Huang G-J, Kobayashi T, Adachi A, Hanado H, Sekizawa S (2005) Rainfall estimation from C-band polarimetric radar in Okinawa, Japan: Comparisons with 2D-video disdrometer and 400 MHz wind profiler. In: Proceedings 32nd Conference on Radar Meteorology. American Meteorological Society, 24–29 October 2005, Albuquerque, NM, USA, 11R.1

Delahaye JY, Vinson JP, Gloaguen C, Gole P (1995) Le Spectropluviomètre CETP. Document de Travail CETP, version 2.0, Juillet 31, 1995, CETP, France

Frank G, Härtl T, Tschiersch J (1994) The pluviospectrometer: Classification of falling hydrometeors via digital image processing. Atmos Res 34:367–378

Gunn R, Kinzer GD (1949) The terminal velocity of fall for water droplets in stagnant air. J Meteorol 6:243–248

Hauser D, Amayenc P, Nutten B, Waldteufel PH (1984) A new optical instrument for simultaneous measurement of raindrop diameter and fall speed distributions. J Atmos Ocean Tech 1:256–269

Hubbert J, Bringi VN, Chandrasekar V, Schönhuber M, Urban HE, Randeu WL (1997) Storm cell intercepts using a mobile 2D-video distrometer in conjunction with the CSU-CHILL radar. In: Preprints 28th Conference on

Radar Meteorology. American Meteorological Society, 7–12 September 1997, Austin, TX, USA, pp 436–437

Ikeda K, Brandes EA, Zhang G, Rutledge SA (2005) Observations of winter storms with a 2D video disdrometer and polarimetric radar. In: Proceedings 32nd Conference on Radar Meteorology. American Meteorological Society, 24–29 October 2005, Albuquerque, NM, USA, P11R.14

Joss J, Schram K, Thams JC, Waldvogel A (1970) On the quantitative determination of precipitation by radar. Wissenschaftliche Mitteilung Nr. 63, Zürich, ETH 1970, City-Druck AG, Zürich, Switzerland

Joss J, Waldvogel A (1967) Ein Spectrograph für Niederschlagstropfen mit automatischer Auswertung. Pure Appl Geophys 68:240–246

Kanofsky LM, Chilson PB, Schuur TJ, Zhang G, Brandes EA (2005) A comparative study of drop size distribution retrieval using two video disdrometers and a UHF wind profiling radar. In: Proceedings 32nd Conference on Radar Meteorology. American Meteorological Society, 24–29 October 2005, Albuquerque, NM, USA, P11R14

Knollenberg RG (1970) The optical array: An alternative to scattering or extinction for airborne particle size determination. J Appl Meteorol 9:86–103

Löffler-Mang M (1997) Development of a low-cost optical disdrometer. MAP-Newsletter #7, October 1997, Zürich, Switzerland

Marshall JS, Palmer W McK (1948) The distribution of raindrops with size. J Meteorol 5:165–166

Moser A, Geigl BC, Steffan H, Bauer A, Paar G, Fromm R, Schaffhauser H, Köck K, Schönhuber M, Randeu WL (2001) Entwicklung eines Meßsystems zur Bestimmung der zeitlichen und örtlichen Schneehöhe in Lawinenhängen als verbesserte Grundlage für die Lawinenprognose und ihre Anwendung am Präbichl. SAMPLE Snow Avalanche Monitoring & Prognosis by Laser Equipment, Endbericht, Forschung Steiermark, 2001, Graz, Austria

Nešpor V, Krajewski WF, Kruger A (2000) Wind-induced error of raindrop size distribution measurement using a two-dimensional video disdrometer. J Atmos Ocean Tech 17:1483–1492

Poiares Baptista JVP (1992) Minutes of radar working group. In: Proceedings 17th Meeting of the Olympus Propagation Experimenters. 3–5 May 1992, Stockholm/Helsinki, pp 36–40

Pruppacher HR, Beard KV (1970) A wind tunnel investigation of the internal circulation and shape of water drops falling at terminal velocity in air. Q J Roy Meteor Soc 96:247–254

Pruppacher HR, Pitter RL (1971) A semi-empirical determination of the shape of cloud and rain drops. J Atmos Sci 28:86–94

Schönhuber M (1998) About interaction of precipitation and electromagnetic waves. Doctoral thesis, Institute of Communications and Wave Propagation, Technical University Graz/Austria, pp 181

Schönhuber M, Urban HE, Randeu WL, Baptista Poiares JPV (2000) Empirical relationships between shape, water content and fall velocity of snowflakes.

In: ESA SP-444 Proceedings of the Millennium Conference on Antennas & Propagation, 9–14 April 2000, Davos, Switzerland

Schönhuber M, Urban HE, Randeu WL, Riedler W (1996a) Technical Reference (Manual). under ESTEC/Contract No. 9949/92/NL/PB(SC) – Work Order No. 02 ('Development and Delivery of a 2D-Video-Distrometer'). Institute of Applied Systems Technology, JOANNEUM RESEARCH, Graz/Austria

Schönhuber M, Urban HE, Randeu WL, Riedler W (1996b) Field Test Report and Call Off Order Report, under ESTEC/Contract No. 9949/92/NL/PB(SC) – Work Order No. 02 and Call Off Order No. 09 under Work Order No. 01 Institute of Applied Systems Technology, JOANNEUM RESEARCH, Graz/Austria

Schönhuber M, Urban HE, Richter EM, Randeu WL, Riedler W (1993) Design Review Report, under ESTEC/Contract No. 9949/92/NL/PB(SC) – Work Order No. 02 ('Development and Delivery of a 2D-Video-Distrometer'). Institute of Applied Systems Technology, JOANNEUM RESEARCH, Graz/Austria

Schuur TJ, Ryzhkov AV, Clabo DR (2005) Climatological analysis of DSDs in Oklahoma as revealed by 2D-video-disdrometer and polarimetric WSR-88D radar. In: Proceedings 32nd Conference on Radar Meteorology. American Meteorological Society, 24–29 October 2005, Albuquerque, NM, USA, 15R.4

Steiner M (1988) Bericht über Vergleichsmessungen mit verschiedenen Niederschlagsmessinstrumenten Oktober-Dezember 1987. Bericht LAPETH-28, 1988, Laboratorium für Athmosphärenphysik, ETH Zürich Switzerland, pp 90

Teschl F, Randeu WL, Schönhuber M (2006) Modelling microwave scattering by solid precipitation particles. In: Proceedings EuCAP 2006. 6–10 November 2006, Nice, France, (ESA SP-262, October 2006)

Thurai M, Bringi VN (2005) Drop axis ratio from a 2D video disdrometer. J Atmos Ocean Tech 22:966–975

Thurai M, Bringi VN, Huang GJ, Hanado H (2005) Drop axis ratios and polarization dependence in rain. From 2-dimensional video disdrometer data. In: Proceedings 3rd International Workshop of COST 280. 6–7 June 2005, Prague

Tokay A, Bringi VN, Schönhuber M, Huang GJ, Hudak D, Wolff DB, Bashor PG, Petersen WA, Skofornick-Jackson G (2007) Disdrometer derived Z-S relations in South Central Ontario, Canada. In: Proceedings 33rd Conference on Radar Meteorology. 6–10 August 2007, Cairns, Queensland, USA

Urban HE, Schönhuber M, Randeu WL, Riedler W (1994) Technical Note, under ESTEC/Contract No. 9949/92/NL/PB(SC) – Work Order No. 02 (Development and Delivery of a 2D-Video-Distrometer) Institute of Applied Systems Technology, JOANNEUM RESEARCH, Graz/Austria

2 Using vibrating-wire technology for precipitation measurements

Claude E. Duchon

School of Meteorology, University of Oklahoma, OK, USA

Table of contents

2.1 Introduction 33
2.2 Principles of operation 35
2.3 Description of field site and data acquisition 36
2.4 Advantages of using three vibrating wires 41
2.5 Calibration-verification 44
2.6 Temperature sensitivity 47
2.7 Rain rate estimation 50
2.8 Very low precipitation events 53
2.9 Summary 56
References 58

2.1 Introduction

Vibrating-wire technology applied to the measurement of precipitation was developed at the Norwegian Geotechnical Institute (NGI), Oslo in the early 1980s (Bakkehøi et al. 1985). This application was an outgrowth of previous work at NGI involving the design and manufacture of instruments employing the vibrating-wire technique to measure strain and loads in concrete in bridges, earth pressures in dams, soil porewater pressure in boreholes under embankments and other geomechanical examples (Tunbridge and Øien 1988). The vibrating-wire precipitation gauge that has been available for many years is called the NGI Geonor T-200B and is the gauge investigated in this Chapter. Figure 1 is an example of a T-200B with and without its case removed. The Geonor T-200B gauge is in the class of a weighing-recording gauge and is widely

Fig. 1. A Geonor T-200B in the field (*upper*) and with its case removed (*lower*). The 12 L bucket rests in a support dish suspended from a circular flange by three vibrating wire transducers. Maximum precipitation capacity of the bucket is 600 mm (24 in)

used at more than 1400 locations world-wide of which over 500 are in Canada and the United States (Brylawski 2007, personal communication).

The purpose of this Chapter is to provide a summary of various field investigations involving the Geonor gauge that have been performed since 2000, mainly at Norman, Oklahoma, USA. The summary begins with a review of the essentials of gauge operation. This is followed by six Sections dealing with, in order, a description of the measurement site and data acquisition, advantages of using three versus two or one vibrating wires, calibration-verification of the wires, sensitivity of vibrating wire frequency to temperature, comparison of rain rates from the Geonor to rain rates from a disdrometer and observations of very low precipitation events. The last Section highlights the findings. This Chapter deals only with liquid precipitation, the primary reason being that frozen precipitation typically occurs only a few times annually in central Oklahoma and usually in the form of sleet or freezing rain. In addition, the gauge is unheated and, as will be seen later, its orifice is at ground level.

2.2 Principles of operation

The basis of operation of the Geonor vibrating wire precipitation gauge is that the fundamental resonant frequency of a wire secured at one end and under tension at the other end is given by

$$f = \frac{1}{2L}(T/u)^{1/2} \qquad (1)$$

where f is frequency, L is length of the wire, T is tension and u is mass per unit length of wire. The derivation of Eq. (1) is given by Raichel (2006, p. 71). Tension is supplied by the weight of the bucket and its contents.

As shown by Bakkehøi et al. (1985), the relationship between the fundamental resonant frequency of the wire and the strain on the wire is

$$f^2 - f_0^2 = K \frac{\varepsilon E g}{4L^2 \rho} \qquad (2)$$

where ε is strain, E is Young's modulus, g is acceleration due to gravity, ρ is density of the wire, f is frequency of vibration at strain ε, f_0 is the frequency at zero strain and K is a constant of proportionality dependent on the design of the gauge.

However, Lamb and Swenson (2005) pointed out that if there is zero strain, the null frequency also must be zero. To remedy this situation, they consider f_0 to be a reference frequency at some reference strain ε_0 determined by the weight of the empty bucket and/or the bucket support. Ignoring a change in length of the wire in Eq. (2) from reference strain to the applied strain, the appropriate expression for frequency related to strain is

$$f^2 - f_0^2 = K\frac{(\varepsilon - \varepsilon_0)Eg}{4L^2\rho} \tag{3}$$

Given that constant E is the ratio of stress T/A to strain, where A is the cross-sectional area of the wire, it follows that

$$f^2 - f_0^2 = K\frac{(T - T_0)g}{4L^2\rho A} = K\frac{(T - T_0)g}{4L^2 u} \tag{4}$$

The differential tension $T-T_0$ is a consequence of the weight W of the contents in the bucket (or contents plus bucket). Recognizing that tension is a reactive force on the wire, i.e., $T-T_0 = Wg$, Eq. (4) becomes

$$f^2 - f_0^2 = K\frac{Wg^2}{4L^2 u} \tag{5}$$

We see that the relation between precipitation (as embedded in W) and frequency of a vibrating wire is nonlinear.

While Eq. (5) is valid, the actual formulation used by Geonor in calibrating vibrating wires captures the nonlinearity in the form of a second-degree polynomial given by

$$P = A(f - f_0) + B(f - f_0)^2 \tag{6}$$

in which P is the depth of the precipitation in appropriate units consistent with the units of constants A and B. Figure 2 shows a plot of each term in Eq. (6) for a typical vibrating wire. The linear term controls the accumulation P near empty bucket while near full bucket the contributions from the linear and square terms are about equal.

2.3 Description of field site and data acquisition

The data used in the various analyses in this Chapter were acquired from a field site located on the north campus of the University of Oklahoma, Norman (97.465° W, 35.236° N) that has good exposure in all directions. Figure 3 shows the pit from which all Geonor measurements were taken.

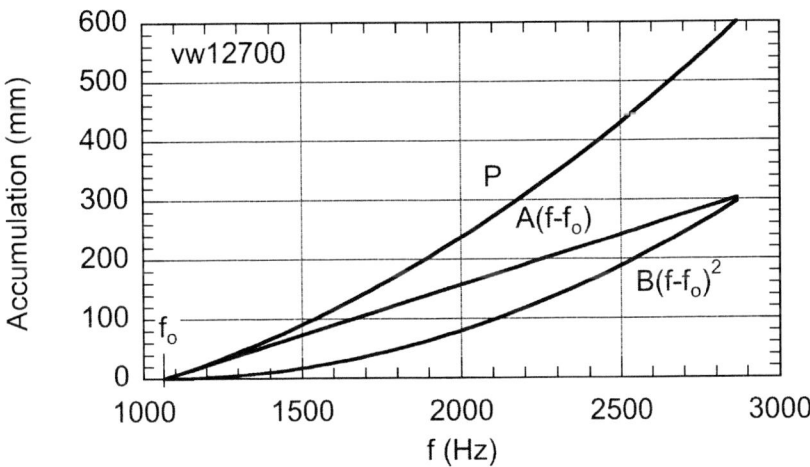

Fig. 2. Typical plot of precipitation accumulation versus frequency from Eq. (6) showing contributions from linear and nonlinear terms

Fig. 3. View of pit with Geonor orifice at far end and 2dvd orifice at near end

The pit has interior dimensions 3.7 m × 1.8 m × 1.4 m deep and is covered by a grill to which is attached a fabric to prevent splashing of raindrops. The Geonor is located at the far end of the pit and its orifice is about 1 cm above the fabric. A two-dimensional video disdrometer (2dvd) is located at the near end of the pit and its orifice is also about 1 cm above the fabric. Figure 4 shows the interior of the pit with the Geonor at the near end and the 2dvd and supporting electronic equipment at the far end. The 2dvd was designed and constructed by Joanneum Research, Graz, Austria. More will be said about the disdrometer in Sect. 2.7.

All data were collected using a Campbell Scientific, Inc., 23X data logger. One-minute averages of wind speed and wind direction near the pit at a height of 2 m, air temperature inside the Geonor case close to a vibrating wire transducer, air temperature inside the pit near the Geonor, data logger temperature, data logger voltage, frequencies from each vibrating wire and their conversion to accumulation of precipitation in units of millimeters were transmitted from the site to the University of Oklahoma and archived on disk.

There are two instructions in the data logger available to determine the frequency of each vw (vibrating wire). The P3 instruction counts the integer number of cycles over a selected time interval using a pulse counting method. The result is that the frequency is accurate to within +/− 1 pulse (cycle) over the given time interval. For 1-minute averaging the inaccuracy is within +/− 1 cycle/60 s = +/−0.0166 Hz which is equivalent to the resolution in frequency and is independent of frequency. However, because of the quadratic relation in Eq. (6), there is a linear relation between error in accumulation with frequency due to error in frequency. For a typical vw the results are: +/−0.0166 Hz = +/−0.003 mm for empty bucket and +/−0.0166 Hz = +/−0.008 mm for full bucket.

The P27 instruction calculates the time required to count a fixed number of cycles so that the ratio of the latter to the former is the frequency of the vw. The P27 instruction was used most often with its control parameters set such that 3000 cycles were required to occur within 3 s for each vw and the procedure repeated every 10 s. Thus, in successive 10 s increments 9 s were available for measurements of frequency from the three vws and 1 s for all other meteorological data. Six successive measurements for each variable were averaged to yield the 1-minute data used in all subsequent analyses.

Chapter 2 - Vibrating-wire technology for precipitation measurements

Fig. 4. View of inside of pit with Geonor in foreground and 2dvd in background

The P27 instruction produces a smoother accumulation time series than the P3 instruction when rain rates are small. Figures 5a and 5b show a low intensity rain event over a 2-hour period beginning about minute 760 and ending at minute 870. Figure 5c shows the subsequent hour in which there is no precipitation. The standard deviation of the differences in Fig. 5c, in which there is no rain, reflects mainly the resolution error in the P3 instruction. Because the standard deviations in Figs. 5a and 5b are approximately the same as the standard deviation in Fig. 5c, the conclusion is that instruction P27 yields a better estimate of rain rate than instruction P3. The larger fluctuations using instruction P3 relative to P27 also can be visually observed in each panel.

Just the opposite conclusion follows when rain rates are high. Figure 6a is an example of very high rain rates concentrated around minute 1410. Note that the P3 time series has been shifted by 1 minute so that the individual rain rates can be easily distinguished. Because of the insensitive scale, the rain rates appear to be virtually identical. However, Fig. 6b, in which the time series of differences is plotted (without the time shift), shows that differences as large as 10 mm/h can be observed. The differences arise because of the nonlinear changes in rain rate during the course of a minute. Instruction P3 samples continuously in time while instruction P27 computes the time needed to count 3000 cycles six times per minute, which, at full bucket, is about 6.3 s (see Fig. 2).

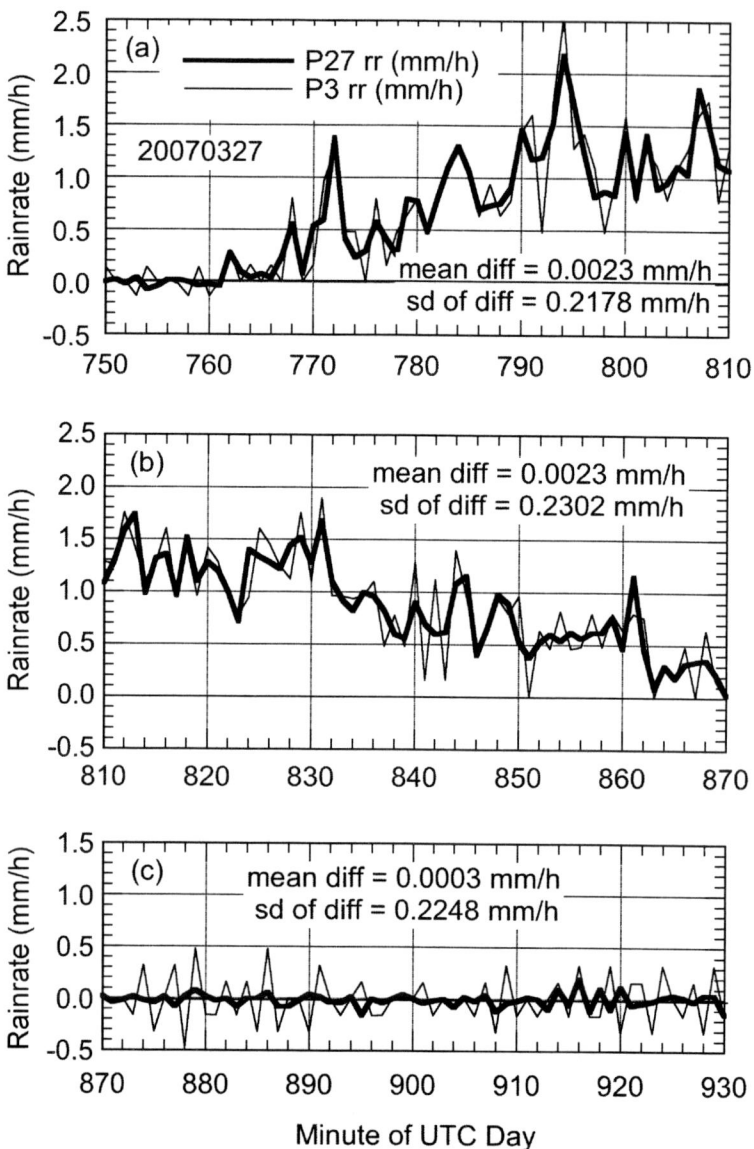

Fig. 5. Comparison of time series from instructions P3 and P27 for low 1-minute rain rates. (**a**) First 60 minutes, increasing rain rate. (**b**) Second 60 minutes, decreasing rain rate. (**c**) Third 60 minutes, zero rain rate

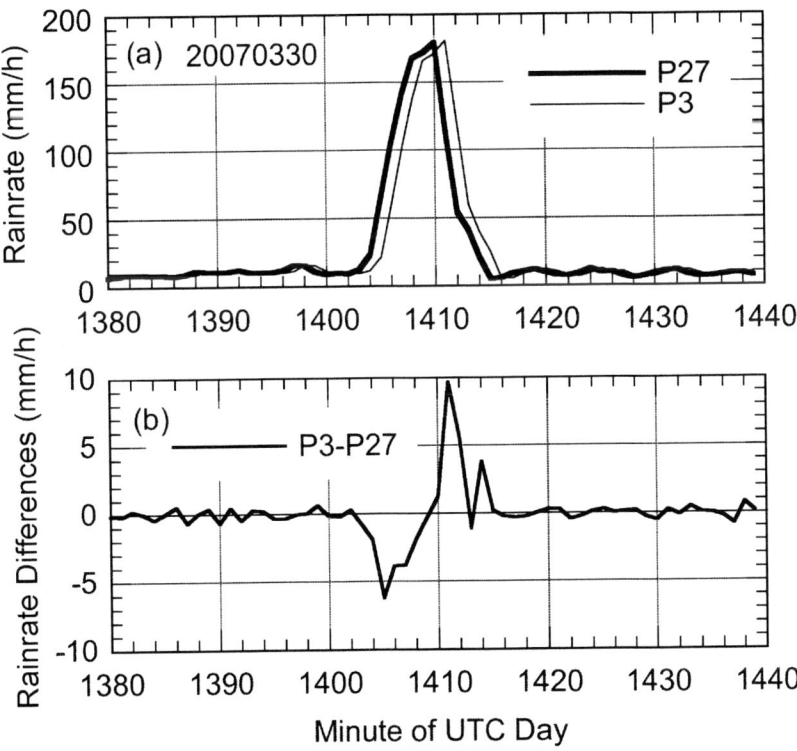

Fig. 6. (a) Rain rates using instruction P27 and P3 for an intense rainfall event. (b) Differences in rain rate using P27 and P3

In summary, instruction P27 is better for estimating low rain rates than instruction P3 because of its higher resolution in frequency, while at higher rain rates P3 is better because it samples continuously with time thereby accommodating nonlinear changes in rain rate during the course of a minute. Both instructions yield the same accumulations of precipitation with time.

2.4 Advantages of using three vibrating wires

The Geonor T-200B can operate with one, two, or three wires. In the case of using one or two wires, the remaining suspension is provided by two chains or one chain, respectively. Apart from the cost of transducers, the advantage lies in using three wires. With three wires, a continuous comparison between pairs of wires is available that can be used to determine if and when the performance of a wire warrants replacement. If that occurs, the remaining two and a chain can be used to provide continuous measurements of accumulation. Also, and particularly for low rain rate estimation and high wind speed, the average of the accumulations from three wires yields better estimates than employing only one wire. As air passes over the inlet orifice, the air pressure inside the case is reduced (the Bernoulli effect). The gustiness in the wind produces a variable pressure on the suspended bucket resulting in variable tension on the wires and fluctuations in their natural frequency. By averaging the three outputs, noise due to the wind is significantly reduced and the best estimate of accumulation retained.

Figure 7 is an example that demonstrates the value of averaging. Each of the five time series (a) through (e) shows the residuals or departures (in mm) from a smooth curve fitted to each time series of accumulation (not shown) over a period of 2 hours with no precipitation. The residuals are equivalent to the fluctuations in 1-minute accumulations. The top three times series show simultaneous residuals for a mean wind speed of 3.94 m/s at 2 m and an accumulation of 98 mm. Their standard deviations vary from 0.00108 mm to 0.00141 mm. If the three accumulation time series are averaged minute-by-minute and a smooth curve fitted to the averages, the residuals are shown in time series (d). The standard deviation is 0.00073 mm, a substantial reduction of noise relative to that in any single wire. If one hypothesizes that the residuals in time series (a), (b) and (c) are white noise, then, theoretically, the variance of the average time series (d) would be 1/3 the average of the variances of the three wires, or a reduction in variance of 67%. The actual reduction is 66%.

Figure 7e shows that the higher the wind speed the greater the noise. The computation of the time series of residuals here is similar to that in time series (d) except that a different date and time have been examined in which the accumulation is only 11 mm greater but the mean wind speed is 9.11 ms^{-1}. We see that the standard deviation is five times larger than in time series (d) in which the mean wind speed is somewhat less than ½ that in (e). Thus wind speed is the primary contributor to noise in accumulation time series in otherwise normally functioning vibrating wires. Accumulation noise is impacted also by the magnitude of the accumulation itself. In a case (not shown) in which there were 585 mm in the bucket and the mean wind speed was 4.04 ms^{-1}, the standard deviation of the residuals or fluctuations in accumulation was 0.00159 mm, more than twice that in comparable time series (d) in Fig. 7 with 98 mm accumulation. While one would expect the greater mass in the bucket to dampen the wind-induced

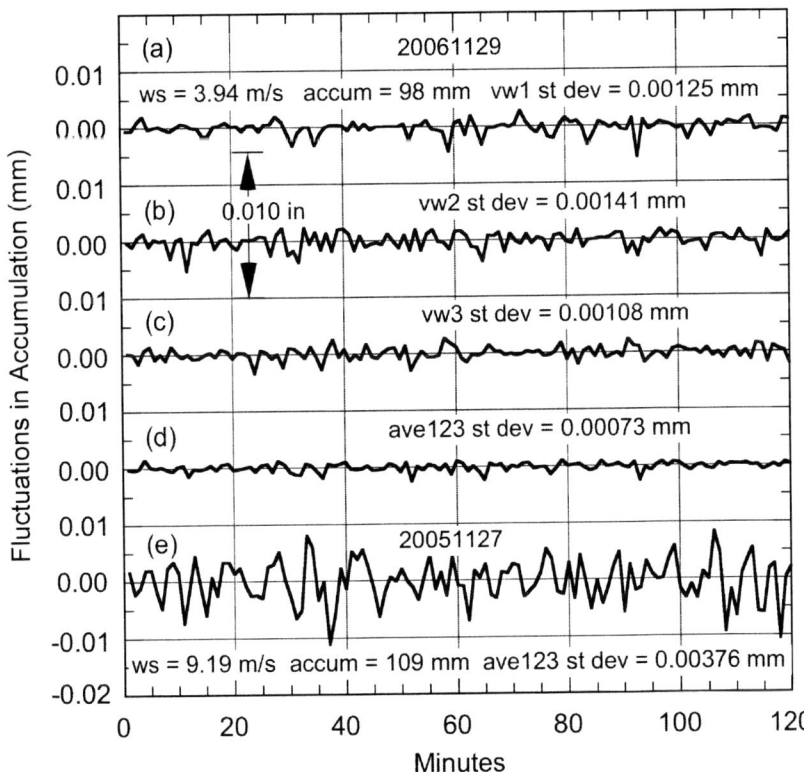

Fig. 7. (**a**) – (**c**) Noise in accumulation due to wind (~4 m/s at 2 m) for each of the three vibrating wires over a 2-hour period. (**d**) Wind noise when the accumulations from the three wires are averaged. (**e**) Wind noise when wind speed is somewhat more than doubled (~9 m/s)

fluctuations in accumulation, the change in accumulation per unit change in frequency at 585 mm accumulation is twice that at 98 mm (see Fig. 2). Apparently, the greater sensitivity of change in accumulation to change in frequency with increasing accumulation overwhelms the effect of increasing mass in the bucket.

It should be emphasized that the above analysis was based on a gauge in a pit, well protected from the wind. Any vibrations on an aboveground gauge due to the force of the wind directly on the gauge can cause fluctuations in accumulation that would be added to the noise described above.

In summary, noise in accumulation due to wind increases with both increasing wind speed and increasing accumulation in the bucket.

While this noise will not significantly affect rain or snow accumulations, it can significantly affect computation of rain and snow rates.

2.5 Calibration-verification

To monitor possible changes in coefficients A and B in Eq. (6) from their factory values, a calibration-verification (cal-ver) is performed in the field. If changes are observed based on the cal-ver, the vibrating wire transducer(s) requires recalibration. In the cal-ver procedure various known weights are placed in the bucket, the equivalent accumulations in mm of water calculated and the observed accumulations based on the frequency-to-accumulation conversion from Eq. (6) compared to the accumulations based on the weights. Fourteen stainless steel weights, each in the form of a disk with diameter 14.3 cm, thickness 0.6 cm and a hole in the center, were machined to weigh 800 g each for which the equivalent water accumulation in the bucket is 40 mm. The disks were weighed using both a precision mechanical balance and a precision electronic balance. Based on independent measurement trials, the inaccuracy of the weight of a disk was found to be within +/−0.5 g and the total weight of all disks within +/−4 g or +/−0.20 mm at 600 mm accumulation. The range in mean disk weight was from 797.5 g to 800.2 g.

The first step in a cal-ver is to sequentially place the disks onto a spindle mounted in the center of another disk made of a polymer machined to snugly fit inside the base of the bucket. Two or three 1-minute measurements were allowed for the empty bucket, spindle and mount and each added disk. Measurements were made until all 14 metal disks were stacked, followed by similar measurements after each disk and spindle and mount were removed and at empty bucket.

Results from a complete cal-ver are shown in Fig. 8. Figure 8a shows the differences between the observed or measured accumulation and the weight (in units of equivalent accumulation) versus weight. That there are two traces for each wire (vw1, vw2, vw3) and their average (ave123) is a result of successively adding all the weights and then reversing the process. This figure shows that the absolute departures from their initial values for the individual wires increase with increasing weight in the bucket, while the average differences are comparatively flat. The explanation for the diverging curves is that the center of mass becomes slightly displaced from the vertical as the disks are stacked one upon another. Thus this method of calibration-verification would not be appropriate when only one or two wires are used because compensation for displacement of the center of mass would not be taken into account.

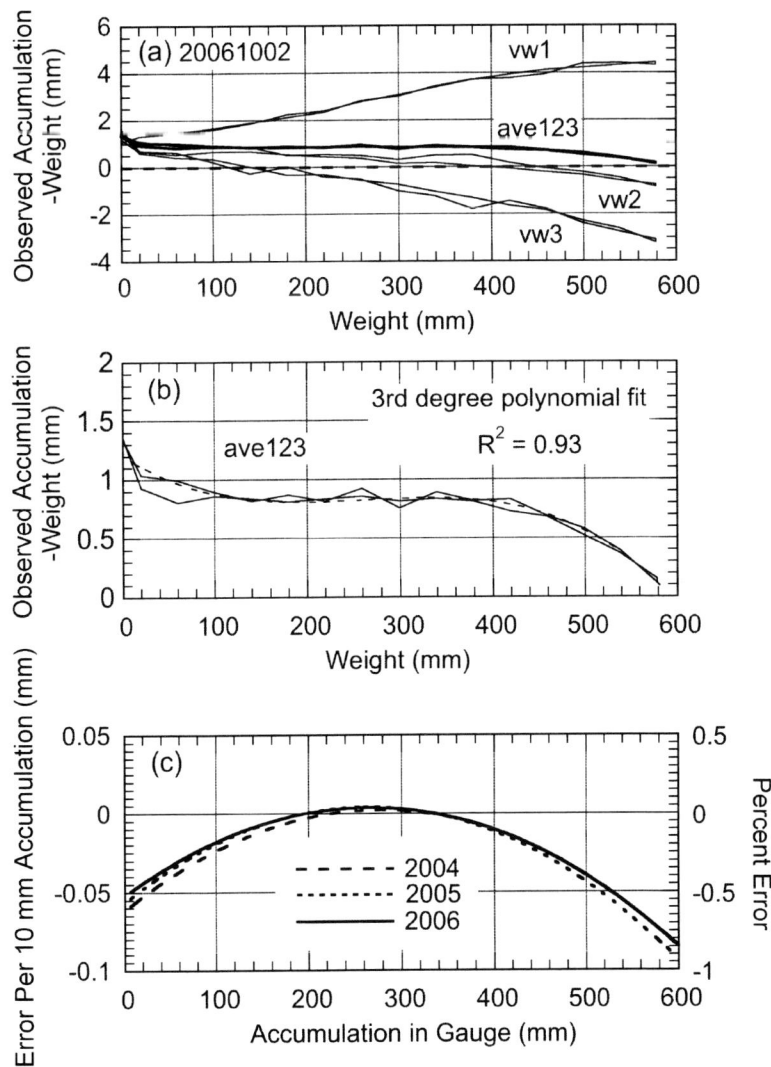

Fig. 8. (a) Differences between observed accumulation and weight in bucket (converted to accumulation) for each wire and their average. (b) As in (a) but with an expanded scale showing the best polynomial fit to the average differences. (c) Derivative of the polynomial in (b) showing error per 10 mm accumulation and percent error with accumulation. Results from previous two years are also shown

It should be noted that the mean curves for cal-vers of the same wires in 2004 and 2005 (not shown) were quite similar to that for 2006 in Fig. 8a. Although the slopes of the individual vibrating wire curves were different, their positions relative to the mean curve were the same. Given that the bucket, spindle disk and vibrating wire transducers have the same emplacement relative to each other for each cal-ver, the conclusion is that the increasing horizontal displacement of the center of mass with increasing number of weights is due to a slight offset of the spindle from vertical.

The next step in the cal-ver is to fit a smooth curve to the average differences. Figure 8b is an expanded view of the ave123 curve in Fig. 8a and includes a minimum least-squares best-fit polynomial. It is believed that the true average differences, in fact, vary smoothly with weight in the bucket and the departures from the fitted curve are measurement noise. The third step is to differentiate the polynomial. The result is shown in Fig. 8c for the polynomial in (b) and for similar polynomials from the previous two years. The curves can be interpreted as the error in accumulation per 10 mm of accumulation or percent error. A 10 mm denominator was chosen because it is a nominal amount for precipitation event totals in the Southern Great Plains. Figure 8c indicates that near empty bucket the error in a 10 mm precipitation event is about –0.05 mm or an undercatch of about 0.5%. When the bucket is around one-half full the error is close to zero and when it is completely full the error for a 10 mm precipitation event approaches –0.1 mm or an undercatch of 1%.

It is apparent that, based on these cal-vers, there has been no need to recalibrate any of the wires. That is, the values of the coefficients A and B in Eq. (6) originally provided by the factory for each wire have not significantly changed. It should be emphasized that the only purpose of a cal-ver is to assess the correctness of the coefficients in Eq. (6). The inaccuracy of a measured or observed precipitation amount depends mainly on other factors, e.g., undercatch due to wind (aboveground gauges), particularly for solid precipitation, wetting loss, evaporation or sublimation loss (negligible with continuous measurements) and splash-in or splash-out (see, for example, Groisman and Legates 1994). This investigation suggests that factory determined coefficients for vibrating wires are accurate and stable with time. The author has performed many other calibration-verifications since 2000 not only of this gauge (with different wires) but also five three-wire gauges at the National Center for Atmospheric Research, Boulder, CO, USA. To date a recalibration was recommended for only one wire and one wire broke.

2.6 Temperature sensitivity

Figure 9a shows an example of the strong inverse relation between accumulation and gauge temperature that can be easily observed on days with a smoothly varying diurnal oscillation in air temperature. The temperature fluctuations between minute 360 and minute 840 (6 pm and 8 am local time) are a consequence of the corresponding fluctuations in 2-m wind speed shown in Fig. 9b with a steep temperature inversion. The data from the two curves in Fig. 9a were used to produce a plot of accumulation versus gauge temperature as shown in Fig. 10a. After applying a least-squares linear fit to these data, the result is the dashed line, the slope of which is the temperature coefficient that, for this example, has the value -0.0901 mm/10°C or approximately 0.1 mm per 10°C.

Plots similar to that in Fig. 10a were created for each suitable day for each wire and their average (ave123) between 17 January 2003 and 7 April 2007. The resulting temperature coefficients are shown in Fig. 10b. The solid line is the minimum least-squares linear fit to the data (the circles) beginning 11 October 2003, the beginning date from which the same wires have been in continuous use. There is considerable scatter of the individual temperature coefficients, but, with an R-squared value of 0.94, the linear change of temperature coefficients with accumulation is clearly significant. The greater the accumulation in the bucket, the more negative the temperature coefficient – by a factor of four from empty bucket to full bucket. Approximately one-third of the available days provided an acceptable accumulation-gauge temperature plot from which a temperature coefficient could be calculated. The remaining days had precipitation, a frontal passage, too small a diurnal temperature range, or other weather phenomena that produced complicated diurnal variations in temperature. Comparatively few days showed the ideal highly eccentric ellipse-like variation in Fig. 10a or the near 1:1 temperature – accumulation relation in Fig. 9a.

Temperature coefficients were calculated also for each wire. The labeled dashed lines in Fig. 10b are their minimum least-squares linear fits. If the supporting data for the dashed lines were plotted, the scatter of points for each wire would be larger than that for the average or solid line. As Fig. 10b shows, each wire has its own characteristic variation of temperature coefficient with accumulation.

The impact of the sensitivity of transducer frequency to temperature occurs during precipitation. Typically, temperature decreases during a precipitation event, particularly at the onset. Assuming a drop in

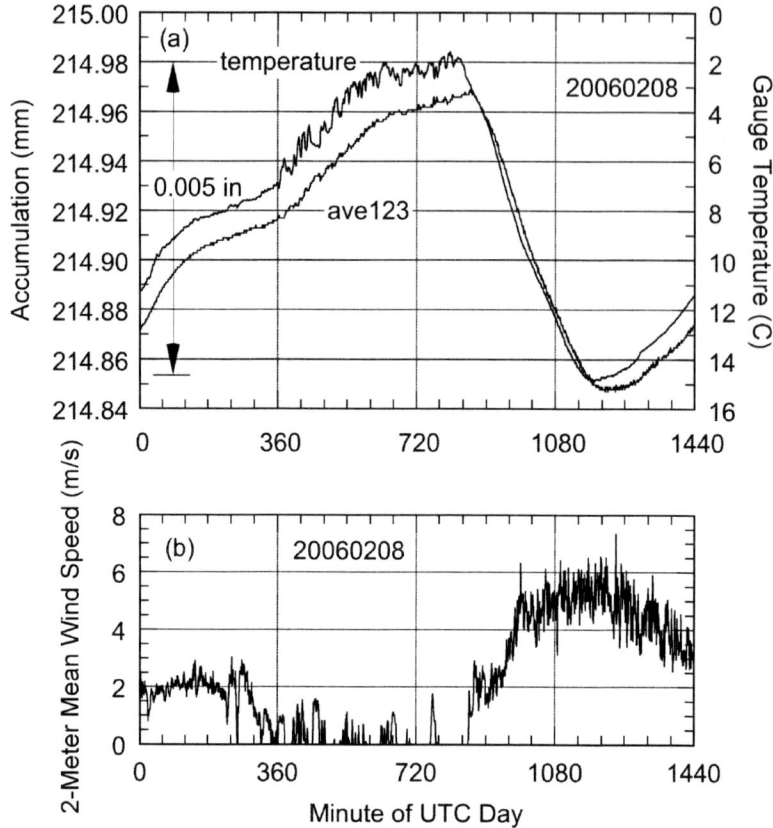

Fig. 9. (a) An example of the relationship between apparent accumulation and gauge temperature. (b) Associated 2-m wind speed

temperature of 10°C, an initial accumulation of 200 mm and a true 10 mm precipitation event, the overestimate would be about 0.1 mm or 1%. For a true 1 mm event, overestimate would be 10%. Near empty bucket, the respective overestimates are one-half the values given and near full bucket twice the values.

Should one correct for overestimation? For a number of reasons, the answer is no. The temperature coefficients found here are unique to the set of vibrating wire transducers that were used. While all coefficients can be expected to be negative, Fig. 10b shows there are substantial differences in magnitude of the temperature coefficients among the three wires for any given accumulation. Thus it would be necessary to perform a similar type of analysis for each three-wire gauge.

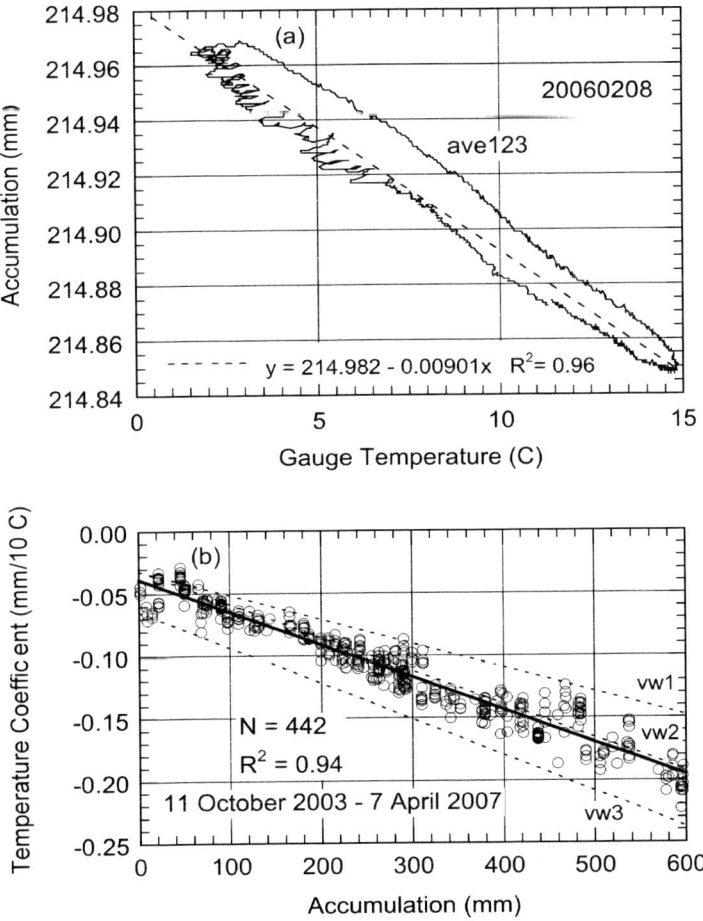

Fig. 10. (a) Plot of accumulation versus temperature in Fig. 9a. The slope of the least–squares linear fit is the temperature coefficient. (b) Temperature coefficients are in units of mm/10°C as a function of accumulation. Dashed lines are least-squares linear fits for individual wires less supporting data. Solid line applies to the average of the three wires with supporting data (open circles)

Undercatch errors due to wind and wetting loss need to be considered, also. Duchon and Essenberg (2001) found undercatch of 4–5% due to wind for both aboveground tipping-bucket and weighing-bucket gauges with and without Alter shields for typical rainfall events across the Southern Great Plains. Clearly, undercatch errors due to wind using an aboveground vibrating wire gauge could easily exceed errors due to temperature.

Golubev et al. (1992) found a 0.03 mm wetting loss for the U.S. standard 8-in rain gauge, but a much higher value for the Tretyakov gauge. Yang et al. (1999) cite a figure of 0.14 mm for the Hellman gauge. In any case, wetting losses depend on both the type of gauge and type of precipitation (Yang et al. 1998). The author is unaware of any such determination for the Geonor gauge.

In conclusion, while there can be a close relationship between accumulation (apparent) and gauge temperature as shown in Fig. 9a, numerically accounting for the magnitude of overestimation in a typical precipitation event is not recommended. Fortunately, this overestimation is counteracted by the undercatch due to wetting loss and wind.

2.7 Rain rate estimation

The availability of a continuous record of gauge-measured 1-min accumulations of precipitation allows for easy calculation of the rate of snowfall or rainfall. In this Section, we examine a heavy rain event in which a two-dimensional video disdrometer or 2dvd, mentioned in Sect. 2.3, provided independent estimates of rain rate. In brief, with the aid of lamps, mirrors and slit plates, two orthogonal video cameras record front and side views of each hydrometeor that falls through a 10 cm × 10 cm plane located below the 25 cm × 25 cm orifice seen in Fig. 4. Software is used to match and process the images from each video camera so that drop shape, size, oblateness and fall speed for each drop can be determined. Rain rate is computed from these variables. Much more detail on the operation of the 2dvd can be found in Kruger and Krajewski (2002) and Schuur et al. (2001).

Figure 11 shows a heavy rain event at the field site as measured by the Geonor gauge, the 2dvd and the Oklahoma Mesonet gauge at Norman (NRMN), located 105 m WNW of the pit. The Geonor recorded the most rainfall, the 2dvd the least. The totals for the day were Geonor 100.20 mm, 2dvd 95.97 mm and NRMN 97.28 mm. Note that both the Geonor gauge and 2dvd yield 1-minute data while NRMN gauge yields 5-minute measurements. Two periods A and B in Fig. 11 were selected for rain rate analysis. The former includes the highest rain rates of the event; the latter, the lowest.

Rain rate for the Geonor was calculated by subtracting the previous from the current 1-minute accumulations. Rain rate from the

Chapter 2 - Vibrating-wire technology for precipitation measurements   51

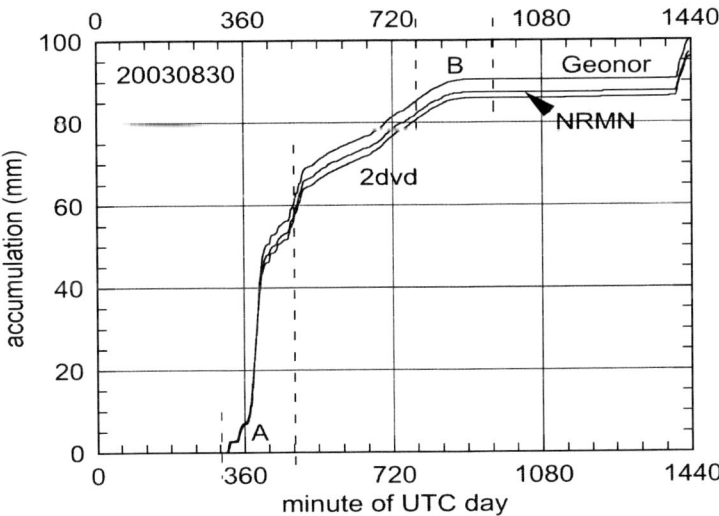

Fig. 11. A heavy rain event. Top accumulation curve is the Geonor, middle curve is the nearby Oklahoma Mesonet gauge (NRMN) and the bottom curve is the two-dimensional video disdrometer (2dvd). The Geonor and 2dvd provide 1-minute data, NRMN 5-minute data. Time period A applies to Fig. 12a,b,c and time period B to Fig. 12d,e,f

2dvd is obtained directly. Figure 12 is a comparison of their rain rates for six 60-minute periods, three in period A and three in period B. With the exception of Fig. 12b, the sensitivities among the rain rate axes are related by an order of magnitude. Including the exception, the range in 1-minute rain rates is more than three orders of magnitude.

Figure 12a covers the initial hour of period A and shows two distinct subperiods of rain and a maximum rain rate of about 42 mm/h. The two curves tend to follow each other quite well. The clocks for the Geonor and 2dvd were not synchronized so there could be a discrepancy of 1-minute or more. The possible discrepancy in time and the fact that the P27 instruction (see Sect. 2.3) was used could account for differences in peak rain rates and/or their displacement in time. These potential limitations apply to all panels in this figure. Figure 12b covers the portion of the rain event with the highest rain rates. Values measured by the Geonor exceed 100 mm/h. A systematic disparity is rain rates can be seen between minutes 380 and 404. Because of the design of the 2dvd wherein the sensing planes are located below the orifice, the trajectories of the raindrops can be at such a low angle relative to the

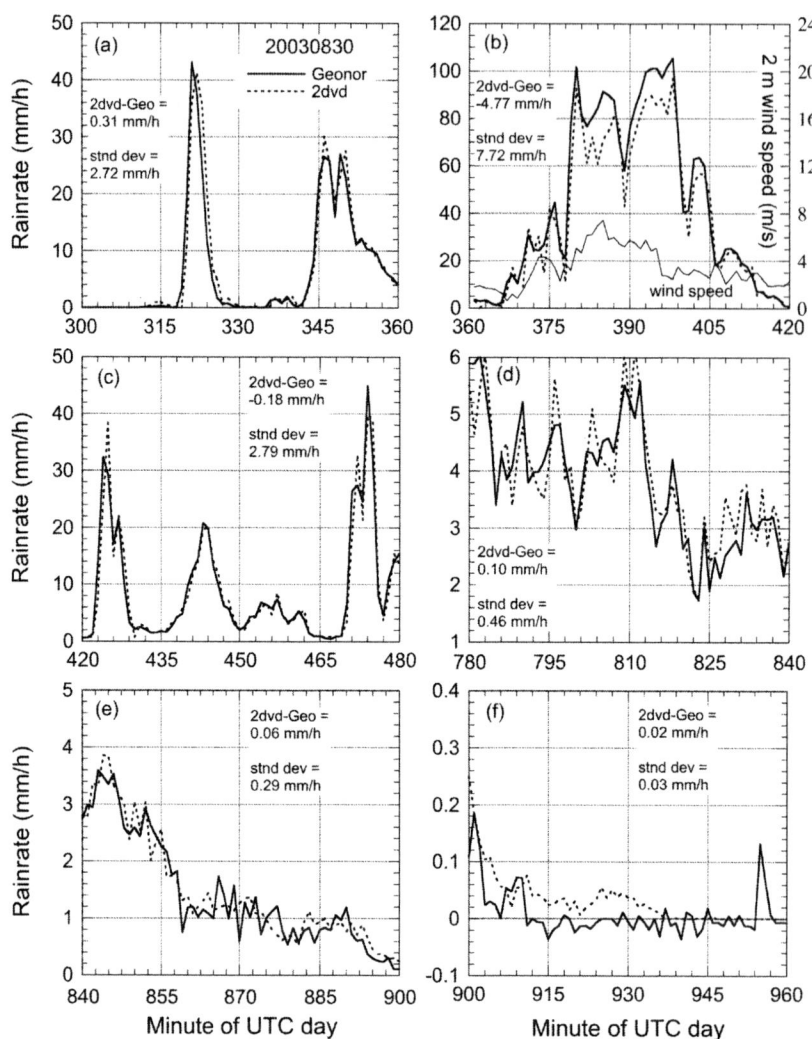

Fig. 12. Comparison of Geonor and 2dvd 1-minute rain rates for periods A and B in Fig. 11. Panels (**a**), (**b**) and (**c**) comprise period A and panels (**d**), (**e**) and (**f**) comprise period B. In (b) the lower curve is wind speed. Note that, with the exception of (b), the sensitivity of rain rate axes can change by an order of magnitude from panel to panel. The mean difference of rain rates and the standard deviation of the differences are shown for each 1-hour period

horizon that there is a 'shadow effect' when the wind speed is sufficiently high.

Due to the low trajectory angle, drops may not pass through part of the sensing planes on the windward side. Godfrey (2002, Sect. 3.2) provides a more detailed explanation. The bottom curve in Fig. 12b shows that the wind speed at 2 m was around or above 4 m/s during this subperiod and may account for the lower rain rates of the 2dvd relative to the Geonor. Figure 12c encompasses the last hour of period A. Correspondence between the two rain rates is good except at the various peaks.

Figure 12d applies to the first hour of period B, the resolution of its ordinate increased by an order of magnitude relative to the previous panel. While the general correspondence between rain rates is satisfactory, differences between their respective peaks and valleys are quite evident.

Rain rates continue to decrease with time in Fig. 12e. When the rain rates decrease to around 1 mm/h, the 2dvd shows a smoother curve than the Geonor. This relationship carries over into Fig. 12f in which the resolution has been increased by another order of magnitude. For values of rain rate less than 0.1 mm/h, the Geonor no longer provides a reliable rain rate as measurement noise dominates beginning about minute 915.

Each figure shows also the mean difference between the 2dvd and Geonor for the associated hour and the standard deviation of the differences. Ignoring Fig. 12b because of the potential bias in the 2dvd due to the wind, a systematic decrease in both the mean differences and standard deviations with decreasing rain rate is observed.

In summary, using the 2dvd as the standard except when wind speed at 2 m exceeds about 4 m/s, the Geonor is capable of measuring 1-minute rain rates with reasonable accuracy from at least 250 mm/h (based on another rain event) down to about 0.1 mm/h (Fig. 12f). With over 1400 gauges world-wide (Sect. 2.1), the opportunity exits to monitor potential changes in rain rates due to a changing climate.

2.8 Very low precipitation events

Because the vibrating wires in the Geonor gauge are sensitive to small changes in tension, it is possible to record trace amounts of precipitation. There are many examples of the occurrence of event totals less than 0.2 mm available in the archived data dating back to 2001. In this Section, four events are discussed that characterize the capability of and the problems associated with estimating trace amounts of rain. No snow events are considered, but, in principle, similar characterizations should apply. The four events are shown in Fig. 13. In each case a

54　C.E. Duchon

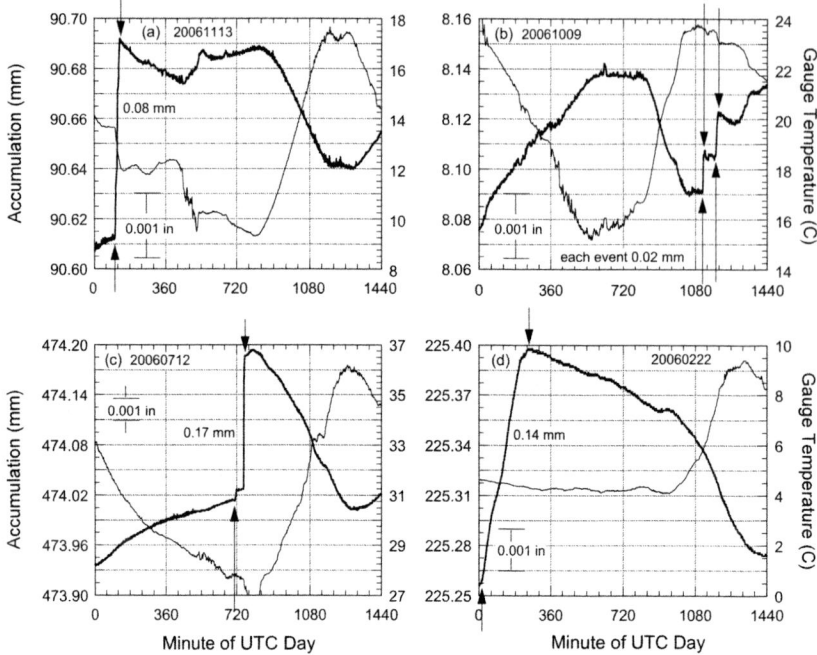

Fig. 13. Accumulation and gauge temperature for four examples of low rainfall events versus minute of a UTC day. Upward and downward pointing arrows define each period of rain. Scale resolution for accumulation in (**c**) is one-third that in (**a**) or (**b**) and in (**d**) is twice that in (**c**). Scale resolution for temperature is the same in all panels

comparison is made with the amount observed from the Mesonet gauge and in one case from the 2dvd, also.

Figure 13a shows an event during a UTC day in which 0.08 mm was recorded over a span of 25 minutes. In each panel the time of rainfall is delineated by the upward and downward pointing arrows. Shortly after the rain ends the accumulation decreases slightly as a consequence of evaporation of droplets that failed to penetrate the oil layer and are lying on the oil surface. There is no significant change in temperature for about 5 hours following the event. After about minute 450 there is an increase in apparent accumulation due to a decrease in temperature. Evaporation likely continues for some time but is masked by the response of the vibrating wires to temperature. The second half of the day is characterized by an increase in temperature of 8°C during which time the accumulation decreases by about 0.05 mm. The decrease in accumulation is in agreement with the nominal temperature coefficient –0.06 mm/10°C seen in Fig. 10b. The NRMN gauge recorded no rainfall, presumably because the rainfall was less than the amount required to tip the bucket, i.e., 0.25 mm.

The first 18 hours of the day in Fig. 13b show the effect of the negative temperature coefficient on (apparent) accumulation. Then beginning at minute 1117 there is a 5-minute shower and at minute 1184 a 10-minute shower, each with a recorded accumulation of 0.02 mm. It is easy to identify even minor shower activity because of the sudden change in accumulation relative to the slowly varying apparent accumulation in response to the daily cycle of temperature. However, the magnitudes of rainfall in both panels (a) and (b) are likely substantial underestimates of the amounts that actually fell due to wetting losses discussed in Sect. 2.6.

Figure 13c shows the results from a day in which three independent estimates of rainfall were available. The accumulation scale is one-third as sensitive as in panels (a) and (b). The Geonor measured 0.17 mm, the Mesonet gauge 0.25 mm and the 2dvd 0.37 mm. The time series of accumulation from the 2dvd matches that of the Geonor along the time axis (not shown), but the amplitudes of the both rain showers are larger. That the Mesonet tipping-bucket gauge shows one tip may be a consequence of residual rain remaining in the bucket from the previous day. Of course, the Mesonet gauge also suffers from wetting loss. Tipping-bucket gauges measure precipitation in discrete amounts and are not useful for measuring very low rainfall events. It seems unlikely that the wetting loss associated with the Geonor could, by itself, account for the difference of 0.20 mm between it and the 2dvd. As cautioned by Kruger and Krajewski (2002, p. 607), the actual size of drops can be overestimated or underestimated depending on whether the 2dvd is properly calibrated. It is the integration of the drop volumes that results in the accumulation estimate. In short, the comparatively large difference between the accumulations from the Geonor and 2dvd remains unexplained.

Following the second rain shower, the gauge temperature reaches a minimum (27°C), then increases to its maximum for the day (36°C), corresponding to an increase of 9°C. During this time the apparent accumulation decreases by 0.18 mm. From Fig. 10b the temperature coefficient is around –0.17 mm/10°C for the given accumulation in the bucket. Thus the observed decrease in accumulation is in line with that expected as a result of the temperature increase; however, evaporation is also likely part of the decrease.

Figure 13d is an unusual case of precipitation in which there is continuous mist for a period of almost 4 hours starting shortly after the day began. The total accumulation was 0.14 mm (the accumulation scale is twice as sensitive as in panel (c)). The Mesonet gauge recorded 0.25 mm (one tip), but is unreliable because the Mesonet time series suggests it is the last tip of a sequence of tips resulting from melting snow in the gauge (it is unheated). No 2dvd data were available. Evaporation began at about minute 240 and continued for at least the next 11 hours. Evaporation is evident because there is negligible temperature change (sky cover is overcast). At around minute 900 the accumulation remains steady for about one and one-half hours and then continues to decline as the sky cover changes from broken to scattered to clear (based on data from a nearby National Weather Service Automated Surface Observing System station). The data record shows that there was a decrease in gauge temperature of 0.3°C from minute 820 to minute 950, which can be seen in the figure, so that, perhaps, decreasing accumulation due to evaporation was being compensated by increasing accumulation (apparent) due to decreasing temperature, thus yielding an essentially flat accumulation. In addition, the increase in temperature from minute 950 to the maximum for the day is approximately 5°C, which, from Fig. 10b, would result in a decrease in apparent accumulation of 0.05 mm. That the actual decrease from minute 950 to the end of the day is more than 0.08 mm suggests evaporation was occurring simultaneously.

In conclusion, the Geonor vibrating wire gauge is capable of easily measuring increases in accumulation as low as 0.01 mm (see Fig. 13c). Interpreting this amount as the true precipitation is confounded by the wetting loss on the interior wall of the collection cylinder and subsequent evaporation. The magnitude of this loss may well exceed the recorded accumulation, but no definitive analysis has been done, insofar as is known, for either liquid or solid precipitation. Thus careful examination of a time series of accumulation can show that a very small amount of precipitation has occurred and when, albeit the amount is likely an underestimate. Occurrence and timing information can be useful for both meteorological and climatological applications. Simultaneous recording of gauge temperature is an important tool in making the proper assessment of very low precipitation events.

2.9 Summary

The Geonor T-200B vibrating wire precipitation gauge is a sensitive instrument for measuring liquid and solid forms. All analyses presented in this research were based on measurements acquired from a Geonor gauge located in a pit at a field site in central Oklahoma, USA. Because frozen precipitation seldom occurs in this area and a pit gauge is not suited to measuring snow, only observations of the liquid form were considered. Various aspects of gauge performance have been investigated including (1) advantages of employing three vibrating wires, (2) stability of vibrating wire calibration, (3) the sensitivity of vibrating wire frequency to temperature and (4) application of vibrating wire measurements to a wide range of rain rates and to very low precipitation events.

The redundancy afforded by using three wires can be exploited to build confidence in measurements or to detect poor performance in one or more wires. In addition, using three wires allows one to average their outputs so that random noise due to wind is reduced. Changes in calibration can be detected by placing a succession of known weights into the bucket and observing the output. No significant change in calibrations of the three wires was found over a three-year period. Temperature coefficients are negative and those for the averaged output or accumulation ranged from around -0.05 mm/$10°C$ at empty bucket to about -0.20 mm/10C at full bucket (600 mm), a four-fold change.

Correcting accumulation for temperature change is not recommended because, in general, the correction is small and also dependent on the specific wires employed. In very low precipitation events, however, it may be necessary to invoke temperature considerations to make sure the measured accumulation change is not a result or partial result of a temperature change. Comparisons of rain rates from a disdrometer with those from the Geonor generally show close correspondence from very low rain rates to very high rain rates.

In conclusion, the Geonor gauge is a state-of-the-art instrument for accurate and reliable measurement of point precipitation continuous in time.

Acknowledgements

The author wishes to express his many thanks to various members of the Oklahoma Climatological Survey under the direction of Kenneth C. Crawford. They include David Grimsley, Chris Fiebrich, David Demko, and Derek Arndt, who graciously provided a wide variety of technical advice and assistance as well as data management resources. Terry Schuur was kind enough to provide various 2dvd data sets and personal support for this project.

References

Bakkehøi S, Øien K, Førland EJ (1985) An automatic precipitation gauge based on vibrating-wire strain gauges. Nord Hydrol 16:193–202

Duchon CE, Essenberg GR (2001) Comparative rainfall observations from pit and aboveground rain gauges with and without wind shields. Water Resour Res 37:3253–3263

Godfrey CM (2002) A scheme for correcting rainfall rates measured by a 2-D video disdrometer. MSc Thesis, University of Oklahoma, Norman, USA, 119 pp

Golubev VS, Groisman PY, Quayle RG (1992) An evaluation of the United States standard 8-in. nonrecording raingauge at the Valdai Polygon, Russia. J Atmos Ocean Tech 9:624–629

Groisman PY, Legates DR (1994) The accuracy of United States precipitation data. B Am Meteorol Soc 75:215–227

Kruger A, Krajewski WF (2002) Two-dimensional video disdrometer: a description. J Atmos Ocean Tech 19:602–617

Lamb HH, Swenson J (2005) Measurement errors using a Geonor weighing gauge with a Campbell Scientific Datalogger. In: Proceedings 16th Conference on Climate Variability and Change. American Meteorological Society, San Diego, CA, Paper P2.5

Raichel DR (2006) The science and applications of acoustics. Springer, New York, 660 pp

Schuur TJ, Ryzhkov AV, Zrnić DS, Schönhuber M (2001) Drop size distributions measured by a 2d video disdrometer: comparison with dual-polarization radar data. J Appl Meteorol 40:1019–1034

Tunbridge LW, Øien K (1988) The advantages of vibrating-wire instruments in geomechanics. Field Measurements in geomechanics In: Sakurai S (ed) Proceedings 2nd International Symposium on Field Measurements in Geomechanics. 6–9 April 1987, Kobe, Japan. Balkema A.A. Publishers, Brookfield, VT pp 3–16

Yang D, Goodison BE, Ishida S (1998) Adjustment of daily precipitation at 10 climate stations in Alaska: application of World Meteorological Organization intercomparison results. Water Resour Res 34:241–256

Yang D, Ishida S, Goodison BE, Gunther T (1999) Bias correction of daily precipitation measurements for Greenland. J Geophys Res 104:6171–6181

3 Measurements of light rain, drizzle and heavy fog

Ismail Gultepe

Cloud Physics and Severe Weather Research Section
Meteorological Research Division, Toronto, Environment Canada, Ontario
M3H 5T4, Canada

Table of contents

3.1 Introduction 59
3.2 FRAM field projects and observations 62
    3.2.1 FD12P measurements 63
    3.2.2 VRG101 measurements 64
    3.2.3 POSS measurements 65
    3.2.4 Total Precipitation Sensor (TPS) measurements ... 67
    3.2.5 FMD and CIP measurements 68
3.3 Analysis 68
3.4 Results 69
    3.4.1 Case studies 69
    3.4.2 Overall comparisons 70
3.5 Discussion 73
    3.5.1 Light precipitation and drizzle measurements 75
    3.5.2 Visibility calculations 75
    3.5.3 Uncertainties 78
3.6 Conclusions 79
References 80

3.1 Introduction

Measurements of light rain, drizzle and settling out (or collision collection at the surfaces) of heavy fog can be important to estimate daily, monthly and annually averaged precipitation amounts. For an example, in northern climates (Stewart et al. 2004), there are no accepted standards for the short-term measurements (on the minute time scale) due to low precipitation rates. This chapter explores the potential of various instruments to measure light precipitation and settling rate under heavy fog conditions.

The World Meteorological Organization (WMO) Guide to Instruments and Methods of Observation (WMO 1983) suggested that precipitation rate should be measured from 0.02 up to 2000 mm h^{-1} and time average should be 1 min. In the guide, required uncertainties were 0.1 mm h^{-1} between 0.2 and 2 mm h^{-1} and 5% above 2 mm h^{-1} (also in WMO 1983). There are several manual instruments that collect precipitation amounts based on various techniques but most of them do not satisfy the criteria stated by WMO (Sevruk and Hamon 1984; Goodison et al. 1998).

Nystuen et al. (1996) provided an extensive work on quality of automatic precipitation measurements. Automatic rain gauges usually provide both accumulated precipitation amount (PA) and precipitation rate (PR). The main types of rain gauge systems include (1) tipping bucket systems, (2) weighing systems and (3) optical systems. In addition to these systems, disdrometers and radar-based systems have also been used for precipitation measurements. The tipping bucket and weighing systems (Humphrey et al. 1997; Nystuen 1999; Ciach 2003) are affected by the flow irregularities occurring within the catchments' basin and flow chambers and also by time response occurring during the tipping process. Rain gauges in general are assumed to underestimate rain due to wind and turbulent effects at the edges of the rain gauges (Yang et al. 1998). Nystuen et al. (1996) stated changes in droplet size spectra could not explain the scatter observed in optical rain gauges. This suggests that rain measurements face large uncertainties that need to be explored.

Tipping buckets sample rain amount differently compared to other instruments. They usually tip when approximately 0.2 mm accumulation occurs. The rainfall rate accuracy of the tipping buckets cannot be better than 12 mm h^{-1} over a minimum interval (Nystuen et al. 1996; Humphrey et al. 1997). This means that it will not measure precipitation rates until this amount is reached and therefore cannot be used for measurements of light precipitation (<0.5 mm h^{-1}). Weighing rain gauges work by weighing the accumulated water amount for a specific time period. They are not useful for light precipitation types e.g., drizzle and heavy fog conditions. Nystuen et al. (1996) suggested that their expected accuracy in precipitation rate can be about 1.8 mm h^{-1} over a 10-s sampling time period but, in reality, this can not be achieved because measurements are not possible during the draining of accumulated water.

Optical gauges have also been often used for measuring precipitation rate and precipitation accumulation. The most common ones are (1) ScTI optical rain gauges (Nystuen et al. 1996) and (2) VAISALA FD12P (Sheppard and Joe 2000; Gultepe and Milbrandt 2007a,b). These instruments are based on scintillation in an optical beam produced by rain/snow drops falling between a light source and a receiver. The change in light intensity due to a drop is related to drop size, fall velocity, optical geometry and the light source. The studies of Wang et al. (1978) and Wang and Clifford (1975) suggested that variations in light intensity are related to rainfall amount and this concept is used to calculate PR and PA values (Nystuen et al. 1996). The typical PR values from optical probes can include an uncertainty between 0.2 and 0.4 mm h^{-1}. The FD12P, however, cannot be used for snow accumulation and melted equivalent of snow total amount if they are not calibrated against a weighing gauge at the field (personal communication, VAISALA Inc., 2007).

The disdrometers have also been used commonly for PR and PA calculations that are based on droplet size distributions (Illingworth and Stevens 1986; Kruger and Krajewski 2002; Tokay et al. 2003). The most common ones used are Joss and Waldvogel (JW) disdrometer (1969), POSS (Sheppard and Joe 1994), OTT Parsivel (Loffler-Mang and Joss 2000) and two-dimensional video disdrometers, 2DVD (Kruger and Krajewski 2002). The OTT Parsivel measurements can be displayed over 32 size bins and stores drop counts for a given time step. The size range is between 0.062 mm and 24.5 mm in diameter. Within high sound/electrical noise environments, this threshold amount for disdrometers increase significantly, usually to more than 0.5 mm.

Nystuen et al. (1996) compared the precipitation rates from various instruments summarized above and found that PR was at an acceptable levels when PR>5 mm h^{-1} whereas the tipping bucket measurements over a 1 min interval can be in error of about 12 mm h^{-1}. They also suggested that the relative error in PR values can reach to +20% for heavy rain, 50% for average rain episodes and +300% for light rain and drizzle. Overall, their results suggested that uncertainties were about 12–14% for PR>5 mm h^{-1} and 38–40% for PR<5 mm h^{-1}. Disdrometer measurements can have large uncertainty at low PR values because of their small sampling area (50 cm^2) and a lower particle threshold size about 300–500 μm.

The PR and PA cannot be obtained accurately when the droplet sizes are less than 500 μm (0.5 mm). It is important to note that the duration of precipitation with small PR and PA values can be much higher than those of high PR events. But, if PR=1 mm h^{-1} occurs over a 10 h time period then it can result in PA=10 mm which can not be measured accurately by various instruments.

The standard procedure for intercomparisons among various precipitation measurements obtained from the instruments is usually done against a standard one e.g. a manual or weighing gauge but, for a light precipitation, this method doesn't work accurately; therefore, relative intercomparisons in this study are made (e.g. instrument to instrument). The measurements collected during the Fog remote sensing and modeling (FRAM) projects (Gultepe et al. 2007) will be summarized to better understand (1) light precipitation events that include heavy fog and drizzle, (2) precipitation rate accuracy and (3) drizzle effect on visibility.

3.2 FRAM field projects and observations

Surface observations collected during the FRAM field projects were conducted at the Center for Atmospheric Research Experiment (CARE) site near Toronto, Ontario during the winter of 2005–06 and in Lunenburg, Nova Scotia, during summers of 2006 and 2007 (Gultepe et al. 2007; Gultepe and Milbrandt 2007a). The FRAM-C and FRAM-L represent observations collected at the CARE and Lunenburg sites, respectively. During the FRAM field projects (FRAM-C; FRAM-L1; FRAM-L2), various measurements related to low precipitation measurements e.g., drizzle and fog were collected by the instruments summarized in Table 1. During FRAM-C project, the YES Inc. TPS and VRG101 were not available.

The measurements of precipitation can be divided into the three types: (1) optical droplet counters that measure droplets with sizes <965 μm (e.g., DMT optical probes), (2) instruments that use either a radiation source (e.g., FD12P) or a conventional technique to measure PR by weighing the mass or measuring volume (e.g., VRG101) and (3) instruments that are based on heating rate to evaporate a droplet or snow particle that falls onto the hot sampling area (e.g., YES total precipitation sensor (TPS)). Figure 1 shows the instruments used to obtain precipitation and drizzle rates and accumulated amounts. Details of the measurements from all three groups are given below.

3.2.1 FD12P measurements

The main observations from the FD12P VAISALA present weather instrument were visibility (Vis) and precipitation rate (PR) for rain and snow. Details on this instrument can be found in Gultepe et al. (2007) and Gultepe and Milbrandt (2007b). The observations made during winter represented mostly snow conditions and during summer, represented rain/drizzle conditions.

The FD12P instrument is a multi-variable sensor designed for automatic weather stations and airport weather observing systems (VAISALA Inc., 2002). The sensor combines the functions of a forward scatter Vis meter and a present weather sensor. This instrument is also used to determine Vis and precipitation type related to weather codes in WMO standard SYNOP and METAR messages. The structural basis of the FD12P is a pole mast that supports a transducer cross-arm (FDC115) which contains the optical units of FDT12B Transmitter and FDR12

 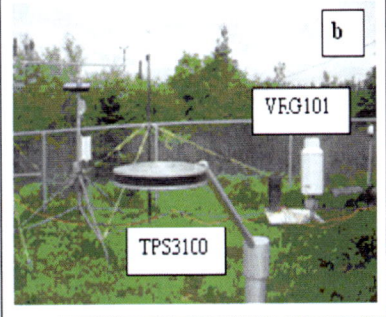

Fig. 1. The FD12P and TPS and VRG101 rain gauges used during FRAM projects are shown in (**a**) and (**b**), respectively. The CIP and FMD instruments are shown in (**c**). The FMD is not directly used in the calculations

Receiver. The DTS14 temperature sensor and the DRD12 Rain Detector (RainCap) is fastened to the cross-arm. The FDT12B Transmitter, tilted 16.5° downward, emits near IR light while the FDR12 Receiver, also tilted 16.5° downward, measures the scattered light from the transmitted beam allowing the receiver to measure light scattered at an angle of 33°. The measured radiance is then converted to a Vis value also called Meteorological Optical Range (MOR). The FD12P detects precipitation droplets from rapid changes in the scattered signal. Then, the droplet data are used to estimate optical precipitation rate and amount. The accuracy of FD12P measurements according to the manufacturer is about 10% for Vis and about 0.05 mm h^{-1} for PR. Measurement uncertainties compared to other instruments can be much larger than this estimate given by the manufacturers' and they are summarized in the results and discussion sections.

In the winter, snow accumulation is usually measured by measuring the thickness of new snow on the ground; however, its water equivalent amount can directly be measured by the FD12P that is based on the forward scatter of light within a sample volume between the arms of the FD12P sensor (VAISALA Inc., 2002). By analyzing the scatter signal, which includes the fall velocity of different precipitation types in conjunction with the RainCap and air temperature sensor measurements, an algorithm is used to identify the type of precipitation. The snowflake has its own characteristics for the scatter signal. Therefore, the snow accumulation can only be estimated if the wetness of snowflakes is known but in many conditions, the 'wetness' of the snow cannot be obtained accurately. For this reason, a weighing gauge (e.g., VRG101) should also be used for calibration and error analysis purposes.

3.2.2 VRG101 measurements

The VRG101 is designed to be a reliable and accurate all-weather precipitation gauge (Turtiainen et al. 2006) when PR> 0.1 mm h^{-1}. The following information is available from the company's technical manual (VAISALA, M210729EN-A, March 2006). The electronics unit includes a processor with embedded algorithms for calculation of cumulative rainfall and intensity. VRG101 utilizes the latest high-accuracy, temperature compensated load cell technology. The single point-type load cell is designed for direct mounting of the weighing platform.

Its wide 400 cm^2 collecting area is advantageous when measuring light rain and its large 650 mm net capacity decreases the risk of overflow (VAISALA, M210729EN-A, March 2006; Turtiainen et al. 2006). The deep container, together with the constriction formed by the inlet funnel decrease the evaporation error and out-blowing of collected snow. Another error source that is eliminated by the advanced mechanics is the underestimation caused by water and snow sticking to the inner surfaces of the gauge inlet funnel. In conventional designs, this mass is not measured and eventually evaporates. In VAISALA's design, the funnel element rests on the collector container. All water and snow on its surface is, therefore, included in the measured mass. The gauge software uses advanced algorithms to filter out noise and spurious signals (e.g., vibration by wind, mechanical impacts and rubbish or other objects entering the collecting container) and to compensate for evaporation. In addition to cumulative rainfall, the gauge data message also includes precipitation intensity, supply voltage, electronics internal temperature, gauge status with error flags and air temperature (if the optional Pt100 sensor is connected). Complete raw data (weight of the container) are also available to be used for diagnostic or research purposes. As stated by Turtiainen et al. (2006), uncertainty in PA can reach to about 33–35% when daily PA<1 mm.

The VRG101 uses 58 s time period to ensure a 1-min sampling time to record the precipitation. Then, the software performs further averaging to the 1 min data to find the 'dry baseline'. When new data are consistently above the base line for more than 0.1 mm, the precipitation sum is added by the increment and the baseline is shifted to the increment value. Then, VRG101 waits for the following 0.1 mm amount. The precipitation rate is then calculated using four last 1 min samples as follows:

$$PR\ [mm\ h^{-1}] = ((S_0-S_3)+ (S_0-S_2) + (S_0-S_1)) * 10, \quad (1)$$

where S with subscripts 0, 1, 2 and 3 (latest) represent intensities. When precipitation rates are less than 0.5 mm h^{-1}, they are recorded as zero values. The resolutions for PA and PR by the manufacturer are given as 0.1 mm and 0.1 mm h^{-1}, respectively (see Table 1).

3.2.3 POSS measurements

The Environment Canada (EC) Precipitation Occurrence Sensor System (POSS) is a bi-static, X-band Doppler radar. The POSS measures a signal in which frequency is proportional to the raindrop Doppler velocity and its amplitude is proportional to the raindrop size distribution (Sheppard 1990; Sheppard and Joe 1994). Sheppard and Joe (2007) stated that the POSS can estimate precipitation rates as low as 0.001 mm h^{-1} for rain and 0.002 mm h^{-1} for snow. Although the uncertainties were given related to quartile values, they can be about 25% when PR< 1 mm h^{-1}. The lowest size range detectable with the POSS is greater than 350 µm. Details on the EC POSS measurements can be found in Sheppard and Joe (1994; 2000; 2007).

Table 1. The instruments used and their characteristics during the FRAM projects to obtain precipitation rate (PR, mm h^{-1}) and accumulated precipitation rate (PA, mm)

Instrument	Model	Measurement	Accuracy	Sensitivity
Vaisala, all weather Precipitation instrument	FD12P	PR (mm h^{-1}) and (PA) (mm); precipitation type; visibility	0.1 mm h^{-1} over a minute	0.1 mm
Vaisala, All Weather Weighing Precipitation Gauge	VRG-101	PR (mm h^{-1}) and PA (mm); collecting area: 400 cm^2, capacity: 650 mm, height: 950 mm, Diameter: 400 mm	>0.5 mm h^{-1} <2000 mm h^{-1} 0.2 mm for precipitation >0.5 mm Resolution: 0.1 mm	0.1 mm h^{-1} >200 µm
YES Inc. Total Precipitation Sensor (TPS)	TPS-3100	PR (mm h^{-1}) and PA (mm)	>0.25 mm h^{-1} over 1 min for snow 0.5 mm h^{-1}	0.1 mm h^{-1} 0.2 mm for PR>0.5 mm
Droplet Measurements Technologies (DMT), fog measuring device	DMT FMD	2–50 µm droplet size spectra	–	>2 µm
Droplet Measurements Technologies (DMT), cloud imaging probe	DMT CIP	15–965 µm droplet size spectra	–	>15 µm

3.2.4 Total Precipitation Sensor (TPS) measurements

Details on the TPS-3100 (hereafter as TPS) can be found in Tryhane et al. (2005). Yankee Environmental Systems, Inc. (YES) markets the TPS. The TPS measures instantaneous total wet deposition, which includes liquid and frozen precipitation during a precipitation event (YES Inc. Manual, 2007). The sensor head consists of two hot plates (HP) about 5 inches in diameter warmed by electrical heaters. During precipitation events, it measures the rate of rain or snow by how much power is needed to evaporate precipitation on the upper plate while maintaining a constant surface temperature. The second lower plate, positioned directly under the evaporating plate, is heated to the same temperature and is used to factor out cooling caused by the wind. The precipitation rate is then derived from the power difference between the two plates, corrected for ambient temperature and wind speed.

The TPS measurement accuracy of the real-time liquid equivalent snowfall rates exceeds other snow gauges because of its unique compact design and no fluids to change during precipitation events and its capability to correct measurements for wind speed variations without the use of wind shielding (Tryhane et al. 2005). It also measures precipitation rates and accumulation amounts. Other instrumentation, such as weighing snow gauges, continuously provide accumulation information but often require 5–10 min of data to derive the corresponding precipitation rate (Rasmussen et al. 2005).

The TPS provides PR measurements instantaneously. On the other hand, a tipping bucket gauge measure PR as low as 0.1 mm h^{-1} but takes a full hour to accumulate it. During that period the rain or snow can evaporate before the tip occurs. Meanwhile, TPS provides output within a minute after rain starts. If the rain exceeds about 25 mm h^{-1}, the TPS may not keep up with evaporating all of the rain and may result in significant differences compared to other instruments. Its measurement range is between 0.1 and 50 mm h^{-1}, time constant 1 min and resolution 0.1 mm h^{-1}. Table 1 provides additional information on the TPS.

The TPS algorithm to calculate precipitation rate uses both 1 and 5 min running averages. The onset of precipitation is based on the 5 min average of instantaneous precipitation. This is done to prevent random variations in wind speed on the top and bottom plates due to turbulence.

3.2.5 FMD and CIP measurements

During the FRAM-L projects, both Droplet Measurement Technologies (DMT) Fog Measuring Device (FMD) and Cloud Imaging Probe (CIP) measurements (Table 1) were collected. These measurements were used to estimate PR based on droplet size distributions. The FMD uses a size range from 1 to 50 μm in diameter and the CIP probe uses a size range from 15 to 965 μm that cannot be resolved by the precipitation probes summarized above. Detailed information on these instruments can be found in Gultepe and Milbrandt (2007a,b) and on the DMT Inc. website. During the field project, the author observed droplet sizes less than 500 μm and PR reaching up to 2–3 mm h^{-1}. This suggests that disdrometers likely miss these droplets with sizes less than 500 μm. While in the field, droplet sizes less than 50 μm also contributed to the rain amount but they were not detected by rain gauges except for the TPS and FD12P instruments.

3.3 Analysis

In the analysis, precipitation rates are used directly from the recorded values by each instrument as specified in the company's manuals. First, PR values in mm h^{-1} for each minute are plotted against each other. Then, the 20 min averages of PR values are used to reduce the noise found at the individual data points. The results for two cases are presented for light rain and moderate rain episodes. The results are also presented using long-time averages of some measurements. For light rain and drizzle conditions, the DMT CIP and FMD and YES TPS measurements are used for PR calculations and better understand the drizzle component of precipitation.

In the present work, PR [mm h^{-1}] from the CIP probe is obtained using the following equation given by Steiner et al. (2004) as:

$$PR = \frac{0.36}{10^5} \sum_{i=1}^{i=64} N(D)D^3 v(D) dD \qquad (2)$$

where N is drop number density [cm^{-3} μm^{-1}], D the diameter [μm], v the terminal velocity [m s^{-1}] and dD=15 μm. The v for small droplets is given by Kunkel (1984) as:

$$v(D) = 1.202 \times 10^{-2} (D/2)^2 \qquad (3)$$

where D is in μm and v is in cm s^{-1}. These types of equations used commonly in weather radar-based PR calculations but they cannot not be valid for droplet diameters less than 500 μm (Steiner et al. 2004). This means that radar based PR inherently will have some error if it is used for assumed light rain or drizzle cases.

The uncertainties, represented by absolute and relative errors, are calculated for comparisons. The absolute error (standard error of mean) is obtained by Wilson (1963) as

$$\sigma_x = \sigma_p / \sqrt{N} \qquad (4)$$

where σ_p is the standard error of the measurements and N is the number of data points (measurements). The relative error is defined similar to the study of Ciach (2003) as

$$\varepsilon_i = (PR_i - \frac{1}{n}\sum_{i=1}^{n} PR_i) / \frac{1}{n}\sum_{i=1}^{n} PR_i \qquad (5)$$

The ε represent the error assigned a specific rain gauge known as i and n represents total number of instruments (for this work as 3). The PR_i represents the PR value measured by each rain gauge. To provide information on the accuracy of the FD12P measurements, its measurements are compared with the YES TPS measurements and the results are shown in Sect. 3.4.2.

3.4 Results

3.4.1 Case studies

Two cases are studied in this analysis: Case 1: 15–16 June 2007 (light rain-drizzle) and Case 2: 10–11 June 2007 (moderate rain). These cases were chosen because they represent low and moderate precipitation values for long durations, respectively.

Case 1: The time series of PR values for each minute from the VRG101, FD12P and TPS are shown in Fig. 2a. Fog, drizzle and rain durations indicated by FD12P measurements (as explained in Sect. 3.2.1) are also shown at the bottom of the plot. This plot suggests that PR due to fog and drizzle cases can be detected by both TPS and FD12P. The VRG101 did not measure PR for the same time intervals. Figure 2b shows TPS (also known as HP: hot plate) versus FD12P PR values for drizzle and rain events. The PR values less than 0.5 mm h^{-1} show more scatter. When the FD12P shows a value of 0.1 mm h^{-1}, the TPS PR reaches 0.3–0.4 mm h^{-1}. This suggests that both instruments responded light rain and drizzle. Differences in the measurements at the low values (<0.5 mm h^{-1}) are likely due to sensitivity of the measurements and/or different measurement techniques. The VRG101 could not measure PR values (at about 04:00 EST) because of its higher measurement threshold. For this case, Fig. 2c shows the scatter plot of VRG101 versus FD12P.

The results suggest that their measurements did not agree with each other for both drizzle and rain events when PR was less than 1 mm h^{-1} and this agrees with manufacturer's requirements.

Case 2: Similar to the Case 1 plot (Fig. 2a), Fig. 3a shows a time series of PR from the 3 instruments. For this case, PR reached up to 30 mm h^{-1} with no fog but drizzle and rain occasionally occurred. Although TPS values showed more scattering among the 3 instruments, their lines had similar trends. When drizzle occurred at 06:00 EST, the TPS clearly responded better to low PR values compared to others. Figure 3b shows the scatter plot of PR obtained from TPS (HP) and FD12P. Similar to Fig. 2b, both TPS and FD12P responded drizzle and low precipitation (<0.5 mm h^{-1}). It is possible that differences between two instruments as in Fig. 2 may arise due to (1) sensitivity of the measurements and (2) measurement techniques. Note that FD12P retrieves precipitation amount and type utilizing an inverse method as described in Sect. 3.2. Figure 3c shows that VRG101 PR values for PR>3–4 mm h^{-1} were more accurate compared to light precipitation conditions but scattering is still large compared to Fig. 3b. A possible reason for the shift in time for VRG101 can likely be due to a recording issue and it will be considered in future studies.

3.4.2 Overall comparisons

In this section, PR values obtained during FRAM-C project were used in the comparisons. PR comparisons are made only between the FD12P and EC POSS.

Chapter 3 - Measurements of light rain, drizzle and heavy fog   71

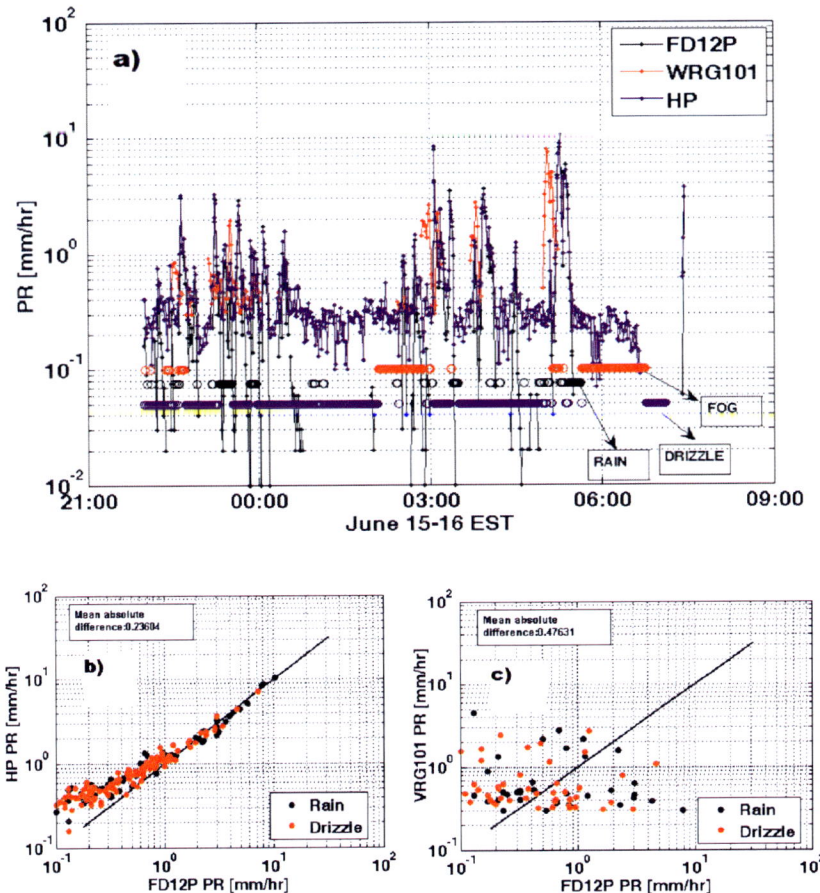

Fig. 2. Time series of 1-min observations of PR for rain and drizzle from three instruments (FD12P, VRG101 and TPS) for the 15–16 June case (**a**). *Horizontal lines* indicating periods of fog (*red*), drizzle (*blue*) and rain (*black*) from the FD12P are also shown in (a). The scatter plot of TPS (HP) PR versus FD12P PR is shown in (**b**). The *solid black line* is for the 1:1 line. The scatter plot of VRG101 PR versus FD12P PR is shown in (**c**)

Figure 4a shows the scatter plot of PR values obtained from FD12P and POSS instruments from November 2005 to April 2006 during FRAM-C. As a condition, wind speed >2 m s^{-1} and < 2 m s^{-1} were used in the scatter plot. This figure shows that the scattering is large when PR <a few mm h^{-1} and becomes smaller for large PR values. The POSS uses a radar technique and the power returned from the target

is proportional to the 6th power of particle size; therefore, small particles cannot be detected accurately. Increasing wind results in increasing POSS PR for a given FD12P value (Sheppard 1990). When temperature (T) is used as a condition, the results are shown in Fig. 4b. There was no clear T effect on the scatter of data points. Both Figs. 4a and 4b suggest that for small PR values, scattering was very large.

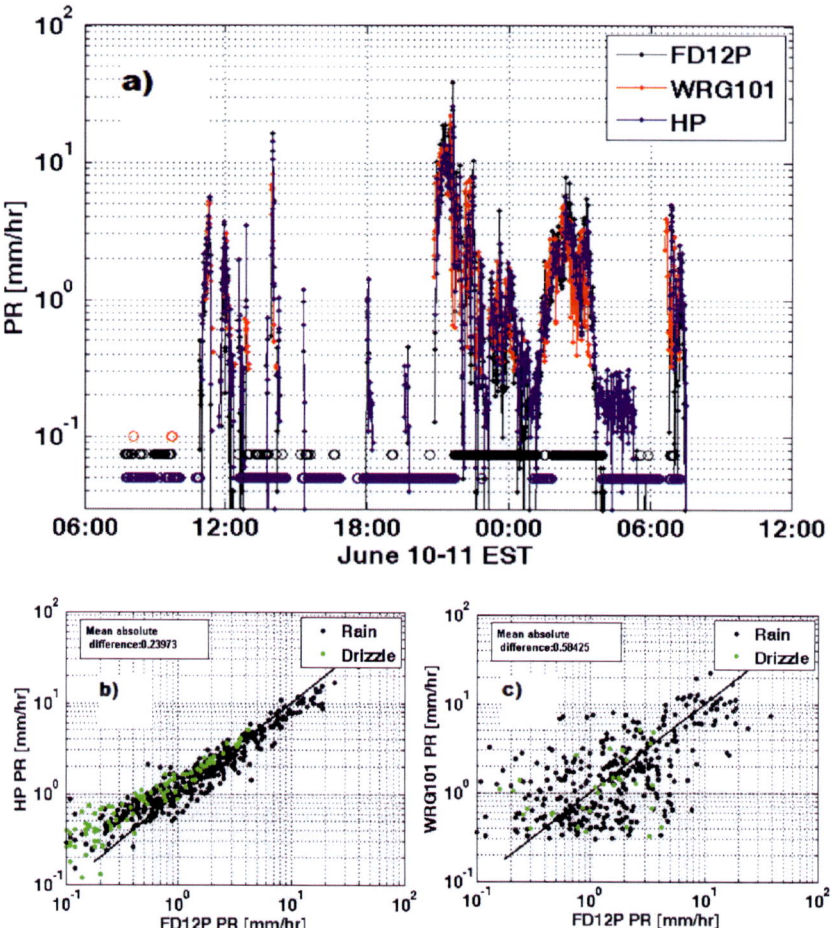

Fig. 3. Time series of 1-min observations of PR for rain and drizzle from three instruments (FD12P, VRG101 and TPS) for the 10–11 June case (**a**). *Horizontal lines* indicating periods of drizzle (*blue*) and rain (*black*) from the FD12P are also shown in (a). The scatter plot of TPS (HP) PR versus FD12P PR is shown in (**b**). The *solid black line* is for the 1:1 line. The scatter plot of VRG101 PR versus FD12P PR is shown in (**c**)

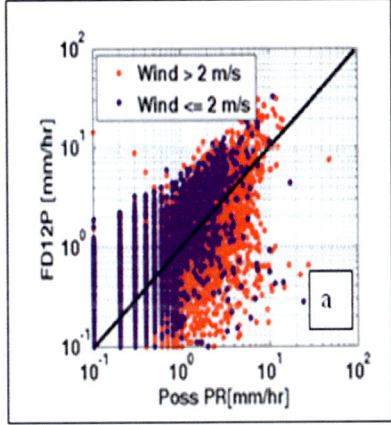
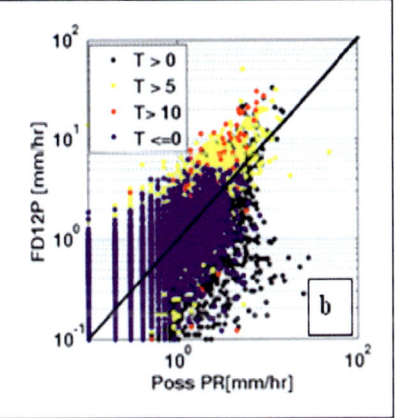
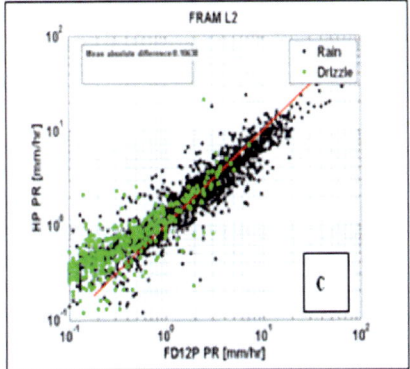

Fig. 4. The scatter plot of FD12P PR rate versus POSS PR with conditions based on wind (**a**) and temperature (**b**) for the observations collected during FRAM-C for about 6 months. Scatter plot of the entire TPS (HP) PR versus FD12P PR observations during FRAM-L2 (**c**). FD12P algorithm is used to obtain precipitation type

3.5 Discussion

In this section, light rain and drizzle measurements obtained using the DMT FMD and CIP probes, VAISALA FD12P, VRG101, TPS, POSS and visibility parameterizations are discussed and the uncertainties related to light precipitation are given.

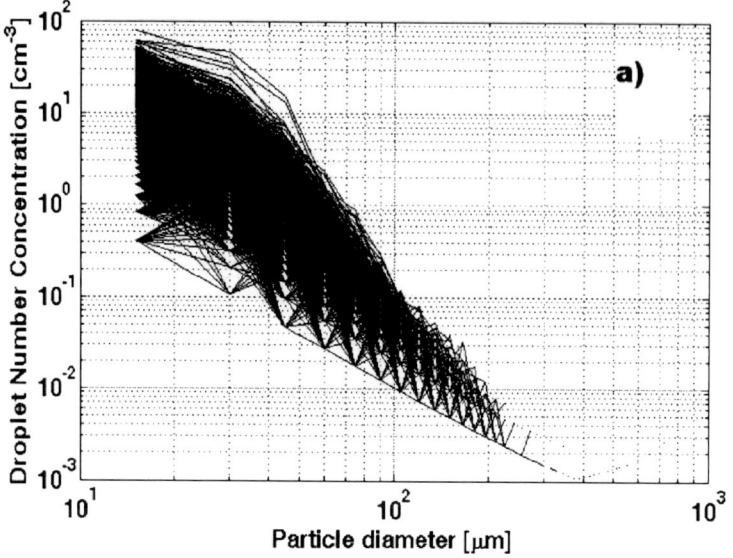

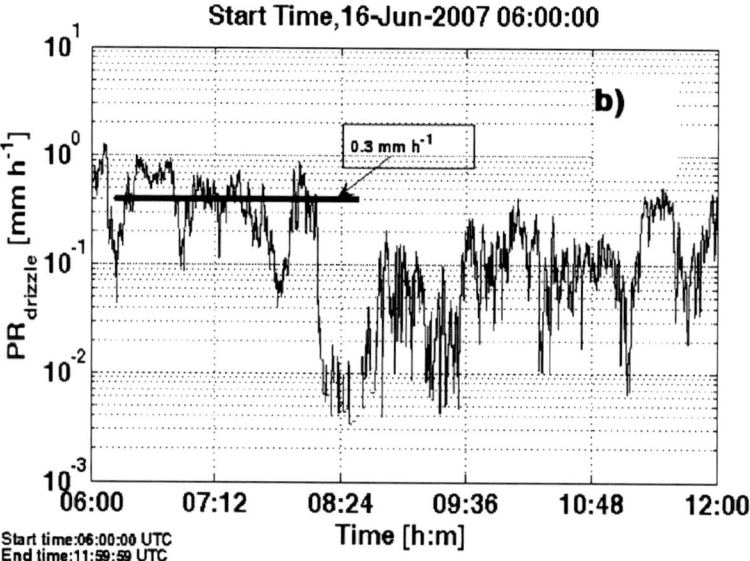

Fig. 5. Droplet spectra for the measurements of the CIP probe during the time period of 06:00–12:00 EST on 16 June 2007 (Box **a**). Box **b** shows the PR values calculated using Eq. (2) that utilized the CIP observations. The horizontal line corresponds to CIP measurements over 2 h time period

3.5.1 Light precipitation and drizzle measurements

Light rain or drizzle cannot be obtained accurately from the traditional precipitation instruments (e.g., tipping buckets and weighing gauges) and from presently available disdrometers. It is possible that a modified version of TPS to improve its sensitivity can help to solve this issue but a detailed research program is needed to guide this improvement. Comparisons between FD12P and YES TPS (Fig. 4c) suggest that when FD12P and TPS are co-located, light precipitation conditions can be obtained confidently because their measurements technologies are different. In addition, the Droplet Measurement Technologies (DMT) Cloud Imaging Probe (CIP) and Fog Measuring Device (FMD) can be adapted to solve light precipitation measurements issues. Fog droplet size is usually found <20 μm in diameter and for sizes >20 μm, some precipitation such as light drizzle can be expected to occur (Gultepe and Milbrandt 2007b).

Using measurements obtained during a light drizzle event, the droplet spectra and PR versus time are obtained from the CIP measurements and are shown in Figs. 5a and 5b, respectively. The mean PR estimated from the CIP is approximately 0.3 mm h^{-1} over ~2 h time period. After calculating the averaged PR from the spectra, it is found that a relative difference between the mean TPS PR (Fig. 2) and CIP PR is estimated at about 25–30%. These results suggest that the TPS measurements can be used for drizzle PR calculations but a more detailed study is needed.

Figure 6 shows the time series of PR obtained from the TPS and the other instruments for the 10 June case. At the end of time series between 04:00 and 06:00 EST, the TPS clearly indicates the light drizzle conditions which are not indicated by the others. PR due to fog settling (droplet size< ~20 μm) is not considered and its contribution (>20 μm) is assumed to be included in the CIP PR calculation.

3.5.2 Visibility calculations

Visibility is usually obtained using the extinction of the visible light over a given distance (Gultepe and Milbrandt 2007a) and it is strongly related to the particle cross-section area and number concentration of particles in a given volume. Based on the definition of visibility, the number concentration is a driving parameter for low visibilities; therefore, visibility can not be considered solely as a function of either T or PR. Figure 7 is obtained using FD12P measurements and shows that when drizzle occurs, Vis usually becomes less than rain visibility alone.

Therefore, Vis for drizzle should be obtained independently and should not be included in rain Vis calculations. Gultepe and Milbrandt (2007b) showed that uncertainty in Vis calculations from various instruments can be high as 1 km in snow conditions and this needs to be reduced to develop better Vis parameterizations for nowcasting applications.

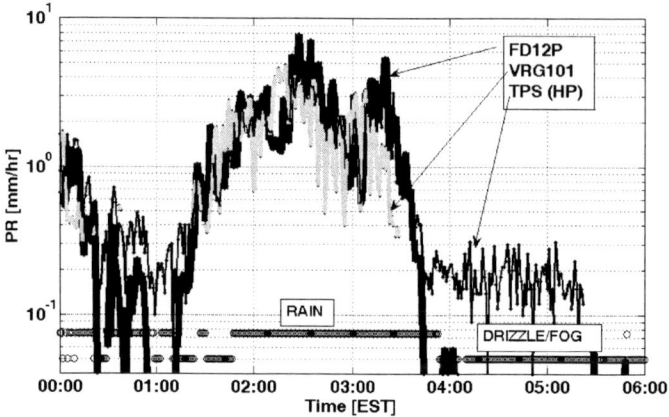

Fig. 6. Time series of precipitation rate for the rain and drizzle case for time period of 06:00–12:00 EST on 10–11 June 2007. Note that when PR is less than about 0.3 mm h^{-1}, only the TPS responses to the drizzle droplets and heavy fog conditions

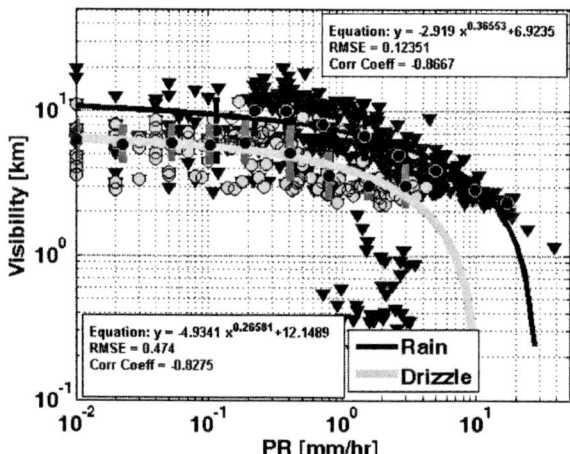

Fig. 7. Visibility versus precipitation rate for rain (*triangles*) and drizzle (*circle*) events for the 10–11 June case. The fitted equations and fits to the 1-min data points are also shown. Drizzle conditions are obtained from FD12P measurements as specified in manufacturer's manual

The Vis (or PR) can be related to temperature for climate change studies but the scatter of data points is so large that any short-term prediction of Vis based on PR and T cannot be used because often low precipitation rates (over the drizzle size range) with large number concentrations results in lower visibility values (Ulbrich and Atlas 1985). Therefore, for nowcasting issues, visibility versus PR (Rasmussen et al. 1999) or Vis versus f(PR;T) relationships should somehow consider the probability curves (Gultepe and Milbrandt 2007a).

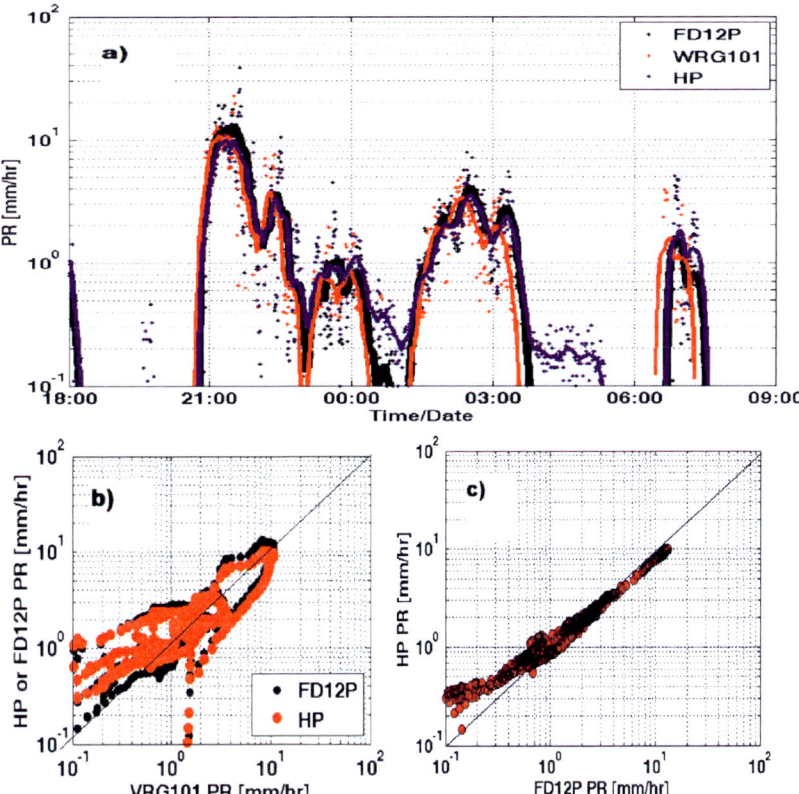

Fig. 8. (**a**) Time series of precipitation rate from three rain gauges for the 10–11 June case. The 1-min PR rates are smoothed out with the 20-point running averages. The 1-min data shows more scatter compared to 20-point smoothed data points, (**b**) scatter plot among FD12P, TPS (HP) and VRG101 PR measurements averaged over 20 min time intervals and (**c**) scatter plot for TPS (HP) versus FD12P PR values averaged over 20 min time interval

3.5.3 Uncertainties

Calculation of PR values as mm h^{-1} over a specific time period can be important for differences in PR values obtained from various instruments. Basically, sampling issues cannot be ignored in the analysis. Figure 8a shows that PR values over 1 min intervals show a large scatter and some time delay occurs in VGR measurements at the end of the time period. After plotting 20-point running averages in a time series, for PR values less than 0.3 mm h^{-1}, only the TPS responded to changes of PR over the drizzle size range (Fig. 8a). Although large PR values follow the same trends, a time gap occurred between VRG101 and others. This figure suggests that PR values in the FRAM data set should be smoothed out for larger time intervals.

Figures 8b and 8c show the scatter plots of VRG101 versus TPS and FD12P and TPS versus FD12P, respectively, for measurements averaged over a 20-min time interval. The large scatter between VRG101 and others is seen but TPS versus FD12P plot doesn't show this. When PR is less than 0.5 mm h^{-1}, TPS measures PR up to 2–3 times more than the FD12P. Surprisingly, both instruments indicate the existence of light precipitation and even the fog.

Table 2 shows the absolute and relative errors for the VRG101, FD12P and TPS. Note that the comparisons between TPS and VRG101 suggest that VRG101 is not useful for light precipitation detection. Absolute error is calculated using Eq. (4). The relative error is calculated using Eq. (5) by getting a difference between a single instrument value and an average value of all three instruments and then, dividing it by their mean values (see Sect. 3.3). Based on Table 2, it is found that large relative and absolute errors up to 44% and 0.31 mm h^{-1}

Table 2. Shows the possible errors calculated using Eqs. 4 and 5 for the measurements of the TPS, VRG101 and FD12P

FRAM-L2	10–11 June 2007 rain case		
instruments	TPS-FD12P [mm h^{-1}]	FD12P-VRG101 [mm h^{-1}]	VRG101-TPS [mm h^{-1}]
Absolute error	0.16	0.28	0.13
Relative error	2%	9%	9%
Mean/SD	1.68/2.62	1.85/3.57	1.56/2.80
	15–16 June 2007 drizzle case		
Absolute error	0.23	0.09	0.31
Relative error	44%	32%	33%
Mean/SD	0.58/0.96	0.35/1.03	0.27/0.78

occur when mean PR obtained over ~12 h time period is about ~0.60 mm h^{-1} for the 15–16 June case. When mean PR is >1.5 mm h^{-1} for the 10–11 June case, corresponding values become ~10% and up to 0.28 mm h^{-1}. It should be noted that both TPS and FD12P responded to low PR values (<0.3 mm h^{-1}) better than VRG101, suggesting that they may also be used for heavy fog detection in the remote areas.

3.6 Conclusions

In the present study, precipitation measurements from three new instruments (VRG101, FD12P and TPS3100) are compared to each other for light precipitation that includes drizzle and heavy fog conditions. The major conclusions are found as:

a) The TPS and FD12P clearly respond to low precipitation and drizzle rate that cannot be measured by weighing instruments e.g. VRG101 or tipping buckets.
b) Large scatter of data points exists when PR<1 mm h^{-1} and uncertainty in PR increases with decreasing precipitation amount
c) Comparisons among the FD12P, TPS and CIP probes suggest that both the TPS and CIP can be used to obtain detailed observations of light precipitation.
d) Measurements of precipitation during drizzle and light precipitation events include large uncertainties.
e) Averaging interval effects can result in large differences in PR values.
f) A time delay occurred between measurements of the VRG and others (FD12P and TPS) and this needs to be considered for PR intercomparisons.
g) Visibility cannot be obtained from PR because Vis is directly related to total number concentration of drops and their cross-sectional area; therefore, the probability curves should be considered for extreme weather applications.

These conclusions were obtained using the measurements collected during the FRAM field programs which were designed for fog studies. Therefore, a detailed field program needs to be designed for light precipitation conditions that are usually represented by PR values less than a few mm/h (e.g. <1 mm h^{-1}). The uncertainty in PR can reach to up to at least 40% but presently many instruments even do not have a capability to measure droplets less than 500 μm where drizzle usually forms.

Acknowledgements

The author would like to thank Dr. G. Isaac for use of one of his total precipitation sensor (TPS), R. Nitu for VRG101 instrument and to Dr. S. Cober for his ongoing support for this work during the FRAM projects. The author also thanks Drs. P. Joe and N. Donaldson for their valuable comments during the internal review process. Funding for this work was provided by the Canadian National Search and Rescue Secretariat and Environment Canada. Some additional funding was also provided by the European COST-722 fog initiative project office. Technical support for the data collection was provided by the Cloud Physics and Severe Weather Research Section of the Science and Technology Branch, Environment Canada, Toronto, Ontario. The author is also thankful to M. Wasey and R. Reed of Environment Canada for technical support during FRAM.

References

Ciach GJ (2003) Local random errors in tipping-bucket rain gauge measurements. J Atmos Ocean Tech 20:752–759

Goodison BE, Louie PYT, Yang D (1998) WMO solid precipitation measurement intercomparison - WMO/TD No 872. World Meteorological Organizatiion, Geneva, Switzerland, pp 211

Gultepe I, Milbrandt J (2007a) Visibility parameterizations for precipitation types for modeling applications. J Appl Meteorol (submitted)

Gultepe I, Milbrandt J (2007b) Microphysical observations and mesoscale model simulation of a warm fog case during FRAM project. Pure Appl Geophys [Special issue on fog, Gultepe I. (ed)] 164:1161–1178

Gultepe I, Cober SG, Isaac GA, Hudak D, King P, Pearson G, Taylor P, Gordon M, Rodriguez P, Hansen B, Jacob M (2007) The fog remote sensing and modeling (FRAM) field project and preliminary results. B Am Meteorol Soc (accepted)

Humphrey MD, Istok JD, Lee JY, Hevesi JA, Flint AL (1997) A new method for automated dynamic calibration of tipping-bucket rain gauges. J Atmos Ocean Tech 14:1513–1519

Illingworth AJ, Stevens CJ (1986) An optical disdrometer for the measurement of raindrop size spectra in windy conditions. J Atmos Ocean Tech 4: 411–421

Joss J, Waldvogel A (1969) Raindrop size distribution and sampling size errors. J Atmos Sci 26:566–569

Kruger A, Krajewski WF (2002) Two-dimensional video disdrometer: A description. J Atmos Ocean Tech 19:602–617

Kunkel BA (1984) Parameterization of droplet terminal velocity and extinction coefficient in fog models. J Clim Appl Meteorol 23:34–41

Loffler-Mang M, Joss J (2000) An optical distrometer for measuring size and velocity of the hydrometeors. J Atmos Ocean Tech 17:130–139

Nystuen JA (1999) Relative performance of automatic rain gauges under different rainfall conditions. J Atmos Ocean Tech 16:1025–1043

Nystuen JA, Proni JR, Black PG, Wilkerson JC (1996) A comparison of automatic rain gauges. J Atmos Ocean Tech 13:62–73

Rasmussen RM, Vivekanandan J, Cole J, Myers B, Masters C (1999) The estimation of snowfall rate using visibility. J Appl Meteorol 38: 1542–1563

Rasmussen RM, Hallet J, Tryhane M, Landolt S, Purcell R, Beaubien MC, Jeffries WQ, Hage F, Cole J (2005) The hotplate snow gauge. In: Preprints 15th Conference on Applied Climatology/13th Symposium on Meteorological Observations and Instrumentation. American Meteorological Society, 19–23 June, Savannah, GA

Sevruk B, Hamon WR (1984) International comparison of national precipitation gauges with a reference pit gauge. Instruments and observing methods – Rep. 17. World Meteorological Organization, Geneva, Switzerland, 86 pp

Sheppard BE (1990) The measurement of raindrop size distributions using a small Doppler radar. J Atmos Ocean Tech 7:255–268

Sheppard BE, Joe PI (1994) Comparison of raindrop size distribution measurements by a Joss-Waldvogel disdrometer, a PMS 2DG spectrometer, and a POSS Doppler radar. J Atmos Ocean Tech 11:874–887

Sheppard BE, Joe PI (2000) Automated precipitation detection and typing in Winter: A two year study. J Atmos Ocean Tech 17:1493–1507

Sheppard BE, Joe PI (2007) Performance of the precipitation occurrence sensor system (POSS) as a precipitation gauge. J Atmos Ocean Tech (in press)

Steiner M, Smith JA, Uijlenhoet R (2004) A microphysical interpretation of radar reflectivity–rain rate relationships. J Atmos Sci 61:1114–1131

Stewart RE, Burford JE, Hudak DR, Currie B, Kochtubajda B, Rodriguez P, Liu J (2004) Weather systems occurring over Fort Simpson, Northwest Territories, Canada during three seasons of 1998/99. Part 2: precipitation features. J Geophys Res 109 D22109, doi: 10.1029/2004JD004929

Tokay A, Wolff DB, Wolff KR, Bashor P (2003) Rain gauge and disdrometer measurements during the Keys area microphysics project (KAMP). J Atmos Ocean Tech 20:1460–1477

Tryhane ML, Landolt SD, Rasmussen RM (2005) Applications of the hotplate snow gauge. In: Preprints 15th Conference on Applied Climatology/13th Symposium on Meteorological Observations and Instrumentation. American Meteorological Society, 19–23 June, Savannah, GA, JP1.33

Turtiainen H, Nylander P, Puura P, Oyj V (2006) A new high accuracy, low maintenance all weather precipitation gauge for meteorological, hydrological and climatological applications. In: Proceedings 22nd International Conference on Interactive Information Processing Systems for Meteorology, Oceanography, and Hydrology. American Meteorological

Society 28 January – 3 February, Atlanta, GA. http:// ams.confex.com/ams/ pdfpapers/ 102118.pdf

Ulbrich CW, Atlas D (1985) Extinction of visible and infrared radiation in rain: Comparison of theory and experiment. J Atmos Ocean Tech 2:331–339

Wang TI, Clifford SF (1975) Use of rainfall-induced optical scintillations to measure path-averaged rain parameters. J Opt Soc Am 65:927–937

Wang TI, Earnshaw KB, Lawrance RS (1978) Simplified optical path-averaged rain gauge. Appl Optics 17:384–390

Wilson JW (1963) Evaluation of precipitation measurements with the WSR-57 Weather radar. J Appl Meteorol 3:164–174

WMO (1983) Guide to Instruments and Methods of Observation – WMO No 8, 7th edn. World Meteorological Organization, Geneva, Switzerland

Yang D, Goodison BE, Metcalfe JR, Golubev VS, Bates R, Pangburn T, Hanson CL (1998) Accuracy of NWS 8" standard non-recording precipitation gauge: Results and application of WMO inter-comparison. J Atmos Ocean Tech 15:54–68

4 The Droplet Spectrometer – a measuring concept for detailed precipitation characterization

Sebastian Glasl, Magnus Anselm

Technische Universität München, Germany

Table of contents

4.1 Introduction 83
4.2 Physical basis 84
    4.2.1 Drop size calculation 84
    4.2.2 Calibration 86
4.3 The measuring concept 87
    4.3.1 The droplet sensor 87
    4.3.2 The software 'Rainalyser' 89
4.4 Discussion and applications 93
    4.4.1 Measuring range 93
    4.4.2 Influence of wind 94
    4.4.3 Drop shapes and drag coefficient 94
    4.4.4 Significance of the impulse of the drops 94
    4.4.5 Application possibilities 95
4.5 Future plans and improvements 96
4.6 Appendix 97
References 99

4.1 Introduction

The climate change issue increasingly attracts the interest of the media and the public. In southern Europe, for example, devastating forest fires are noted more often after long dry periods; also reports for intense storms and damaging floods are more frequent. The winter in Germany in 2006–2007 recorded about two degrees warmer mean temperatures than for the average winter and in some regions the deviation was even higher. The climate change is endorsed by several International bodies. Hence, it becomes more and more important to be able to observe and analyze in detail the changing climate and weather with suitable measuring instruments.

Focusing on the observation of precipitation and its effects, a simple information like '10 mm of precipitation were recorded in two hours' is already much imprecise. In order to be able to carry out a detailed precipitation analysis, size and speed is of special interest; even more the momentum of the hydrometeors is. To measure these parameters some concepts have already been available for some time. Optical and mechanical measuring instruments exist which are used for the local determination of precipitation intensity or the drop size distribution. Some rain detectors are able to calculate these parameters, like laser disdrometers (e.g., OTT's Parsivel) or the Joss-Waldvogel Disdrometer (by Distromet LTD), by measuring the size of single raindrops (see also Tokay et al. 2005). However, there is no compact measuring instrument in the market which records information about detection time, size and momentum of every single drop directly without carrying out a projection on more inexact parameters (e.g., intensity or whole amount of precipitation).

Hence, the authors saw the need to develop a measuring instrument which stores the raw data and allows a very detailed characterization of precipitation. The early development process begun in 2005 and the first prototypes are already in use. The so-called 'Droplet Spectrometer' is a three-part measuring system basically consisting of a droplet sensor, a measuring amplifier and a computer with the software 'Rainalyser' for analysis and calculation of the momentum of the individual drops and for displaying the results (see Fig. 1).

4.2 Physical basis

4.2.1 Drop size calculation

For the calculation of drop size and drop mass the measured mechanical impulse, p, is mathematically transformed by using also the terminal fall speed into the mass m (the procedure for the calculation of droplet mass from momentum is given in the Appendix; for all references to equations

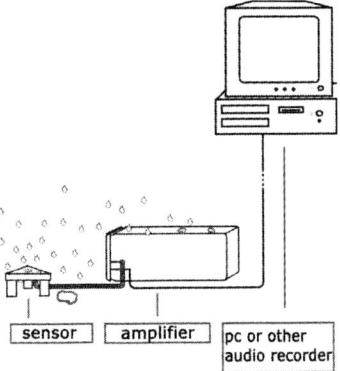

Fig. 1. The instrument consisting of three parts

that follow, the reader should refer to this Appendix). The terminal fall speed is reached when the air friction F_r equals weight F_g. As long as the terminal fall speed is not reached, the drops are accelerated during the fall by F_g. At the same time, the air friction F_r increases and the drops' velocity approaches asymptotically the terminal fall speed which is dependent on the drops' mass and which is nearly reached after a few seconds of flight.

From Eq. (4), for the terminal fall speed, it becomes evident that the speed of the drops depends on their mass m, their front surface A and the air drag coefficient. The term 'front surface' is defined as the drop's resisting surface. For simplification, the front surface of drops is assumed to have the form of a circle. To calculate the front surface, A, knowledge of the radius of the sphere or the circle is necessary. This is calculated with the aid of the volume in Eq. (5). From Eq. (8) it is seen that the front surface A depends only on the mass of the drops.

For calculation of the mass the following constants are needed:

- Circle constant, π: 3.14159
- Acceleration of gravity, g: 9.80665 m/s^2
- Air density of the gas in which the drop falls, ρ_{air}: 1.293 kg/m^3
- Water density of the liquid of which the drop consists of, ρ_{water}: 998 kg/m^3
- Air drag coefficient for a sphere in air, c_w: ~0.45

4.2.2 Calibration

It must be note that the detector of the droplet spectrometer does not deliver directly the impulse of the drops, but a tension value which is proportional to this impulse. This tension is stored by recording the signal as amplitude values in AIFF files. To be able to calculate the impulse depending on the measured amplitude, the following approach helps: (Eq. (15))

$$\gamma = \frac{Y}{p}$$

With:
    Y = amplitude value in the AIFF file
    γ = sensibility factor
    p = impulse (Eq. (14))

The sensibility of the device γ must be determined with the help of calibration measurements. Therefore, drops of known size are dropped onto the sensor from a defined height. A drop with a diameter of 2 mm e.g., reaches 30% of its terminal fall speed and approximately 10% of its terminal air friction after 0.5 m fall height and 2 s fall time (see Fig. 2).

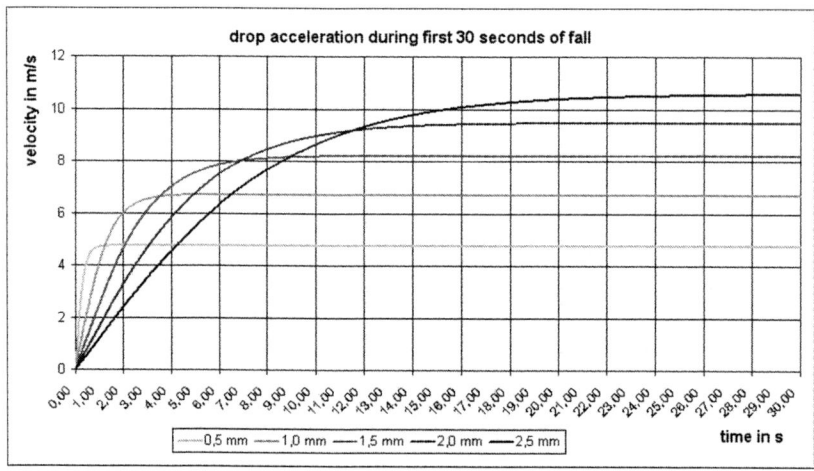

Fig. 2. Drop acceleration during the first 30 seconds of fall

Hence, the impulse of the drops can be calculated without more precise consideration of the aerial friction and can be combined with the measured amplitude value in order to calculate the sensibility factor gamma, γ. Nevertheless, this method of the calibration provokes some uncertainties. First of all, it is assumed that the sensor measures linearly, which means that the sensibility factor gamma can be applied for the entire measuring range. In addition, the aerial friction strength is neglected during the calibration and consequently a systematic measurement error is caused. During the experimental tests, it appeared that the linearity of the sensor can be considered roughly as given. Since this assumption is not verified exactly, the authors are currently working on a function that describes the correlation between drops size and amplitude values. The disregard of the air friction strength during the calibration process influences the calculation exactness. If a higher fall height is selected, the influence of the air friction becomes increasingly larger. To avoid this fact, the calibration should be performed with a relatively low fall height. Nevertheless, this calibration method is not quite satisfactory and is also under revision.

Also, other calibration methods have already been considered. For example, a calibration with plasticine balls or other steady materials. This would have the advantage that the mass can be weighed out with high precision before the impulse measurement. Using water droplets is much more challenging, because they burst with the impact. A production of drops with exactly defined mass is complex and tricky. However, regarding the transfer of impulse during the calibration, the behavior of the steady materials is expected to be different compared to water drops. Raindrops apply an inelastic push and splash-off to the sides. Thereby secondary droplets can emerge. Indeed, a plasticine ball would likewise carry out an inelastic push but does not burst in several parts. It would rather be compressed and change its shape but remains in one piece, implying that the momentum that is induced might be larger and transferred more rapidly.

4.3   The measuring concept

4.3.1 The droplet sensor

In order to measure the detailed characteristics of rain, a piezoelectric droplet sensor has been developed (see Fig. 3) which transforms the mechanical impulse of each single drop into an electrical voltage. It

Fig. 3. The piezoelectric droplet sensor (3rd generation)

essentially consists of a base plate with piezoelectric elements on it, a cover plate and a spring which connects base and cover plate. When a drop hits the measuring surface the piezoelectric elements are compressed and generate a voltage which is amplified by the measuring amplifier. Subsequently, the signal is transmitted via cable and can be recorded by a computer.

The piezoelectric discs are made of modified lead zirconate-titanate (PZT) type PIC 155 provided by the company 'PI Ceramic'. These possess properties which are suitable especially for application in microphones and oscillation receivers combined with preamplifiers. With the development of the sensor, a lot of attention has been paid to a quick attenuation of the construction. This is important because after every strike of a drop the analysis software 'Rainalyser' has to apply a certain 'dead time' (in the following also called the event length) which prevents the treatment of post oscillations as individual drops. Hence, it is aimed to reach a very strong damping of the oscillating sensor surface; a high resonance frequency can be also improving the attenuation. To succeed in these requirements, light and well subdued aluminum PVC sandwich plates of the types 'Alucobond ®' and 'Dibond ®' were chosen as coverage. In addition, a tension spring is integrated which pulls the cover plate to the basis construction performing as a damping unit. The total mass of the coverage is only raised by approximately five grams, but the post oscillations are considerably damped.

The measuring amplifier strengthens the signal tensions to enable higher cable lengths of up to 100 m and to diminish the background noise. For the sensitivity adjustment the amplification can be set between 0.5- to 100-fold. Typically, the tension is amplified from a few mV up to 0.5V and limited to protect the computer against over voltage.

The signals created by the measuring amplifier are recorded via the soundcard's line-in as an audio stream and stored as AIFF-files. Thereby no significant signal information data gets lost. Furthermore, before the analysis, longer recording periods can be cut into shorter terms to advance the investigation. The original measuring data is yet preserved due to the strict separation of recording and analysis.

4.3.2 The software 'Rainalyser'

Analysis of the raw signals

In the first step, the software 'Rainalyser' analyses the AIFF-files. During this process, 'Rainalyser' scans the audio files for amplitudes which possess higher values than a certain lower threshold. As the sensor works comparable to a microphone, it is evident that a background noise exists that has to be disregarded, especially while operating under pluvial-noise conditions. Consequently, a threshold is adjusted which has a slightly higher value than the background noise and which can be overstepped by the smallest amplitude value desired to measure.

When the software detects an amplitude value which is greater than the minimum amplitude, it searches in a defined period for the biggest peak value. This time period is called event length. It should be chosen at least so long for all oscillations of the sensor to have decreased below the threshold within the event length. During this time period, only one drop can be registered. This leads to a dead time which defines the minimum time interval needed to distinguish individual drops. In the case of using the second sensor generation, this value lies between 10 and 20 ms for ordinary drop sizes (~1.5 mm in diameter). Subsequently, 'Rainalyser' starts its scans again until the end of the audio file is reached. The amplitude peaks and their moment of impinge are saved in an analysis file (see Fig. 4). The analysis needs a certain time depending on the AIFF-file size and the analysis parameters. The recording is performed with the following parameters: sampling frequency: 8000 Hz, bit-depth 16 bits, number of channels: one. It has,

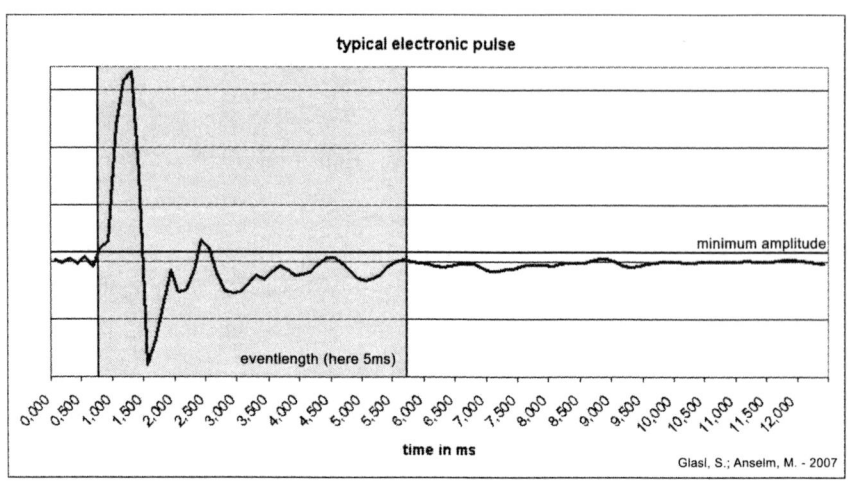

Fig. 4. Typical electronic pulse

therefore, a data rate of 16 KB/s (~55 megabytes and 28.8 million frames per hour).

The two important evaluation parameters, namely, event length and minimum amplitude, limit decisively the measuring area of the droplet spectrometer. The greater the event length is, the fewer the drops that can be registered per second are and when facing high drop rates, a lot of drops can be suppressed. The minimum amplitude defines the size of the smallest measurable drops (realistic value approximately 0.5 mm of diameter). The smaller the minimum amplitude the measured drops can also be the smaller. Indeed, then the risk exists to register a post oscillation as a drop if the event length was put too brief.

It is important to achieve a balance between both parameters in order to receive the best possible results. The analytic evaluation method which conserves the raw data makes this adjustment very well feasible. For example, analysis files with different evaluation parameters can be generated and compared. To simplify the adjustment of the evaluation parameters, it is examined whether it is possible to define the event length dynamically which means that the software automatically adjusts the event length to the size of the drop. Small drops could thereby receive a shorter event length and big drops, which stimulate the system stronger, a longer one.

Besides the two main parameters for the analysis, a maximum amplitude can be defined. In addition, all physical variables which are relevant for the calculation of the drop size can be adjusted manually (see also Sect. 4.2.1).

Display options

Besides the analysis of the sound files, 'Rainalyser' offers also the possibility to display the results in tabular form and graphically. In a value table all measured values of the single drop events, like mass, diameter, momentum, kinetic energy and point in time can be read. This table can be exported as a comma delimited file and be processed by other programs. An overview page allows a quick overview about parameters like whole amount of precipitation, duration of the recording, time with and without precipitation, average drop size etc.

A probably unique display option might be the time-drop size diagram (see Fig. 5). The software is able to create a point diagram, with time in the x-direction and drop size in y-direction. Every single drop is shown as a dot. This allows drawing conclusions on the temporal development of the precipitation and its intensity. Especially, variations in intensity and breaks are easy to identify. The diagram is adjustable in detail; for example, the axis can be described, a grid net can be drawn or a subsection can be marked and enlarged for precise consideration. Also, the time and dimension axes are arbitrarily scaleable.

The second graphic display form is the droplet spectrum (see Fig. 6): the measured drops are grouped in terms of their size into up to 100 classes and these classes are shown in a histogram. The droplet size distribution makes it possible to see in one view of which kind the precipitation has been and the characteristic trait of every precipitation event. For example, the bandwidth of the drop classes admits conclusions on the intensity.

Combining the information from the different display possibilities and the specialist knowledge, a detailed characterization of the measured precipitation event is feasible. The measured results can be tuned to the needs of the user by numerous settings and export possibilities and may be adapted on the concrete measuring situation.

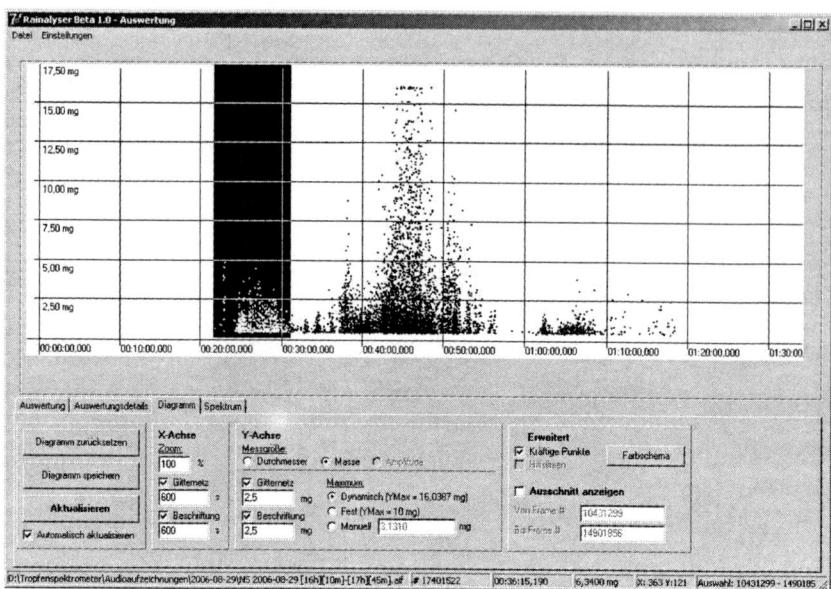

Fig. 5. The time-drop size diagram

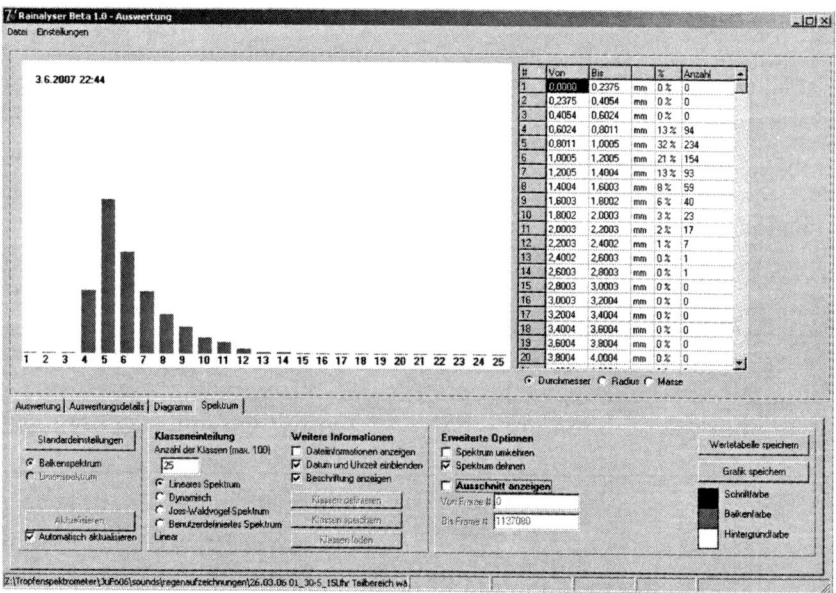

Fig. 6. The droplet spectrum

4.4 Discussion and applications

4.4.1 Measuring range

Lower limit

The size of the smallest rain droplets possible to measure depends on the evaluation parameters of 'Rainalyser', the amplification factor and signal-to-noise-ratio of the measurement amplifier. With low background noise and short post oscillations, the minimum amplitude can be also reduced. Using the first operative prototype in 2006, we were already able to measure droplets smaller than 1.0 mm in diameter. Indeed, the sensor was stimulated to long lasting post oscillations. Consequently, short drop intervals lead to a disregard of individual drops. Therefore, the successor was optimized especially for short post oscillations and a quick response to stimulation. This was realized by implementing high quality material, primarily X5CrNi18-10 (1.4301) stainless-steel, and professional piezo-ceramics which replace the fire igniters of the predecessor.

However, physical effects fix a limit, too. Droplets with a diameter smaller than 0.3 mm possess an enormous surface in comparison to their weight. Consequently, very low air-streams can be sufficient to manipulate their fall behavior distinctly. This effect is well known especially from fog droplets which almost levitate in the air. Even if these droplets hit the surface, both the mass and speed are so low that no recognizable signal amplitude is induced.

Nevertheless, if the volume contribution to the rainfall of small, undetected droplets is relatively small, neither the radar reflectivity nor the precipitation rate are influenced significantly by this drop class (Grimbacher 2002).

Upper limit

The size of drops with diameters greater than 5 mm can only be determined with decreasing exactness when measuring their momentum. Due to their high velocity, these drops become increasingly unstable and behave rather unpredictably. The physical principles which are applied for small drops are only partially valid for the large ones. In practice, these large drops can occur especially in thundershowers. But the larger these drops are the higher is the probability that they diverge during their fall into smaller secondary drops which can be measured easily. Nevertheless, all drops with a momentum that causes amplitude values higher than the minimum amplitude are registered. In order to minimize the possible inaccuracies, we are working on a more precise consideration of the physical principles.

4.4.2 Influence of wind

Upstream impinging wind influences the terminal fall speed of the rain drops and can lead to distorted drop dimension values. Such a measuring error can be noticed by a shift or distortion of the drop size spectrum. In general, it can be said that the smaller the raindrops are, the stronger they are susceptible to wind and measurement errors. The reason for this is the fact that the decrease of the mass of the drops is disproportional compared to the surface when the diameter is reduced.

Vertical winds influences only the vertical speed component of the falling drops which is not measured by the system and therefore can be disregarded. It would be considered to mount a droplet sensor in a vertical position to measure this speed component , but this idea has not been tested yet.

4.4.3 Drop shapes and drag coefficient

While droplets up to 1 mm diameter are nearly perfectly round and drops up to 2–3 mm diameter still are unambiguously spherical, bigger drops flatten and buckle increasingly on the underside. These drops with larger diameter change their shape from a sphere to an oblate spheroid with a dent on the bottom side. This change in shape causes an increasing drag coefficient (Vössing 2001). While for smaller drops a drag coefficient of 0.45 is assumed, this would rise up to 1.35 for a kidney-shaped spheroid.

This shape change causes a modified fall behavior: drops with diameters greater than 3 mm become increasingly unstable and start to oscillate in their shape (Vössing 2001, p. 90). This fact entails that big drops adopt a lower, due to the oscillation, even varying fall speed which can diverge up to 30% from the calculated final speed (Vössing 2001, p. 91).

4.4.4 Significance of the impulse of the drops

One is not only interested in the drop size. The physical/mechanical effects of raindrops are put into the focus of, for example, the erosion research. In this area, the impulse and the kinetic energy become the key sizes. A lot of research projects use measuring instruments and methods which first determine the size and/or speed of the drops. In a second step, the impulse is computed on the basis of mass and speed. In contrast to this method, the Droplet Spectrometer can measure the impulse the drops directly mechanically and display it without any calculation (only the sensitivity factor is needed). An unpredictable behavior of, for example, larger drops or hail does not affect the impulse measurement.

4.4.5 Application possibilities

Application in the climatology

Our comprehension of the future climate, but also partially of the climate of the presence and the past, comes from computer aided model calculations. These climate models could be extended or confirmed by precipitation measurements with the Droplet Spectrometer, so that this measuring system would be helpful to draw a more sharply outlined picture of our climatic future.

Microphysical research

The Droplet Spectrometer could be used to investigate the precipitation origin and development. For example, the microphysical models, which are used to simulate the processes in clouds, occasionally require evidence for the suitability of their spectra (Grimbacher 2002).

Erosion modeling or simulation

To investigate erosion under lab conditions, one uses rain simulators which generate precipitation with the desired properties. For being able to interpret the erosion behavior correctly, one needs to know the characteristics of the artificially generated precipitation. Besides the precipitation structure, one is interested particularly in the kinetic energy or the impulse of the single drops (Hassel and Richter 1991). For the operation in the erosion simulations, the Droplet Spectrometer owns the great advantage that the determining parameter, the impulse is measured directly.

4.5 Future plans and improvements

The Droplet Spectrometer is still in a very early phase of its development. During the past two years, the basis of the concept has been created and it has been shown that the piezo-electric sensor and the software are working properly; nevertheless, there are still many isues that have to be carefully considered.

For example, there are plans to compare the signals produced by sensors with different cover plates and varying piezo-ceramic types. The first prototype had a triangle-like form, but we have also built a rectangular and a round one. Besides the sensor, also the measuring amplifier has recently been redesigned in order to provide a better signal quality with less noise and a more precise frequency handling. However, the amplifier can be improved further. For example, a digital version with wireless communication would be a great advance to extend the applications of the system.

When thinking about the calculation algorithms, further enhancement possibilities come out. For example, the linearity of the system has to be analyzed in detail to be able to include this relation into the formula for the droplet size calculation. Also, the calibration should be worked over to improve the measuring accuracy.

To reduce the dead time of the system after a drop hit the surface, the characteristics of the post oscillations are studied. By using Fourier transformation, it could be possible to compute out the post oscillations and to reduce the dead time to a minimum.

To verify the correct operation of the system it is planned to compare with other measuring instruments, for example, the Joss-Waldvogel disdrometer (JWD). Some facilities have already offered their assistance in the form of providing these instruments and their knowledge.

With the development of the Droplet Spectrometer a precipitation measuring system has been emerged which is suitable for a great spectrum of applications. Although there is still a lot of work to do in terms of improvements, we believe that the system has a solid basis and possesses a high potential. Through its simple and cheap construction it could compete with commercial rain gauges and established instruments (e.g., the Joss-Waldvogel disdrometer).

4.6 Appendix

Calculation of droplet mass from momentum

The air friction is defined:
c_w = air drag coefficient
A = front surface
ρ_{air} = density of the air
v = velocity of the drop

$$F_r = c_w \cdot A \cdot \frac{\rho_{air} \cdot v^2}{2} \quad (1)$$

The weight is defined as:
m = drop mass
g = acceleration of gravity

$$F_g = m \cdot g \quad (2)$$

The terminal fall speed is reached when air friction and gravity force values are equal:

$$F_g = F_r \quad (3)$$

Now Eqs. (1), (2) are inserted into Eq. (3) and the formula is solved for the terminal fall speed

$$v = \sqrt{\frac{2 \cdot m \cdot g}{c_w \cdot A \cdot \rho_{air}}} \quad (4)$$

Radius calculation with the aid of the volume

$$V = \tfrac{4}{3} r^3 \pi \quad (5)$$

Volume and mass are associated with density (where r = drop radius, ρ_{water} = density of water):

$$V = \frac{m}{\rho_{water}} \quad (6)$$

Summarizing and solving for radius:

$$r = \sqrt[3]{\frac{3 \cdot m}{4 \cdot \rho_{water} \cdot \pi}} \quad (7)$$

Substituting the radius by Eq. (7), one receives:

$$A = \left(\sqrt[3]{\frac{3 \cdot m}{4 \cdot \rho_{water} \cdot \pi}}\right)^2 \cdot \pi \quad (8)$$

Eq. (8) inserted into Eq. (4):

$$v = \sqrt{\frac{2 \cdot m \cdot g \cdot 4^{\frac{2}{3}} \cdot \rho_{water}^{\frac{2}{3}}}{c_w \cdot \rho_{air} \cdot \pi^{\frac{1}{3}} \cdot 3^{\frac{2}{3}} \cdot m^{\frac{2}{3}}}} \quad (9)$$

The detector delivers the value of the impulse, expressed by the amplitude. The former is the product of mass and velocity:

$$p = m \cdot v \qquad (10)$$

Substituting the velocity by Eq. (9) and pulls m under the radical:

$$p = \sqrt{\frac{2 \cdot m^3 \cdot g \cdot 4^{\frac{2}{3}} \cdot \rho_{water}^{\frac{2}{3}}}{c_w \cdot \rho_{air} \cdot \pi^{\frac{1}{3}} \cdot 3^{\frac{2}{3}} \cdot m^{\frac{2}{3}}}} \qquad (11)$$

If one extracts the root and cancels out:

$$p = \frac{2^{\frac{1}{2}} \cdot 4^{\frac{1}{3}}}{3^{\frac{1}{3}} \cdot \pi^{\frac{1}{6}}} \cdot \frac{g^{\frac{1}{2}} \cdot \rho_{water}^{\frac{1}{3}}}{c_w^{\frac{1}{2}} \cdot \rho_{air}^{\frac{1}{2}}} \cdot m^{\frac{7}{6}} \qquad (12)$$

The impulse is directly proportional to the mass to the power of seven sixths:

$$p \sim m^{\frac{7}{6}} \qquad (13)$$

Solving for mass, the mass is:

$$m = \frac{1}{2} \cdot \sqrt[7]{\frac{9 \cdot \pi \cdot c_w^3 \cdot \rho_{air}^3}{g^3 \cdot \rho_{water}^2}} \cdot \sqrt[7]{p^6} \qquad (14)$$

Acknowledgements

We want to express our deep gratitude to Dr. Silas Michaelides for his invitation to contribute to this book and his support at the European Geosciences Union General Assembly 2007 in Vienna, Austria.

We also want to thank the GSF – National Research Centre for Environment and Health for award the 'Carl Friedrich von Martius Environmental Price' to us and the trust in our measuring instrument. Great thanks to Frank Moeller of the company PI Ceramic for donating us the piezo ceramics, as well as to the companies Aluform and Gleich for the material samples. Furthermore, we are indebted to Matthias Kuhnert of the Geo Research Center, Potsdam, Karl Auerswald of the Technische Universität München and Ali Tokay of the Goddard Space Flight Center for making available to us literature related to our project.

References

Grimbacher T (2002) http://www.iac.ethz.ch/en/groups/richner/cd/index.html (Updated at 1 June 2007). Institute for Atmospheric and Climate Science, Swiss Federal Institute of Technology, Zürich, Switzerland

Hassel J, Richter G (1991) Ein Vergleich deutscher und schweizerischer Regensimulatoren nach Regenstruktur und kinetischer energie. Zeitschrift für Pflanzenernährung und Bodenkunde 155:185–190

Tokay A, Bashor PG, Wolff KR (2005) Error characteristics of rainfall measurements by collocated Joss-Waldvogel disdrometers. J Atmos Ocean Technol 22:513–527

Vössing H-J (2001) In-situ messung großer hydrometeore mit hilfe der in-line-holographie. Johannes Gutenberg-Universität Mainz, pp 90–91

5 Quality control of precipitation data

Thomas Einfalt[1], Silas Michaelides[2]

[1]Hydro & meteo GmbH & Co. KG, Lübeck, Germany
[2]Meteorological Service, Nicosia, Cyprus

Table of contents

5.1 Introduction 101
5.2 Quality Control of rain gauge data 102
    5.2.1 Gaps in the data 103
    5.2.2 Physically impossible values 103
    5.2.3 Constant values 103
    5.2.4 Values above set thresholds 103
    5.2.5 Improbable zero values 104
    5.2.6 Unusually low daily values 104
    5.2.7 Unusually high daily values 104
    5.2.8 Data check time series 104
    5.2.9 Station data quality 105
    5.2.10 Generalization and future work 106
    5.2.11 Conclusion: what can we do automatically? ... 106
5.3 Quality Control of radar data 106
    5.3.1 Data Quality report of COST 717 107
    5.3.2 Error sources 108
    5.3.3 Data Quality Index 111
    5.3.4 Correction methods 114
5.4 Future developments 123
References 123

5.1 Introduction

Valuable information about natural processes is acquired from measurements and the data they produce. However, such sources of information are subject to disturbances of different kinds: device specific errors, uncertainties over data storage, transmission problems and data transformation effects – to name but a few. Therefore, data acquired from measurements do not always represent the real behavior of the observed processes. To read the data and identify their reliable and unreliable components, normally, requires experience and a lot of effort. Nevertheless, for many applications, from pipe design to flood warning, a good quality of data is required, usually reflecting an economic value.

5.2 Quality Control of rain gauge data

Quality Control (QC) of rain gauge data has been an important topic since the beginning of data collection. Attempts to formalize this task have been started in several countries (see Einfalt et al. 2000, for an overview). However, the conclusion has always been that the check of point rainfall measurements can only be done by human eye with a reliably good result (Jörgensen et al. 1998; Maul-Kötter and Einfalt 1998).

Two aspects may prevent such manual procedures: real-time data to be further processed for flood warning and large amounts of data to be investigated.

A quality check approach starting from simple cases before tackling the difficult ones yields surprisingly good results. This attitude has been proposed by the Bavarian Agrometeorological Service (Vaitl 1988) and refined by MeteoSwiss (Musa et al. 2003).

Basically, the rain gauge quality check starts with items that are easy to check, followed by more complex ones. In practice, this means that firstly, existing gaps in the data are excluded from further treatment, then features on the data of one station only are investigated and finally an intercomparison between the data from several rain gauges is performed.

The structure from simple check elements at the top, down to more complex ones, as proposed above, comprises the following steps (Einfalt et al. 2006):

1. Detection of gaps in the data;
2. Detection of physically impossible values;
3. Detection of constant values;
4. Detection of values above set thresholds;
5. Detection of improbable zero values;
6. Detection of unusually low values (which may be real, though);
7. Detection of unusually high values (which may also be real, though).

5.2.1 Gaps in the data

The detection of gaps in the data mainly serves to exclude the gap interval of a station to be used for comparison to other stations. Furthermore, a gap statistic can be derived which is an indicator of the reliability of the data from this station. Experience shows that well maintained stations with a good data quality rarely have gaps in their measurement series.

5.2.2 Physically impossible values

Physically impossible data consist of negative rainfall values and very high intensities, e.g., more than 5 mm per minute in a moderate climate. Such values should be excluded from further evaluation.

As a function of the underlying data and software, very high intensities may be a side effect of digitized paper charts. Such values can be further used if the time step for further analysis is large enough. For example, a value of 5 mm in 1 min can, in reality, be representative for 5 or 10 min when analyzed with additional information (e.g., the basic paper charts). In such a case, the evaluation on a 5 or 10 min time grid is the correct further treatment.

5.2.3 Constant values

Constant values over a certain time are an indicator of unusable data which may be either due to bad digitization or to missing values in case of digital registration. The 'certain time' for digital data with a time step of 1–5 min is around 15 min for intensities above 1 mm/h. For digitized data derived from paper charts, this time interval is a function of the paper chart resolution and may be as long as 60 min for old paper charts.

5.2.4 Values above set thresholds

For predefined durations, it is useful to indicate when the measurements are above statistically rare values, e.g., higher than an event occurring every 5 or 10 years. Useful durations comprise 5 min, 15 min, 60 min and 1440 min (1 day).

Such thresholds were defined in order to identify 'interesting' events, i.e., events which should be carefully checked before accepting the measurements.

5.2.5 Improbable zero values

While all of the above checks are performed on one station only, this check and the following two use the spatio-temporal rainfall structure as seen by several gauges. Improbable zero values can be detected at one station if all surrounding stations have significant rainfall and stations are close enough to each other.

5.2.6 Unusually low daily values

A more sophisticated check is the check on too low values where the daily sum of a selected station is compared to the daily sums of the neighboring stations. If the surrounding stations recorded significantly more rainfall than the selected one, the measurement of this station has to be considered as doubtful.

5.2.7 Unusually high daily values

A season dependent check is the one on too high values, where the daily sum of a selected station is compared to the daily sums of the neighboring stations. If the surrounding stations recorded significantly less rainfall than the selected one, the measurement of this station has to be considered as doubtful if a convective rainfall can be excluded. This is usually the case in Germany between October and March.

5.2.8 Data check time series

As a result, check time series were produced which present a non-zero check value for each observation in the data of a time series (Fig. 1). With the help of these check series, the data were screened, assuming that the main source of error had been detected by the automatic checks.

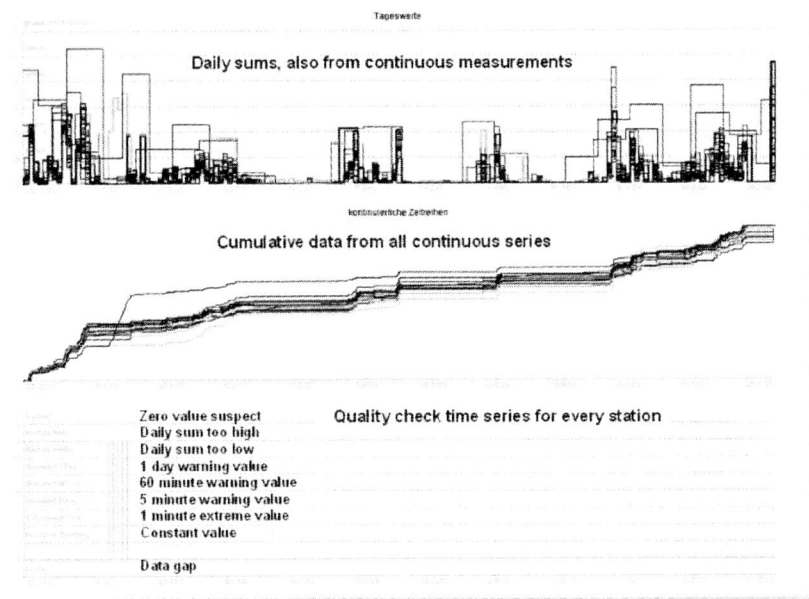

Fig. 1. Time series of daily values (*upper graph*), cumulated values (*middle graph*) and data quality (*lower graph*)

5.2.9 Station data quality

In a project where monthly values of areal rainfall for more than 5000 areas over 25 years had to be calculated (Einfalt et al. 2006), this approach was tested on a large amount of data. The quality check of the initially more than 20,000 station years resulted in a total of nearly 15,000 station years of plausible, quality controlled data. Main results for the difference of more than 25% of all station years were:

- Recognized inconsistencies in the data (8%);
- Marked gaps for diverse reasons;
- Data gaps due to stations only in service during summer (in early years);
- Data gaps due to missing data from the most recent years (mainly DWD stations: not delivered).

For continuous data, a clear temporal trend was visible: the data became more reliable with less gaps and the whole was based on a growing network.

5.2.10 Generalization and future work

QC of data is a task which needs a lot of experience. Although many methods have been developed to perform these checks automatically, they depend on the climate, the density of the network and the quality of the data; the latter is very important, since only with good quality data erroneous data can be detected, from a spatial perspective.

The presented approach does not use any additional information such as topographical height, temperature, radar or satellite measurements, or results from numerical weather models as it is done elsewhere (e.g., Musa et al. 2003). However, an application to mountain areas requires at least the knowledge of topography and temperature.

The use of radar data or other concurrent data sources are highly interesting for online data (historical data do not exist from these sources).

5.2.11 Conclusion: what can we do automatically?

In Einfalt et al. (2006) it proved to be extremely useful to help the human investigator with clearly defined automatic checks which were represented by time series and could be viewed in parallel with the original data of a whole region. Thus observations by the automatic check could be swiftly verified and subsequently accepted or rejected.

In particular, the 'simple' checks can be done automatically – there was nearly complete agreement with the human observer. More complex methods like the spatial consistency check need a well founded parameterization and can only serve as assistance to the visual inspection, still. This assistance is extremely valuable because is saves a lot of time by attracting the attention of the observer to the cases that are suspicious.

After the automatic checks have verified the 99% of the data, the final decision for the correctness of the data rests with the human observer who performs a cross-check with the remaining 1%.

5.3 Quality Control of radar data

Rainfall measurements by radar are subject to uncertainties, due to measurement errors and due to the indirect measurement by the radar as an instrument. These uncertainties and their consequences for further applications have not yet been systematically evaluated. There are two main reasons for this: on the one hand, many successfully presented pilot projects inherently take care of these uncertainties when transforming measurement results into useful information (e.g., Browne et al. 1998); on the other hand, it is difficult to assess the many different uncertainties which are partly dependent on the radar location and on the rainfall situation. Therefore, Fabry (2004) made the statement that 'no radar specialist can compute with any certainty what is the expected range-dependent accuracy of a rainfall estimate in a specific weather situation'. Krajewski and Ciach (2004) go even further by postulating: 'we acknowledge the fact that in practice it is impossible to delineate and estimate these errors separately based on the available measured quantities'. This contribution does not comprise an attempt to contradict the above statements but to contribute to a pragmatic way to improve knowledge about the uncertainties in radar measurement and to ways to communicate them.

5.3.1 Data Quality report of COST 717

The COST 717 Action 'Use of Radar Observations in Hydrological and Numerical Weather Prediction Models' (http://www.smhi.se/cost717) has been active in three working areas, organized along the application fields of radar data. A cross cutting activity in this context was the production of a status report on 'Radar Data Quality in Europe' (Michelson et al. 2005).

The report summarizes the activities in Europe related to radar data quality, both country-wise and project-wise. It starts with definitions, objectives and error sources and terminates by presenting ways to formulate radar data quality as well as open issues and forthcoming challenges for radar data providers and users.

The definitions provided are worth outlining here:

- *Data quality:* attribute of the data, which is inverse to uncertainties and errors, i.e., error-free data with few uncertainties are of high data quality and data with errors or large uncertainties are of low quality;
- *Data error:* wrong measurement contained in the data (e.g., clutter);
- *Data uncertainty*: measurement where the validity is not guaranteed (e.g., Z-R relationship).

It is interesting to note that there is no direct definition given for data quality but only the opposite: to the error laden or uncertain data. A differentiation is made between errors and uncertainties; this issue will be discussed in the following section.

5.3.2 Error sources

There are several sources of error which affect the ability of radars to measure precipitation and which influence the accuracy of the measurements. Such errors are discussed by Browning (1987) and Joe (1996), among others. Since it is not easy to differentiate which error sources affect reflectivity measurements and which affect surface rainfall estimates, the two are combined in this section. Figure 2 illustrates some of the most important factors. This section is influenced by the list of error sources given by Joe (1996) and it touches upon the most important ones starting with the radar system itself and continuing out to various interactions with targets.

Electronics stability: Modern components vary slightly with time and with temperature. A monitoring system can keep the stability to within one dB or warn when tolerances are exceeded.

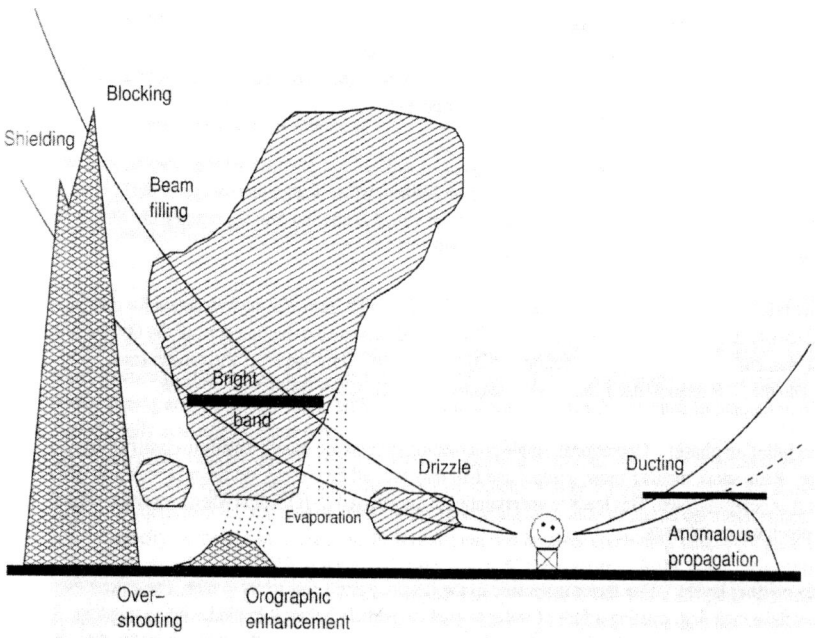

Fig. 2. Factors affecting the radar rainfall measurement (from Michelson et al. 2005)

Antenna accuracy: The antenna should be designed to minimize side lobes. If its orientation accuracy is not regularly checked, data will be inaccurately navigated which will subsequently result in inaccurate location of measurements.

Signal processing accuracy: The combination of the sampling capabilities of the radar hardware together with the performance of the signal processor will define the ability of the system to process data to derive the most accurate radar observables and treat known errors. Regardless of the signal processor's performance, it must accurately interpret the radar equation in order to reduce the risk of error.

Electromagnetic interference: Other radars, microwave links, the sun and military jamming, all can cause interference which can result in errors.

Attenuation due to a wet or snow/ice covered radome: In heavy rain, a thin film of water will cover the radome, causing signal attenuation. In cold conditions, snow and ice may build up on top of the radome, also causing attenuation and limiting the quantitative use of reflectivity measurements.

Clutter: Ground Clutter (GC) is usually strong due to the relative radar cross-section of the ground being much greater than that from meteorological targets and despite echoes from the ground being generated from side lobe radiation which is much weaker than that from the main lobe. Ground clutter can be minimized through intelligent radar siting, Doppler suppression and through the use of post-processing methods, such as static clutter maps.

Anomalous propagation (anaprop): Specific atmospheric temperature and/or moisture gradients will cause part of the radar beam to propagate along a non-normal path. If the fraction of the beam that is refracted downward (super-refraction) is refracted sufficiently, the radiation will illuminate the surface and return signals to the radar from distances further than are normally associated with ground clutter targets.

Shielding: If the radar siting is inappropriate, nearby topographic features, but also trees, buildings and other structures can block the radar beam in whole or in part, causing shielding of sectors of interest. Regardless of the siting quality, anomalous propagation can still cause problems at distant ranges, although this is a less significant problem in terms of the radar's ability to detect meteorological targets.

Other non-precipitation echoes: Such echoes can originate from birds, insects, chaff (strips of metal foil used by the military) and refractive inhomogeneities, known as clear air echoes. These echoes are often not static in space which means that they cannot be effectively treated using

Doppler techniques. They are, however, often easily identifiable by an operator.

Attenuation by precipitation: Heavy rain, graupel and hail can attenuate energy, leading to strong underestimation of precipitation intensities. Especially in hail, where scattering takes place in the Mie region, the scattered energy can be attenuated to the point of virtual extinction over the return path. Shorter wavelengths (X and C bands) are more seriously affected.

Z-R relation: This relation, expressed in the form $Z = AR^b$, provides the foundation for relating radar reflectivity Z to rainfall intensity R. Z and R are usually assumed to be functions of the 6th and roughly 4th moment of the drop size distribution (DSD), respectively. Thus, the DSD will fundamentally influence Z, thus the Z-R relation itself can be very sensitive to the choice of coefficients A and b.

Precipitation phase: Operational, single polarization systems are usually unable to classify hydrometeor type. Rain, snow, melting snow, graupel and hail may thus all be present, yet the radar can only consider them as being of one type. This leads to uncertainties in the selected Z-R relation when converting reflectivity to precipitation intensity.

The melting layer: This factor is specific to the region where snow melts into rain. The extremities of a snowflake melt first, causing a film of water to coat the particle before it implodes into a raindrop. Since water is a much more conductive medium than ice, this causes strong reflectivities in radar data, leading to an effect known as the Bright Band (BB); its region is found at more-or-less uniform heights/ranges. In southern Europe, the melting layer exists throughout the year and can reach up to above 4 km during summer. It is often absent or very close to the surface during the winter in northern Europe and seldom reaches above 3 km during summer.

Beam filling and overshooting: These two effects are problems which increase in severity with increasing range from the radar, as the beamwidth increases. Beam filling occurs where the scale of precipitation is small relative to the pulse volume, as for example in convection. Overshooting, in whole or in part, occurs where the precipitation is shallow in relation to the pulse volume. Overshooting is thus a greater problem in cold climates, as winter snow is usually considerably shallower than summer rain.

Non-uniformly vertically distributed precipitation: Several of the factors mentioned above can combine and lead to problems interpreting the observable as being valid as a surface measurement. This leads to problems applying the radar equation which must usually be neglected.

These errors also lead to representativeness problems, if the objective is to achieve a measurement which is applicable as a surface estimate; such problems are related to the Vertical Profile of Reflectivity (VPR) and its characteristics.

5.3.3 Data Quality Index

A *data quality index* takes values between 0 (bad quality) and 1 (good quality) and is attributed to each pixel of the radar data. Initial ideas in this field came from the German DLR (Friedrich et al. 2006), the Italian ARPA-SMR (Fornasiero et al. 2004) and Météo France (Lamarque et al. 2004). The data quality index uses a confidence value function for each investigated error source (e.g., Fig. 3) and combines the different information to create one number per pixel (Fig. 4).

The most comprehensible advantages of the *data quality index* are that it presents the quality of radar data

- in very high detail (one value per pixel); and
- in just one number per investigated data point.

The main drawbacks of a *data quality index* are:

- the index describes only the analyzed errors and uncertainties;
- its value may be application dependent (e.g., a warning scheme may not care about ground clutter); and
- a method how to evaluate the radar data after the correction has not yet been determined.

There are several approaches of how to calculate such an index. The main attempts have been initiated by DLR (Germany), ARPA SMR (Italy) and Météo France. Nevertheless, other institutions, such as the UK Met. Office, German Weather Service and Polish Weather Service have formulated their own index, also.

Data correction plays a growing role in the current use and preparation of radar data for practical applications (e.g., Lamarque et al. 2004; Jessen et al. 2005). The reasons for this are – among others – that users start to look very much into radar data details, that different applications need different data and that radar products are not frequently delivered in a way that they can be used directly.

However, the true rainfall not being known, any modification made to the radar data can improve or deteriorate them – the quality of the result of the intended correction is not known beforehand and can not be easily assessed after data modification, either. A reference

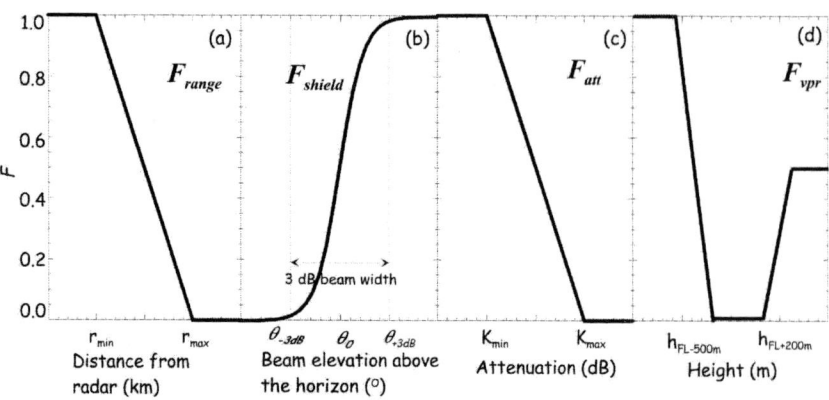

Fig. 3. Example for individual index functions (from Friedrich et al. 2006)

proclaiming 'the truth' does not exist. This is why a *data quality index* for *corrected data* is difficult to achieve and its conclusion is doubtful.

Bearing in mind the above, a number of questions are still open today:

- Which evaluation method can determine that a correction improved the data?
- What are the look and the statistical parameters of an *ideal* radar image?
- How *durable* is a quantification scheme, i.e., if you apply a quantification to an original image, after what time has science progressed so much that a different classification of the same image is likely?

Furthermore, different users and data providers use different correction schemes which result in different data quality. The development of correction algorithms has just started and improved ones based on new knowledge, additional (multisource) data information and new measurement technologies (e.g., double polarization) will multiply the number of applicable procedures. Thus, data correction is always a temporal repair of data, working as long as the data are doing their job (producing reliable results) or until new tools are providing one step forward.

For these reasons, probably the best information to provide to the user is the raw data, the corrected data (together with proper documentation) and a *data quality index* field based on the original data.

Questions to be answered when producing a data quality index field include the following ones:

- How should the number of functions included in the data quality scheme be determined? The data quality can only be evaluated on the basis of the predefined control functions and these functions are specific to given errors or uncertainties. It is impossible to check for all possible errors, because some data may not allow checking for certain types of errors (not all errors are detectable) and the effort required for some tests may inhibit their use in real-time.
- How should the quality functions be defined? A proper definition would require a completely known error distribution statistics for a given radar station and each individual error or uncertainty. This is impossible to achieve because it cannot be guaranteed how long this distribution will be valid. Therefore, a rough approximation is likely to be used for data quality index fields.

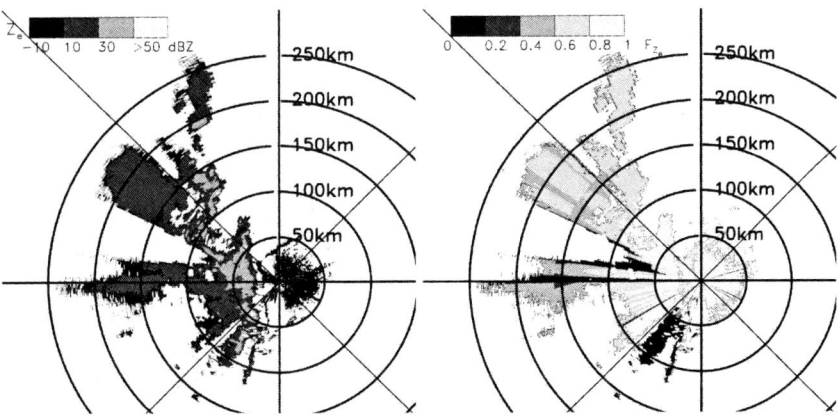

Fig. 4. Example for a reflectivity image and the corresponding gradient image (from Friedrich et al. 2006)

The following conclusions can be drawn:
- A quantification of radar measurement errors and uncertainties is possible.
- It should be well documented which modifications have been applied to the data.
- Any quantification scheme used should be documented.
- A harmonization of meta-data to be used in the radar data correction context is required.
- Many errors can only be detected, not corrected.
- There is no objective criterion to tell whether a data modification is an improvement.
- There is (yet) no objective approach to a data quality index calculation *after* image correction procedures are implemented.

5.3.4 Correction methods

The choice of data correction methods is strongly dependent on the available kind of data, the so-called *data product*. All corrections which are applied *before* creating a product (e.g., Doppler based clutter removal) are *not* presented in this chapter.

Furthermore, it is important to clarify whether a radar product is a measured radar volume, a polar scan or a Cartesian radar image. For each of these basic product types, some methods are applicable, others are not. In general, more sophisticated methods yielding a high quality of corrected data are the ones working on more sophisticated data, e.g., volume data are more reliable to correct than polar data and the latter are better than Cartesian ones.

Ground clutter and speckle

Ground clutter is unavoidable in radar measurements with a low elevation angle, as they are required for hydrological applications. Such echoes may be mostly recognized as stationary and are presented as pixels with high values. Doppler radars can detect clutter more easily through the comparison of potential clutter pixels with the corresponding movement speed (i.e., clutter does not move). Speckles are radar reflections on a limited very small area, usually variable in time, e.g., from airplanes.

Hannesen (2001) presented an overview of aspects to be considered for rainfall rate derivation. He named different possible error sources and available algorithms to overcome such errors.

To filter ground clutter, different techniques may be used, such as Doppler filter or statistical filter applied in the signal processor, clutter maps or combination of both.

None of the clutter filters works perfectly because the clutter intensity changes over time. Additionally, other clutter (e.g., from birds) cannot be removed by real-time algorithms. Isolated speckles or points with abnormally large values can be replaced by interpolating surrounding measurements.

Clutter map: For permanent ground clutter it is usual to define a fixed clutter map. Such a map is not variable, which can lead to problems (e.g., if signals of water vapor from power plants are detected by the radar).

Texture-based algorithm: The 'texture-based' algorithm by Gabella and Notarpietro (2002) detects small areas like speckles which have higher gradients to the neighborhood. This filter can also help to remove artifacts which arise, for example, from hardware interferences.

Speckle Filter: Peura (2002) presented eight different filters based on pattern recognition techniques for radar data QC. The filter BIOMET works well in detecting the existence of birds and insects. Biometeors are characterized by low-intensity speckled pattern near the radar. SPECK computes the segment size and sets a size threshold to filter speckle.

Segment size and reverse speckle filters: The algorithm 'segment size' (Golz et al. 2006) computes the number of connected image pixels with values greater than zero, constituting a 'segment'. With a defined segment size threshold, also depending on the pixel size, it is possible to eliminate speckle areas of a few pixels on a radar image. In contrast to the 'texture-based' algorithm, this method works only on 'segments' surrounded by zero-values (see Fig. 5).

The reverse method of 'segment size' is called 'reverse speckle'. This filter helps to fill single zero-pixels surrounded by pixels with non-zero values. This can be useful before using other correction methods (e.g., filter for radial anomalies).

Vertical and horizontal substitution: This filter combines the vertical and horizontal substitution methodologies as a function of the spatial variability of the rainfall field: vertical for convective and horizontal for stratiform. The first part of the filter works like a clutter map and replaces horizontally in one elevation. If these substitutes exceed a convective threshold (45 dBZ, proposed by Sánchez-Diezma et al. 2001), these pixels are again substituted by the first non-contaminated value of the vertical (second part of the algorithm). The substitute of the vertical (in higher elevations) is accepted, if it is higher than the result of the horizontal substitute; it is assumed to use correct data of a vertically extended convective cell.

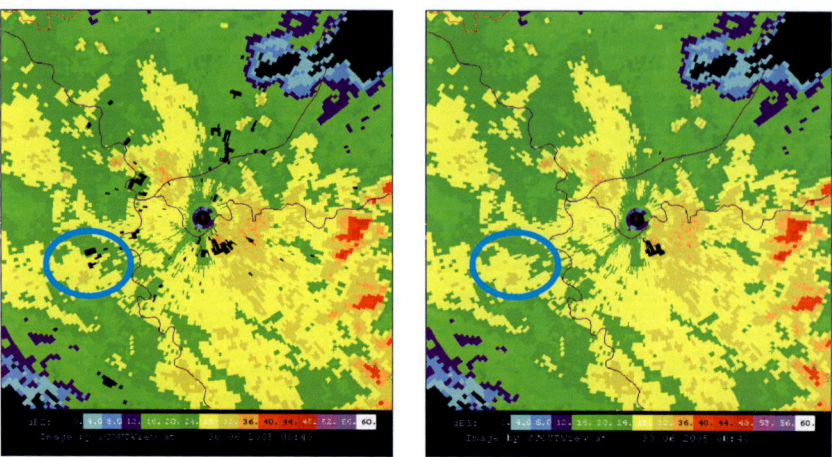

Fig. 5. The segment size and reverse speckle filters reduce clutter areas (*in black*)

Attenuation

Attenuation is the weakening process along the radar beam. The higher precipitation intensities, the more the radar beam is attenuated. In particular, convective cells have a large influence on the radar reflection. In extreme rainfall events the whole radar signal may be lost. Moreover, the wavelength of the radar is important for the degree of attenuation. Beams of smaller wavelength radars (e.g., X-Band radar) get attenuated more rapidly than the beams of long wavelength radar (e.g., S-Band radar). With regard to QC algorithms for attenuation, it must be concluded that the detection of attenuation is much easier than its correction.

Cumulative gate-by-gate algorithm: As a correction algorithm for the attenuated areas behind high dBZ values of convective cells, a 'cumulative gate-by-gate' algorithm can be used (Harrison et al. 2000). Such an algorithm has to be capped (e.g., at the factor of two for the increase in rain rate), because it tends to become unstable in the case of severe attenuation.

Mountain – return method: By using stable ground clutter measurements, it is possible to correct attenuation, especially radome attenuation, with the 'mountain – return' method (Sempere-Torres et al.

2001). With this method, an averaged clutter map is used, as a reference for the analysis and to determine the correction factor.

Iterative attenuation correction: Kraemer et al. (2006) investigated a microwave link and an X-band radar in order to derive an estimate of the Path Integrated Attenuation (PIA) and to assess the true rainfall after correction. The analysis was based on a variation scheme for the parameters A and b of the Z-R relationship. With the microwave link data and data from a disdrometer, the above authors concluded that an attenuation correction scheme is variable in time and space and needs to be readjusted for each measurement. They propose an iterative scheme whereby they fix the b value, define an initial guess for the A value which is likely to be too large and stepwise reduce the value of A after the full computation of the potential correction effects. A stable correction scheme is one possible solution.

Radome attenuation: Attenuation by the radome is still an important problem. A first step towards the assessment of radome attenuation has been performed by Kurri and Huuskonen (2006) who have established curves of attenuation factors as function of rainfall intensity. However, the state of the radome is very important (e.g., clean or dirty) as are its material and structure. Waxing may reduce attenuation effects.

Radial anomalies: Radial anomalies are radially different radar reflections on the track of a radar beam. Radial anomalies may be caused by hardware or software problems during the radar measurement or beam blockage through obstacles, like towers, high buildings and mountains. The 'radial filter' works better on single blocked rays (Golz et al. 2006). The 2-D filters 'beamblock' and 'visibility map' work on partly blocked radar areas.

Beamblock: A data-driven method to correct for beam blockage in polar PPI (Plan Position Indicator) radar data has been developed (Jessen et al. 2005) which does not require the existence of a DEM (Digital Elevation Model) or precise knowledge of the radar parameters, as other methods do (e.g., Bech et al. 2003).

Although being based on the disregard of physical properties of the radar beam, a careful analysis of the radar data leads to the determination of beam specific correction factors and results in acceptable to good correction results.

The analysis of the PPI radar data is performed for each angle radially: the sum of the reflectivity values along the ray is computed for every radar image and compared to the surrounding values. This analysis resulted in potential factors for correction of the different angles of the radar data. The corresponding factor, applicable to the reflectivity values, is then determined for the full extent of the radar angle.

The approach developed appears to be useful where DEM data are not available, where partial beam blockage cannot be explained by DEM information or where the reason for partial beam blockage is unknown.

The correction algorithm can also be used to correct for areas behind a known obstacle (e.g., ground clutter), treating only the part of the beam that is behind this known location. The location, as well as the correction factors, are parameters specific to the method.

Visibility map: The 'visibility map' filter works with provided visibility information from a radar operating agency. It requires geometric visibility information in percents of the pulse volume in the form of a polar matrix for each elevation and each pixel. By applying the visibility map, it is possible to correct by a small amount the blockage by mountains.

Emitter: The 'Emitter' filter (see Peura 2002) replaces radial anomalies in low elevations with values of higher elevations.

Differentiation into convective and stratiform precipitation: Two tested filters are used for the identification of convective or stratiform areas. The first one called '3 criteria' by Ehret (2003) can be used for 2D data and gives the information for a whole image. The result can be convective, stratiform or 'undefined'. The second one called '2 methods' combines horizontal and vertical methods. The horizontal part (Steiner and Houze 1995) identifies convective cells and the vertical part (Sánchez-Diezma et al. 2000) searches for BB signatures and identifies with it stratiform areas. When the data base includes no event with a BB signature, this latter part cannot be used. For this reason only the first part has been compared to the results of the first filter.

Vertical profile

Maximum method: The effects of BB are difficult to detect and to correct for, if no radar volume data are available. Temperature data at the site can help, but the sole use of ground temperature as additional information is not sufficient because:

- the temperatures on the ground are not always homogeneous in space to estimate the height of the zero degree level in the atmosphere with sufficient reliability,
- the transformation of ground-based temperature to the zero degree level intersected by the radar beam is leaving a certain band width

of uncertainty if a mean correction factor (e.g., 0.6° C per 100 m height difference) is used.

For the above reasons, Golz et al. (2006) propose a combination of the use of ground-based temperature measurements, which are transformed to estimated temperatures at the zero degree level and an image analysis approach.

Such an analysis is based on the following assumptions:

- a mean temperature difference of 0.6° C per 100 m height difference (and, thus, no effects of temperature inversion, for example);
- a straight propagation of the radar beam in the atmosphere (excluding effects like ducting or anomalous beam propagation); and
- the zero degree level is the top of the BB signature and the peak is in the middle of the BB (see Fig. 6).

A key assumption for the image analysis approach is that a BB effect is characterized by increased reflectivities at a constant distance to the radar (equivalent to a constant height of the radar beam and of a stratiform character of the observed precipitation).

Firstly, the height of the zero degree level was estimated by using the average of the ground-based temperatures, normalized by their respective height above sea level. The resulting height above sea level was intersected with the height of the radar beam, using the beam elevation and the beamwidth in order to obtain the distance for which BB effects can be expected. The thickness of BB was determined using the rain intensity dependency after Fabry (1997):

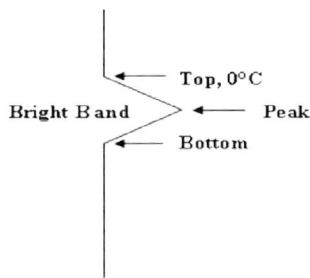

Fig. 6. Bright Band signature in an idealized and simplified vertical profile

$$B[m] = \begin{vmatrix} 140\, Z_e^{0.17}, \text{ for rain} > 100\,\text{mm}^6/\text{m}^3 \\ \\ 210\, Z_e^{0.085}, \text{ for rain} < 100\,\text{mm}^6/\text{m}^3 \end{vmatrix} \quad (1)$$

with the equivalent reflectivity factor Z_e in dBZ. As an alternative to the use of ground-based temperatures, existing measured freezing levels that are available in the radar data can be used.

In a second step, the circularly computed mean reflectivities (i.e., mean reflectivities for the same distance from the radar) were analyzed for every image. If there was a pronounced maximum which was persistent in time and distance (Fig. 7), the presence of a BB was assumed. The detected maximum is assumed to be equivalent to the peak of the BB and with the calculated BB thickness the top of the BB, the zero degree level can be determined (Fig. 6).

The third step is to combine the two zero degree levels. The heights of the temperature and of the image analysis approach are averaged in the ratio of two thirds to one third, so that the result of the temperature approach gets more weight. For example, if the zero degree level of the temperature approach is 800 m and the zero degree level of the image analysis approach is 500 m, the result would be 700 m.

The correction for the BB was performed based on an idealized simplified vertical profile (Fig. 6) similar to an idealized vertical profile

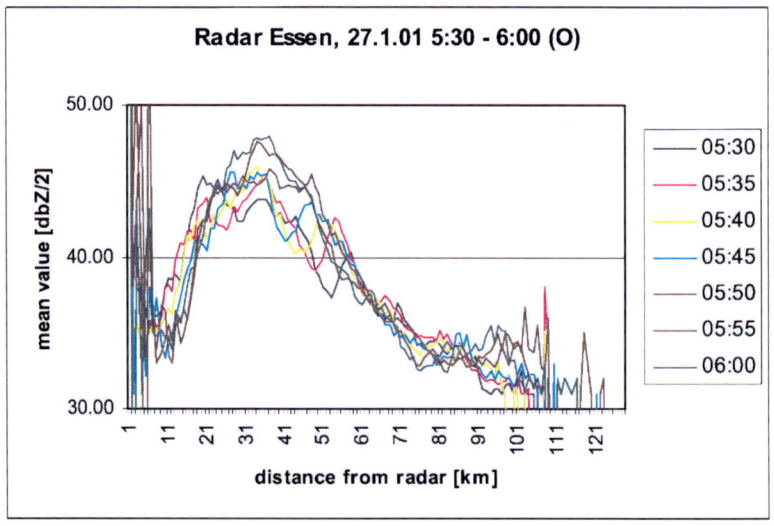

Fig. 7. Peak of reflectivities at constant distance from radar

without orographic enhancement (Kitchen et al. 1994). Knowing the heights and the values of the top, the peak and the bottom of the BB, a linear function is calculated between top and bottom. This function determines how much the value of the peak is 'too high' in comparison to the average of the top and bottom values of the BB. If one of these values is zero, the next pixel in the direction of the BB peak is taken. For the areas below and above the peak, linearly interpolated factors are computed which are applied to the pixels concerned. Such a correction is not always possible (e.g., if the BB is over the radar and the peak cannot exactly be determined).

Mean Apparent Vertical Profile of Reflectivity (MAVPR): Franco et al. (2002) used the Mean Apparent Vertical Profile of Reflectivity to obtain rain rates on the ground at different distances from the radar. The MAVPR is estimated near the radar using volume data; subsequently, it is adjusted to the reflectivity values measured at different ranges, thus improving the rain rate estimations at these distances by utilizing the lowest elevation scan.

Franco et al. (2006) improved the classification part of the algorithm. Their method combines criteria relative to the vertical development of the precipitation with criteria relative to its horizontal pattern. First results showed that it improves the Steiner algorithm (Steiner and Houze 1995) by avoiding the false convective detection produced when the lowest PPI is contaminated by the BB.

Anomalous propagation (anaprop): To reduce anaprop, Borga et al. (2002) used a procedure called 'tilt test'. This is a vertical echo continuity check, which includes the knowledge that the areal extent of anaprop often decreases rapidly, as the antenna elevation steps up to higher angles. The test had been implemented in the ratio curve (which is the ratio of reflectivities from scans taken at two different elevation angles and at discrete ranges from the radar) computation for VPR correction and exhibited good results.

Several of the above algorithms have a proven usefulness and perform well for online applications, whereas, others are only applicable for offline use (see Table 1). Limited experience with algorithms in the VOLTAIRE project (see Chap. 20 of this book) does not exclude their successful application elsewhere (e.g. the MAVPR is in operational use at Universitat Politècnica de Catalunya in Barcelona, Spain). Finally, it should be noted that some observations, as for example the attenuation, can often only be diagnosed, but not corrected.

Table 1. Overview of QC algorithms applied in the VOLTAIRE project (for abbreviations see text)

Algorithms	Reference	Data type	Tested on data from			
			Cyprus	Switzerland	Germany	Spain
Attenuation: cumulative gate-by-gate	Harrison et al. (2000)	2D			x	some tests
Attenuation: mountain return	Sempere-Torres et al. (2001)	2D	some tests			
GC: static map	Own development	2D	x		x	
GC: texture-based	Gabella and Notarpietro (2002)	2D	x	x	x	x
Ground clutter: segment size	Own development	2D	x	x	x	x
GC: vertical and horizontal substitution	Sánchez-Diezma et al. (2001)	3D	some tests			
Convection/ stratiform: 3 criteria	Ehret (2003)	2D		some tests	some tests	
Convection/ stratiform: 2 methods	Sánchez-Diezma et al. (2000)	3D	some tests			
VPR: maximum method	Own development/ Golz et al. (2006)	2D			x	
VPR: MAVPR	Franco et al. (2002)	3D	some tests			
Radial: emitter	Peura 2002	3D	some tests			
Radial: radial filter	Own development	2D			x	
Radial: beamblock	Own development/ Jessen et al. (2005)	2D		some tests	x	
Radial: visibility map	Own development	2D		some tests		
Anaprop: tilt-test	Borga et al. (2002)	3D	some tests			

5.4 Future developments

Further improvements of data quality can be expected by cross referencing independent measurement parameters, such as Doppler information or dual polarization information. Research has started into this promising direction; however, methods will become more complex.

The online feasibility and operational applicability of new methods is of very high importance for a routine employment of methods. Therefore, simple methods have an advantage over complex ones – as long as they provide results of sufficient quality.

References

Bech J, Codina B, Lorente J, Bebbington D (2003) The sensitivity of single polarization weather radar beam blockage correction to variability in the vertical refractivity gradient. J Atmos Ocean Tech 20:845–855

Borga M, Tonelli F, Moore RJ, Andrieu H (2002) Long-term assessment of bias adjustment in radar rainfall estimation. Water Resour Res 38:1226

Browne O, Auriaux G, Idier F, Delattre JM (1998) A decision aid system for real-time operation of seine saint-denis sewer network. In: Proceedings 3ème Conférence Internationale sur les Nouvelles Technologies en Assainissement Pluvial (Novatech98). Lyon, GRAIE, pp 147–154

Browning KA (1987) Towards a more effective use of radar and satellite imagery in weather forecasting. In: Collinge VK, Kirby C (eds) Weather Radar and Flood Forecasting. John Wiley and Sons, pp 239–269

Ehret U (2003) Rainfall and flood nowcasting in small catchments using weather radar. University of Stuttgart, Institute of Hydrology, 121 pp, ISSN 0343–1150

Einfalt T, Arnbjerg-Nielsen K, Spies S (2000) Rainfall data measurement and processing for model use in urban hydrology. In: Proceedings 5th International Workshop on Precipitation in Urban Areas. 10–13 December, Pontresina

Einfalt T, Jessen M, Quirmbach M (2006) Can we check raingauge data automatically? In: Proceedings 7th International Workshop on Precipitation in Urban Areas. 7–10 December, St. Moritz, Switzerland, ISBN 3-909386-65-2

Fabry F (1997) Vertical profiles of reflectivity and precipitation intensity. UNESCO: Weather Radar Technology for Water Resources Management. Montevideo, Uruguay, pp 137–145

Fabry F (2004) Obstacles to the greater use of weather radar information. In: Proceedings 6th International Symposium on Hydrological Applications of Weather Radar. 2–4 February, Melbourne, Australia

Fornasiero A, Amorati R, Alberoni PP, Ferraris L, Taramasso AC (2004) Impact of combined beam blocking and anomalous propagation correction algorithms on radar data quality. In: Proceedings 3rd European Conference on Radar in Meteorology and Hydrology. 6–10 September, Visby, Sweden. http://www.copernicus.org/erad/2004/

Franco M, Sempere-Torres D, Sánchez-Diezma R (2006) Correction of the error related to the vertical profile ofreflectivity: previous partitioning of precipitation types. In: Proceedings 4th European Conference on Radar in Meteorology and Hydrology. 18–22 September, Barcelona, Spain. http://www.erad2006.org

Franco M, Sempere-Torres D, Sánchez-Diezma R, Andrieu H (2002) A methodology to identify the vertical profile of reflectivity from radar scans and to estimate the rainrate at ground at different distances. In: Proceedings 2nd European Conference on Radar Meteorology. 18–22 November, Delft, The Netherlands, pp 299–304. http://www.copernicus.org/erad/index2002.html

Friedrich K, Hagen M, Einfalt T (2006) A quality control concept for radar reflectivity, polarimetric parameters, and doppler velocity. J Atmos Ocean Tech 23:865–887

Gabella M, Notarpietro R (2002) Ground clutter characterization and elimination in mountainous terrain. In: Proceedings 2nd European Conference on Radar Meteorology. 18–22 November, Delft, The Netherlands, pp 305–311. http://www.copernicus.org/erad/index2002.html

Golz C, Einfalt T, Galli G (2006) Radar data quality control methods in VOLTAIRE. Meteorol Z 15:497–504

Hannesen R (2001) Quantitative precipitation estimation from radar data – a review of current methodologies: Gematronik (MUSIC), http://www.geomin.unibo.it/orgv/hydro/music/reports/D4.1_QPE-revi.pdf

Harrison DL, Driscoll SJ, Kitchen M (2000) Improving precipitation estimates from weather radar using quality control and correction techniques. Meteorol Appl 6:135–144

Jessen M, Einfalt T, Stoffer A, Mehlig B (2005) Analysis of heavy rainfall events in North Rhine-Westphalia with radar and raingauge data. Atmos Res 77:337–346

Joe P (1996) Precipitation at the ground: radar techniques. In: Raschke E (ed) Radiation and water in the climate system. Springer, NATO ASI Series, Chap. 12, pp 277–321

Jörgensen HK, Rosenörn S, Madsen H, Mikkelsen PS (1998) Quality control of rain data used for urban runoff systems. Water Sci Technol 37(11)

Kitchen M, Brown R, Davies AG (1994) Real-time correction of weather radar data for the effects of bright band, range and orographic growth in widespread precipitation. Q J Roy Meteor Soc 120:1231–1254

Krajewski WF, Ciach GJ (2004) Towards operational probabilistic quantitative precipitation estimation using weather radar. In: Proceedings 6th International Symposium on Hydrological Applications of Weather Radar. 2–4 February, Melbourne, Australia

Kraemer S, Verworn H-R., Hartung A, Holt AR., Upton GJG, Becker M (2006) Uncertainty quantification of operational X-band weather radar rainfall measurements. In: Proceedings 7th International Workshop on Precipitation in Urban Areas. 7–10 December, St. Moritz, Switzerland, ISBN 3-909386-65-2

Kurri M, Huuskonen A (2006) Measurement of the transmission loss of a radome at different rain intensities. In: Proceedings 4th European Conference on Radar in Meteorology and Hydrology. 18–22 September, Barcelona, Spain. http://www.erad2006.org

Lamarque P, Tabary P, Desplat J, Do Khac K, Eideliman F, Parent du Châtelet J (2004) Improvement of the french radar rainfall accumulation product. In: Proceedings 3rd European Conference on Radar in Meteorology and Hydrology. 6–10 September, Visby, Sweden. http://www.copernicus.org/erad/2004/

Maul-Kötter B, Einfalt T (1998) Correction and preparation of continuously measured raingauge data: A standard method in North Rhine-Westphalia. Water Sci Technol 37(11)

Michelson D, Einfalt T, Holleman I, Gjertsen U, Friedrich K, Haase G, Lindskog M, Jurczyk A (2005) Weather radar data quality in Europe quality control and characterisation. Review, COST Action 717 – Use of radar observations in hydrological and NWP models. Luxembourg

Musa M, Grüter E, Abbt M, Häberli C, Häller E, Küng U, Konzelmann T, Dössegger R (2003) Quality control tools for meteorological data in the meteoSwiss data warehouse system. In: Proceedings ICAM/MAP 2003. 19–23 May, Brig, Switzerland

Peura M (2002) Anomaly detection and removal in radar images (ANDRE) – Final Project Report. Finnish Meteorological Institute, Helsinki, Finland

Sánchez-Diezma R, Sempere-Torres D, Delrieu G, Zawadzki I (2001) An improved methodology for ground clutter substitution based on a pre-classification of precipitation types. In: Preprints 30th International Conference on Radar Meteorology. American Meteorological Society. Munich, Germany, pp 271–273

Sánchez-Diezma R, Zawadzki I, Sempere-Torres D (2000) Identification of the bright band through the analysis of volumetric radar data. J Geophys Res 105(D2):2225–2236

Sempere-Torres D, Sánchez-Diezma, Cordoba MA, Pascual R, Zawadski I (2001) An operational methodology to control radar measurements stability from mountain returns. In: Preprints 30th International Conference on Radar Meteorology. American Meteorological Society. Munich, Germany, pp 264–266

Steiner M, Houze RA (1995) Sensitivity of monthly convective rain fraction to the choice of Z-R relation. In: Preprints 27th International Conference on Radar Meteorology. American Meteorological Society, Vail, CO

Vaitl W (1988) Beschreibung der Prüfkriterien für die Qualitätskontrolle stündlicher bzw. 10-minütiger Daten von automatischen agrarmeteorologischen Stationen der Bayerischen Landesanstalt für Bodenkultur und Pflanzenbau. München-Freising, 16 pp

Part II. Estimation of precipitation

i. Space estimation

6 Global precipitation measurement

Arthur Y. Hou[1], Gail Skofronick-Jackson[1], Christian D. Kummerow[2], James Marshall Shepherd[3]

[1] NASA Goddard Space Flight Center, Greenbelt, MD, USA
[2] Colorado State University, Fort Collins, CO, USA
[3] University of Georgia, Athens, GA, USA

Table of contents

6.1 Introduction 131
6.2 Microwave precipitation sensors 135
6.3 Rainfall measurement with combined use of active
    and passive techniques 140
6.4 The Global Precipitation Measurement (GPM) mission 143
    6.4.1 GPM mission concept and status 145
    6.4.2 GPM core sensor instrumentation 148
    6.4.3 Ground validation plans 151
6.5 Precipitation retrieval algorithm methodologies 153
    6.5.1 Active retrieval methods 155
    6.5.2 Combined retrieval methods for GPM 157
    6.5.3 Passive retrieval methods 159
    6.5.4 Merged microwave/infrared methods 160
6.6 Summary 162
References 164

6.1 Introduction

Observations of the space-time variability of precipitation around the globe are imperative for understanding how climate change affects the global energy and water cycle (GWEC) in terms of changes in regional precipitation characteristics (type, frequency, intensity), as well as extreme hydrologic events, such as floods and droughts. The GWEC is driven by a host of complex processes and interactions, many of which are not yet well understood. Precipitation, which converts atmospheric water vapor into rain and snow, is a central element of the GWEC. Precipitation regulates the global energy and radiation balance through coupling to clouds and water vapor (the primary greenhouse gas) and shapes global winds and atmospheric transport through latent heat release. Surface precipitation directly affects soil moisture and land hydrology and is also the primary source of freshwater in a world that is facing an emerging freshwater crisis. Accurate and timely knowledge of global precipitation is essential for understanding the multi-scale interaction of the weather, climate and ecological systems and for improving our ability to manage freshwater resources and predicting high-impact weather events including hurricanes, floods, droughts and landslides.

In terms of measurements of precipitation, it is critical that data be collected at local scales over a global domain to capture the spatial and temporal diversity of falling rain and snow in meso-scale, synoptic-scale and planetary-scale events. However, given the limited weather station networks on land and the impracticality of making extensive rainfall measurements over oceans, a comprehensive description of the space and time variability of global precipitation can only be achieved from the vantage point of space.

The spatial and temporal scales required to resolve the impact of precipitation for different hydrometeorological processes are illustrated in Fig. 1. This figure shows that surface water can vary on the order of minutes and meters; measurements at these scales are relevant for landslide and flooding conditions. Short-term (<~1 day) weather related events include for example, flood warnings, urban drainage and hydropower optimization. Seasonal to inter-annual (~1 day to several decade) hydrological scale events include management of irrigation and water supply reservoirs, land use decisions and culvert operations. Oceanic processes near coastlines have fine resolution requirements, while open ocean processes can span decades or hundreds of years and thousands of kilometers. On the climate scale for long-term planning over 50 years to centuries, hydrologists must anticipate minor and major dam needs and assess environmental impacts of water resources. As can be expected, satellite observations cannot measure to all the spatial and temporal scales required for hydrometeorological applications.

Nevertheless, satellites can provide certain types of data at high spatial and temporal scales. The first images of clouds in relationship to meteorological processes were provided by the Television and Infrared Observation Satellite (TIROS-1), which was launched in April 1960. These early investigations noted the importance of satellite observation

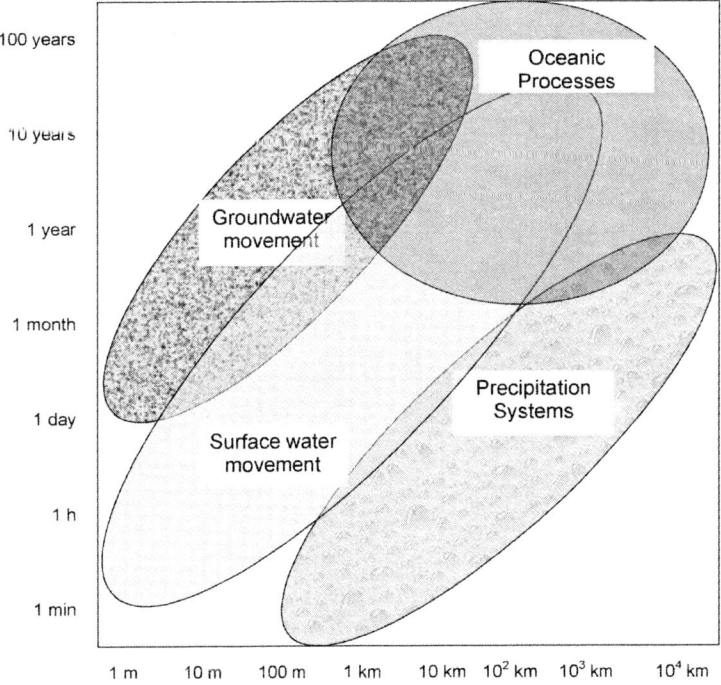

Fig. 1. Schematic of selected atmospheric, surface and subsurface hydrologic processes and their temporal and spatial scales of occurrence (from Bloschl and Sivapalan 1995)

of clouds since precipitation is inherently linked with clouds (Kidder 1981) although properly resolving the spatio-temporal precipitation from space would prove to be a challenging task.

Currently, observations of cloud tops using visible and infrared sensors from geostationary orbits such as the Geostationary Operational Environmental Satellites (GOES) spacecraft are done with near continuous (fine temporal) scans at footprint resolutions of 1–8 km. Kidd (2001) summarizes other geostationary satellites and reviews various approaches inferring precipitation from visible and infrared sensors. Measurements of rainfall rate inferred from cloud top data do not probe into the cloud nor provide information on the vertical structure and microphysics of clouds. Active radars at Ku, Ka and W band (~14, 35 and 95 GHz, respectively), for example, can measure profiles of precipitating hydrometeor characteristics (e.g., size) within clouds.

Passive precipitation radiometers (~10–89 GHz) can measure the integrated cloud water and ice paths and are used to estimate rain rate (Barrett and Beaumont 1994; Petty and Krajewski 1996; Smith et al. 1998). Passive radiometers in the 1990's and 2000's typically had horizontal surface footprints of 5–50 km, while radar footprints were on the order of 1–10 km. While there are a few active and several passive precipitation sensors in orbit, none are currently in geostationary orbit and thus the temporal resolution is limited to the number of overpasses per day.

Wideband multifrequency passive radiometers can provide microphysical information about both liquid and frozen hydrometeors in clouds. Passive microwave sounders with multiple channels centered around oxygen and water vapor absorption lines provide vertically-resolved information on the temperature and water vapor profiles of clear air atmospheres and the sounder channels are also sensitive to hydrometeors for retrievals of cloud properties (Chen and Staelin 2003; Kidder et al. 2000; Spencer 1993). Current active microwave satellite radars (at Ku and W-band) provide fine-scale vertical profile structure information about atmospheric clouds (Meneghini et al. 2000; Stephens et al. 2002). Combined radar-radiometer systems, such as the Tropical Rainfall Measuring Mission (TRMM) (Kummerow et al. 2000; Simpson et al. 1988) are particularly important for studying and understanding the microphysical processes of precipitating clouds and for accurate estimates of rainfall rate. Since TRMM is a single satellite in a non-Sun-synchronous 35° orbit, it cannot provide fine temporal resolution alone. A generation of blended, 3-hourly rainfall products has emerged to exploit the temporal resolution of geosynchronous techniques, the improved accuracy of passive microwave techniques and the direct rainfall measurement from active microwave sensors (See Ebert et al. 2007, for a review of these multi-sensor techniques and past intercomparison activities). The next stage in the evolution of precipitation observations from space is the Global Precipitation Measurement (GPM) Mission, which is designed to unify a constellation of research and operational satellites to provide integrated, uniformly-calibrated precipitation measurements at every location around the globe every 2–4 h.

This Chapter begins with a brief history and background of microwave precipitation sensors, with a discussion of the sensitivity of both passive and active instruments, to trace the evolution of satellite-based rainfall techniques from an era of inference to an era of physical measurement. Next, the highly successful Tropical Rainfall Measuring Mission will be described, followed by the goals and plans for the GPM

Mission and the status of precipitation retrieval algorithm development. The Chapter concludes with a summary of the need for space-based precipitation measurement, current technological capabilities, near-term algorithm advancements and anticipated new sciences and societal benefits in the GPM era.

6.2 Microwave precipitation sensors

Satellite-based remotely-sensed visible and infrared imagery provides high spatial resolution from instruments of moderate aperture size (<1m), even at geosynchronous distances. However, due to the large hydrometeor extinction at infrared and visible wavelengths, such sensors are unable to probe through most cloud cover. In contrast, microwave, millimeter-wave and sub-millimeter-wave remote sensing provides the capability of probing through clouds and precipitation while retaining useful sensitivity to hydrometeors (Staelin 1981; Njoku 1982; Ishimaru 1991). Three common types of microwave sensors exist: active radars and passive radiometric imagers and sounders. Cloud and precipitation radars are those that observe the direct backscatter from hydrometeors. Imagers operate primarily in the window regions of the microwave spectrum away from oxygen and water vapor absorption lines. The atmosphere tends to be relatively transparent in these window regions so that a robust signal can often be obtained as a function of total water in the atmospheric column. Sounders operate in microwave absorption lines in order to profile the atmospheric temperature and water vapor contents but have recently been found to have some uses in the detection and quantification of cold season precipitation and are expected to provide indirect information about light rain over land. Recent reviews by Kidd (2001) and Levizzani et al. (2007) are excellent resources complementary to this discussion.

Passive precipitation radiometers measure brightness temperature (T_B) which is a function of the upwelling electromagnetic radiation intensity I (z, θ, f) where f is the radiometer frequency, θ is the angle of observation and z is the height of observation. This radiation intensity is a measure of the vertically integrated emission, reflection and scattering of passively-generated thermal radiation from the Earth's surface, atmospheric gases and cloud and precipitation hydrometeors. The radiation intensity is typically converted to a brightness temperature using the Rayleigh-Jeans approximation to Planck's Law. Microwave window radiometers are designed to operate in the electromagnetic spectrum away from strong absorption lines of oxygen and water vapor. With relatively little attenuation from oxygen or water vapor in these 'window' regions, microwave radiometers can probe through cloud layers to provide information about precipitation to near the Earth's surface. In the absence of large hydrometeors and away from absorption lines, the radiative transfer equation may be written as (Olson et al. 2001):

$$T_B \approx \varepsilon\, T_S \exp(-\tau) + T_{atm}[1 - \varepsilon \exp(-\tau) - (1 - \varepsilon)\exp(-2\tau)] \quad (1)$$

From this, it is immediately evident that window channel radiometers can be designed to retrieve T_s (the surface temperature), ε (the surface emissivity which is closely related to the wind speed over oceans and the vegetation cover over land) and τ, the atmospheric absorption. Absorbing constituents in the media are the residual water vapor effects, cloud water and rain water. Water vapor is usually observed using a weak absorption line near 22.235 GHz. Cloud water and rain water can then be estimated from the residual absorption but distinguishing one from the other is difficult unless scattering is sufficiently large so as to distinguish the two signals.

For large raindrops and frozen hydrometeors, scattering by microwave radiation cannot be ignored. The degree of sensitivity depends on the frequency of observation, the hydrometeor phase (e.g., cloud water, rain droplets, ice, snow, graupel, and/or hail), the hydrometeor density and particle size distribution (Gasiewski 1993). For example, frequencies below ~20 GHz respond to only the strongest liquid precipitation, while frequencies above ~220 GHz respond to even light non-precipitating ice clouds such as cirrus. Equation (1) can be expanded to include the contribution from scattering as would be observed at height z and observation angle θ. For a horizontally planar-stratified atmosphere, the radiative transfer equation with frequency dependence assumed is (Gasiewski 1993):

$$\cos\theta\, \frac{dT_B(z,\theta)}{dz} = -K_e(z)T_B(z,\theta) + K_a(z)T(z)$$

$$+ K_s(z)\int_0^\pi \overline{P}(z,\theta,\theta')T_B(z,\pi-\theta')\sin\theta'd \quad (2)$$

In this equation, $T(z)$ is the atmospheric temperature profile at height z, while K_a, K_e, K_s are the bulk layer absorption, extinction and scattering coefficients for the layer dz and the assumed frequency f. The reduced phase matrix, $\overline{P}$ describes the fraction, magnitude and polarization of energy scattered from angle θ' to angle θ at z and f. The bulk asymmetry parameter G is used to define the phase matrix and the asymmetry provides a measure of the direction of scattering (e.g., mostly forward, backward, or isotropic). In performing the integration to solve Eq. (2) for T_B, the contributions from the boundary conditions (surface and cosmic background) are incorporated.

Fundamental to interpreting Eq. (2) is the concept that scattering increases very rapidly with frequency across the microwave domain. Adding higher frequency channels (freq >40 GHz) to radiometers is thus useful to exploit this scattering signature. While absorption continues to be important for liquid particles at these frequencies, ice particles become almost pure scatterers at these frequencies due to their dielectric properties. Figure 2 shows computed brightness temperatures from 1 to 1000 GHz for representative cloud types as would be observed over a calm ocean surface. The representative cloud types shown here include convective rain with ice aloft, falling snow, non-precipitating anvil ice and clear air with low and high water vapor. The saturation at the water vapor absorption sounding channels (e.g., 23, 183, 380, 448, 556, 752,

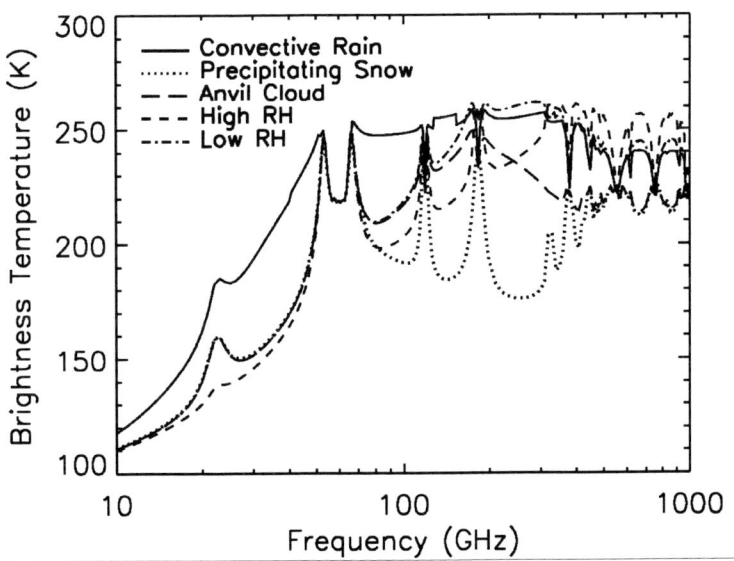

Fig. 2. Representative brightness temperatures from 10 to 1000 GHz for five different atmospheric and cloud conditions (from Skofronick-Jackson 2004)

987 GHz) is apparent in the figure, while the evidence of the oxygen channels (e.g., 50–60, 118, 368, 424, 487, 715, 834 GHz) is limited to the lower frequency channels. For precipitation sensing over oceans, the window channels with center frequencies near 10, 18, 36 and 89 GHz have proved to be the most useful. To capture the heaviest precipitation rates (and more information about the surface features during the absence of precipitation), 6 GHz can be added. Over land, where the highly variable surface temperature and emissivity contaminate the brightness temperature signal for lower frequencies $<\sim 90$ GHz, precipitation estimates can be obtained by relating the scattering from ice aloft to the rain at the surface using channels with center frequencies near 89, 150 (or 166) GHz and at multiple offsets from the 183 GHz water vapor absorption line (e.g., 183±1, 183±3, 183±7 GHz). These higher frequency channels are sensitive to the ice particles in clouds (causing an increase in scattering and hence a reduction in the brightness temperature). Thus the 150 and 183GHz channels have also been shown to be useful in estimating falling snow characteristics (e.g., Ferraro et al. 2005; Skofronick-Jackson et al. 2004). Frequencies above ~200 GHz have been proven to be useful for estimating information about ice in clouds through aircraft instrumentation (Evans et al. 2005), but no dedicated cloud ice satellite missions at these frequencies exist at this time.

Another important aspect of instrument sensitivity is that if designed to do so, radiometers can receive electromagnetic energy in a fully polarimetric mode (i.e., measuring the four Stokes parameters, Gasiewski 1993; Skou and Le Vine 2006). Typically, satellite precipitation sensors measure only the vertical (V) and/or horizontal (H) polarizations that are related to the Stokes parameters, though ground-based radars are moving toward fully-polarimetric measurements. Differences in measured V and H can emanate, for example, from oceanic wave patterns and their foam crests or from oriented ice particles in clouds. The information content of V versus H polarization is exploited to improve retrieval accuracy of various parameters.

Multi-frequency window radiometers such as the Scanning Multichannel Microwave Radiometer (SMMR) launched in 1978 (Njoku et al. 1980) and the Defense Meteorological Satellite Program (DMSP) Special Sensor Microwave/Imager (SSM/I) first launched in 1987 (Hollinger et al. 1990) make use of window channels to retrieve surface wind speed, column water vapor, cloud water and rainfall over the oceans. Lower frequencies (e.g., 10 and 6 GHz) are needed to retrieve parameters such as sea surface temperature. These lower frequency channels were first flown on TRMM's Microwave Imager (TMI) launched in 1997 (Kummerow et al. 1998) and the Advanced Microwave Scanning Radiometer-Earth Observing System (AMSR-E) launched in 2002 (Kawanishi et al. 2003), respectively. The WindSat instrument on the Coriolis satellite launched in 2003 (Gaiser et al. 2004) further adds wind direction capabilities by measuring all four Stokes parameters at the low frequencies that are most sensitive to the surface state. These window channel passive precipitation sensors typically operate in a conically-scanning mode at ~53° inclination angle that is fixed throughout the scanning operation such that the V and H polarized signals are not mixed as can occur in cross-track scanning systems (Ulaby et al. 1981).

Passive microwave sounders such as the Microwave Sounding Unit (MSU), first launched in 1978 (Kidder and VonderHaar 1995), followed by the Advanced MSU (AMSU-B) first launched in 1998 aboard National Oceanic and Atmospheric Administration (NOAA)-15 satellite, the Humidity Sounder of Brazil launched on the Earth Observing System (EOS) Aqua spacecraft in 2002 and the Microwave Humidity Sounder (MHS) instruments aboard the European Meteorological Operational (MetOp) satellite launched in 2006 and also on the NOAA-18 launched in 2005, all operate near the oxygen (60 and 118 GHz) bands and/or the water vapor (183 GHz) bands. These radiometers are designed to derive profiles of temperature and water vapor by sounding the atmosphere at multiple frequencies around the absorption lines. The relatively high frequencies employed by these radiometers, however, make them sensitive to scattering by liquid and frozen hydrometeors. The atmosphere becomes more opaque as the waver vapor increases, particularly closest to the center of the absorption lines and this reduces the channel sensitivity to surface emissivity. This can be advantageous when the surface emissivity is not well known as is the case over land surfaces, especially over most frozen surfaces.

For active remote sensing, radars transmit and receive signals. The signal returned to the receiver provides a measure of the interacting media through the backscattering from that media. Depending on its design, radars can measure the distance from the media and the amount and relative size of the particles in the media. The media probed by radars includes, for example, clouds, vegetation and soil. The advantage of precipitation radars is their ability to sense information on the location and size distribution of cloud particles. Essentially, they provide a detailed vertical distribution of the precipitation particles in the cloud. The radar equation (see Atlas 1990 for derivation) shows that the intensity of the signal received by the radar is dependent on the size of the particle, r, to the sixth power. There is also sensitivity to the liquid versus frozen particles in the cloud. In fact, in the melting layer of clouds, radars often exhibit what is called the bright band. The bright band is caused by the exterior melting of large frozen particles that makes them appear as large raindrops to the radar. Unfortunately, depending on their operating frequency, radars suffer from attenuation (higher frequencies saturate at shorter distances from the transmitter as the cloud optical depth increases). Techniques (Meneghini et al. 2000; Iguchi and Meneghini 1994) have been developed to address and remove the attenuation so that the full vertical picture from radars can be analyzed. Even though these attenuation correction techniques can be further improved to reduce the mismatch between what the radar retrieves and the actual conditions in the cloud, the utility of precipitation radars has been proven with the first precipitation radar in space on TRMM, launched in 1997. In addition, TRMM has shown that combining active and passive sensor measurements can provide a powerful tool for investigating precipitation and cloud particle microphysics.

6.3 Rainfall measurement with combined use of active and passive techniques

While passive microwave radiometers can provide information about precipitating liquid and/or ice particles, the inference of rain rates from microwave brightness temperatures requires additional information and assumptions. One way to vouch for the validity of passive microwave retrieval techniques for rainfall estimation is to compare results with coincident estimates from an active sensor such as precipitation radar. The combined use of active and passive microwave sensors also provides complementary information about the macro and microphysical processes of precipitating clouds, which can be used to reduce uncertainties in combined radar/radiometer retrieval algorithms. TRMM is a joint satellite precipitation mission between the United States National Aeronautics and Space Administration (NASA) and the Japan Aerospace and Exploration Agency (JAXA) that uses both radar and radiometer instrumentation to provide more accurate rain rate estimates than what can be accomplished by either sensor alone (Simpson et al. 1988). Launched in 1997, TRMM quickly became the world's prototype satellite for the study of precipitation and climate processes in the tropics (Kummerow et al. 2000). The orbit for TRMM was designated as an inclined non-Sun-synchronous processing orbit extending between 35°N and 35°S in order to focus efforts on tropical rainfall and hurricanes. TRMM, which continues to provide data for nearly ten years after launch, has been a tremendous success both in terms of advancing scientific understanding of the global water cycle and practical societal applications (NRC 2007).

The success of TRMM can be traced back to a complement of carefully designed precipitation sensor instrumentation (Table 1). The satellite includes the first rain radar instrument in space along with a radiometer and other instrumentation. The five instruments on TRMM were designed to provide information and products independently, as well as be linked for joint product deliverables. The conically-scanning (53° inclination) TRMM Microwave Imager (TMI) serves as the radiometer with frequencies at 10.7, 19.3, 21.3, 37.0 and 85.5 GHz, with V and H polarization on all channels except 21.3 GHz, which has vertical polarization only. The TMI had a swath width of 760 km at a 350 orbital altitude and footprint resolutions ranging from 36 km × 60 km for 10.7

Table 1. TRMM sensor summary – Rain package (derived from Kummerow et al. 2000)

Instrument	Radiometer (TMI)	Radar (PR)	Visible and Infrared radiometer (VIRS)
Channels	10.7, 19.3, 21.3, 37.0 and 85.5 GHz (dual-polarized except for 21.3: vertical only)	13.8 GHz	0.63, 1.61, 3.75, 10.8 and 12 µm
Resolution	10 km × 7 km field of view at 37 GHz	4.3-km footprint and 250-m vertical resolution	2.2-km resolution
Scanning Mode	Conically scanning (53° inc.)	Cross-track scanning	Cross-track scanning
Swath Width	760 km	215 km	720 km

GHz to 4 km × 7 km for 85.5 GHz. The Precipitation Radar (PR) operates at 13.8 GHz with a 4.3 km footprint and 250 m vertical resolution at the 350 km orbital altitude. The PR operates in cross-track scanning mode, having a 215 km swath at the 350 km orbit. The other instruments include the Visible and Infrared Radiometer (VIRS), the Lightning Imaging Sensor (LIS) and the Clouds and the Earth's Radiant Energy System (CERES). Mission changes since launch include the failure of the CERES in mid-1998 and in August of 2001 TRMM's orbit was boosted to ~400 km to conserve fuel that had been required at the lower orbit for station keeping.

The unique function of the PR is to provide the three-dimensional structure of rainfall, obtaining high quality rainfall estimates over ocean and land and improve TRMM rainfall estimates through combined radar-radiometer retrieval algorithms (Tao et al. 2000). The PR instrument was designed and built by JAXA. The TMI design was based on the SSM/I onboard the DMSP satellites since 1987. The TMI, built in the United States, has added V and H polarized 10.7 GHz channels and a slightly different water vapor channel at 21.3 GHz. The TMI provides increased swath coverage, sensitivity to higher rain rates and a link to passive precipitation radiometers on other satellites. The VIRS uses visible and infrared channels to provide additional information related to precipitation. VIRS also provided a connection between TRMM data and visible/infrared data sets from geostationary satellites.

TRMM has a series of measured and estimated products that are used to fulfill the scientific objectives of the mission. There are three major levels of these TRMM products: the Level 1 Earth-located and calibrated radiance/reflectivity swath data; the Level 2 physical retrieval swath-format products; and the Level 3 gridded products. The primary Level 2 operational products associated with TRMM include: (1) radar surface scattering cross-section and total path attenuation; (2) classification of rain (convective/stratiform) and height of bright band; (3) surface rainfall and three-dimensional (3D) structure of hydrometeors and heating over the TMI swath; (4) surface rainfall and 3D structure of hydrometeors over PR swath; and (5) surface rainfall and 3D structure of hydrometeors derived from TMI and PR simultaneously. Major operation products at Level 3 include: (1) 5° gridded TMI-only monthly rain-ocean; (2) 5° gridded PR monthly average; (3) PR-TMI monthly average; (4) TRMM multi-satellite (3-hourly, 0.25° resolution); and (5) TRMM multi-satellite precipitation merged with ground-based gauge measurements.

Ground validation has been an important component of TRMM. Techniques to produce quality controlled ground radar data sets and estimated surface rainfall rates based on ground radar have been developed. These efforts have led to validation at monthly and instantaneous time scales. The ground validation efforts have been instrumental in verifying the accuracy of the TRMM rain estimates and have led to improvements in calibrating the ground radars.

The validation of satellite products is classically defined as a ground-based observing strategy intended to assess whether or not the satellite products meet their stated accuracy requirements and objectives. In the case of TRMM, this philosophy was translated to quasi-continuous operation of four ground radar sites for which TRMM and ground-based rainfall products were compared. Findings from these four sites (Houston, Texas; Melbourne, Florida, Darwin, Australia; and Kwajelin Atoll) revealed that products were indeed generally within the stated objectives. However, direct comparison between rainfall estimates from the TRMM PR and TMI revealed that differences between the satellite estimates had regional and seasonal components leading to questions about the representativeness of the fixed validation sites. In addition, it is now clear that the nature of the errors themselves is very important. Small, but systematic rainfall errors over a large domain may be difficult to detect at individual ground validation sites. Yet, these errors are critical for climate studies and precipitation process studies. On the other hand, larger random errors are that assumed to cancel in climatologies may have significant consequences on hydrologic applications. Future precipitation missions need to work toward improving instrument accuracies, retrieval capabilities and validation of retrieved products.

6.4 The Global Precipitation Measurement (GPM) mission

The GPM Mission is an international satellite mission to unify and advance global precipitation measurements from a constellation of research and operational microwave sensors. The goal of this upcoming mission is to provide uniformly calibrated precipitation observations at every location around the world every 2–4 h to advance the understanding of the Earth's water and energy cycle and to improve the monitoring and prediction of weather, climate, freshwater availability, as well as high-impact natural hazard events such as hurricanes, floods and landslides. The GPM Mission is a primarily science mission to better understand the microphysics and the space-time variability of global precipitation. At the same time, by making precipitation observations available in the near real-time to wide segments of the user community, GPM has tremendous potential for practical benefits to society. The GPM Science Objectives thus embrace both fundamental research and application-oriented research (see Table 2). Each of the five high-level objectives listed in the table represents a key science driver for the measurement and sampling strategies for the mission.

Table 2. Scientific Objectives of GPM

1. **Advancing Precipitation Measurement Capability from Space**
   - Measurements of microphysical properties and vertical structure information of precipitating systems using active remote-sensing techniques.
   - Combination of active and passive remote-sensing techniques to provide a calibration standard for unifying and improving global precipitation measurements by a constellation of dedicated and operational passive microwave sensors.

2. **Improving Knowledge of Precipitation Systems, Water Cycle Variability and Freshwater Availability**
   - Four-dimensional measurements of space-time variability of global precipitation to better understand storm structures, water/energy budget, freshwater resources and interactions between precipitation and other climate parameters.

3. **Enhancing Climate Modeling and Prediction**
   - Estimation of surface water fluxes, cloud/precipitation microphysics and latent heat release in the atmosphere to improve Earth system modeling and analysis.

4. **Advancing Weather Prediction and 4-D Reanalysis**
   - Accurate and frequent measurements of precipitation-affected microwave radiances and instantaneous precipitation rates with quantitative error characterizations for assimilation into numerical weather prediction systems.

5. **Improving Hydrometeorological Modeling and Prediction**
   - High resolution precipitation data through downscaling and innovative hydrological modeling to advance predictions of high-impact natural hazard events (e.g., flood/drought, landslide and hurricanes).

The GPM scientific program requires measurements of the four-dimensional distribution of precipitation and its variability from diurnal to inter-annual time scales, quantitative estimates of the associated latent heat release and detailed information on bulk precipitation microphysics including the particle size distribution (PSD) information. The integrated application goals of GPM ensure that the knowledge gained by advanced precipitation measurement capabilities from space is transferred to meeting the extended goals in monitoring freshwater availability, climate modeling, weather prediction and hydrometeorological modeling.

Global precipitation measurements provide the necessary framework for understanding changes in the global water cycle and the context in which to interpret causes and consequences of local trends in water-related variables. Within the United States, GPM is envisioned to be the first in a series of Earth science missions in the coming decade to improve the understanding of the Earth's water and energy cycle (NRC, 2007). Such improvements will in turn improve decision support systems in broad societal applications identified by international communities (e.g., water resource management, agriculture, transportation, energy, health, etc.). In terms of international programs, GPM serves as a cornerstone for the development of a unified satellite constellation for monitoring global precipitation under the Committee on Earth Observation Satellites (CEOS) within the Global Earth Observing System of Systems (GEOSS) Program to provide comprehensive, long-term and coordinated observations of the Earth. During its mission life, GPM will be a mature realization of a multi-national CEOS Precipitation Constellation.

6.4.1 GPM mission concept and status

The GPM concept centers on deploying a Core spacecraft carrying both active and passive microwave sensors to serve as a 'precipitation physics observatory' and a 'calibration reference' for a constellation of dedicated and operational passive microwave sensors (most of which are in Sun-synchronous polar orbits) to produce accurate, uniform global precipitation products within a consistent framework. The GPM core spacecraft will carry the first Ku/Ka-band Dual-frequency Precipitation Radar (DPR) and a multi-channel GPM Microwave Imager (GMI) with high-frequency capabilities in a non-Sun-synchronous orbit at 65° inclination. The GMI is specifically designed to serve as a reference standard for constellation radiometers by employing a state-of-the-art calibration system. The DPR will provide detailed microphysical measurements including particle size distribution information and vertical structure of precipitating cloud systems, which will be used in conjunction with cloud-resolving models to provide a common cloud/hydrometer database for precipitation retrievals from both the Core and Constellation radiometers. The combination of the DPR with Ku and Ka bands and the GMI with 10–183 GHz channels will enable GPM to take on the new science of estimating falling snow and light rain characteristics over both ocean and land surfaces. It is expected that the GPM core instrumentation will provide estimated rain rates from ~0.2–110 mm/h.

The role of the constellation satellite system is to provide the best possible global and temporal coverage with a partnership of international space agencies over the GPM mission life. The constellation build-up will follow a 'rolling wave' strategy with a flexible architecture to capitalize on 'satellites of opportunity'. Each constellation member may have its own unique scientific mission, while participating in the partnership via sensors with precipitation measurement capabilities, such as a conically scanning radiometer and/or cross-track humidity sounder.

GPM is a currently a partnership between NASA and JAXA, with opportunities of additional partnerships with U.S. and international space agencies. NASA will provide the Core Spacecraft with a GMI instrument. JAXA will provide the DPR and launch service for the Core Satellite. In addition, NASA plans to provide a constellation satellite to be flown in an orbit that optimizes the GPM constellation coverage based on available partner assets over the mission life. This NASA Constellation Spacecraft will also augment the sampling provided by the Core Observatory for improved cross-satellite calibration of constellation radiometers, as well as enhanced capabilities for near real-time weather (e.g., hurricane) monitoring and prediction. The Core is planned for a ~400 km orbital altitude, while the NASA constellation is expected to be at ~650 km altitude with a non-Sun-synchronous orbit at ~40° inclination. Both spacecrafts are designed for a prime mission of 3 years, with consumables for a minimum of 5 years of operation. The GPM Core Spacecraft and instruments are under development by NASA and JAXA for an anticipated launch date of June 2013 with the NASA Constellation Spacecraft to be launched in 2014.

The current constellation partnership plans (see Fig. 3) include conical-scanning microwave imagers; e.g., Japan's Global Change Observation Mission – Water (GCOM-W) (Shimoda 2005), French and Indian Megha-Tropique (Aguttes et al. 2000), United States DMSP satellites (Hollinger et al. 1990) augmented by microwave temperature/humidity sounders over land including Advanced Technology

Microwave Sounder (ATMS) on the National Polar-orbiting Operational Environmental Satellite System (NPOESS) Preparatory Project (NPP) and NPOESS-C1 (Bunin et al. 2004) and Microwave Humidity Sounder (MHS) on NOAA-N' and the European MetOp satellite (Edwards and Pawlak 2000). The inclusion of microwave humidity sounders in the baseline GPM sampling reflects recent advances of precipitation retrievals from sounder instruments such as AMSU-B (with channels at ~89, 150, 183±1, 183±3, 183±7 GHz), which have been shown to be comparable in quality to those from conical-scanning radiometers over land (Lin and Hou 2007). These sounder instruments make it possible for GPM to provide a precipitation estimate every 1–2 h at every location on land over the prime mission life. Over oceans, the sounder retrievals are not as accurate due their inability to detect warm rain systems and thus not included in the baseline GPM sampling. Without the sounders, the average revisit time over oceans for the GPM constellation ranges from 2 to 4 h over the first three years of the mission. Overall, the GPM constellation can provide better than 3-hour sampling over as much as 90% of the globe if all anticipated partner assets are available. The GPM sampling is designed for graceful degradation of performance if partner assets are not available or launched on schedule.

The GPM mission is supported on the ground by (1) a NASA-provided mission operations system for the operation of the Core and NASA Constellation Spacecrafts, (2) a Ground Validation (GV) System consisting of an array of ground calibration and validation sites, provided by NASA, JAXA and international and U.S. domestic partners and (3) a NASA-provided Precipitation Processing System (PPS) in coordination with GPM partner data processing sites to produce near-real-time and standard global precipitation products. The PPS will have the data processing and communications capacity to process the full quantity of input data from the space segment and ancillary GPM sources as it is generated and to create science products in three categories: immediate real-time, outreach and research data. The PPS will process Level 1, 2 and 3 products (similar to TRMM). The PPS shall be sized to handle all data from the NASA Core Observatory, NASA Constellation Spacecraft and partner assets. In addition, the NASA Precipitation Measurement Missions (PMM) Science Team together with its JAXA counterpart work to ensure the scientific success of the GPM mission in the sensor design, retrieval algorithm development and validation, as well as innovative methodologies for data utilization in applications that range from numerical weather prediction to hydrological modeling and prediction.

Fig. 3. The GPM Mission configuration with Core Spacecraft (*upper right*), the NASA Constellation Spacecraft (*middle left*) and partner constellation satellites providing global coverage

6.4.2 GPM core sensor instrumentation

For the GPM Mission, NASA and JAXA will design, develop and operate two spaceborne instruments – the GMI and DPR. The GMI provides measurements of precipitation intensity and distribution, while the DPR provides three-dimensional estimates of cloud and microphysical properties. The GMI and DPR are designed to work together to provide better precipitation estimates than either sensor alone. The sampling strategy of the DPR and GMI is shown in Fig. 4. As illustrated in the figure, the DPR Ka-band has a 125 km swath with a vertical resolution of 250 or 500 m, the Ku-band has a 245 km swath with 500 m vertical resolution, while the GMI extends measurements beyond the DPR domain to the larger swath width of 885 km.

NASA will provide two GMIs, one for the Core Spacecraft and one for the NASA Constellation Spacecraft. GMI is being designed to make simultaneous measurements of a range of precipitation rates, including

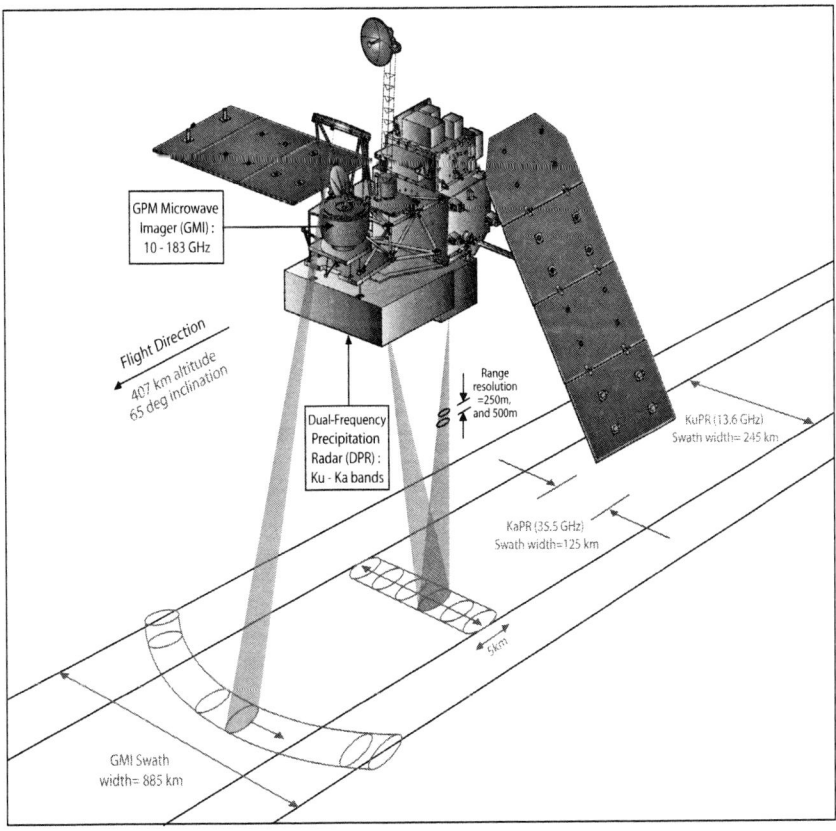

Fig. 4. The GPM Core Spacecraft with the DPR and GMI instruments

light rain and snowfall often found at the Earth's higher latitudes. These measurements are key to understanding the precipitation processes and storm structures of mid-latitude and high-latitude systems, both over land and water. The frequency and footprint characteristics of the GMI for the Core Observatory and Constellation Observatory are provided in Table 3 and the GMI instruments operate in a conically-scanning mode as shown in Fig. 4.

The GMI is being designed to have independent calibration measurements to ensure accuracy. GMI's calibration is achieved through the use of the standard methodology of hot and cold load gain measurements that provide an instrument count (measure of intensity) to brightness temperature value (Ulaby et al. 1981). In addition, both the hot and cold load measurement ports will have injected noise diode inputs. This will provide hot+noise and a cold+noise calibration points. If there is a detected hot load failure due to sun impinging on the hot load, the measurements from the cold and cold+noise can be used to calibrate the GMI radiometer. Further, the noise diodes provide a method to track the non-linearity drift in the radiometer signals over time. It is especially important to quantify this drift when using GMI radiometer measurements to estimate parameters to be used for climate warming studies, such as sea surface temperature.

The DPR under development by JAXA will provide three-dimensional measurements of cloud structure, precipitation particle size distribution (PSD) and precipitation intensity and distribution while serving as an orbiting reference system for the passive microwave-based precipitation estimations. The dual frequency design of the DPR will utilize the differential attenuation of the returned signals to infer information about the bulk characteristics of the particle size distribution (e.g., the diameter that divides the rain water content into two equal parts) and hydrometeor category (e.g., rain, snow, mixed, wet graupel/hail). Particularly at 35 GHz,

Table 3. The frequencies and footprints for the GMI instrument

Frequency (GHz)	V/H Polarization	Core footprint (km) (at 405 km altitude)	Constellation footprint (km) (at 650 km altitude)
10.65	V & H	19.4 × 32.2	30.8 × 51.7
18.7	V & H	11.2 × 18.3	18.0 × 29.4
23.8	V	9.2 × 15.0	14.8 × 24.1
36.5	V & H	8.6 × 14.4	13.8 × 23.1
89.0	V & H	4.4 × 7.3	7.1 × 11.7
165.5	V & H	4.4 × 7.3	7.1 × 11.7
183.31± 3	V	4.4 × 7.3	7.1 × 11.7
183.31±8	V	4.4 × 7.3	7.1 × 11.7

cloud drops may contribute to the integrated attenuation of precipitating clouds as a function of cloud depth and possibly the characteristics cloud drop spectra (e.g., marine versus continental clouds). The DPR operates in a cross-track scanning mode as shown in Fig. 4. The frequencies are Ku (13.6 GHz) and Ka (35 GHz) and have a footprint of 5km at nadir for the Core Observatory.

6.4.3 Ground validation plans

The GPM validation program is being developed with a somewhat modified paradigm with respect to the methodology used for TRMM. Aside from verifying GPM products through statistical comparisons with ground-based measurements, GPM requires that certain validation sites be equipped to additionally diagnose the underlying causes of any retrieval algorithm discrepancies. This diagnosis component, when framed in the context of meteorological conditions is intended to: (a) provide invaluable information regarding the algorithm's expected performance in other regions; and (b) provide information that algorithm developers typically need in order to improve algorithms. The metrics to be used include the ability to predict the success or failure of the algorithms, based upon meteorological circumstances, as this is a quantitative measure of our understanding.

While it is important to learn from prior and current satellite-precipitation measurement missions such as TRMM, GPM must also be forward-looking. It is becoming apparent that the future of precipitation research is probably not one in which satellite data is used in isolation. Instead, integration of satellite precipitation measurements with ground observations and cloud resolving models is likely to replace satellite-only precipitation products, particularly for applications such as hydrology that require precipitation as input.

For GPM, the GV strategy follows a three-prong approach to focus on the different needs of validation. There will be Statistical Validation sites, Precipitation Process sites and Integrated Application sites. The surface precipitation statistical validation sites will be used for direct assessment of GPM satellite data products. These will be co-located with existing or upgraded national network (e.g., Weather Surveillance Radar 88 Doppler (WSR-88D), etc.) and dense gauge networks. Their primary purpose will be to validate the satellite estimates using statistical and other procedures. There will need to be sites over ocean and land surfaces. Complications due to different sampling volumes of the satellite versus the ground sensors will need to be addressed. Contributions to the error in the comparisons and validation will be investigated and analyzed.

The precipitation process sites are used for improving understanding of precipitation physics, modeling and satellite retrieval algorithms and provide valuable observations both pre and post launch. These sites will focus on tropical, mid- and high-latitude precipitation studies. The sites will include orographic/coastal sites and targeted sites for resolving discrepancies between satellite algorithms. Sites will be selected for validating the newer products of GPM, namely light rain and falling snow. These sites may include aircraft for in situ measurements and mobile instrument assets that can be moved to different locations, for example, to observe snow in cold seasons and tropical rain in warm seasons. These process sites will be used to improve and validate the physics of cloud resolving models, land and hydrology models and coupled land-atmosphere models.

The integrated hydrological sites will focus on improving hydrological applications. These sites will be co-located with existing watersheds maintained by other US agencies and international research programs. The idea is to integrate or assimilate the precipitation estimates into hydrological and/or climate models. The expectation is that precipitation will improve the model predictions, yet part of the process is to assess the impacts of errors in the precipitation estimates on errors in the predicted hydrology or climate application.

In advance of GPM, ground validation is occurring to provide data for the falling snow algorithm developers. Data sets are needed to (1) develop and validate models that convert the physical properties (shape, size distribution, density, ice-air-water ratio) of single snowflakes to their radiative properties (asymmetry factor and absorption, scattering and backscattering coefficients); and (2) relate the bulk layer radiative properties (summation of the single particle radiative properties over a discrete vertical layer) to calculated and observed passive microwave radiances and radar reflectivities. These models are central to physically-based snowfall retrieval methods, as well as the characterization of likely retrieval uncertainties. In the past, the microwave community has used a number of approximate models all giving different results based on choices of parameters and assumptions. The snowfall retrieval algorithm community cannot make significant advances without a better understanding of the relationships between non-spherical snowflakes and their radiative properties.

In the winter of 2006–2007, GPM participated in the Canadian CloudSat/CALIPSO Validation Program field campaign (C3VP, http://c3vp.org/) held near Toronto, Canada, to measure data for retrieval algorithm development. This provided an opportunity to collect data in cold-latitudes needed for falling snow retrieval algorithm development and provide a basis for improving retrieval accuracy. The C3VP field campaign ground-based assets collected high resolution dual frequency (DPR matched) polarimetric radar measurements of snow/rainfall rate, particle type and mass-content coincident to aircraft sampling, precipitation particle size distributions and shapes measured near the ground under cover of radar and aircraft and a comprehensive set of measurements of environmental conditions. The C3VP field campaign used aircraft to collect in situ microphysics in snow and mixed-phase precipitation events of lake effect snow bands and in synoptic snow conditions. Overpasses by the CloudSat radar (94 GHz), the NOAA AMSU-A and AMSU-B and the AMSR-E radiometers provided microwave satellite data. These measurements will lead to improved information for defining relationships between the physical properties of frozen precipitation and its active and passive radiative signatures. Since international partnership is key to ensuring the success of the GPM GV program, similar collaborations on ground validation are underway between NASA and other international research organizations.

6.5 Precipitation retrieval algorithm methodologies

It is not enough to collect observed satellite data and ground measurement data sets. Reliable and accurate retrieval algorithms must be developed, tested, validated and improved as our understanding of the underlying physical processes or the retrieval mechanics is enhanced or updated. For passive radiometers, four retrieval methodologies have been used in the past: empirical algorithms, neural network approaches, maximum likelihood methods and Bayesian techniques. Because the Bayesian techniques permit real-time estimation as well as a link to the understanding of the physical state and underling processes, they have been the cornerstone of the TRMM passive algorithms and are being considered for the GPM algorithms. Due to the greater sampling of the precipitation column by spaceborne radars and the less ill-posed nature of the radar retrieval problem in general, analytical inversion methods have been at the heart of most TRMM-era algorithms. However, in recognition of the inherent uncertainties in radar reflectivity measurements, variational techniques incorporating additional constraints on the inversion of radar data have also been employed with success.

Retrieval algorithms for satellite-based precipitation estimates from both active and passive sensors are complicated by the fundamentally underconstrained nature of the inversion problem. Precipitating clouds simply have more free parameters, including their size, shape and internal distribution of hydrometeors along with their associated sizes, shapes and compositions, than can realistically be retrieved from a finite set of satellite observables. Assumptions regarding the composition of clouds are, therefore, necessary. These assumptions and how they are represented in algorithms, are the chief source of discrepancies among existing rainfall products (Stephens and Kummerow 2007). The key to building a consistent framework within GPM is thus to merge these physics assumptions across active and passive and combined active-passive algorithms. This can be accomplished using the GPM core satellite that is designed explicitly to observe precipitating cloud microphysical structure with greater accuracy than ever before. The ability to create an inventory of naturally occurring cloud states with the details afforded by the GPM core satellite greatly simplifies the construction of Bayesian algorithms, which in turn can be used to derive products from the constellation of radiometers. Consistent retrievals from the constellation satellites would then simplify the construction of merged microwave/IR products for very high space/time resolution products and allow for precipitation data assimilation into General Circulation Models (GCMs) and Numerical Weather Prediction (NWP) models.

The algorithm development efforts for GPM reflect a hierarchy of products, as described above. Most critical in this hierarchy are the dual frequency radar algorithms, which serve as the basis for the merged radar-radiometer algorithm implemented on the core satellite. The merged algorithm is applied to the core satellite data streams to create a priori databases representing our best estimate of observed cloud structures. These representative databases are then incorporated in the construction of the constellation algorithm(s), which should then be applicable to any spaceborne radiometer irrespective of the sensor details. The consistent rainfall estimates derived from the radiometer constellation serve as a calibration reference in combined microwave/infrared techniques designed to provide high resolution hourly rainfall.

6.5.1 Active retrieval methods

The GPM core satellite will carry a dual frequency radar operating at 13.6 and 35.5 GHz. From this dual frequency radar one can, in principle, determine two parameters of the drop size distribution at each range bin in the vertically sampled profile. While this is a significant step forward from TRMM and most surface radars that rely on measurements at a single frequency, it is not enough to unambiguously determine rainfall rate. For a general gamma size distribution of raindrop sizes one needs three independent measurements to specify the three free parameters of the gamma distribution. Practical considerations such as radar calibration uncertainty, sub-pixel variability and gaseous and particle attenuation complicate the inversion. Because the three free parameters of the gamma distribution cannot be unambiguously determined from two radar frequency measurements, robust solutions require that one assumes plausible relationships between drop size distribution parameters, their radiative properties and the observations.

The reflectivity and total beam attenuation of a precipitation radar target can be related directly to the sum of the radiative backscattering and extinction cross-sections of individual particles in the target volume. It follows that the effective radar reflectivity $Z_{e\lambda}$ and extinction k_λ at wavelength λ characterizes the backscattering and extinction, respectively, from a unit volume of the precipitating atmosphere. While the relations between drop size and backscattering and extinction are well specified by Mie formulas for individual spherical particles, the rainfall rate as well as the reflectivity and extinction depend not upon individual drops, but upon the size distribution of these drops in a unit volume. This size distribution, N(D), has multiple degrees of freedom and cannot be uniquely characterized by $Z_{e\lambda}$ or even by $Z_{e\lambda}$ combined with k_λ. The general strategy adopted for the dual frequency radar algorithm is thus one that describes the size distribution as having two free parameters that are then determined by the two radar frequencies. Even within this framework, a number of potential solutions exist, especially in frozen precipitation cases where multiple particle habits with distinct size distributions may occur.

There are four approaches currently being considered for retrievals from the DPR on GPM. In the first approach, particle sizes follow a gamma distribution of the form $N(D) = N_0 D^\mu \exp(-\Lambda D)$ in which μ is assumed fixed and then the classical solution developed by Hitschfeld and Bordan (1954) for attenuating radars is employed. This technique assumes power law relations hold between the extinction k_λ and $Z_{e\lambda}$ at each of the frequencies in order to correct for the intervening attenuation.

While the single frequency Hitschfield-Bordan solution is known to be unstable, the dual frequency method has the additional constraint that the rainfall rate at each range gate should be the same for each of the frequencies employed. This method does not require knowledge of the total beam attenuation from a surface reference technique, but it has the disadvantage that the accuracy of rain estimates is limited by the accuracy of the assumed PSD model.

A second technique relies on differential or integral equations for the parameters of the PSD. For a fixed μ, one can directly obtain a set of coupled differential equations for N_0 and Λ (or more commonly the mass weighted median diameter, D_0, which is related by $\Lambda = (\mu + 3.67)/D_0$) as a function of range, r, that can be solved numerically. Once N_0 and D_0 are estimated, the rain rate and equivalent water content profiles can be derived. These N_0, D_0 equations can be solved either in a forward-going (from the storm top downward) or backward-going (from the surface upward) direction. Although the forward-going solutions do not require an independent estimate of path attenuation, the estimates tend to become unstable as the rain rate increases. The backward solutions are more stable and, in the rain, are independent of cloud water and mixed-phase particles above the rain layer. However, the procedure requires accurate estimates of the total path attenuation at both frequencies.

When the total attenuation is unknown or has a high degree of uncertainty, however, a different approach is needed. The third approach requires one to assume that N_0 and D_0 are somehow related. While there are many possible methods of implementing this constraint, a well known method is that of Marzoug and Amayenc (1994), who assume that the normalized intercept parameter, N_0^*, is constant along the radar path. This approach can be used both for single and dual-frequency radars and with or without an attenuation constraint. In the dual-frequency case, a relationship between the specific attenuation profiles at the two frequencies is assumed and various parameters are adjusted to minimize the root mean squared difference along the radar path.

The fourth proposed method is one in which the difference between the measured reflectivities for the two frequencies at two ranges is utilized. This 'difference of differences' provides an estimate of the differential path attenuation over a range interval. The attenuation can then be related directly to rainfall rate, taking advantage of the fact that the two quantities are nearly linearly related. While this method has historically been applied to systems in which one wavelength suffers little or no attenuation, it can in principle be applied to any wavelength combination. The major source of error arises from non-Rayleigh scattering at the higher frequency in the presence of large drop sizes, an effect that biases the estimate of attenuation.

For intermediate rain rates from several mm/h to about 15 mm/h and where accurate estimates of path attenuation can be obtained, the DPR can provide size distribution information along the full column including the snow, rain and mixed-phase regions. In the more general situation, the most robust approach will depend upon the signal provided by each of the wavelengths, uncertainty in total attenuation estimates from the surface reference technique in providing the total attenuation as absolute calibration and system noise considerations. The performance of the various methods and how best to apply them are areas of active research in the GPM community. Merely having two frequencies instead of the single frequency available on TRMM will, however, significantly reduce the uncertainties irrespective of the final method (or combination of methods) chosen for GPM. A study designed to provide a better understanding of the improvement in spaceborne radar estimates from TRMM to GPM was recently performed (Haddad et al. 2006). This analysis used cloud model results as well as storm 'snapshots' synthesized from high resolution airborne radar measurements and from TRMM overpasses. The study showed that uncertainty in estimates of the surface rain rate should be substantially smaller with the GPM core suite of instruments than it was with the TRMM radar and radiometer. Figure 5 summarizes the main conclusion: GPM-core's inner-swath algorithm should yield rain rates estimates with an uncertainty less than 20% for rain rates between 1.5 and 12 mm/h (as opposed to the TRMM radar's 40%), though we will be hard-pressed to achieve such a small uncertainty at lighter rain rates (because the corresponding 14-GHz and 35-GHz signatures are not sufficiently different) or at higher rain rates (because significant attenuation at 35-GHz will affect the results).

6.5.2 Combined retrieval methods for GPM

The dual frequency radar rain estimates may be further improved by using the GMI on board the core spacecraft. Unlike the two radar frequencies that sample a common volume, merging satellite radar and radiometer information is considerably more challenging. Additional difficulties arise due to the varying resolutions of the radiometer channels, the different view direction of the radiometer and radar and ultimately the different aspects of the cloud and underlying surface that result from the different viewing geometries. A number of methods have been developed for the TRMM satellite starting with the operational

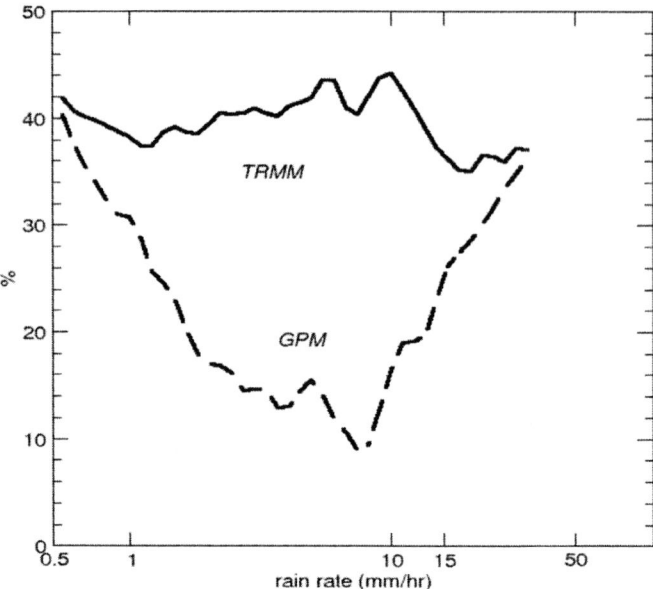

Fig. 5. Uncertainties of GPM-Core and TRMM rain rate estimates based on simulations (derived from Haddad et al. 2006)

algorithm developed by Haddad et al. (1997), followed by Grecu et al. (2004) and Masunaga and Kummerow (2005). While each approach is somewhat different, they all share the common goal of using the radiometer signal as an integral constraint on the column attenuation seen by the radar. This is particularly important in lighter rainfall cases where the surface reference estimates of attenuation are too noisy, leaving the radar with only an assumed drop size distribution for interpreting the surface rainfall. While each technique is different, the solutions each produce a hydrometeor profile, particle size distribution and surface parameters for which simulated brightness temperatures and reflectivities are consistent with the actual measurements.

Since a number of formulations for the combined radar/radiometer retrieval for TRMM have already been established, the GPM algorithm is being developed along the same lines – by varying the cloud and background constituents until a solution consistent with both the DPR and the GMI observations is obtained. While limited to the narrow swath of the dual frequency radar, this solution is nonetheless critical for the GPM concept in that it forms the basis for the constellation radiometer algorithm discussed in the next section.

6.5.3 Passive retrieval methods

A number of different approaches have been taken for deriving rainfall rates from passive microwave sensors, including empirical and neural network methodologies. However, a class of physically based algorithms to retrieve the precipitation profile (e.g., Evans et al. 1995; Kummerow and Giglio 1994; Kummerow et al. 1996, 2001; Marzano et al. 1999; Smith et al. 1994a, b) is especially well suited for the general framework needed by GPM. These physically-based algorithms retrieve precipitation rate, but also provide information that will lead to a better understanding of the relationships between the cloud state and the observations. This framework accommodates not only the core and constellation sensors currently envisioned for GPM, but any future sensors that have yet to be specified. The combined radar/radiometer algorithm discussed in the previous section thus serves not only to establish the most complete product from the GPM core satellite itself, but it also can be used to produce an a priori Bayesian database of candidate solutions for the constellation radiometer algorithms. The a priori database consists of vertical profiles of precipitating clouds with associated computed brightness temperatures for each profile. Since one can compute T_Bs at any sensor frequency, the same microphysical cloud profiles remain consistent across all constellation members.

The radiometer algorithms in the TRMM era were dominated by schemes using cloud resolving models (CRMs) to produce a priori databases of cloud profiles. The ability of explicit cloud resolving models to faithfully reproduce and fully represent the actual microphysics structure of observed storms is one of the essential conditions for most of these algorithms to give reliable results. While CRMs were approaching this condition in the TRMM era, the use of these CRMs also created problems that complicated the inversion scheme. Most important was the lack of representativeness that was introduced when a finite and typically small number of these models were used to represent all raining environments. The a priori database generated by the GPM core satellite, whether it uses ancillary information from CRMs or not, will overcome this representativeness problem by creating a priori databases of hydrometeor profiles as they were observed in nature. Any radar profile can serve as the observational basis of the a priori database; for example, the database can be initialized with TRMM and CloudSat results and updated as microphysical profiles from the DPR become available.

With regard to ice-phase precipitation, creating representative a priori databases will require some effort. Ice-phase hydrometeors occur above the freezing level in most precipitation systems and therefore radiative transfer calculation through the entire precipitation column must account for raindrops as well as ice-phase and mixed-phase particles in the column at the higher microwave frequencies (37–200 GHz). There will also be northern-latitude cold-season events where there is no melting layer and the precipitation falls in the form of frozen hydrometeors. Therefore, not only do the cloud profile microphysics need to be appropriate for these precipitation systems, but also the models to compute high frequency (37–200 GHz) radiative properties from physical models of ice-phase and mixed-phase particles need to be developed with care. Liquid precipitation hydrometeors are most commonly modeled as homogeneous dielectric spheres, so that standard Mie codes may be utilized to compute their radiative properties. For most ice-phase hydrometeors (snow crystals and aggregates, rimed graupel particles, etc.), the spherical assumption is typically motivated more by convenience and by the lack of practical alternatives, than by realism. At frequencies at or above ~60 GHz, the spherical approach is no longer accurate for highly nonspherical hydrometeors, such as dendrites and aggregates (Liu 2004; Kim 2006). The computation of the radiative properties of nonspherical snow and mixed-phase particles at higher frequencies requires computationally intense numerical solutions, such as the Finite-Difference Time Domain (Yang and Liou 1995; Sun et al. 1999), Conjugate Gradient (Meneghini and Liao 2000), the Discrete-Dipole Approximation (DDA) methods (Purcell and Pennypacker 1973), and/or the Generalized Multiparticle Mie model (Xu 1997). Inclusion of meltwater in these calculations only adds to their complexity. In order to compute the radiative properties of ice- and mixed-phase particles, one must have representative habits, densities and PSDs of the particles for use in the rigorous calculation of their radiative properties. Field campaigns and investigations are underway to provide such data for GPM era algorithms and to determine the best methodology for computing the radiative properties of these particles.

6.5.4 Merged microwave/infrared methods

Having a consistent set of rainfall products from each of the passive microwave radiometers in the GPM constellation is an essential step forward in producing the desired 3-hour rain rate estimates at high space/time resolution. In order to achieve consistent precipitation estimates from all constellation members for the desired temporal coverage, GPM scientists are developing techniques to inter-calibrate the radiometer brightness temperatures and precipitation rates from constellation member instruments using GPM Core sensor measurements as a reference. This inter-calibration will provide common standards for GPM data streams without interfering with the calibration requirements of the individual constellation members. An initial goal would be to ensure common file formats among the brightness temperature data sets. Later goals might include common microphysical databases that can be used to physically link the observations, via forward radiative transfer models, among the different sensors.

While consistent 3-hour precipitation estimate coverage is important, there are and always will be, applications such as hydrology that require even greater temporal and spatial resolution. In addition to downscaling, a number of techniques have been developed that use the passive microwave estimates as anchors for geostationary based infrared (IR) rainfall estimates (see Ebert et al. 2007). These estimates can, in principle, be made at 15-minute intervals with resolution equaling that of the IR data itself. The techniques themselves vary, but perhaps the simplest to envision is a simple morphing algorithm such as the one developed by Joyce et al. (2004) in which the rainfall derived by two passive microwave overpasses of a given scene is morphed (in an image sense) between one overpass and another. The morphing is guided by cloud motions derived from the more frequent IR images, bracketed by the microwave overpasses. The technique itself is, therefore, not dependent upon the nature of the microwave algorithm itself – merely the instantaneous rainfall product at the time of the overpass. The objective is the creation of a rain product with the temporal and spatial resolution of the IR data, but bias-corrected by the microwave rain estimates. The merged microwave/infrared methods will thus benefit immediately from the constellation algorithms being developed for GPM in that GPM will take advantage of coincident measurements by the Core Observatory in a non-Sun-synchronous orbit intersecting the constellation satellites to ensure that the microwave products being used for bias correction are consistent with one another – an essential ingredient to make these algorithms perform optimally. The complementary information provided by the DPR and the GMI on the GPM Core satellite is the key to integrating multiple satellite precipitation estimates within a consistent framework to improve the accuracy of global 3-hour precipitation products.

6.6 Summary

Water cycling and the future availability of fresh water resources are immense societal concerns that impact all nations on Earth as it affects virtually every environmental issue. Precipitation is also a fundamental component of the weather/climate system for it regulates the global energy and radiation balance through coupling to clouds, water vapor, global winds and atmospheric transport. Accurate and comprehensive information on precipitation is essential for understanding the global water/energy cycle and for a wide range of research and applications with practical benefits to society. However, rainfall is difficult to measure because precipitation systems tend to be random in character and also evolve and dissipate very rapidly. It is not uncommon to see a wide range of rain amounts over a small area; and in any given area, the amount of rain can vary significantly over a short time span. These factors together make precipitation difficult to quantify, yet measurements at such local scales are needed for many hydrometeorological applications such as flood and landslide forecasting.

Historical, multi-decadal measurements of precipitation from surface-based rain gauges are available over continents, but oceans remained largely unobserved prior to the beginning of the satellite era. Early visible and infrared satellites provided information on cloud tops and their horizontal extent; however, wide-band microwave frequencies proved extremely useful for probing into the precipitating liquid and ice layers of clouds. It was only after the launch of the first SSM/I on the DMSP satellite series in 1987 that precipitation measurements over oceans from passive microwave radiometers have become available on a regular basis. Recognizing the potential of satellites as a vital tool for measuring global precipitation from the vantage point of space, NASA and the predecessor of JAXA launched in 1997 the Tropical Rainfall Measuring Mission (TRMM) satellite carrying the first precipitation radar and a multi-frequency microwave imager to confirm the validity of space-based precipitation measurements from passive microwave radiometers. The success of TRMM has led the widespread use of merged multi-satellite precipitation products in a broad range of scientific research and practical applications (NRC 2005).

Encouraged by the success of TRMM, NASA and JAXA are jointly planning a new international satellite mission named the Global Precipitation Measurement (GPM) mission to develop the next-generation of global precipitation measurements. The GPM mission consists of the Core Spacecraft, a NASA-provided Constellation

Spacecraft and multiple U.S. and international partner precipitation satellites in order to provide global rain rate estimates with footprint resolutions from 5 to 50 km and temporal resolutions of 2–4 h. GPM's Core Spacecraft will have measurement capabilities beyond that of TRMM by carrying a dual-frequency (Ku and Ka) radar and a wide-band (10–183 GHz) radiometer with extra calibration hardware to serve as a reference standard for cross-calibrating rainfall estimates from a constellation of passive microwave imagers and sounders. This GPM Core Spacecraft sensor package is designed to observe precipitating cloud microphysical structure with greater accuracy than ever possible, leading to improved retrieval algorithms and a better understanding of precipitation processes through estimations of not only moderate and high rain rates, but also light rain and falling snow over both land and oceans.

Operations, data processing and ground validation are built into the GPM mission framework. Ground validation ensures that the satellite estimates are statistically accurate, that precipitation process knowledge is gained at the macroscopic and microphysical scales and that integrated application goals, especially in terms of hydrology, are addressed and validated as part of GPM. The data processing system is designed to process the Core and constellation data streams to produce real-time estimates, research products and outreach information. Much effort is spent and will continue to be spent, on algorithm development. The algorithm heritage from TRMM will lead the initial directions for the radar-only, radiometer-only and combined radar-radiometer retrieval methodologies. With the additional channels on the DPR and GMI, algorithm performance is expected to improve such that rain rate estimates from ~0.2 to 110 mm/h will be available from the Core Spacecraft. Further, information gained from the Core Spacecraft algorithm development will greatly enhance retrievals from the constellation members.

GPM precipitation estimates will be available globally and this availability is perhaps most important to the developing nations where freshwater resources are critical. In 2002, GPM was identified by the United Nations as an outstanding example of peaceful uses of space. The GPM concept is currently serving as the scientific basis for the formulation of an international Precipitation Constellation by the Committee on Earth Observation Satellites (CEOS) under the auspices of the Global Earth Observing System of Systems (GEOSS). GEOSS is an inter-governmental effort to provide coordinated, comprehensive and long-term observations of the Earth. During its mission phase, the GPM

Mission will be a mature realization of the CEOS Precipitation Constellation for the benefits of many nations.

The anticipated societal benefits of GPM are manifest in the integrated science plans of the mission. Given the central importance of precipitation in the global water and energy cycle, GPM measurements will make significant contributions to the understanding of the detailed microphysics and the space-time variability of precipitation. The precipitation estimates and knowledge gained from GPM will also be useful for weather forecasting through four-dimensional data assimilation and climate forecasting through better estimates of soil moisture and freshwater fluxes into the oceans. GPM's integrated application goals will support improvements in climate prediction at seasonal to inter-annual scales. This is possible, in part, because variations in precipitation patterns are traceable to cycles in global atmospheric dynamics such as the El Nino/Southern Oscillation and the Madden Julian Oscillation. A majority of these patterns are driven by oceanic processes affecting atmospheric and precipitation processes that will be measured at temporal resolutions of 2–4 h over the oceans by GPM constellation satellites. Further, GPM's observations continue the multi-decadal history of satellite precipitation estimates. While these are a few of GPM's integrated application goals, stakeholders such as those in public health, agriculture and urban planning will find GPM's global high resolution, accurate precipitation data useful for their decision support systems and operational requirements. GPM represents a truly significant milestone in international partnership in providing state-of-the-art global precipitation estimates for both scientific research and societal applications.

Acknowledgments

The authors thank Dr. Robert Adler for reference materials on TRMM and Drs. Robert Meneghini and William Olson for reviewing algorithm sections of this work. The support of this work by NASA's Precipitation Measurement Missions (PMM) Program managed by Dr. Ramesh Kakar is gratefully acknowledged.

References

Aguttes JP, Schrive J, Goldstein J, Rouze C, Raju G (2000) MEGHA-TROPIQUES, a satellite for studying the water cycle and energy exchanges in the tropiques. In: Proceedings Geoscience and Remote Sensing Symposium, IGARRS 2000, pp 3042–3044

Atlas D (ed) (1990) Radar in meteorology: Battan memorial and 40th anniversary. In: Proceedings Radar Meteorology Conference, American Meteorological Society, Boston, pp 806

Barrett EC, Beaumont MJ (1994) Satellite rainfall monitoring: An overview. Rem Sens Rev 11:23–48

Bloschl G, Sirapalan M (1995) Scale issues in hydrological modeling – A review. Hydrol Process 9:251–290

Bunin SL, Holmes D, Schott T, Silva HJ (2004) NOAA/NESDIS Preparation for the NPOESS era. In: Preprints 20th International Conference on Interactive Information and Processing Systems (IIPS) for Meteorology, Oceanography, and Hydrology. American Meteorological Society, Seattle, Washington.

Chen FW, Staelin DH (2003) AIRS/AMSU/HSB precipitation estimates. IEEE T Geosci Remote 41:410–417

Ebert EE, Janowiak J, Kidd C (2007) Comparison of near real time precipitation estimates from satellite observations and numerical models. B Am Meteorol Soc 88:47–64

Edwards PG, Pawlak D (2000) MetOp: The space segment for EUMETSAT's polar system. ESA Bulletin 102:6–18

Evans KF, Turk J, Wong T, Stephens GL (1995) A Bayesian approach to microwave precipitation profile retrieval. J Appl Mcteorol 34:260–279

Evans KF, Wang JR, Racette PE, Heymsfield G, Li L (2005) Ice cloud retrievals and analysis with the compact scanning submillimeter imaging radiometer and the cloud radar system during CRYSTAL FACE. J Appl Meteorol 44:839–859

Ferraro RR, Weng F, Grody N, Zhao L, Meng H, Kongoli C, Pellegrino P, Qiu S, Dean C (2005) NOAA operational hydrological products derived from the AMSU. IEEE T Geosci Remote 43:1036–1049

Gaiser PW, St Germain KM, Twarog EM, Poe GA, Purdy W, Richardson D, Grossman W, Linwood Jones W, Spencer D, Golba G, Cleveland J, Choy L, Bevilacqua RM, Chang PS (2004) The WindSat spaceborne polarimetric microwave radiometer: Sensor description and early performance. IEEE T Geosci Remote 42:2347–2361

Gasiewski AJ (1993) Microwave radiative transfer in hydrometeors. In: Janssen MA (ed) Atmospheric remote sensing by microwave radiometry. John Wiley and Sons, New York, NY, pp 91–144

Grecu M, Olson WS, Anagnostou EN (2004) Retrieval of precipitation profiles from multiresolution, multifrequency, active and passive microwave observations. J Appl Meteorol 43:562–575

Haddad ZS, Smith EA, Kummerow CD, Iguchi T, Farrar MR, Durden SL, Alves M, Olson WS (1997) The TRMM 'day-1' radar/radiometer combined rain-profiling algorithm. J Meteorol Soc Jpn 75:799–809

Haddad ZS, Meagher JP, Durden SL, Smith EA, Im E (2006) Drop size ambiguities in the retrieval of precipitation profiles from dual-frequency radar measurements. J Atmos Sci 63:204–217

Hitschfeld W, Bordan J (1954) Errors inherent in the radar measurement of rain fall at attenuating wavelengths. J Meteorol 11:58–67

Hollinger JP, Pierce JL, Poe GA (1990) SSM/I instrument evaluation. IEEE T Geosci Remote 28:781–790

Iguchi T, Meneghini R (1994) Intercomparison of single-frequency methods for retrieving a vertical rain profile from airborne or space borne radar data. J Atmos Ocean Tech 11:1507–1516

Ishimaru A (1991) Electromagnetic wave propagation, radiation, and scattering. Prentice Hall, Englewood Cliffs, NJ

Joyce RJ, Janowiak JE, Arkin PA, Xie P (2004) CMORPH: A method that produces global precipitation estimates from passive microwave and infrared data at high spatial and temporal resolution. J Hydrometeorol 5:487–503

Kawanishi T, Sezai T, Ito Y, Imaoka K, Takeshima T, Ishido Y, Shibata A, Miura M, Inahata H, Spencer RW (2003) The advanced microwave scanning radiometer for the Earth observing system (AMSR-E), NASDA's contribution to the EOS for global energy and water cycle studies. IEEE T Geosci Remote 41:184–194

Kidd C (2001) Satellite rainfall climatology: A review. Int J Climatol 21: 1041–1066

Kidder SQ (1981) The measurement of precipitation frequencies by passive microwave radiometry. In: Precipitation measurements from space, Workshop Report. October 1981. NASA Goddard space flight center: Greenbelt, MD

Kidder SQ, VonderHaar TH (1995) Satellite meteorology: An introduction. Academic Press, San Diego, CA

Kidder S, Goldberg M, Zehr R, DeMaria M, Purdom JFW, Velden CS, Grody NC, Kusselson SJ (2000) Satellite analysis of tropical cyclones using the advanced microwave sounding unit (AMSU). B Am Meteorol Soc 81:1241–1259

Kim M-J (2006) Single scattering parameters of randomly oriented snow particles at microwave frequencies. J Geophys Res vol 111, D14201, doi:10.1029/2005JD006892

Kummerow C, Giglio L (1994) A passive microwave technique for estimating rainfall and vertical structure information from space. Part I: Algorithm description. J Appl Meteorol 33:3–18

Kummerow C, Olson WS, Giglio L (1996) A simplified scheme for obtaining precipitation and vertical hydrometeor profiles from passive microwave sensors. IEEE T Geosci Remote 11:125–152

Kummerow C, Barnes W, Kozu T, Shiue J, Simpson J (1998) The tropical rainfall measuring mission (TRMM) sensor package. J Atmos Ocean Tech 15:809–817

Kummerow C, Simpson J, Thiele O, Barnes W, Chang ATC, Stocker E, Adler RF, Hou A, Kakar R, Wentz F, Ashcroft P, Kozu T, Hong Y, Okamoto K, Iguchi T, Kuriowa K, Im E, Haddad Z, Huffman G, Ferrier B, Olson WS, Zipser E, Smith EA, Wilheit TT, North G, Krishnamurti T, Nakamura K (2000) The status of the Tropical Rainfall Measureing Mission (TRMM) after two years in orbit. J Appl Meteorol 39:1965–1982

Kummerow C, Hong Y, Olson WS, Yang S, Adler RF, McCollum J, Ferraro R, Petty G, Shin D-B, Wilheit TT (2001) The evolution of the Goddard profiling algorithm (GPROF) for rainfall estimation from passive microwave sensors. J Appl Meteorol 40:1801–1820

Levizzani V, Bauer P, Turk FJ (eds) (2007) Measuring precipitation from space: EURAINSAT and the future, vol 28. Springer, Netherlands, pp 722

Lin X, Hou AY (2007) Evaluation of coincident passive microwave rainfall estimates using TRMM PR and ground measurements as references. J Appl Meteorol (in review)

Liu G (2004) Approximation of single scattering properties of ice and snow particles for high microwave frequencies. J Atmos Sci 61:2441–2456

Marzano FS, Munai A, Panegrossi G, Pierdicca N, Smith EA, Turk J (1999) Bayesan estimation of precipitating cloud parameters from combined measurements of spaceborne microwave radiometer and radar. IEEE T Geosci Remote 37:596–613

Marzoug M, Amayenc P (1994) A class of single- and dual-frequency algorithms for rain-rate profiling from a spaceborne radar. Part I: Principle and tests from numerical simulations. J Atmos Ocean Tech 11:1480–1506

Masunaga H, Kummerow CD (2005) Combined radar and radiometer analysis of precipitation profiles for a parametric retrieval algorithm. J Atmos Ocean Tech 22:909–929

Meneghini R, Liao L (2000) Effective dielectric constants of mixed-phase hydrometeors. J Atmos Ocean Tech 17:628–640

Meneghini R, Iguchi T, Kozu T, Liao L, Okamoto K, Jones JA, Kwiatkowski J (2000) Use of the surface reference technique for path attenuation estimates from the TRMM Precipitation Radar. J Appl Meteorol 39: 2053–2070

Njoku EG (1982) Passive microwave remote sensing of the Earth from Space- A review. In: Proceedings of the IEEE 70:728–750

Njoku E, Christensen E, Cofield R (1980) The Seasat Scanning Multichannel Microwave Radiometer (SMMR): Antenna pattern corrections development and implementation. IEEE J Oceanic Eng 5:125–137

NRC (National Research Council) (2005) Assessment of the benefits of extending the tropical rainfall measuring mission, National Academies Press, Washington, DC, pp 103

NRC (National Research Council) (2007) Earth science and applications from space: National imperatives for the next decade and beyond. National Academies Press, Washington, DC

Olson WS, Bauer P, Kummerow CD, Hong Y, Tao WK (2001) A melting layer model for passive/active microwave remote sensing applications Part II: Simulation of TRMM observations. J Appl Meteorol 40:1164–1179

Petty GW, Krajewski WF (1996) Satellite estimation of precipitation over land. Hydrolog Sci J 41:433–451

Purcell EM, Pennypacker CR (1973) Scattering and absorption of light by nonspherical dielectric grains. Astrophys J 186:705–714

Shimoda H (2005) GCOM Missions. In: Proceedings IEEE International Geoscience and Remote Sensing Symposium. 25–29 July, Seoul, Korea, vol 6 pp 4201–4204

Simpson JR, Adler RF, North GR (1988) A proposed tropical rainfall measuring mission (TRMM) satellite. B Am Meteorol Soc 69:278–295

Skofronick-Jackson G (2004) Brightness temperature sensitivity to variations in solid precipitation cloud profiles (2004) In: Proceedings of IGARSS. vol 2, Anchorage, Alaska, pp 1378–1381

Skofronick-Jackson G, Kim M, Weinman JA, Chang D (2004) A physical model to determine snowfall over land by microwave radiometry. IEEE T Geosci Remote 42:1047–1058

Skou N, Le Vine D (2006) Microwave radiometer systems: Design and analysis, 2nd edn. Artech House, Norwood, MA

Smith EA, Xiang X, Mugnai A, Tripoli GJ (1994a) Design of an inversion-based precipitation profile retrieval algorithm using an explicit cloud model for initial guess microphysics. Meteorol Atmos Phys 54:53–78

Smith EA, Kummerow C, Mugnai A (1994b) The emergence of inversion-type profile algorithms for estimation of precipitation from satellite passive microwave measurements. Rem Sens Rev 11:211–242

Smith EA, Lamm JE, Adler R, Alishouse J, Aonashi K, Barrett EC, Bauer P, Berg W, Chang A, Ferraro R, Ferriday J, Goodman S, Grody N, Kidd C, Kniveton D, Kummerow C, Liu G, Mozano F, Mugnai A, Olson W, Petty G, Shibato A, Spencer R, Wentz F, Wilheit T, Zipser E (1998) Results of the WetNet PIP-2 project. J Atmos Sci 55:1483–1536

Spencer RW (1993) Global oceanic precipitation from the MSU during 1979–92 and comparisons to other climatologies. J Climate 6:1301–1326

Staelin DH (1981) Passive microwave techniques for geophysical sensing of the Earth from satellites. IEEE Trans Antennas Propag AP-29(4):683–687

Stephens GL, Kummerow C (2007) The remote sensing of clouds and precipitation from space: A review. J Atmos Sci 64:3742–3765

Stephens GL, Vane DG, Boain RJ, Mace GG, Sassen K, Wang Z, Illingworth AJ, O'Connor EJ, Rossow WB, Durden SL, Miller SD, Austin RT, Benedetti A, Mitrescu C, CloudSat Science Team (2002) The cloudsat mission and the a-train. B Am Meteorol Soc 83:1771–1790

Sun W, Fu Q, Chen ZZ (1999) Finite-difference time domain solution of light scattering by dielectric particles with a perfectly matched layer absorbing boundary condition. Appl Optics 38:3141–3151

Tao WK, Adler R, Braun S, Einaudi F, Ferrier B, Halverson J, Heymsfield G, Kummerow C, Negri A, Kakar R (2000) meeting summary: Summary of a symposium on cloud systems, hurricanes, and TRMM: Celebration of Dr. Joanne Simpson's career-The first fifty years. B Am Meteorol Soc 81:2463–2474

Ulaby FT, Moore RK, Fung AK (1981) Microwave Remote Sensing: Active and passive, vol I. Addison-Wesley publishing company, Reading, Massachusetts

Xu Y-L (1997) Electromagnetic scattering by an aggregate of spheres: Far field. Appl Optics 36:9496–9508

Yang P, Liou KN (1995) Light scattering by hexagonal ice crystals: Comparison of finite-difference time domain and geometric optics models. J Opt Soc Am A 12:162–176

7 Operational discrimination of raining from non-raining clouds in mid-latitudes using multispectral satellite data

Thomas Nauss[1], Boris Thies[1], Andreas Turek[1], Jörg Bendix[1], Alexander Kokhanovsky[2]

[1]Laboratory of Climatology and Remote Sensing, University of Marburg Germany
[2]Institute of Remote Sensing, University of Bremen, Germany

Table of contents

7.1 Introduction 171
7.2 Conceptual model for the discrimination of raining from non-raining mid-latitude cloud systems 172
7.3 Retrieval of the cloud properties using multispectral satellite data 173
7.4 Application of the conceptual model to Meteosat Second Generation SEVIRI data 175
    7.4.1 The daytime approach 175
    7.4.2 The night-time approach 178
7.5 Evaluation of the new rain area delineation scheme 183
    7.5.1 Evaluation study using daytime scenes 184
    7.5.2 Evaluation study using night-time scenes 186
7.6 Conclusions 188
References 190

7.1 Introduction

The detection of rainfall by geostationary (GEO) weather satellites has a long tradition as they provide area-wide information about the distribution of this key parameter of the water cycle in a very high temporal and high spatial resolution (e.g., Adler and Negri 1988). Most retrieval techniques developed so far for GEO systems are based on the relationship between cloud top temperature in the infrared channel and rainfall probability. Such retrievals which are often referred to as IR retrievals are appropriate for the tropics where precipitation is generally linked with deep convective clouds that can be easily identified in the infrared and/or water vapor channels (e.g., Levizzani et al. 2001; Levizzani 2003) but show considerable drawbacks in the mid-latitudes (e.g., Ebert et al. 2007; Früh et al. 2007) where great parts of the precipitation originates from clouds preferably formed by spatially extended frontal lifting processes in extra-tropical cyclones (hereafter denoted as advective/stratiform precipitation).

To overcome this drawback, some authors have suggested to use the effective cloud droplet radius (a_{ef}) defined as the ratio of the third to the second power of the cloud droplet spectrum (Hansen and Travis 1974) which can be retrieved from multispectral satellite data. They propose to use values of a_{ef} of around 14 µm as a fixed threshold value (THV) for precipitating clouds (e.g., Rosenfeld and Gutman 1994; Lensky and Rosenfeld 1997; Ba and Gruber 2001) but these studies have mainly focused on convective systems and a fixed THV seems to be not applicable for a reliable differentiation between frontal induced raining and non-raining stratiform clouds over large parts of Europe. In this context, Nauss and Kokhanovsky (2006, 2007) recently proposed a new scheme for the discrimination of raining and non-raining cloud areas applicable to mid-latitudes using daytime multispectral satellite data. Similarly, Thies et al. (2008) introduced a new technique for rain area delineation in the mid-latitudes using night-time multispectral satellite data. In the following sections, the conceptual model of this new approach as well as its application to geostationary MSG (Meteosat Second Generation) SEVIRI (Spinning Enhanced Visible and InfraRed Imager) data will be presented. Since the final technique is different for day- and night-time scenes, the two algorithms will be presented separately.

7.2 Conceptual model for the discrimination of raining from non-raining mid-latitude cloud systems

Due to the very homogenous spatial distribution of cloud-top temperature T for (warm) clouds with values of T differing not significantly between raining and non-raining regions, the advective/stratiform precipitating cloud area is generally underestimated or even not detected by some of the advanced infrared temperature threshold techniques like the Convective-Stratiform-Technique CST (Adler and Negri 1988) or the Enhanced Convective-Stratiform-Technique ECST (Reudenbach 2003). Therefore, the authors propose to use the cloud liquid water path to identify raining clouds in optical satellite data. This idea is based on the conceptual model that rainfall is favored by both, cloud droplets with sufficiently large diameters where terminal velocity can over-compensate updraft wind fields and a vertical cloud extent large enough to allow droplets to grow and preventing them from evaporating below the cloud base (which in turn has an influence on the required droplet size; see Lensky and Rosenfeld 2003a). Consequently, precipitating clouds in the new conceptual model must be characterized by a specific combination of droplet size and the cloud thickness, both large enough to form rain droplets. Since neither the droplet spectrum nor the geometrical thickness of a cloud can be computed from optical data without additional theoretical assumptions, the effective droplet radius (a_{ef}) and the cloud optical thickness (τ) is used as a proxy for the particle size and the cloud thickness. Multiplying both parameters according to:

$$\text{lwp} = \frac{2}{3} \cdot \tau \cdot a_{ef} \qquad (1)$$

one gets the liquid water path (lwp) which again is related to the rainfall probability of a cloud so that raining clouds can finally be characterized by a sufficiently large lwp.

The new proposed scheme shows an improvement in rain area delineation compared to existing techniques using only a threshold for cloud top infrared temperature especially for advective/stratiform precipitation clouds.

7.3 Retrieval of the cloud properties using multispectral satellite data

Values of lwp (i.e., values of a_{ef} and τ) can be retrieved on a pixel basis during daytime using a combination of two solar channels (e.g., Nakajima and Nakajima 1995; Kawamoto et al. 2001; Kokhanovsky et al. 2003; Kokhanovsky et al. 2005; Platnick et al. 2003; Nauss et al. 2005). This is due to the fact that the reflection of solar light by a cloud in a non-absorbing wavelength (i.e., a visible channel between 0.4 and 0.8 μm) is strongly correlated to the optical thickness, while the reflection of solar light in a slightly absorbing wavelength (i.e., a near-infrared channel between 1.6 and 3.9 µm) is mainly a function of the cloud effective droplet radius.

To proof the conceptual model presented above within an initial test study, Nauss and Kokhanovsky (2006, 2007) utilize the Semi-Analytical CloUd Retrieval Algorithm (SACURA, Kokhanovsky et al. 2003; Kokhanovsky et al. 2005; Nauss et al. 2005) to compute a_{ef}, τ and finally lwp using data from NASA's Terra-MODIS sensor (Moderate Resolution Imaging Spectroradiometer, http://modis.gsfc.nasa.gov/, last access 2007/07/30). SACURA is based on asymptotic solutions and exponential approximations of the radiative transfer theory valid for weakly absorbing media (Kokhanovsky and Rozanov 2003, 2004), which are applicable for cloud retrievals up to a wavelength of around 2.2 µm. For a single scattering albedo (ϖ_0) equal to one, the equations coincide with more general asymptotic formulae valid for all values of ϖ_0 (Germogenova 1963; van de Hulst 1980; King 1987) and differ only insignificantly from general equations as $\varpi_0 \rightarrow 1$. However, the exponential approximation provides much simpler final expressions, which can be used as a basis for a high-speed cloud retrieval algorithm necessary for near-real-time applications (Kokhanovsky et al. 2003). SACURA has been validated over sea and land surfaces against the commonly used but computer-time expensive look-up table approaches of the Japanese Space Agency JAXA (Nakajima and Nakajima 1995; Kawamoto et al. 2001) and the NASA MODIS cloud property product MOD06 (Platnick et al. 2003) showing good agreement for optically thick (e.g., raining) cloud systems (Nauss et al. 2005). However, as SACURA is only valid for water clouds it does not consider the ice phase which leads to inaccuracies concerning precipitating clouds in the mid-latitudes as efficient precipitation processes are mainly connected to the ice phase and the so-called Bergeron-Findeisen process (e.g., Houze 1993). Recently, Kokhanovsky and Nauss (2005, 2006) showed that a fast and accurate calculation of the effective cloud particle radius (a_{ef}) and the cloud optical thickness (τ) is possible for water and ice clouds by using again a non-absorbing visible and an absorbing near infrared channel (e.g., 0.8 µm and 1.6 µm).

Since the cloud microphysical and optical properties are strongly related to the reflection of solar light but not to the thermal emission of the cloud, there is no retrieval at hand that can explicitly compute a_{ef} and τ during night-time. Anyhow, several case studies have shown that implicit information about a_{ef} and τ is available in the emissive channels during night-time. Stone et al. (1990), Ou et al. (1993), González et al. (2002), Ou et al. (2002) and Hutchison et al. (2006), used a 3.7 μm channel and a 11 μm channel combination to infer microphysical and optical cloud properties. The studies of Inoue (1985), Wu (1984) and Baum et al. (1994), have shown that both the brightness temperature differences (ΔT) between a 3.7 μm channel and a 11 μm channel ($\Delta T_{3.7-11}$) and between a 11 μm channel and a 12 μm channel (ΔT_{11-12}) are sensitive to the cloud's microphysical and optical properties. Baum et al. (1994) stated that both brightness temperature differences used in combination provide more information regarding cloud properties than either ΔT alone. Ackerman et al. (1998a) and Huang et al. (2004) demonstrated the sensitivity of the ΔT between a 8.5 μm and 11 μm channel ($\Delta T_{8.5-10.8}$) and ΔT_{11-12} to values of a_{ef}. Lensky and Rosenfeld (2003a) utilized $\Delta T_{3.7-11}$ to check a passing criteria indicating the actual cloud geometrical depth and particle size combination is large enough for the pixel to be considered as precipitating.

7.4 Application of the conceptual model to Meteosat Second Generation SEVIRI data

With the availability of the SEVIRI sensor aboard the new European GEO system Meteosat Second Generation (Aminou 2002; Schmetz et al. 2002; Levizzani et al. 2001), a system is in orbit which provides a sufficient spectral resolution to infer information about the liquid water path and the ice water path (hereafter both referred to as cloud water path (cwp)) as well as about the cloud phase. Furthermore it offers a high temporal (15 min) and spatial (3 by 3 km at sub-satellite point) resolution necessary for a continuous area-wide monitoring of the rainfall distribution which is essential for nowcasting purposes. Therefore, the authors chose that system for implementing a new operational technique for the rain area delineation in mid-latitudes on a 15 min basis for daytime and night-time data.

7.4.1 The daytime approach

As stated in the previous Chapter, SACURA is only applicable to water clouds. Concerning the rain area delineation in the mid-latitudes this represents a shortcoming as effective precipitation processes in these regions are mainly connected to the ice phase and the so-called

Bergeron-Findeisen process. As a consequence, Kokhanovsky and Nauss (2006) have already presented the fast and accurate forward radiative transfer scheme CLOUD which enables the computation of the cloud properties for water and ice clouds using one non-absorbing and one absorbing band available on MSG SEVIRI. However, a fast inverse radiative transfer scheme is required for the operational retrieval of cloud properties which is currently under final evaluation. Because this scheme (called SLALOM) is not yet finally approved, the authors decided to use the original reflections of the 0.56–0.71 μm ($VIS_{0.6}$) and 1.5–1.78 μm ($NIR_{1.6}$) SEVIRI channels for this study, instead of computed values of a_{ef} and τ.

Information about the cloud phase are incorporated by means of $\Delta T_{8.7-10.8}$ and $\Delta T_{10.8-12.1}$ (refer to Strabala et al. 1994; Ackerman et al. 1998b). The differentiation is based on the observation that the increase of water particle absorption is greater between 11 and 12 μm than between 8 and 11 μm. The ice particle absorption increases more between 8 and 11 μm than between 11 and 12 μm (Strabala et al. 1994). Therefore, $\Delta T_{10.8-12.1}$ of water clouds are greater than $\Delta T_{8.7-10.8}$. On the other hand, $\Delta T_{8.7-10.8}$ of ice clouds are greater than coincident $\Delta T_{10.8-12.1}$.

To use the information about the cwp and the cloud phase for a proper detection of potentially precipitating cloud areas (i.e., a large enough cwp and ice particles in the upper part of the cloud) the rainfall confidence is calculated as a function of the value combinations of the four variables $VIS_{0.6}$, $NIR_{1.6}$, $\Delta T_{8.7-10.8}$ and $\Delta T_{10.8-12.1}$ (e.g., Bellon et al. 1980; Cheng et al. 1993; Kurino 1997; Nauss and Kokhanovsky 2007). The computation of the pixel based rainfall confidence is realized by a comparison of these combinations with ground-based radar data from the German Weather Service (DWD 2005) for daytime precipitation events from January to August 2004 (altogether 850 scenes).

Figure 1 shows the calculated rainfall confidence as a function of $VIS_{0.6}$ and $NIR_{1.6}$ (a), as well as a function of $\Delta T_{8.7-10.8}$ and $\Delta T_{10.8-12.1}$ (b). Equation (2) shows the calculation of the rainfall confidences as a function of two different variables.

$$\text{RainConf}(x_1, x_2) = \frac{N_{\text{Rain}}(x_1, x_2)}{N_{\text{Rain}}(x_1, x_2) + N_{\text{NoRain}}(x_1, x_2)}, \qquad (2)$$

where N_{Rain} and N_{NoRain} are the raining and the non-raining frequencies, respectively and x_1 and x_2 denote the channel or channel

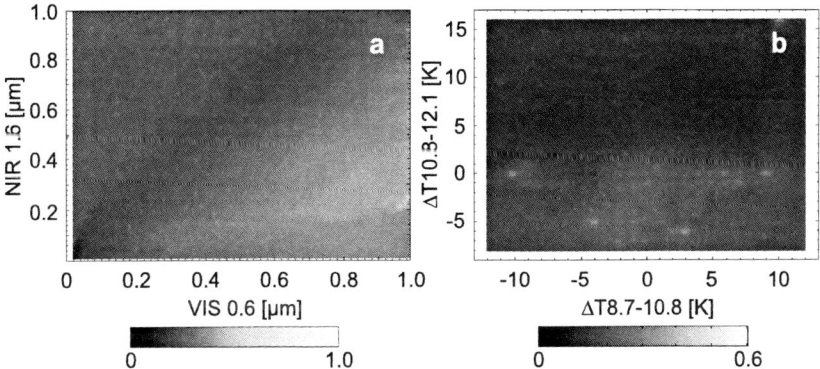

Fig. 1. The rainfall confidence as a function of $VIS_{0.6}$ and $NIR_{1.6}$ (**a**), as well as a function of $\Delta T_{8.7-10.8}$ and $\Delta T_{10.8-12.1}$ (**b**) calculated with Eq. (2)

difference ($VIS_{0.6}$, $NIR_{1.6}$, $\Delta T_{8.7-10.8}$, $\Delta T_{10.8-12.1}$) combined for the calculation of the rainfall confidence.

As can be seen in Fig. 1a high values of the rainfall confidence coincide with high values of $VIS_{0.6}$ and low values of $NIR_{1.6}$, indicating a large cwp. High values of $VIS_{0.6}$ indicate a high optical thickness and low values of $NIR_{1.6}$ indicate large cloud particles as the absorption increases with increasing particle size. Figure 1b indicates that ice clouds ($\Delta T_{8.7-10.8} > \Delta T_{10.8-12.1}$) possess high rainfall confidences and water clouds ($\Delta T_{8.7-10.8} < \Delta T_{10.8-12.1}$) are characterized by lower rainfall confidences.

To make use of the combined information content in each channel difference for rain delineation, the rainfall confidence is computed as a function of the combined values of the four variables as shown in Eq. (3) using the above mentioned 850 scenes:

$$\text{RainConf}(x_1, x_2, x_3, x_4) = \frac{N_{\text{Rain}}(x_1, x_2, x_3, x_4)}{N_{\text{Rain}}(x_1, x_2, x_3, x_4) + N_{\text{NoRain}}(x_1, x_2, x_3, x_4)}, \quad (3)$$

where N_{Rain} and N_{NoRain} are the raining and the non-raining frequencies, respectively and x_1, x_2, x_3 and x_4 denote the channel or channel difference ($VIS_{0.6}$, $NIR_{1.6}$, $\Delta T_{8.7-10.8}$, $\Delta T_{10.8-12.1}$) combined for the calculation of the rainfall confidence.

The threshold of the calculated rainfall confidence appropriate for rain area delineation is determined by optimizing the equitable threat score (ETS) which is based on the number of pixels that have been identified by the satellite (S) and radar (R) techniques as raining

(S_Y, R_Y) or non-raining (S_N, R_N). It indicates how well the classified rain pixels correspond to the rain pixels observed by the radar, also accounting for pixels correctly classified by chance ($S_Y R_{Y\,Random}$). Its value can range from −1/3 to 1 with the optimum value 1. The ETS is calculated according to

$$\text{ETS} = \frac{S_Y R_Y - S_Y R_{Y\,Random}}{S_Y R_Y + S_N R_Y + S_Y R_N - S_Y R_{Y\,Random}} \qquad (4)$$

with

$$S_Y R_{Y\,Random} = \frac{(S_Y R_Y + S_N R_Y) \cdot (S_Y R_Y + S_Y R_N)}{T_{SR}}, \qquad (5)$$

where T_{SR} denotes the total number of pixels.

Different rainfall confidence threshold values between 0.1 and 0.7 were used to delineate the satellite-based rain area. The ETS for the delineated rain areas based on the different rainfall confidence levels were calculated again in comparison with ground-based radar data. The delineated rain area using a rainfall confidence threshold of 0.34 yields to the optimized ETS of 0.24. Therefore, the rainfall confidence of 0.34 is chosen as the minimum threshold for precipitating clouds during daytime.

7.4.2 The night-time approach

As already mentioned above, no operational technique is currently at hand to compute the cloud water path based on the cloud emissions during night-time. However, based on the findings mentioned in Sect. 7.3, the brightness temperature differences between the following SEVIRI channel differences are considered to gain implicit information on the cloud water path as well as on the cloud phase to detect potentially precipitating cloud areas:

- $\Delta T_{3.9-10.8}$: ΔT between the 3.9 μm channel (3.48–4.36 μm) and the 10.8 μm channel (9.8–11.8 μm);
- $\Delta T_{3.9-7.3}$: ΔT between the 3.9 μm channel and the 7.3 μm channel (6.85–7.85 μm);
- $\Delta T_{8.7-10.8}$: ΔT between the 8.7 μm channel (8.3–9.1 μm) and the 10.8 μm channel;
- $\Delta T_{10.8-12.1}$: ΔT between the 10.8 μm channel and the 12.1 μm channel (11–13 μm).

Regarding $\Delta T_{3.9-10.8}$, a large cwp is the product of a large effective particle radius and a high optical thickness. Large particles have a higher emission in the 3.9 µm channel compared to smaller particles. This is due to the decreased absorption of smaller particles which reduces the cloud emissivity. As a result, the brightness temperature in the 3.9 µm channel is higher for larger particles. This dependence on particle size is much less distinct in the 10.8 µm channel. Therefore, $\Delta T_{3.9-10.8}$ is higher for larger particles. For optically thin clouds the emission in the 3.9 µm channel is less than in the 10.8 µm channel. As a result, the 3.9 µm transmittance is larger than the 10.8 µm transmittance, which implies a larger transmissivity of below-cloud radiance of the former wavelength (see Lensky and Rosenfeld 2003b). Thus, for optically thin clouds consisting of small or large particles (small or medium cwp), the brightness temperature of the 3.9 µm channel is larger than that of the 10.8 µm channel and $\Delta T_{3.9-10.8}$ reaches the highest values. Large particles together with a high optical thickness (large cwp) result in medium to high difference values but these differences are always lower than for optically thin clouds. Thick clouds with small particles (medium cwp) lead to small $\Delta T_{3.9-10.8}$.

In general, $\Delta T_{3.9-7.3}$ should show similar characteristics as $\Delta T_{3.9-10.8}$. Because of the diminishing effect of the water vapor absorption and emission in mid- to low tropospheric levels on the brightness temperature (BT) in the 7.3 µm channel ($BT_{7.3}$) (Schmetz et al. 2002), $\Delta T_{3.9-7.3}$ should be generally higher than $\Delta T_{3.9-10.8}$. Therefore, $\Delta T_{3.9-7.3}$ is expected to provide additional information about the cloud water path. For thin clouds with small or large particles, respectively (small or medium cwp), $BT_{3.9}$ is larger than $BT_{7.3}$ and $\Delta T_{3.9-7.3}$ reaches the highest values. Large particles together with a high optical thickness (high cwp) result in medium to high difference values which are lower than for optically thin clouds. Thick clouds with small particles (medium cwp) lead to small $\Delta T_{3.9-7.3}$.

Concerning $\Delta T_{8.7-10.8}$ cloud radiative properties in both channels are dependent upon the cloud particle size. Scattering processes and the dependence on particle size are stronger in the 8.7 µm channel relative to the 10.8 µm channel (Strabala et al. 1994). Therefore, for larger particles $\Delta T_{8.7-10.8}$ increases. The water vapor absorption in the 8.7 µm channel is higher relative to the 10.8 µm channel (Soden and Bretherton 1996; Schmetz et al. 2002). This is why $\Delta T_{8.7-10.8}$ is lower for low optical thicknesses. For higher optical thicknesses, $\Delta T_{8.7-10.8}$ increases. As a result, $\Delta T_{8.7-10.8}$ reaches high values for large effective particle radii and large optical thicknesses (large cwp). A low optical thickness in combination with small effective particle radii (small cwp) lead to minimum $\Delta T_{8.7-10.8}$. A low optical thickness with large particles (medium cwp) and a large optical thickness with small particles (medium cwp) result in medium values of $\Delta T_{8.7-10.8}$.

$\Delta T_{10.8-12.1}$ is positive at low optical thicknesses due to the increased water vapor absorption in the 12.1 µm channel relative to the 10.8 µm channel. For higher optical thicknesses $\Delta T_{10.8-12.1}$ decreases as the transmittance and the influence of water vapor emission from beneath diminish (Inoue 1987; Baum et al. 1994). Particle absorption differences at 10.8 µm and 12.1 µm decrease with increasing effective radius. An increase in particle size acts to decrease $\Delta T_{10.8-12.1}$ (Baum et al. 1994). As a result, $\Delta T_{10.8-12.1}$ reaches lowest values for large particles and large optical thicknesses (large cwp). Highest values for $\Delta T_{10.8-12.1}$ are characteristic for low optical thicknesses together with small effective particle radii (small cwp). Medium values for $\Delta T_{10.8-12.1}$ are reached for high optical thicknesses together with small effective particle radii (medium cwp) as well as for low optical thicknesses in combination with a large effective particle radius (medium cwp).

The effect of the cwp on the respective channel difference is summarized in Table 1. High $\Delta T_{3.9-10.8}$ together with high $\Delta T_{3.9-7.3}$ correspond to small cwp. Highest values for both ΔT correspond to medium cwp. Medium $\Delta T_{3.9-10.8}$ and $\Delta T_{3.9-7.3}$ are indicative for large cwp. Low $\Delta T_{3.9-10.8}$ and low $\Delta T_{3.9-7.3}$ are the result of a medium cwp. High (highest) $\Delta T_{3.9-10.8}$ together with low $\Delta T_{8.7-10.8}$ correspond to small (medium) cwp. Medium $\Delta T_{3.9-10.8}$ and high $\Delta T_{8.7-10.8}$ refer to large cwp. Low $\Delta T_{3.9-10.8}$ and medium $\Delta T_{8.7-10.8}$ are indicative for medium cwp. The same statements hold true for $\Delta T_{3.9-7.3}$ and $\Delta T_{8.7-10.8}$ in combination. High (highest) $\Delta T_{3.9-10.8}$ and high $\Delta T_{10.8-12.1}$ are the result of small (medium) cwp. Medium $\Delta T_{3.9-10.8}$ together with low to medium $\Delta T_{10.8-12.1}$ refer to large cwp. Low $\Delta T_{3.9-10.8}$ and medium $\Delta T_{10.8-12.1}$ correspond to medium cwp. The same features are characteristic for $\Delta T_{3.9-7.3}$ and $\Delta T_{10.8-12.1}$ in combination. Low $\Delta T_{8.7-10.8}$ together with high $\Delta T_{10.8-12.1}$ are indicative for small cwp. High $\Delta T_{8.7-10.8}$ and low $\Delta T_{10.8-12.1}$ correspond to large cwp. Medium $\Delta T_{8.7-10.8}$ and medium $\Delta T_{10.8-12.1}$ refer to medium cwp.

The rainfall confidence as a function of two different channel differences calculated with Eq. (2) is depicted in Fig. 2. The computation of the pixel based rainfall confidence is analogous to the daytime scheme and is done by a comparison of the SEVIRI channel differences with ground-based radar data for night-time precipitation events from January to August 2004 (altogether 709 scenes).

Table 1. The effect of the cloud water path (cwp) on the respective channel difference

	$\Delta T_{3.9-10.8}$	$\Delta T_{3.9-7.3}$	$\Delta T_{8.7-10.8}$	$\Delta T_{10.8-12.1}$
small cwp (small τ with small a_{ef})	high	high	low	high
medium cwp (small τ with large a_{ef})	highest	highest	medium	medium
medium cwp (large τ with small a_{ef})	low	low	medium	medium
large cwp (large τ with large a_{ef})	medium	medium	high	low

For the combination of $\Delta T_{3.9-10.8}$ with $\Delta T_{3.9-7.3}$ (Fig. 2a) high rainfall confidences can be found for small $\Delta T_{3.9-10.8}$ and small $\Delta T_{3.9-7.3}$ as well as for medium $\Delta T_{3.9-10.8}$ and medium $\Delta T_{3.9-7.3}$. These intervals coincide with those for medium to large cwp (refer to Table 1). Low rainfall confidences are characterized by high $\Delta T_{3.9-10.8}$ and high $\Delta T_{3.9-7.3}$ which correspond to low cwp (refer to Table 1).

Regarding the combination of $\Delta T_{3.9-10.8}$ with $\Delta T_{8.7-10.8}$ (Fig. 2b), high rainfall confidences are indicated for small $\Delta T_{3.9-10.8}$ and medium $\Delta T_{8.7-10.8}$ as well as for medium $\Delta T_{3.9-10.8}$ and large $\Delta T_{8.7-10.8}$. Both value intervals correspond to medium and large cwp (refer to Fig. 1b). Low rainfall confidences can be found for high $\Delta T_{3.9-10.8}$ and small $\Delta T_{8.7-10.8}$ which coincide with low cwp (refer to Table 1). Concerning the combination of $\Delta T_{3.9-10.8}$ and $\Delta T_{10.8-12.1}$ (Fig. 2c) high rainfall confidences are indicated for small $\Delta T_{3.9-10.8}$ and medium $\Delta T_{10.8-12.1}$ as well as for medium $\Delta T_{3.9-10.8}$ and small $\Delta T_{10.8-12.1}$ which coincide with medium to large cwp (refer to Table 1). Low rainfall confidences can be found for high $\Delta T_{3.9-10.8}$ and high $\Delta T_{10.8-12.1}$ which correspond to low cwp (refer to Table 1).

To summarize, it can be stated that intervals of the channel differences representative for high rainfall confidences correspond with the intervals indicative for medium to large cwp. This corroborates our conceptual model that clouds with a large enough cwp together with ice particles in the upper parts possess a high probability to produce precipitation.

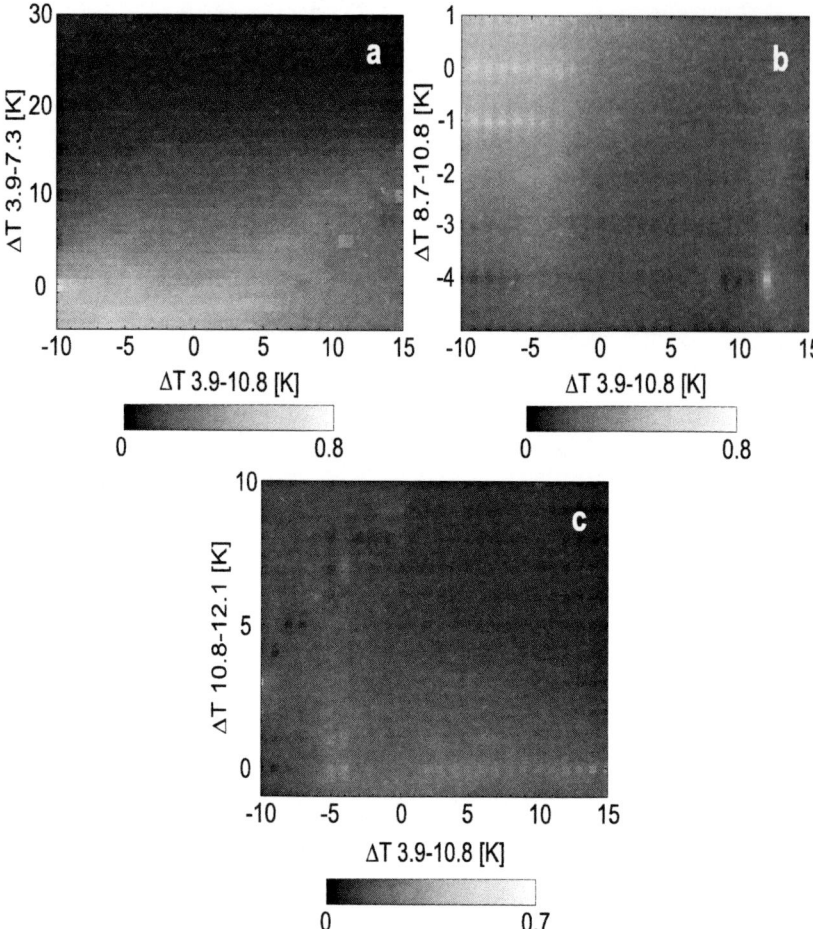

Fig. 2. The rainfall confidence as a function of $\Delta T_{3.9-10.8}$ versus $\Delta T_{3.9-7.3}$ (**a**), $\Delta T_{3.9-10.8}$ versus $\Delta T_{8.7-10.8}$ (**b**), $\Delta T_{3.9-10.8}$ versus $\Delta T_{10.8-12.1}$ (**c**) calculated with Eq. (2)

To make use of the combined information content in each channel difference for rain delineation, the rainfall confidence is computed as a function of the combined values of the four channel differences as shown in Eq. (3) using the above mentioned 709 scenes. The threshold of the rainfall confidence appropriate for rain area delineation is determined analogously to the daytime scheme by optimizing the ETS. The delineated rain area using a rainfall confidence threshold of 0.35 yields to the optimized ETS of 0.25. Therefore, the rainfall confidence of 0.35 is chosen as the minimum threshold for precipitating clouds during night-time.

7.5 Evaluation of the new rain area delineation scheme

In order to get an idea of the reliability of the new Rain Area Delineation Scheme for daytime and night-time (RADS-DN), 720 day- and 676 night-time precipitation scenes between January and August 2004 have been evaluated. The rainfall events within these scenes are independent of the above mentioned precipitation events used for the algorithm development.

To evaluate the potential improvement by the new RADS-DN, the validation scenes were also classified by the Enhanced Convective Stratiform Technique (ECST, Reudenbach 2003; Reudenbach et al. 2001) which is similar to the Convective Stratiform Technique (CST) of Adler and Negri (1988) but additionally includes the water vapour channel temperature for a more reliable deep convective/cirrus clouds discrimination (see also Tjemkes et al. 1997). The ECST which was first transferred from Meteosat-7 MVIRI (Meteosat Visible and InfraRed Imager radiometer) to MSG SEVIRI (Thies et al. 2007) is used for the identification of convective rain areas since these regions approximately represent the performance of many present IR rainfall retrievals.

Standard verification scores following the suggestions of the CGMS International Precipitation Working Group (IPWG, Turk and Bauer 2006) were calculated on a pixel basis for each scene in comparison with corresponding ground-based radar data from the German Weather Service (S_Y and R_Y represent the sum of pixels identified as raining in the satellite and radar product, respectively; S_N and R_N represent the sum of pixels identified as non-raining). Thereby, the bias describes the ratio between S_Y and R_Y, the probability of detection (POD) the ratio between $S_Y R_Y$ and the sum of $S_Y R_Y$ and $S_N R_Y$, the probability of false detection (POFD) the ratio between $S_Y R_N$ and the sum of $S_Y R_N$ and $S_N R_N$ and the false alarm ratio (FAR) the ratio between $S_Y R_N$ and the sum of $S_Y R_Y$ and $S_Y R_N$. The critical success index (CSI), which encloses all pixels that have been identified as raining by either the radar network or the satellite technique, describes the ratio between $S_Y R_Y$ and the sum of $S_Y R_Y$, $S_N R_Y$ and $S_Y R_N$. All scores except the bias range from 0 to 1 and the optimum value for the POD, CSI and bias is 1, while it is 0 for the POFD and FAR. Since the POD can be increased by just increasing the satellite rainfall area (i.e., by reducing the rainfall confidence threshold), it has to be analyzed in connection with corresponding values of the FAR and the POFD since both measure the fraction of the satellite pixels that have been incorrectly identified as raining. The verification scores were calculated on a pixel basis for each single scene without any spatio-temporal aggregation. For a detailed discussion of the verification scores, see Stanski et al. (1989) or the web site of the World Weather Research Program/Working Group on Numerical Experimentation Joint Working Group on Verification (see WWRP/WGNE 2007)

7.5.1 Evaluation study using daytime scenes

The verification scores calculated for the 720 daytime validation scenes are summarized in Table 2. RADS-D slightly overestimates the rain area detected by the radar network which is indicated by the bias of 1.15 (see Table 2). In contrast to this, the rain area is strongly underestimated by the ECST (bias of 0.22). Sixty-one percent of the radar observed raining pixels are also identified by RADS-D. This indicates a much better performance compared to the POD of 9% for the ECST, even if this coincide with a higher POFD of 0.18 for RADS-D in comparison to 0.04 for the ECST. Anyhow, the FAR indicates that a lower fraction of the pixels where misclassified as rain by RADS-D (0.46) than by the ECST (0.51). Altogether, the good performance of the new RADS-D is further supported by the CSI (0.39) and the ETS (0.25). Compared to ECST (CSI: 0.1; ETS: 0.06) this signifies a marked improvement concerning the delineated rain area.

An overview of the performance of RADS-D in comparison to the ECST is given by the relative operation characteristic (ROC) plot in Fig. 3. The visual impression additionally supports the good and improved performance of the new developed scheme. The combination of medium to high values for POD together with low to medium values for POFD which is valid for the main part of the classified scenes underlines the overall good skill of the new scheme. In contrast, for scenes classified by the ECST the POD and POFD indicate much lower or even no skills.

Table 2. Results of the standard verification scores applied to the rain-area identified by RADS-D and ECST on a pixel basis. The scores are based on 676 precipitation scenes with 24,914,160 pixels of which 5,872,220 have been identified as raining by RADS-D

Test	RADS-D				ECST			
	Mean	StDev	Min	Max	Mean	StDev	Min	Max
Bias	1.15	0.38	0.16	2.17	0.22	0.27	0.0	2.82
POD	0.61	0.21	0.12	0.98	0.12	0.17	0.0	0.97
POFD	0.18	0.09	0.02	0.54	0.04	0.05	0.0	0.78
FAR	0.46	0.12	0.03	0.84	0.51	0.27	0.0	1.00
CSI	0.39	0.14	0.1	0.77	0.10	0.14	0.0	0.64
ETS	0.25	0.11	−0.04	0.53	0.06	0.09	−0.05	0.39

To gain a visual impression of the performance of the new developed rain area delineation scheme, the classified rain area for a scene from 12 January 2004 12:45 UTC is depicted in Fig. 4. Figure 4a shows the brightness temperature in the 10.8 μm channel ($BT_{10.8}$), Fig. 4b the rain area delineated by RADS-D as well as by ECST and Fig. 4c the rain area detected by RADS-D in comparison to the radar data.

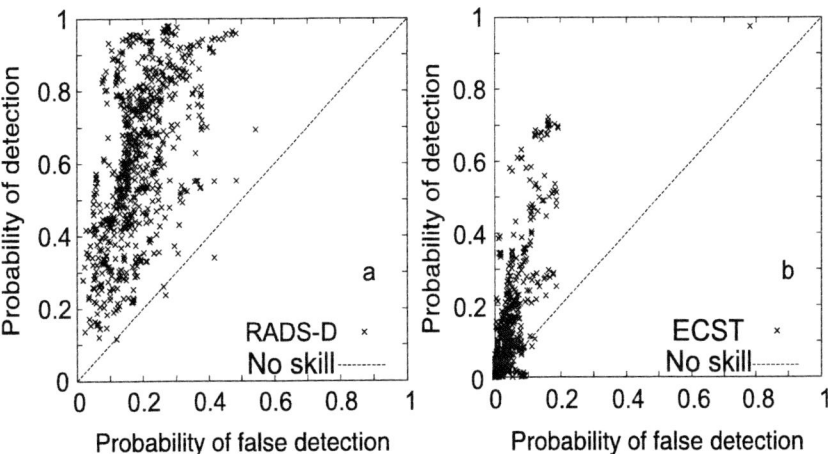

Fig. 3. ROC curves for the comparison between RADS-D and ground-based radar (**a**) and ECST and ground-based radar (**b**). The calculated probability of detection (POD) and probability of false detection (POFD) are based on the 720 scenes mentioned in the text

7.5.2 Evaluation study using night-time scenes

The verification scores calculated for the 676 night-time validation scenes are summarized in Table 3. Compared to the strong underestimation of the rain area by the ECST, RADS-N moderately overestimates the rain area detected by the radar network which is indicated by the bias of 0.21 for the ECST and of 1.4 for RADS-N (see Table 3). The POD shows that 68% of the radar observed raining pixels are also identified by RADS-N which points to a much better performance compared to 9% for the ECST, even if this coincide with a higher POFD of 0.24 for RADS-N in comparison to 0.04 for the ECST. However, the false alarm ratio shows that a lower fraction of the pixels where wrongly classified as rain by RADS-N (0.52) than by the ECST (0.57). The overall good performance of RADS-N, indicated by the good range of the verification scores is further supported by the CSI (0.37) and the ETS (0.22) which outperform the results of the ECST (CSI: 0.07; ETS: 0.03).

The relative operation characteristic (ROC) plot in Fig. 5 gives an overview of the performance of RADS-N in comparison to the ECST. It underlines again the good performance of the new developed scheme and the improvement in comparison to the ECST. For the main part of the classified scenes the POD and POFD indicate a good skill with medium to high values for POD together with low to medium values for POFD. In contrast, for scenes classified by the ECST the POD and POFD indicate much lower or even no skills.

Table 3. Results of the standard verification scores applied to the rain-area identified by RADS-N and ECST on a pixel basis. The scores are based on 676 precipitation scenes with 23,392,304 pixels of which 4,746,069 have been identified as raining by RADS-N

Test	RADS-N				ECST			
	Mean	StDev	Min	Max	Mean	StDev	Min	Max
Bias	1.42	0.67	0.16	4.97	0.21	0.36	0.0	4.28
POD	0.62	0.18	0.12	0.97	0.09	0.14	0.0	0.95
POFD	0.24	0.13	0.01	0.84	0.04	0.08	0.0	0.94
FAR	0.52	0.14	0.11	0.88	0.57	0.32	0.0	1.00
CSI	0.37	0.13	0.10	0.74	0.07	0.10	0.0	0.53
ETS	0.22	0.12	–0.03	0.57	0.03	0.06	–0.06	0.35

Chapter 7 - Discrimination of raining from non-raining clouds   187

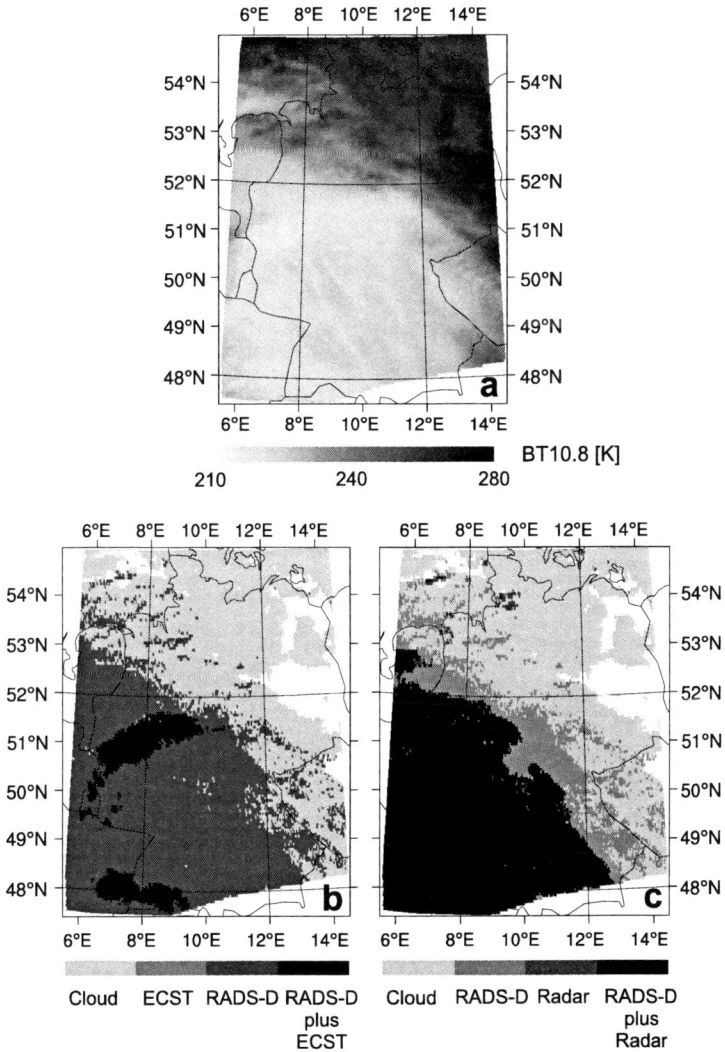

Fig. 4. Delineated rain area for the scene from 12 January 2004 12:45 UTC. Figure 4a shows the $BT_{10.8}$ image, Fig. 4b the rain area delineated by RADS-D as well as by ECST and Fig. 4c the rain area detected by RADS-D in comparison to the radar data

To gain a visual impression of the performance of the new developed rain area delineation scheme, the classified rain area for a scene from 31 May 2004 00:45 UTC is depicted in Fig. 6. Figure 6a shows the brightness temperature in the 10.8 μm channel ($BT_{10.8}$), Fig. 6b the rain area delineated by RADS-N as well as by ECST and Fig. 6c the rain area detected by RADS-N in comparison to the radar data.

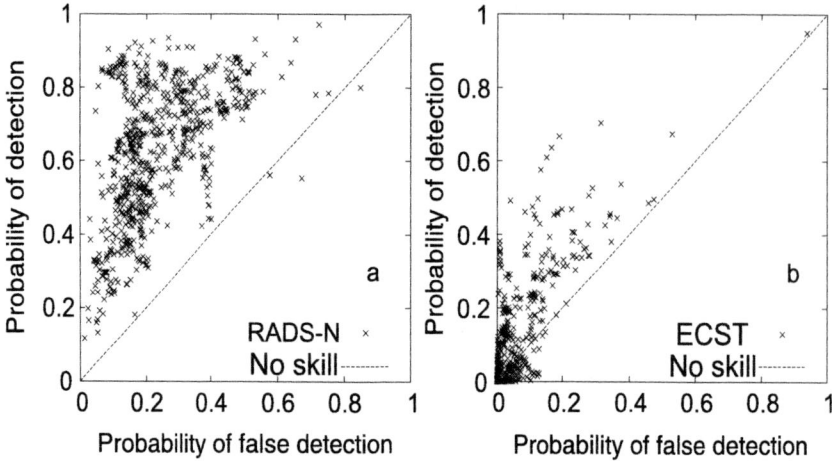

Fig. 5. ROC curves for the comparison between RADS-N and ground-based radar (**a**) and ECST and ground-based radar (**b**). The calculated probability of detection (POD) and probability of false detection (POFD) are based on the 676 scenes mentioned in the text

7.6 Conclusions

A new algorithm for rain area delineation during day- and night-time using multispectral optical and thermal IR satellite data of MSG SEVIRI was proposed. The method allows not only a proper detection of mainly convective precipitation by means of the commonly used connection between infrared cloud top temperature and rainfall probability but also enables the detection of advective/stratiform precipitation (e.g., in connection with mid-latitude frontal systems). It is based on the new conceptual model that precipitation is favored by a large cloud liquid or ice water path and the presence of ice particles in the upper part of the cloud.

The daytime technique considers the $VIS_{0.6}$ and the $NIR_{1.6}$ channel to gain information about the cloud water path. The night-time technique considers information about the cloud water path inherent in the channel differences $\Delta T_{3.9-10.8}$, $\Delta T_{3.9-7.3}$, $\Delta T_{8.7-10.8}$ and $\Delta T_{10.8-12.1}$. Additionally, both techniques utilize the channel differences $\Delta T_{8.7-10.8}$ and $\Delta T_{10.8-12.1}$ to gain information about the cloud phase.

Chapter 7 - Discrimination of raining from non-raining clouds   189

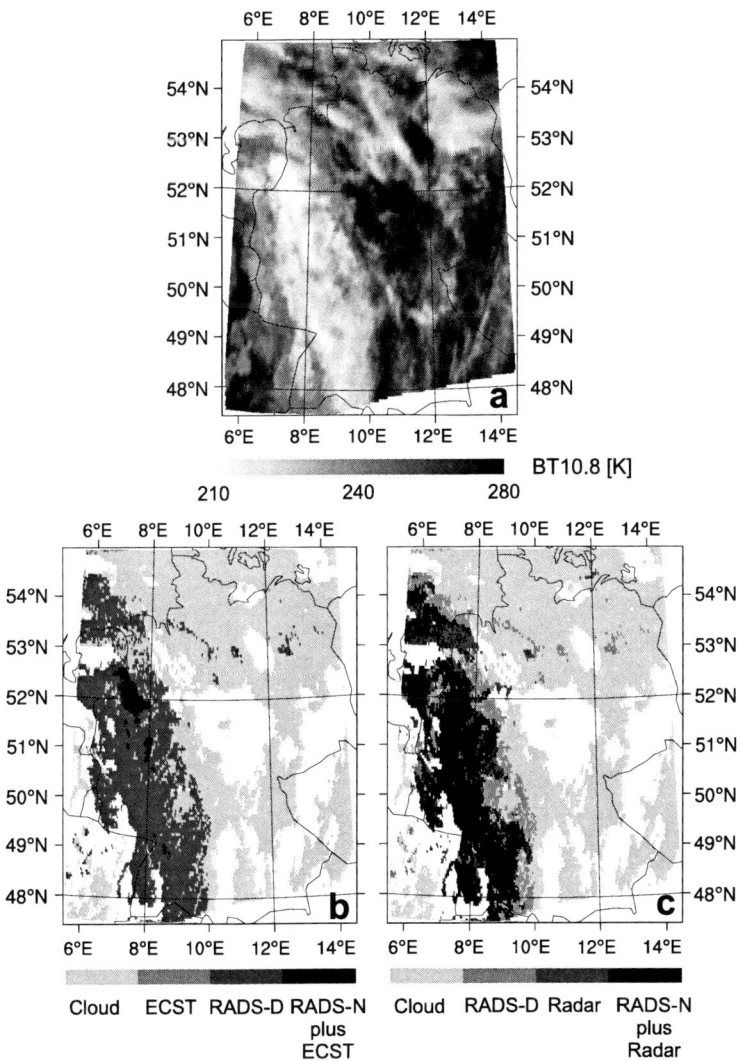

Fig. 6. Delineated rain area for the scene from 31 May 2004 00:45 UTC. Figure 6a shows the $BT_{10.8}$ image, Fig. 6b the rain area delineated by RADS-N as well as by ECST and Fig. 6c the rain area detected by RADS-N in comparison to the radar data

The information about the cwp and the cloud phase inherent in the four variables is merged and incorporated into the new developed rain area delineation algorithm. Rain area delineation is accomplished by using the pixel based rainfall confidence as a function of the respective value combination of the four variables. The calculation of the rainfall confidence is based on a comparison of the value combinations of the four variables with ground-based radar data. A minimum threshold for the rainfall confidence of 0.34 for the daytime scheme and of 0.35 for the night-time scheme was determined as appropriate for rain area delineation.

The results of the algorithm were compared with corresponding ground-based radar data. The proposed technique performs better than existing retrieval techniques using only IR thresholds for cloud top temperature.

The new developed algorithm shows encouraging performance concerning precipitation delineation during daytime and night-time in the mid-latitudes using MSG SEVIRI data and offers the great potential for a 24 h technique for rain area delineation with a high spatial and temporal resolution.

Acknowledgements

The current study is partly funded by the German Ministry of Research and Education (BMBF) in the framework of GLOWA-Danube project (G-D/2004/TP-10, precipitation/remote sensing) as well as by the German Research Council DFG (BE 1780/18-1) within the SORT project.

The authors are grateful to the German weather service (DWD) for providing the radar datasets within the Eumetsat/DWD Advanced Multisensor Precipitation Experiment (AMPE).

References

Ackerman SA, Moeller CC, Strabala KI, Gerber HE, Gumley LE, Menzel WP, Tsay SC (1998a) Retrieval of effective microphysical properties of clouds: A wave cloud case study. Geophys Res Lett 25:1121–1124

Ackerman SA, Strabala KI, Menzel WP, Frey RA, Moeller CC, Gumley LE (1998b) Discriminating clear sky from clouds with MODIS. J Geophys Res-Atmos 103:32141–32157

Adler RF, Negri AJ (1988) A satellite technique to estimate tropical convective and stratiform rainfall. J Appl Meteorol 27:30–51

Aminou DMA (2002) MSG's SEVIRI instrument. ESA Bulletin 111:15–17

Ba MB, Gruber A (2001) GOES Multispectral Rainfall Algorithm (GMSRA). J Appl Meteorol 40:1500–1514

Baum BA, Arduini RF, Wielicki BA, Minnis P, Tsay SC (1994) Multilevel cloud retrieval using multispectral HIRS and AVHRR data: Night-time oceanic analysis. J Geophys Res-Atmos 99:5499–5514

Bellon A, Lovejoy S, Austin GL (1980) Combining satellite and radar data for the short range forecasting of precipitation. Mon Weather Rev 108: 1554–1556

Cheng M, Brown R, Collier CG (1993) Delineation of precipitation areas by correlation of Meteosat visible and infrared data in the region of the United Kingdom. J Appl Meteorol 32:884–898

DWD (2005) Weather radar network. http://www.dwd.de/en/Technik/Datengewinnung/Radarverbund/Radarbroschuere_en.pdf (11 May 2007)

Ebert EE, Janowiak JE, Kidd C (2007) Comparison of near-real-time precipitation estimates from satellite observations and numerical models. B Am Meteorol Soc 88:47–64

Früh B, Bendix J, Nauss T, Paulat M, Pfeiffer A, Schipper JW, Thies B, Wernli H (2007) Verification of precipitation from regional climate simulations and remote-sensing observations with respect to ground-based observations in the upper Danube catchment. Meteorol Z 16:275–293

Germogenova TA (1963) Some formulas to solve the transfer equation in the plane layer problem. In: Stepanov BI (ed) Spectroscopy of Scattering Media. Academy of Sciences of BSSR, Minsk, pp 36–41

González A, Pérez JC, Herrera F, Rosa F, Wetzel MA, Borys RD, Lowenthal DH (2002) Stratocumulus properties retrieval method from NOAA-AVHRR data based on the discretization of cloud properties. Int J Remote Sens 23:627–645

Hansen JE, Travis LD (1974) Light scattering in planetary atmospheres. Space Sci Rev 16:527–610

Houze RA (1993) Cloud Dynamics. International Geophysics Series, Vol. 53. Academic Press, San Diego

Huang HL, Yang P, Wei HL, Baum BA, Hu YX, Antonelli P, Ackerman SA (2004) Inference of ice cloud properties from high spectral resolution infrared observations. IEEE T Geosci Remote 42:842–853

Hutchison K, Wong E, Ou SC (2006) Cloud base heights retrieved during night-time conditions with MODIS data. Int J Remote Sens 27:2847–2862

Inoue T (1985) On the temperature and effective emissivity determination of semi-transparent cirrus clouds by bi-spectral measurements in the 10 μm window region. J Meteorol Soc Jpn 63:88–98

Inoue T (1987) An instantaneous delineation of convective rainfall areas using split window data of NOAA-7 AVHRR. J Meteorol Soc Jpn 65:469–481

Kawamoto K, Nakajima T, Nakajima TY (2001) A global determination of cloud microphysics with AVHRR remote sensing. J Climate 14:2054–2068

King MD (1987) Determination of the scaled optical thickness of clouds from reflected solar radiation measurements. J Atmos Sci 44:1734–1751

Kokhanovsky AA, Nauss T (2005) Satellite-based retrieval of ice cloud properties using a semi-analytical algorithm. J Geophys Res-Atmos 110/D19:D19206, doi:10.1029/2004JD005744

Kokhanovsky AA, Nauss T (2006) Reflection and transmission of solar light by clouds: asymptotic theory. Atmos Chem Phys 6:5537–5545

Kokhanovsky AA, Rozanov VV (2003) The reflection function of optically thick weakly absorbing turbid layers: A simple approximation. J Quant Spectrosc Ra 77:165–175

Kokhanovsky AA, Rozanov VV (2004) The physical parameterization of the top-of-atmosphere reflection function for a cloudy atmosphere-underlying surface system: The oxygen A-band case study. J Quant Spectrosc Ra 85: 35–55

Kokhanovsky AA, Rozanov VV, Nauss T, Reudenbach C, Daniel JS, Miller HL, Burrows JP (2005) The semianalytical cloud retrieval algorithm for SCIAMACHY. I: The validation. Atmos Chem Phys 6:1905–1911

Kokhanovsky AA, Rozanov VV, Zege EP, Bovensmann H, Burrows JP (2003) A semi-analytical cloud retrieval algorithm using backscattered radiation in 0.4–2.4 micrometers spectral range. J Geophys Res-Atmos 108:4008, doi:10.1029/2001JD001543

Kurino T (1997) A satellite infrared technique for estimating 'deep/shallow' precipitation. Adv Space Res 19:511–514

Lensky IM, Rosenfeld D (1997) Estimation of precipitation area and rain intensity based on the microphysical properties retrieved from NOAA AVHRR data. J Appl Meteorol 36:234–242

Lensky IM, Rosenfeld D (2003a) A night-time delineation algorithm for infrared satellite data based on microphysical considerations. J Appl Meteorol 42:1218–1226

Lensky IM, Rosenfeld D (2003b) Satellite-based insights into precipitation formation processes in continental and maritime convective clouds at night-time. J Appl Meteorol 42:1227–1233

Levizzani V (2003) Satellite rainfall estimations: New perspectives for meteorology and climate from the EURAINSAT project. Ann Geophys-Italy 46:363–372

Levizzani V, Schmetz J, Lutz HJ, Kerkmann J, Alberoni PP, Cervino M (2001) Precipitation estimations from geostationary orbit and prospects for Meteosat Second Generation. Meteorol Appl 8:23–41

Nakajima TY, Nakajima T (1995) Wide-area determination of cloud microphysical properties from NOAA AVHRR measurements for FIRE and ASTEX regions. J Atmos Sci 52:4043–4059

Nauss T, Kokhanovsky AA (2006) Discriminating raining from non-raining clouds at mid-latitudes using multispectral satellite data. Atmos Chem Phys 6:5031–5036

Nauss T, Kokhanovsky AA (2007) Assignment of rainfall confidence values using multispectral satellite data at mid-latitudes: First results. Adv Geosci 10:99–102

Nauss T, Kokhanovsky AA, Nakajima TY, Reudenbach C, Bendix J (2005) The intercomparison of selected cloud retrieval algorithms. Atmos Res 78:46–78

Ou SC, Liou KN, Gooch WM, Takano Y (1993) Remote sensing of cirrus cloud parameters using advanced very-high-resolution radiometer 3.7- and 10.9-μm channels. Appl Optics 32:2171–2180

Ou SC, Liou KN, Takano Y, Higgins G, Larsen N, Slonaker R (2002) Cloud Effective Particle Size and Cloud Optical Thickness; Visible/Infrared Imager/Radiometer Suite. Algorithm Theoretical Basis Document

Platnick S, King MD, Ackerman SA, Menzel WP, Baum BA, Riédi JC, Frey RA (2003) The MODIS cloud products: Algorithms and examples from Terra. IEEE T Geosci Remote 41:459–473

Reudenbach C (2003) Konvektive Sommerniederschläge in Mitteleuropa. Eine Kombination aus Satellitenfernerkundung und numerischer Modellierung zur automatischen Erfassung mesoskaliger Niederschlagsfelder. Bonner Geographische Abhandlungen, p 109

Reudenbach C, Heinemann G, Heuel E, Bendix J, Winiger M (2001) Investigation of summertime convective rainfall in Western Europe based on a synergy of remote sensing data and numerical models. Meteorol Atmos Phys 76:23–41

Rosenfeld D, Gutman G (1994) Retrieving microphysical properties near the tops of potential rain clouds by multispectral analysis of AVHRR data. Atmos Res 34:259–283

Schmetz J, Pili P, Tjemkes S, Just D, Kerkmann J, Rota S, Ratier A (2002) An introduction to Meteosat Second Generation (MSG). B Am Meteorol Soc 83:977–992

Soden BJ, Bretherton FP (1996) Interpretation of TOVS water vapor radiances in terms of layer-average relative humidities: Method and climatology for the upper, middle and lower troposphere. J Geophys Res- Atmos 101:9333–9343

Stanski HR, Wilson L, Burrows W (1989) Survey of common verification methods in meteorology, World Weather Watch Technical Report No.8. WMO. Geneva. WMO/TD No. 358

Stone RS, Stephens GL, Plant CMR, Banks S (1990) The remote sensing of thin cirrus cloud using satellites, lidar and radiative transfer theory. J Appl Meteorol 29:353–366

Strabala KI, Ackerman SA, Menzel WP (1994) Cloud Properties Inferred from 8–12-μm Data. J Appl Meteorol 33:212–229

Thies B, Nauss T, Bendix J (2007) Detection of high rain clouds using water vapour emission – transition from Meteosat First (MVIRI) to Second Generation (SEVIRI). Adv Space Res (under revision)

Thies B, Nauss T, Bendix J (2008) Discriminating raining from non-raining cloud areas at mid-latitudes using Meteosat Second Generation SEVIRI night-time data. Meteorol Appl (in press)

Tjemkes SA, van de Berg L, Schmetz J (1997) Warm water vapour pixels over high clouds as observed by Meteosat. Contr Atmos Phys 70:15–21

Turk J, Bauer P (2006) The International Precipitation Working Group and its role in the improvement of quantitative precipitation measurements. B Am Meteorol Soc 87:643–647

Van de Hulst HC (1980) Multiple Light Scattering: Tables, Formulas and Applications. Academic Press, San Diego

Wu MC (1984) Radiation properties and emissivity parameterization of high level thin clouds. J Clim Appl Meteorol 23:1138–1147

WWRP/WGNE (2007) World Weather Research Program/Working Group on Numerical Experimentation Joint Working Group on Verification: Forecast Verification – Issues, Methods and FAQ. http://www.bom.gov.au/bmrc/wefor/staff/eee/verif/verif_web_page.html

8 Estimation of precipitation from space-based platforms

Itamar M. Lensky[1], Vincenzo Levizzani[2]

[1]Department of Geography and Environment, Bar-Ilan University, Ramat-Gan, Israel
[2]Institute of Atmospheric Sciences and Climate, National Research Council, Bologna, Italy

Table of contents

8.1	Introduction	195
8.2	Estimating rainfall from space	196
	8.2.1 VIS/IR	197
	8.2.2 Passive microwave	198
	8.2.3 Active sensors	200
	8.2.4 Blended techniques	202
8.3	Retrieval of precipitation formation processes using microphysical data	205
	8.3.1 Rain estimates using microphysical considerations	205
	8.3.2 Retrieval of precipitation formation processes	207
	8.3.3 Future developments	212
8.4	Abbreviation	212
References		213

8.1 Introduction

Measuring precipitation intensity from spaceborne sensors is a highly difficult problem whose solution is yet to be completely reached. While the problems of physical and space-time representativeness of ground-based measurements are to some extent typical, also of the space-based ones, spaceborne sensing adds a few more issues that need to be considered when trying to make a quantitative use of data. The indirect character of retrievals from ground-based radars, for example, is even more exacerbated from the satellite passive remote sensing perspective, which deals with radiation scattered or emitted from the clouds in the visible (VIS), infrared (IR) and passive microwave (PMW) spectral bands. These retrievals of precipitation characteristics and intensity from space became significantly less indirect when the Tropical Rainfall Measuring Mission (TRMM) was launched in 1997 (Kummerow et al. 1998, 2000) with the first radar for precipitation ever in space. It is not the scope of the present Chapter to go in depth into the field of rainfall measurements from space and the reader is referred to the book edited by Levizzani et al. (2007), which represents the most recent and perhaps complete overview of the state of the art in the field. Other notable reviews were compiled with a general perspective (Levizzani et al. 2001) and with a focus on over land applications (Petty 1995) and climatology (Kidd 2001). An historical perspective of the field is offered by Barrett and Martin (1981).

Note that the problem of measuring precipitation from space can be decomposed into more than one step to be necessarily tackled and solved before pretending to obtain any quantitative result: (1) assess the physical content of the radiance measurements with respect to cloud hydrometeor content and precipitation formation mechanisms, (2) identify cloud type in terms of precipitation content, and (3) delineate precipitation areas. A crude simplification could delimit the problem to two major aspects: (a) delimit rain areas, and (b) quantitatively estimate precipitation. In fact, before making any attempt to estimate the amount of rain falling from a particular cloud seen from a satellite sensor, we must first make sure that the cloud is indeed precipitating and this is far from being an easy task.

The first part of the Chapter will briefly examine the status of precipitation estimates from space using the available passive and active sensors and give a perspective on possible improvements from advances in sensor technology and better physical understanding of cloud vertical structure. The second part will discuss the potential of multispectral observations for improving the knowledge of the physical status of a cloud thus allowing a step forward in estimating rainfall.

8.2 Estimating rainfall from space

Rainfall estimation from spaceborne sensors has a relatively long history dating back to the 1970s when the first VIS/IR methods were conceived for an indirect retrieval of precipitation from geostationary

(GEO) and low-earth (LEO) orbit. Some of the early methods are still in use for global climatological applications, but the field has evolved very rapidly in recent times. This Section will give a brief account of the main types of rainfall retrievals from space using VIS/IR, PMW and active sensors. The emphasis will be on recent developments trying also to provide a glimpse of the trends in developing new sensor technology and algorithm concepts.

8.2.1 VIS/IR

Observations in VIS/NIR/IR spectral bands allow for measuring scattering or emission from cloud top or near cloud top (e.g., Rosenfeld et al. 2004) and thus they generally cannot be exploited to directly estimate precipitation intensities. However, such radiation measurements are very instrumental to assess cloud microphysical properties and show great potential for raincloud classification (e.g., Cattani et al. 2007) and for improving rainfall intensity retrieval as shown in the second part of the present Chapter. For the purpose of this book we will mention two classes of algorithms that target applications at the extremes of the space-time resolution scale, nowcasting and climate.

The original IR algorithm for the Geostationary Operational Environmental Satellite (GOES), named Auto-Estimator (AE, Vicente et al. 1998) computes convective rain rates from 10.7 μm brightness temperatures (BT) using a relationship derived from more than 6000 collocated radar-satellite pixels. It is adjusted for moisture and dynamic growth, orography, an equilibrium level for warm tops and parallax. Non-raining areas are identified using the spatial gradients in the temperature of the top and changes from the previous image. Then the adjustments are applied using ancillary input data. However, since the AE is highly dependent on radar data for the identification of cold cloud pixels, another version of the AE, the Hydro-Estimator (HE) was developed trying to expand its use in regions where radar/rain gauge data are not necessarily available (Scofield and Kuligowski 2007). The technique is now operational at the US National Weather Service (NWS) for the analysis of extreme precipitation events.

At the other extreme of the space-time scale is the GOES Precipitation Index (GPI, Arkin and Meisner 1987). The GPI technique estimates tropical rainfall using cloud-top temperature as the sole predictor. The estimation procedure is rather simple:

Precipitation (mm) = FRAC × RATE × TIME   (1)

where FRAC is the fractional coverage of IR pixels associated with BT < 235 K over a reasonably large domain (50 × 50 km^2 and larger), RATE = 3 mm h^{-1} and TIME the number of hours over which FRAC was compiled. Numerous studies have shown that the GPI yields useful results in the tropics and warm-season extratropics. The major advantage of the technique is that it is based on IR data which is available frequently over most areas of the globe from GEO and LEO satellites, while the obvious major weakness is that estimation of precipitation from cloud-top temperature is relatively far removed from the physics of precipitation generation processes. Monthly precipitation estimates for the 40N – 40S belt for the period January 1986 through the present are available from the National Oceanic and Atmospheric Administration (NOAA) Climate Prediction Center (CPC) ftp server (ftp://ftp.cpc.ncep.noaa.gov/precip/gpi) together with daily and pentad products.

It is worth mentioning another aspect of the research on the improvement of IR rain retrieval methods, i.e., the use of cloud-to-ground lightning detection as additional information on heavy rainfall areas. Grecu et al. (2000) have conducted an analysis of the correlation between satellite PMW and IR rainfall estimates and on the number of strikes in 'contiguous' areas with lightning, where the contiguity is defined as a function of the distance between strikes. They found that lightning data contain useful information for IR rainfall estimation resulting in a reduction of about 15% in the root-mean-square error of the estimates of rain volumes defined by convective areas associated with lightning.

8.2.2 Passive microwave

PMW frequencies have been used for rain retrieval for about 25 years and the techniques that have been developed and refined in time rely on the emission signal of rain drops over the ocean at frequencies at or below 37 GHz and the scattering signal of ice particles in the precipitation layer over land at frequencies at or above 85 GHz (Ferraro 2007; an overview of operational algorithms). Perhaps the most widespread algorithm for operational use is the one of Ferraro (1997) who introduced into the original NOAA algorithm an emission component over ocean to expand detection of the oceanic rainfall. Continuous improvement is being undertaken at NOAA including new sensors like the Advanced Microwave Scanning Radiometer (AMSR) leading into the next generation of PMW operational algorithms (McCollum and Ferraro 2003).

More complex rainfall retrieval strategies are based on the early work of Smith et al. (1992) and Mugnai et al. (1993) who laid the foundations of statistical-physical algorithms for rain retrieval using PMW data and cloud modeling. The most recent and widespread of these algorithms is the Goddard Profiling (GPROF) technique (Kummerow et al. 2001) whose recent improvements for TRMM are also applicable to other sensors. GPROF retrieves the instantaneous rainfall and the rainfall vertical structure using the response functions for different channels peaking at different depths within the raining column. There are, however, more independent variables within raining clouds than there are channels in the observing system and this requires additional assumptions or constraints. Radiative transfer calculations can be used to determine a BT vector, **Tb**, given a vertical distribution of hydrometeors represented by **R**. An inversion procedure, however, is needed to find the hydrometeor profile, **R**, given a vector **Tb**. The GPROF retrieval method uses the Bayes' theorem, which writes the probability of a particular profile **R**, given **Tb** as

$$\Pr(\mathbf{R} \mid \mathbf{Tb}) = \Pr(\mathbf{R}) \times \Pr(\mathbf{Tb} \mid \mathbf{R}) \qquad (2)$$

where $\Pr(\mathbf{R})$ is the probability with which a certain profile **R** will be observed and $\Pr(\mathbf{Tb} \mid \mathbf{R})$ is the probability of observing the BT vector, **Tb**, given a particular rain profile **R**. The first term on the right hand side of Eq. (2) is derived using cloud-resolving models (CRM).

However, most of the available PMW rainfall retrieval algorithms have been optimized for the corresponding satellite sensor and intercomparisons have shown that each algorithm has its own strengths and weaknesses related to the specific application it was designed for. None of them appears to be universally better than the other. Recently, Kummerow et al. (2007) have addressed the need for a transparent, parametric algorithm for ensuring uniform rainfall products across all available sensors. This is especially needed in view of the Global Precipitation Measurement (GPM) mission (Hou et al., Chap. 6 in this book), which will be composed by a constellation of different sensors with varying frequency ranges and scanning geometries. The need is for a parametric structure that will avoid the algorithm to be conceived for a specific sensor and specific frequencies. The community is now at work and Kummerow et al. (2007) have defined a framework for the non-raining simulations, the raining scene identification, the a priori database from CRM simulations, the retrieval and finally the error model, which is essential for applications.

Recent research on PMW sensor technology has brought about new developments in the direction of exploiting higher frequencies for rainfall retrievals. In particular, higher frequencies should help mitigating the effects of the poor knowledge of emission and scattering properties of land surfaces that cause substantial misclassifications when retrieving rain and snow over land. Staelin and Chen (2000) have conceived a new retrieval method based on simultaneous passive observations at 50–191 GHz from the Advanced Microwave Sounding Unit (AMSU) on the NOAA-15 satellite. Comparisons of the retrieved rain rates with NEXRAD data over the Continental United States have demonstrated the potential of such high-frequency channels for operational global retrievals. Chen and Staelin (2003) have extended the original algorithm to 17 channels of the Atmospheric Infrared Sounder/Advanced Microwave Sounding Unit/Humidity Sounder for Brazil (AIRS/AMSU/HSB) with a range reaching as high as 100 mm h^{-1}.

Note that, however, PMW-based retrieval methods, while being very much tested and strongly linked to the physics of precipitation formation, are still performing much better in heavy rain convective conditions over the ocean and often perform poorly in light rain. A problem for all algorithms is also represented by snow detection that is a fundamental chapter of research nowadays (e.g., Mugnai et al. 2007). One more problem is represented by the diffraction, which limits the ground resolution for a given satellite PMW antenna. PMW sensors are, consequently, only mounted at present on LEO satellites and this greatly limits the time resolution of observations. The advent of higher frequency channels opens up the possibility to board the next generation of PMW sensors on GEO orbit spacecrafts for the long awaited GEO PMW rainfall retrieval (e.g., the Geostationary Observatory for Microwave Atmospheric Sounding, GOMAS, described by Bizzarri et al. 2007).

8.2.3 Active sensors

The history of precipitation retrieval from space using active sensors started in November 1997 with the launch of TRMM, which hosts for the first time a Precipitation Radar (PR) at 13.8 GHz. Since then several algorithms have been conceived and retrieval techniques run operationally (e.g., Iguchi et al. 2000).

Apart from the high value of the precipitation products from the PR, it is safe to say that the PR has become a sort of 'truth' against which all other products are compared and evaluated. The obvious limitations of the sensor are (1) attenuation, (2) TRMM data limited to the 35S – 35N latitude belt, and (3) PR's relatively narrow swath (215 km). However, the groundbreaking characteristics of the instrument in terms of quality and resolution of the products are undisputed and data usage has grown in time quite steadily. Furuzawa and Nakamura (2005) have, for example, investigated the performance of the TRMM Microwave Imager (TMI) in rain retrieval algorithm using collocated PR estimations depending on storm height, demonstrating the value of data from radar in space to improve PMW rain retrieval. Figure 1 shows an example of rain retrieval over Africa using the PR compared with the wide swath retrieval of the PMW radiometer onboard the same satellite. Note the higher spatial resolution of the PR estimate.

The GPM core satellite will host onboard a next generation dual-wavelength precipitation radar (DPR) at 13.6 and 35.5 GHz, which is

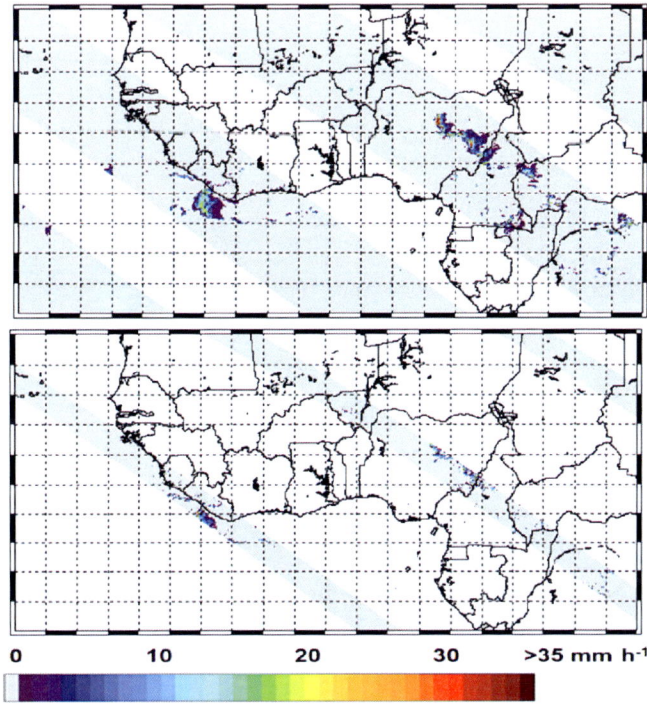

Fig. 1. 1 June 2006. Example of precipitation retrievals and sensor swaths from several TRMM orbits over Western Africa: TRMM Microwave Imager (TMI, *top*) and Precipitation Radar (PR, *bottom*), (courtesy of NASA and F. Torricella, ISAC-CNR, Bologna)

conceived with (1) a high sensitivity to detect light rain and snow, (2) capability to discriminate between liquid and solid precipitation, and (3) better accuracy in rain retrieval with respect to the PR (Nakamura and Iguchi 2007). The swath width would remain the same at 13.6 GHz while it would decrease to 100 km at 35.5 GHz. Note that the beam matching at the two radar channels is essential. The algorithms for liquid and solid precipitation retrieval of the DPR are currently a subject of intense research activity.

The launch of the CloudSat satellite on 28 April 2006 has started a new chapter of clouds and precipitation research from space. CloudSat is an essential centerpiece of the A-Train constellation (Stephens et al. 2002) designed to provide the vertical structure of clouds combining the CloudSat cloud profiling radar data (CPR, Im et al. 2005) at 94 GHz and radiance data obtained from the other sensors of the constellation. This is the first attempt to carefully retrieve the microphysics of cloud particles both liquid and solid using passive radiometers, lidar and radar. While the CPR is not specifically designed for rain retrieval itself, data analyzed after more than one year in orbit have shown that there exists a potential also for rain estimation at these frequency (L'Ecuyer et al. 2007).

8.2.4 Blended techniques

The wide variety of sensors in orbit suggests that their combined use could in principle help alleviate some of the deficiencies of a single-sensor method by using data obtained from another sensor. This kind of strategy is also instrumental in creating global rainfall datasets for which space-time coverage is crucial. Moreover, as pointed out by Stephens and Kummerow (2007), cloud and precipitation retrievals are most often constructed around very unrealistic layered atmosphere models. The retrievals thus become too sensitive to the unobserved parameters of those layers and the atmosphere above and below. A better definition of the atmospheric state and the vertical structure of clouds and precipitation are needed to improve the information extracted from satellite observations. This is why the combination of active and passive measurements offers much scope for improving cloud and precipitation retrievals.

There are several ways of combining passive and active sensor data in a final blended rainfall product, depending on the particular combination of IR, PMW, radar, gauge, lightning and wind data used. Levizzani et al. (2007) give a precise account of most of them. Here we will detail three of them to exemplify the concept of blending.

Turk et al. (1999) have proposed a blending method for real-time rainfall estimation at global scale, the method of the Naval Research Laboratory (NRL), which relates rain retrievals in the PMW to IR BTs from GEO satellites. The constantly evolving temporal and spatial characteristics of precipitation and its relation to satellite observations require that any statistical tuning or calibration to IR BT follow the rain characteristics. The method saves time- and space-coincident PMW and GEO IR data each time a Special Sensor Microwave Imager (SSM/I) or TRMM orbit pass intersects with any of the operational GEO satellites. Every three hours, an update cycle starts and locates the most recent 24 h of past coincident data. Separate histograms of IR BTs and the associated PMW-based rain rate are built in 15° × 15° global boxes. The SSM/I rain rate is computed via the NOAA-NESDIS (National Environmental Satellite, Data and Information Service) operational scheme (Ferraro 1997), which separates land- and ocean-based components, based upon a scattering index test. Figure 2 shows an example of successful rainfall retrieval for a hailstorm in 2006 in Villingen-Schenningen, Germany.

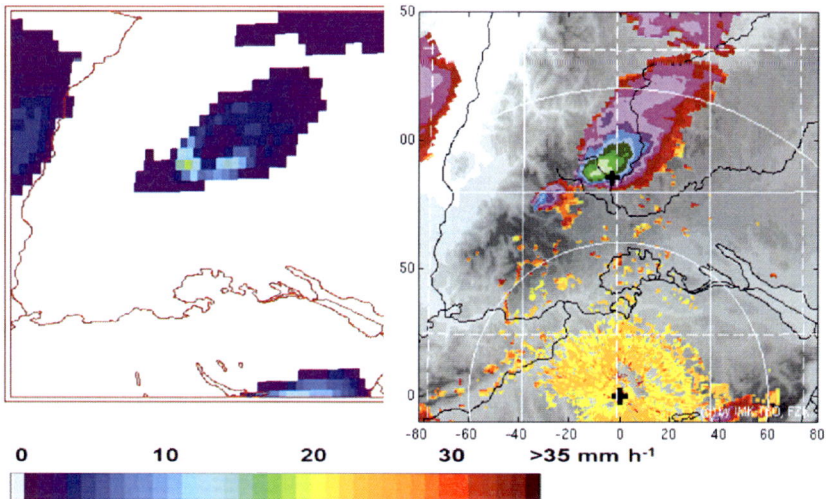

Fig. 2. 28 June 2006 17:20 UTC. Satellite rainfall retrieval using the NRL blended PMW-IR method (MSG-SEVIRI+SSM/I) (left) and radar reflectivity (right). The southern cross in the radar map represents the radar location in Albis, Germany, and the northern one that of the Villingen-Schenningen village where a substantial hail fall was registered, (courtesy of F. Torricella, ISAC-CNR, Bologna)

Huffman et al. (2007) have conceived the TRMM Multisatellite Precipitation Analysis (TMPA), which provides a calibration-based sequential scheme for combining precipitation estimates from multiple satellites, as well as gauge analyses at fine scales (0.25° × 0.25° and 3 hourly). The TMPA shows reasonable performance at monthly scales, while at finer scales it reproduces the surface observation–based histogram of precipitation, as well as reasonably detecting large daily events. Note that, however, TMPA has lower skill in correctly specifying moderate and light event amounts on short time intervals, in common with other fine scale estimators.

The Climate Prediction Center morphing method (CMORPH, Joyce et al. 2004) uses motion vectors derived from half-hourly interval GEO IR satellite imagery to propagate the relatively high quality precipitation estimates derived from PMW data for a global product. The shape and intensity of the precipitation features are modified (morphed) during the time between PMW sensor scans by performing a time-weighted linear interpolation. The process yields spatially and temporally complete PMW-derived precipitation analyses, independent of the IR BT field thus avoiding the problems generated by the validity of the PMW-IR histograms of the NRL method. Note that, however, CMORPH performs well if the PMW rainfall retrieval method performs well and this brings us back to the problem of improving the scores of PMW methods in all conditions and over all surfaces.

Climate and, in general, global applications require long and consistent datasets at daily, pentad and monthly time scales. The Global Precipitation Climatology Project (GPCP) offers global merged daily and monthly rain rates (Huffman et al. 1997; Adler et al. 2003) at 2.5° × 2.5° latitude-longitude from January 1979 to the present. It is a merged analysis that incorporates precipitation estimates from LEO satellite PMW data, GEO satellite IR data and surface rain gauge observations. The merging approach utilizes the higher accuracy of the LEO PMW observations to calibrate, or adjust, the more frequent GEO IR observations.

Xie and Arkin (1997) constructed a dataset, named the CPC Merged Analysis of Precipitation (CMAP), consisting of gridded fields (analyses) of global monthly precipitation on the same 2.5° × 2.5° grid from 1979 to 1995 by merging several kinds of information sources with different characteristics, including gauge observations, estimates inferred from a variety of satellite observations and the NCEP–NCAR (National Center for Environmental Predictions, National Center for Atmospheric Research) reanalysis. More recently, Xie et al. (2003)

constructed analyses of pentad precipitation over the GPCP global grid 1979 to 2001 by adjusting the pentad CMAP against the monthly GPCP-merged analyses. The adjustment was essential to align the two products in terms of input data sources and merging algorithms.

8.3 Retrieval of precipitation formation processes using microphysical data

In this Section, we will give a short overview on rain estimates and retrieval of precipitation formation processes using microphysical data in the VIS/IR. As it will be demonstrated, these retrievals are very instrumental for the improvement of rain estimation techniques.

8.3.1 Rain estimates using microphysical considerations

Remote sensing of cloud microphysics has begun more than two decades ago. Arking and Childs (1985) proposed a method to extract cloud cover parameters: cloud fraction within the field of view, optical thickness (τ), cloud top temperature and a microphysical model parameter, which was an index representing the properties of cloud particles (size, shape and thermodynamic phase). They used three NOAA Advanced Very High Resolution Radiometer (AVHRR) channels: the VIS channel at 0.65 µm, the NIR channel at 3.7 µm and the thermal IR channel at 11 µm. Shortly after, Pilewskie and Twomey (1987) used reflected solar radiation at several wavelengths in the NIR portion of the solar spectrum to discriminate cloud ice from water and pointed out that additional information about microphysics and rain processes near cloud top can be gained by using the radiative information at these wavelengths.

The use of cloud microphysics concepts to investigate precipitation forming processes from satellite started a decade later based on the notion that precipitation processes in clouds with warm tops are very sensitive to the cloud's microphysical structure. More specifically, precipitation processes are more efficient when water droplets and/or ice particles grow to larger sizes. This process cannot be detected by the cloud top temperature alone. Rosenfeld and Gutman (1994) retrieved properties of potentially precipitating cloud tops using NOAA AVHRR data by considering as candidate precipitating clouds only those optically thick in the VIS and filling the field of view, thus avoiding the complications of the effects of emitted and reflected radiation from below the clouds, which are important in semi-transparent or broken clouds (Rosenfeld et al. 2004). They showed that optically thick clouds with effective radius (r_e) greater than about 14 μm match well areas with radar echoes, indicating the existence of precipitation size particles. Their findings are in agreement with the notion that the existence of drops with radius of at least 12 μm is required for efficient precipitation formation in clouds with relatively warm tops via warm rain processes as well as ice multiplication processes.

A first attempt to estimate rain area and rain intensity using microphysical information during daytime was done by Lensky and Rosenfeld (1997). They used the effective radius of cloud particles derived from the AVHRR 3.7-μm window channel to detect warm rain clouds. In addition to the microphysical information, the fraction of rain cloud coverage and cloud spatial structure (convective and stratiform) were used for the rain estimation algorithm. Ba and Gruber (2001) applied this principle to the operational GOES Multispectral Rainfall Algorithm (GMSRA) using microphysical information only at daytime. They used the spatial gradient of cloud top temperature to screen non-raining cirrus clouds and r_e during daytime. During night-time, only cloud tops with BT < 230 K were considered for the screening. At daytime, all clouds having a visible reflectance greater than 40% were considered for the screening, using a r_e = 15 μm threshold for raining clouds. A rain rate was obtained by the product of probability of rain (Pb) and mean rain rate and adjusted by a moisture factor that was designed to modulate the evaporation effects on rain below cloud base for different moisture conditions.

Inoue and Aonashi (2000) compared cloud information from the TRMM Visible and InfraRed Scanner (VIRS) with rain detection and intensity as derived by the TRMM PR. Four radiative parameters were selected to describe the cloud: (1) the ratio of the reflected solar radiance at 0.6 and 1.6 μm (Ch1/Ch2) – higher for dense ice clouds, (2) the brightness temperature difference (BTD) between 11 and 12 μm – smaller for thick clouds, (3) the BTD between 3.7 and 11 μm (BTD34) – only at night-time, and (4) the BT at channel 4 (BT4). The parameters that displayed the highest skill score in the comparison for clouds with BT4 < 260 K were Ch1/Ch2 > 25 at daytime and BTD34 < 8 K at night-time. Lensky and Rosenfeld (2003a) used the BTD34 parameter to extract information on the microstructure and precipitation potential of clouds at night-time. They showed that the two factors that contribute to large BTD34s are the particle size at cloud top and the τ of the cloud layer, which appear to have contradictory effects on precipitation. Simulations with a radiative transfer model were conducted to weigh the respective contributions and results were compared with TRMM observations. Based on these findings, the authors developed a method to use the distribution of BTD34 with cloud top temperature for retrieving cloud microstructure and precipitation properties and implemented this method into a precipitation delineation algorithm (Lensky and Rosenfeld 2003b). The delineation algorithm performs well also in cases of warm clouds over land for which PMW algorithms fail.

Recently, Nauss and Kokhanovsky (2006) proposed a new method for the assignment of rainfall confidences on a pixel basis using cloud properties derived from optical satellite data during daytime. This approach is based on the conceptual model that precipitating clouds must have both a sufficient vertical extent and large enough droplets. They retrieved functions for the computation of an auto-adaptive threshold value of r_e with respect to the corresponding τ, which links these cloud properties with rainfall areas on a pixel basis. This approach enables the detection of stratiform precipitation (e.g., in connection with mid-latitude frontal systems). A first evaluation against ground-based radar data during March 2004 showed good skills (Nauss and Kokhanovsky 2007).

8.3.2 Retrieval of precipitation formation processes

In the previous Section, we saw that delineation algorithms based on microphysical considerations at daytime (Nauss and Kokhanovsky 2006) and night-time (Lensky and Rosenfeld 2003b) are based on the conceptual model that precipitating clouds must have both a sufficient vertical extent and large enough droplets to have terminal fall velocity of a few m s^{-1} so that they can reach the ground before evaporating.

Rosenfeld and Lensky (1998) took this approach one step forward. They investigated the evolution of r_e with temperature in convective cloud, inferring information about precipitation forming processes in the clouds. The Rosenfeld-Lensky Technique (RLT) takes as input r_e (Fig. 3b) and cloud top temperature (T) (Fig. 3a) of all cloudy pixels ($r_e > 0$) in a predefined area. The pixels are grouped into 1°C intervals and all r_e in each interval are sorted (Fig. 3c), then the median and other percentiles are used to build a T-r_e curve (Fig. 3d).

The retrieval of precipitation formation processes is done by analyzing the slope of the T-r_e curve and the values of T and r_e. Five microphysical zones are defined: diffusional growth, coalescence growth, rainout, mixed phase and glaciated. Not all five zones need to appear in a given cloud system. Application of the RLT to maritime clouds showed, from base to top, zones of coalescence, rainout, a shallow mixed-phase region and glaciation starting at −10°C or even warmer.

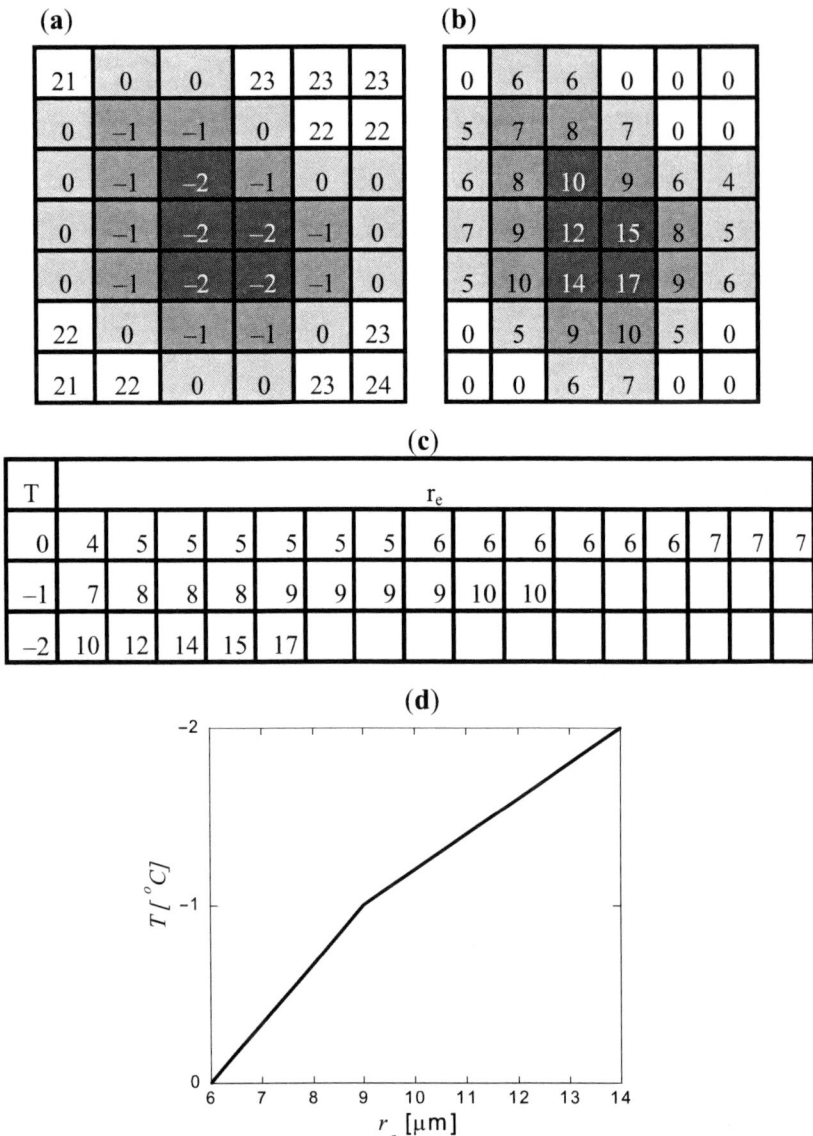

Fig. 3. The input for the RLT is (a) T cloud top temperature [°C] and (b) the effective radius [μm] r_e. (c) The pixels are grouped into 1°C intervals and all r_e in each interval are sorted; (d) the median (and other percentiles) are used to build a T-r_e curve

In contrast, continental clouds showed a deep diffusional growth zone above their bases, followed by coalescence and mixed phase zones and glaciation at −15° to −20°C. Highly continental clouds showed a narrow or no coalescence zone, a deep mixed-phase zone and glaciation occurring between −20° and −30°C. Substantial transformation in the microphysical and precipitation forming processes was observed by the RLT in convective clouds developing in air masses moving from the sea inland. These changes appear to be related to the modification of the maritime air mass as it moves inland and becomes more continental. Further transformations are observed in air masses moving into areas affected by biomass burning smoke or urban air pollution, such that coalescence and thus precipitation, is suppressed even in deep tropical clouds.

The RLT is based on two assumptions:
(a) The evolution of r_e with height (or T), observed by the satellite at a given time t_0 (snapshot) for a cloud ensemble over an area (C_1, C_2, C_3) is similar to the T-r_e time evolution (t_1, t_2, t_3) of a given cloud at one location (C_0). This is the ergodicity assumption, which means exchangeability between the time and space domains (Fig. 4a).
(b) The r_e near cloud top is similar to that well within the cloud at the same height as long as precipitation does not fall through that cloud volume (Fig. 4b).

The second assumption was verified using in situ aircraft measurements (Rosenfeld and Lensky 1998; Freud et al. 2005). To address the ergodicity assumption, Lensky and Rosenfeld (2006) used rapid scan data (imagery at three minute intervals) of the Spinning Enhanced Visible and InfraRed Imager (SEVIRI) on board Meteosat Second Generation (MSG) – the European geostationary satellite (Schmetz et al. 2002).

The RLT was widely accepted and was followed by numerous papers by Rosenfeld and co-authors. However, it seems that it is still hard for other researchers to use this technique. This situation may be a result of three shortcomings of the RLT. First, it does not give a large-scale view but rather describes the precipitation formation processes in a cloud cluster in a user-defined area. Second, the usage of the RLT demands some skills to obtain an informative T-r_e curve from an area that has to contain clouds in different stages of their life: from young through mature to dissipating. Especially the tops of the young clouds must be exposed to the sensor. Third, the user must know how to interpret the resulting T-r_e plot. Lensky and Drori (2007) proposed a method that overcomes these three shortcomings.

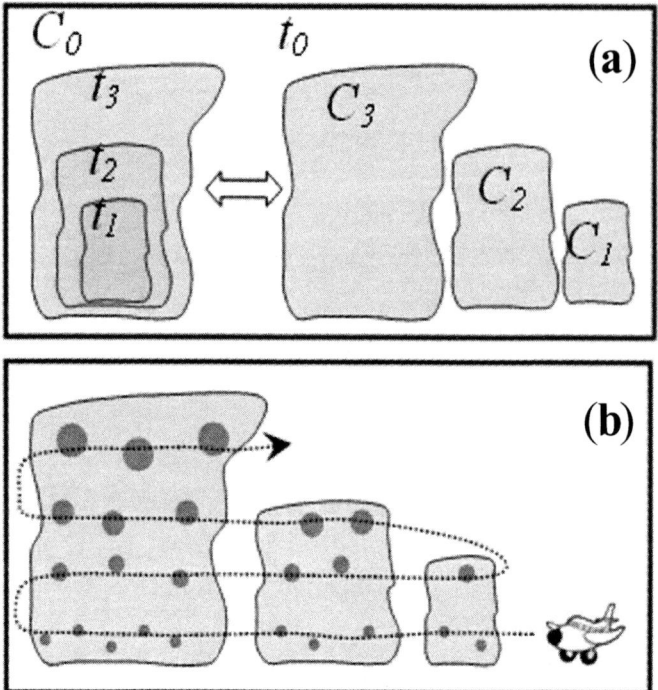

Fig. 4. Schematic representation of the two assumptions underlying the RLT. (**a**) The ergodicity assumption (exchangeability between time and space domains) says that the T-r_e observed by the satellite at a given time (t_0) for a cloud ensemble (C_1 C_2 C_3) over an area, is similar to the time evolution (t_1 t_2 t_3) of the T-r_e of a given cloud (C_0), at one location. (**b**) The r_e near cloud top is similar to that well within the cloud at the same height as long as precipitation does not fall through the cloud volume

This method analyzes all of the available satellite data (full swath) rather than the limited area of the RLT, it is objective (it does not require a skilful user) and the results are easy to interpret. Lensky and Drori (2007) followed the RLT approach and defined the temperature of the onset of precipitation (T_{15}), as the temperature where the median r_e exceeds a precipitation threshold of 15 μm (Lensky and Rosenfeld 1997; Ba and Gruber 2001) and D_{15} as the temperature difference (depth) between T_{15} and the cloud base temperature (T_{base}). They used D_{15} to monitor the nature and spatial extent of the impacts of forest fires and highly polluting mega-cities on cloud microstructure and precipitation. Larger D_{15} is a manifestation of the precipitation suppression effect of small cloud condensation nuclei (CCN) that act to increase the altitude where effective precipitation processes are initiated. Also, warmer land surface with greater sensible heat flux that increase the updraft velocity at cloud base can have the same effect. Therefore, the D_{15} is larger for clouds that develop over more polluted and/or warmer surfaces, due to smoke and urban pollution and/or urban heat island, respectively

The RLT was used also to investigate pyro-clouds (Andreae et al. 2004; Rosenfeld et al. 2007). Heavy smoke from forest fires in the Amazon was observed to reduce cloud droplet size and thus delay the onset of precipitation from 1.5 km above cloud base in pristine clouds to more than 5 km in polluted clouds and more than 7 km in pyro-clouds. Suppression of low-level rainout and aerosol washout allows transport of water and smoke to upper levels, where the clouds appear 'smoking' as they detrain much of the pollution. Elevating the onset of precipitation allows for an invigoration of the updrafts, causing intense thunderstorms, large hail and increased likelihood of overshooting cloud tops into the stratosphere.

Following the insights gained by the RLT, Lensky and Rosenfeld (2003b) looked for a method to retrieve information on precipitation formation processes at night-time. Night-time microphysical retrievals can be obtained to a much lower accuracy than those during daytime, because of lack of the solar radiation. However, the BTD between a thermal IR channel (11 µm) and a NIR channel (3.7 µm) contains information about the microstructure and precipitation potential of clouds at night-time. The BTD is very sensitive to particle size at cloud top and optical depth of the cloud layer. On daytime, the usage of the effective radius is relatively simple because large effective radii indicate that the particle size is large or that the phase of the particles is ice, both contributing to efficient precipitation formation processes. At night-time, the situation is more complex. Smaller BTDs of an optically thick cloud ($\tau_{0.55} > 10$) are related to smaller particle sizes and therefore less efficient precipitation processes. Larger BTDs are related to larger particle sizes or ice clouds and therefore more efficient precipitation processes. However, very large BTDs are related to semi-transparent clouds ($\tau_{0.55} \sim 2$), which are obviously not precipitating. Lensky and Rosenfeld (2003b) demonstrated these principles on a large TRMM dataset of precipitating and non-precipitating clouds according to the PR, in maritime and continental environments and applied these principles to case studies, using a similar approach as the RLT. A clear distinction of cloud microstructure and precipitation potential of clouds in specific areas residing in maritime and continental environments was shown.

8.3.3 Future developments

The T-r_e curve of the RLT is very informative as regards to the impacts of aerosols on cloud microstructure and precipitation forming processes and for the detection of severe weather events such as pyro-clouds. Intended seeding signatures are detectable at much smaller scale by the same satellite technology. Rosenfeld (2007) claims that they are different manifestations of the same sensitivity of precipitation- forming processes to the role of aerosols in the rate of conversion of cloud droplets into precipitation and the dynamic response of the clouds, which result in changes of the amount and distribution of precipitation.

Lensky and Drori (2007) suggested the D_{15} parameter as candidate for a large scale inspection of the impact of aerosols on clouds. A fully automated operative algorithm that will give short prediction for severe storms is surely the next stage needed in this direction.

Another development is the CLAIM-3D (3-Dimensional Cloud Aerosol Interaction Mission) satellite concept by Martins et al. (2007) who propose to combine several techniques to simultaneously measure the vertical profile of cloud microphysics, thermodynamic phase, brightness temperature and aerosol amount and type in the neighborhood of the clouds. The wide wavelength range and the use of multi-angle polarization measurements would allow estimating the availability and characteristics of aerosol particles acting as CCN and their effects on the cloud microphysical structure. These results would provide unprecedented details on the response of cloud droplet microphysics to natural and anthropogenic aerosols in the size scale where the interaction really happens.

8.4   Abbreviation

AE	Auto-Estimator
AIRS	Atmospheric Infrared Sounder
AMSR	Advanced Microwave Scanning Radiometer
AMSU	Advanced Microwave Sounding Unit
AVHRR	Advanced Very High Resolution Radiometer
BT	Brightness Temperature
BTD	Brightness Temperature Difference
CCN	Cloud Condensation Nuclei
CLAIM-3D	3-Dimensional Cloud Aerosol Interaction Mission
CMAP	CPC Merged Analysis of Precipitation
CMORPH	CPC Morphing method

CPC	Climate Prediction Center (NOAA)
CPR	Cloud Profiling Radar (CloudSat)
CRM	Cloud-Resolving Model
DPR	Dual-wavelength Precipitation Radar
GEO	Geostationary Orbit
GMSRA	GOES Multispectral Rainfall Algorithm
GOES	Geostationary Operational Environmental Satellite
GOMAS	Geostationary Observatory for Microwave Atmospheric Sounding
GPCP	Global Precipitation Climatology Project
GPI	GOES Precipitation Index
GPM	Global Precipitation Measurement mission
GPROF	Goddard Profiling algorithm
HE	Hydro-Estimator
HSB	Humidity Sounder for Brazil
IR	Infrared
LEO	Low Earth Orbit
MSG	Meteosat Second Generation
NCAR	National Center for Atmospheric Research
NCEP	National Center for Environmental Predictions
NESDIS	National Environmental Satellite, Data and Information Service (NOAA)
NIR	Near Infrared
NOAA	National Oceanic and Atmospheric Administration
NRL	Naval Research Laboratory
NWS	National Weather Service
PMW	Passive Microwave
PR	Precipitation Radar (TRMM)
RLT	Rosenfeld-Lensky Technique
SEVIRI	Spinning Enhanced Visible and InfraRed Imager
SSM/I	Special Sensor Microwave Imager
TMI	TRMM Microwave Imager
TRMM	Tropical Rainfall Measuring Mission
VIRS	Visible and InfraRed Scanner (TRMM)
VIS	Visible

References

Adler RF, Huffman GJ, Chang A, Ferraro R, Xie P, Janowiak J, Rudolf B, Schneider U, Curtis S, Bolvin D, Gruber A, Susskind J, Arkin P, Nelkin E (2003) The version-2 Global Precipitation Climatology Project (GPCP) monthly precipitation analysis (1979–present). J Hydrometeor 4:1147–1167

Andreae MO, Rosenfeld D, Artaxo P, Costa AA, Frank GP, Longo KM, Silva-Dias MAF (2004) Smoking rain clouds over the Amazon. Science 303:1337–1342
Arking A, Childs JD (1985) Retrieval of cloud cover parameters from multispectral satellite images. J Appl Meteor 24:322–333
Arkin PA, Meisner BN (1987) The relationship between large-scale convective rainfall and cold cloud over the Western Hemisphere during 1982–84. Mon Weather Rev 115:51–74
Ba MB, Gruber A (2001) GOES Multispectral Rainfall Algorithm (GMSRA). J Appl Meteor 40:1500–1514
Barrett EC, Martin DW (1981) The use of satellite data in rainfall monitoring. Academic Press, p 340
Bizzarri B, Gasiewski AJ, Staelin DH (2007) Observing rain by millimeter-submillimetre wave sounding from geostationary orbit. In: Levizzani V, Bauer P, Turk FJ (eds) Measuring precipitation from space. Springer, pp 675–692
Cattani E, Torricella F, Levizzani V (2007) Rain areas delineation combining MW and VIS-NIR-IR instruments onboard TRMM. In: Proceedings 3rd IPWG Workshop, 23–27 October 2006, Melbourne [http://www.isac.cnr.it/~ipwg/]
Chen FW, Staelin DH (2003) AIRS/AMSU/HSB precipitation estimates. IEEE T Geosci Remote 41:410–417
Ferraro RR (1997) SSM/I derived global rainfall estimates for climatological applications. J Geophys Res 102:715–735
Ferraro RR (2007) Present and future of microwave operational rainfall algorithms. In: Levizzani V, Bauer P, Turk FJ (eds) Measuring precipitation from space. Springer, pp 189–198
Freud E, Rosenfeld D, Andreae MO, Costa AA, Artaxo P (2005) Robust relations between CCN and the vertical evolution of cloud drop size distribution in deep convective clouds. Atmos Chem Phys Discuss 5:10155–10195
Furuzawa FA, Nakamura K (2005) Differences of rainfall estimates over land by Tropical Rainfall Measuring Mission (TRMM) Precipitation Radar (PR) and TRMM Microwave Imager (TMI)—Dependence on storm height. J Appl Meteor 44:367–382
Grecu M, Anagnostou EN, Adler RF (2000) Assessment of the use of lightning information in satellite infrared rainfall estimation. J Hydrometeor 1:211–221
Huffman GJ, Adler RF, Arkin P, Chang A, Ferraro R, Gruber A, Janowiak J, McNab A, Rudolf B, Schneider U (1997) The Global Precipitation Climatology Project (GPCP) combined precipitation dataset. B Am Meteorol Soc 78:5–20
Huffman GJ, Adler RF, Bolvin DT, Gu G, Nelkin EJ, Bowman KP, Hong Y, Stocker EF, Wolff DB (2007) The TRMM Multisatellite Precipitation Analysis (TMPA): Quasi-global, multiyear, combined-sensor precipitation estimates at fine scales. J Hydrometeorol 8:38–55

Iguchi T, Kozu T, Meneghini R, Awaka J, Okamoto K (2000) Rain-profiling algorithm for the TRMM precipitation radar. J Appl Meteor 39:2038–2052

Im E, Wu C, Durden SL (2005) Cloud profiling radar for the CloudSat Mission. IEEE Aero El Sys Mag 20:15–18

Inoue T, Aonashi K (2000) A comparison of cloud and rainfall information from instantaneous visible and infrared scanner and precipitation radar observations over a frontal zone in East Asia during June 1998. J Appl Meteor 39:2292–2301

Joyce RJ, Janowiak JE, Arkin PA, Xie P (2004) CMORPH: A method that produces global precipitation estimates from passive microwave and infrared data at high spatial and temporal resolution. J Hydrometeor 5:487–503

Kidd C (2001) Satellite rainfall climatology: A review. Int J Climatol 21:1041–1066

Kummerow CD, Barnes W, Kozu T, Shiue J, Simpson J (1998) The Tropical Rainfall Measuring Mission (TRMM) sensor package. J Atmos Oceanic Technol 15:809–817

Kummerow CD, Simpson J, Thiele O, Barnes W, Chang ATC, Stocker E, Adler RF, Hou A, Kakar R, Wentz F, Ashcroft P, Kozu T, Hong Y, Okamoto K, Iguchi T, Kuroiwa H, Im E, Haddad Z, Huffman G, Ferrier B, Olson WS, Zipser E, Smith EA, Wilheit TT, North G, Krishnamurti T, Nakamura K (2000) The status of the Tropical Rainfall Measuring Mission (TRMM) after two years in orbit. J Appl Meteor 39:1965–1982

Kummerow CD, Hong Y, Olson WS, Yang S, Adler RF, McCollum J, Ferraro R, Petty G, Shin D-B, Wilheit TT (2001) The evolution of the Goddard Profiling Algorithm (GPROF) for rainfall estimation from passive microwave sensors. J Appl Meteor 40:1801–1820

Kummerow CD, Masunaga H, Bauer P (2007) A next-generation microwave rainfall retrieval algorithm for use by TRMM and GPM. In: Levizzani V, Bauer P, Turk FJ (eds) Measuring precipitation from space. Springer, pp 235–252

L'Ecuyer T, Miller S, Mitrescu C, Haynes J, Kummerow C, Turk FJ (2007) A first look at the CloudSat precipitation dataset. In: Proceedings 3rd IPWG Workshop. 23–27 October 2006, Melbourne [http://www.isac.cnr.it/~ipwg/]

Lensky IM, Drori R (2007) A satellite based parameter to monitor the aerosol impact on convective clouds. J Appl Meteor Clim 45:660–666

Lensky IM, Rosenfeld D (1997) Estimation of precipitation area and rain intensity based on the microphysical properties retrieved from NOAA AVHRR data. J Appl Meteor 36:234–242

Lensky IM, Rosenfeld D (2003a) A night-rain delineation algorithm for infrared satellite data based on microphysical considerations. J Appl Meteor 42:1218–1226

Lensky IM, Rosenfeld D (2003b) Satellite-based insights into precipitation formation processes in continental and maritime convective clouds at nighttime. J Appl Meteor 42:1227–1233

Lensky IM, Rosenfeld D (2006) The time-space exchangeability of satellite retrieved relations between cloud top temperature and particle effective radius. Atmos Chem Phys 6:2887–2894

Levizzani V, Schmetz J, Lutz HJ, Kerkmann J, Alberoni PP, Cervino M (2001) Precipitation estimations from geostationary orbit and prospects for METEOSAT Second Generation. Meteor Appl 8:23–41

Levizzani V, Bauer P, Turk FJ (eds) (2007) Measuring Precipitation from Space. Springer, Dordrecht, 722 pp

Martins JV, Marshak A, Remer LA, Rosenfeld D, Kaufman YJ, Fernandez-Borda R, Koren I, Zubko V, Artaxo P (2007) Remote sensing the vertical profile of cloud droplet effective radius, thermodynamic phase, and temperature. Atmos Chem Phys Discuss 7:4481–4519

McCollum JR, Ferraro RR (2003) The next generation of NOAA/NESDIS SSM/I, TMI and AMSR-E microwave land rainfall algorithms. J Geophys Res 108:8382–8404

Mugnai A, Smith EA, Tripoli GJ (1993) Foundations for statistical-physical precipitation retrieval from passive microwave satellite measurements. Part II: Emission-source and generalized weighting-function properties of a time-dependent cloud-radiation model. J Appl Meteor 32:17–39

Mugnai A, Di Michele S, Smith EA, Baordo F, Bauer P, Bizzarri B, Joe P, Kidd C, Marzano FS, Tassa A, Testud J, Tripoli GJ (2007) Snowfall measurements by proposed European GPM mission. In: Levizzani V, Bauer P, Turk FJ (eds) Measuring precipitation from space. Springer, pp 655–674

Nakamura K, Iguchi T (2007) Dual-wavelength radar algorithm. In: Levizzani V, Bauer P, Turk FJ (eds) Measuring precipitation from space. Springer, pp 225–234

Nauss T, Kokhanovsky AA (2006) Discriminating raining from non-raining clouds at mid-latitudes using multispectral satellite data. Atmos Chem Phys 6:5031–5036

Nauss T, Kokhanovsky AA (2007) Assignment of rainfall confidence values using multispectral satellite data at mid-latitudes: First results. Adv Geosci 10:99–102

Petty GW (1995) The status of satellite-based rainfall estimation over land. Remote Sens Environ 51:125–137

Pilewskie P, Twomey S (1987) Discrimination of ice from water in clouds by optical remote sensing. Atmos Res 21:113–122

Rosenfeld D (2007) New insights to cloud seeding for enhancing precipitation and for hail suppression. J Weather Modification 39:61–69

Rosenfeld D, Gutman G (1994) Retrieving microphysical properties near the tops of potential rain clouds by multispectral analysis of AVHRR data. Atmos Res 34:259–283

Rosenfeld D, Lensky IM (1998) Satellite-based insights into precipitation formation processes in continental and maritime convective clouds. B Am Meteorol Soc 79:2457–2476

Rosenfeld D, Cattani E, Melani S, Levizzani V (2004) Considerations on daylight operation of 1.6 μm vs 3.7 μm channel on NOAA and Metop satellites. B Am Meteorol Soc 85:873–881

Rosenfeld D, Fromm M, Trentmann J, Luderer G, Andreae MO, Servranckx R (2007) The Chisholm firestorm: Observed microstructure, precipitation and lightning activity of a pyro-Cb. Atmos Chem Phys 7:645–659

Schmetz J, Pili P, Tjemkes S, Just D, Kerkmann J, Rota S, Ratier A (2002) An Introduction to Meteosat Second Generation (MSG). B Am Meteorol Soc 83:977–992

Scofield RA, Kuligowski RJ (2007) Satellite precipitation algorithms for extreme precipitation events. In: Levizzani V, Bauer P, Turk FJ (eds) Measuring precipitation from space. Springer, pp 485–495

Smith EA, Mugnai A, Cooper HJ, Tripoli GJ, Xiang X (1992) Foundations for statistical-physical precipitation retrieval from passive microwave satellite measurements. Part I: Brightness-temperature properties of a time-dependent cloud-radiation model. J Appl Meteor 31:506–531

Staelin DH, Chen FW (2000) Precipitation observations near 54 and 183 GHz using the NOAA-15 satellite. IEEE Trans Geosci Remote Sens 38:2322–2332

Stephens GL, Kummerow CD (2007) The remote sensing of clouds and precipitation from space: A review. J Atmos Sci 64:3742–3765

Stephens GL, Vane DG, Boain RJ, Mace GG, Sassen K, Wang Z, Illingworth AJ, O'Connor EJ, Rossow WB, Durden SL, Miller SD, Austin RT, Benedetti A, Mitrescu C, CloudSat Team (2002) The CloudSat mission and the A-Train- A new dimension of space-based observations of clouds and precipitation. B Am Meteorol Soc 83:1771–1790

Turk FJ, Rohaly GD, Hawkins J, Smith EA, Marzano FS, Mugnai A, Levizzani V (1999) Meteorological applications of precipitation estimation from combined SSM/I, TRMM and infrared geostationary satellite data. In: Pampaloni PP, Paloscia S (eds) Microwave radiometry and remote sensing of the Earth's surface and atmosphere. VSP Int. Sci. Publ., pp 353–363

Vicente GA, Scofield RA, Menzel WP (1998) The operational GOES infrared rainfall estimation technique. B Am Meteorol Soc 79:1883–1898

Xie P, Arkin PA (1997) Global precipitation: A 17-year monthly analysis based on gauge observations, satellite estimates, and numerical model outputs. B Am Meteorol Soc 78:2539–2578

Xie P, Janowiak JE, Arkin PA, Adler RF, Gruber A, Ferraro R, Huffman GJ, Curtis S (2003) GPCP pentad precipitation analyses: An experimental dataset based on gauge observations and satellite estimates. J Climate 16:2197–2214

9 Combined radar–radiometer retrievals from satellite observations

Mircea Grecu[1], Emmanouil N. Anagnostou[2]

[1]Goddard Earth Sciences and Technology Center, University of Maryland Baltimore County and NASA Goddard Space Flight Center Greenbelt, MD, USA
[2]Civil and Environmental Engineering, University of Connecticut, Storrs, CT, USA; Hellenic Center for Marine Research, Institute of Inland Waters, Anavissos-Attikis, Greece

Table of contents

9.1 Introduction 219
9.2 Background 220
9.3 General formulation 223
9.4 Concluding remarks 228
References 228

9.1 Introduction

A major challenge in precipitation estimation from satellite microwave observations stems from the fact that the distributions of precipitation particle sizes are highly variable in time and space. Mathematically, the particle size distribution (PSD) variability can be expressed through analytical functions (such as Euler's gamma function) of at least two or three independent parameters. Even for short time periods and atmospheric volumes small enough to be characterized by constant values of these parameters, precipitation retrieval is subject to uncertainties because the information necessary to estimate the parameters associated with the PSD within such volumes is incomplete. For example, single frequency spaceborne radars provide only reflectivity observations, while the PSDs on which these observations depend are functions of three variables. The problem of determining three independent variables from a single observation is mathematically ill-posed (i.e., it does not have a unique solution, insensitive to small variations in the input data). In practice, it is customary to determine the sensitivity of the observations with respect to the PSD parameters and solve for the parameter that has the largest impact on the observations, while setting the other parameters equal to constant values determined from independent observations. When independent observations are available, it is possible to derive more accurate solutions simultaneously solving for more parameters and reducing the numbers of parameters that have to be set to 'a priori' values. This kind of approach can be applied to dual-frequency radar observations or to combined radar radiometer observations.

Recent developments in the area of precipitation retrieval from combined radar and radiometer observations have been motivated by the deployment of the Tropical Rainfall Measuring Mission (TRMM) satellite. TRMM features a Precipitation Radar (PR) operating at 13.8 GHz and a nine channel TRMM Microwave Imager (TMI) (Kummerow et al. 1998). Inconsistencies in the early version of TRMM products (Kummerow et al. 2000) suggested that both the PR-only and TMI-only estimates might be subject to systematic errors. Combined radar radiometer retrievals are deemed to be less prone to systematic errors than retrievals from individual instruments because the number of assumptions that need to be made to make the retrieval problem mathematically well-defined is smaller. This is the reason why combined retrievals drew considerable attention in the years immediately preceding and following the TRMM's launch in 1997.

Here, the most recent developments in the area of satellite combined and radiometer retrievals are described. Strengths and limitations of combined retrievals are discussed. Conclusions and recommendations on further work are presented as well.

9.2 Background

Many combined radar radiometer algorithms have their origins in radar profiling algorithm. Airborne and spaceborne radars observations are subject to attenuation that in some cases can be quite severe (up to 30 dB in convective rain for a radar operating at 13.8 GHz). The methods used to correct for attenuation, although based on a rigorous mathematical analysis (Hitschfeld and Bordan 1954) may become unstable. Therefore, various adjustment techniques have been devised to keep the path integrated attenuation (PIA) within reasonable bounds (Iguchi and Meneghini 1994). Estimates of the PIA can be derived using surface reference methods (Iguchi and Meneghini 1994). That is, the ratio of the surface return power measured in rain to the ratio of surface return power measured in adjacent rain-free areas may be used to estimate the PIA. When an independent PIA estimate exists, the relationships between the attenuation and effective reflectivity may be adjusted such that the PIA determined from the analytical Hitschfeld Bordan (HB) approach is equal to the independent PIA estimate. Weinman et al. (1990) were among the firsts to eloquently show that estimates of PIA can be also derived from passive microwave radiometer observations and therefore combined algorithms potentially more accurate than radar-only algorithms can be formulated. At the same time, it was recognized that the changes in the assumed attenuation reflectivity relationships have to be associated with changes in the precipitation reflectivity relationships (Marzoug and Amayenc 1994) such that the retrieval be physically consistent. That is, assuming a known distribution of PSDs, a power law specific attenuation-reflectivity relation, $k = \alpha(r)Z^{\beta}$, where r is the range from the radar, can be derived. Similarly, a consistent (based on the same PSDs) precipitation-reflectivity relationship of the type $R = a(r)Z^{b(r)}$ can be derived. The analytical (HB) PIA is given by

$$PIA_{HB} = \left[1 - q\beta \int_0^{r_{surf}} \alpha(s) Z_m^{\beta}(s) ds \right]^{1/\beta} \tag{1}$$

where q=0.2ln(10) and Z_m are the observed radar reflectivity. To make the PIA_{HB} equal to an independent PIA estimate (like the one derived from radiometer observations), $\alpha(r)$ is modified to a different value $\alpha(r)\delta\alpha$ where $\delta\alpha$ can be exactly derived from condition $PIA_{HB} = PIA_{est}$ where PIA_{est} is the independent PIA estimate. For the approach to be physically consistent, it is necessary that the precipitation reflectivity relationship change to $R = a(r)\delta a Z^{b(r)}$. The relationship between $\delta\alpha$ and δa can be easily determined especially using the normalized gamma distribution concept advocated by Testud et al. (2000). A normalized gamma distribution can be expressed as

$$N(D) = N_0^* f(\mu) \left(\frac{D}{D_o}\right)^\mu \exp[-(3.67+\mu)D/D_o] \qquad (2)$$

$$f(\mu) = \frac{6}{3.67^4} \cdot \frac{(3.67+\mu)^{(4+\mu)}}{\Gamma(3.67+\mu)} \qquad (3)$$

where N(D) is the density of particles of diameter D, N_0^* is a generalized intercept, D_0 is the mean diameter and μ is a shape parameter. Both theoretical and observational evidence (Testud et al. 2000) indicate that most precipitation-related relationships (i.e., radar reflectivity versus rain rate, radar reflectivity versus attenuation, absorption versus precipitation content) strongly depend on N_0^* and only weakly on μ. Moreover, the N_0^* dependence can be readily into power law relationships of the type $Y = a N_0^{*(1-b)} X^b$. For example, the Marshall-Palmer relationship, $Z=200R^{1.6}$, which holds for $N_0^*=0.08$ cm^{-4}, can be extrapolated to $Z=200(N_0^*/0.08)^{-0.4}R^{1.6}$ and then applied for $N_0^* \neq 0.08$ cm^{-4}. This concept applied to the adjustment technique described above leads to the relationship $\delta a = \delta \alpha^{\frac{1-b(r)}{1-\beta}}$. Moreover, $\delta \alpha^{\frac{1}{1-\beta}} = \delta N_0^*$ is the change in the PSD intercept relative to the initial value. It follows that an independent estimate of the PIA does not only make the attenuation correction more stable, but also provides additional insight into the PSD properties. These types of adjustments (i.e., α and a) have been exploited in many radar profiling algorithms (Ferreira et al. 2001) including the official TRMM PR algorithm (Iguchi et al. 2000), although in the PR algorithm the multiplicative factors in the power law relationships are not calculated explicitly as a function of δN_0^* but retrieved from a priori relationships. δN_0^* is thus determined from the condition that the PIA determined from the reflectivity profile is equal to the independent estimate of the PIA (using surface reference methods or radiometer based estimates as proposed by Weinman et al. (1990) and Smith et al. (1997) subsequently). Although convenient, the estimation of the PIA from radiometer-only observations (independently of the actual radar observations) may lead to erroneous PIA estimates, which can negatively affect the results obtained through the adjustment technique. Therefore, more complex combined approaches have been developed. These are presented next.

9.3 General formulation

A general combined retrieval methodology can be derived in a Bayesian estimation framework. That is, a solution is determined to maximize the conditional probability of PSDs, given the radiometer and radar observation. If one denotes by **X** the set of PSD related variables needed to calculate the radiometer observations, $\mathbf{t}_m$, and the radar observations, $\mathbf{z}_m$, a solution is determined by the minimization of the following a posteriori probability

$$p = p(\mathbf{X} \mid \mathbf{t}_m, \mathbf{x}_m) \tag{4}$$

which, given Bayes theorem, can be expressed as

$$p(\mathbf{X} \mid \mathbf{t}_m, \mathbf{x}_m) = p(\mathbf{z}_m \mid \mathbf{X}) p(\mathbf{t}_m \mid \mathbf{X}) p(\mathbf{X}) / p(\mathbf{t}_m, \mathbf{x}_m) \tag{5}$$

Assuming Gaussian errors in models and observations the maximization of (5) is equivalent to the minimization of the following functional

$$F = \frac{1}{2}(\mathbf{t}_m - \mathbf{t}(\mathbf{X}))^T \mathbf{W}_T^{-1}(\mathbf{t}_m - \mathbf{t}(\mathbf{X})) + \frac{1}{2}(\mathbf{z}_m - \mathbf{z}(\mathbf{X}))^T \mathbf{W}_Z^{-1}(\mathbf{z}_m - \mathbf{z}(\mathbf{X})) + \frac{1}{2}(\mathbf{X} - \mathbf{m}_X)^T \mathbf{W}_X^{-1}(\mathbf{X} - \mathbf{m}_X)) \tag{6}$$

where $\mathbf{t}(\mathbf{X})$ and $\mathbf{z}(\mathbf{X})$ are radar and radiometer models, **W** are covariance matrices accounting for uncertainties in models and observations, while $\mathbf{m}_X$ is an 'a priori' estimate of **X**. Although conceptually simple, the formulation of Eq. (6) is extremely challenging from the numerical standpoint. This is the reason why various combined algorithms have been developed. The numerical difficulties stems in the complexity of operator $\mathbf{t}(\mathbf{X})$ and in the large number of variables upon which F depends.

Haddad et al. (1997) developed an approach that is the core of the TRMM combined facility algorithm based on parameterized calculations of brightness temperatures. Although mathematically consistent and physically reasonable, this algorithm may induce estimate uncertainties that are hard to interpret, given the simplification in the calculation of brightness temperatures. Quite general approaches, but restricted to airborne observations were developed by Olson et al. (1996) and Marzano et al. (1999). These approaches are based on cloud resolving model (CRM) simulations. That is, CRM simulations are used to create 'a priori' databases of precipitation profiles and associated reflectivity profiles brightness temperatures. In the retrieval phase, these databases are explored and a combination of precipitation profiles in the database is determined as a solution. The application of the CRM-based algorithms to real satellite data is cumbersome because the satellite radiometer footprints are large and encompass multiple profiles in the CRM database. Thus, the solution is based on searches of combinations of profiles in the CRM database. Nevertheless, such approaches exist (Di Michele et al. 2003; Masunaga and Kummerow 2005).

Other approaches are based on the direct minimization of functional F. Such approaches have been derived by Grecu et al. (2004) and Jiang and Zipser (2006). The approach of Grecu et al. (2004), which is computationally very efficient, is in many respects a direct generalization of the adjustment technique presented in the previous section. That is because in order to minimize the computational effort, the covariance matrix W_Z in Eq. (6) is consider small (close to zero). This enforces a solution that precisely matches the reflectivity observations, exactly as in radar profiling algorithms based on Eq. (1). Moreover, the variables to be determined through the minimization of Eq. (6) are not the precipitation contents but a set of generalized intercepts, N_0^*. An N_0^* is assigned to each radar profile. The minimization is achieved through gradient based search. The radiometer observations are calculated using a radiative transfer model based on the Eddington approximation (Grecu et al. 2004). However, the radiometer observations considered in the approach are the normalized polarizations rather than the initial brightness temperatures. This was done in an attempt to reduce the uncertainties caused by cloud water and variability in the surface wind speed on the calculation of radiometer brightness temperatures. For low frequency channels (10, 19 and 37 GHz) the differences between vertically and horizontally polarized brightness temperatures are due mainly to differences in the emissivity of vertically and horizontally polarized radiation. As the amount of precipitation in the atmospheric column sampled by the radiometer increases, the surface effects become smaller and the differences between vertically and horizontally polarized brightness temperature decrease. Therefore, the polarization difference (defined as the difference between the vertically and horizontally polarized brightness temperatures) decreases with the amount of precipitation within the sampling volume. In the combined retrievals of Grecu et al. (2004) the polarization differences are normalized by the clear sky values (Petty 1994).

Shown in Fig. 1 are the TRMM near surface attenuation corrected reflectivity and the 19 GHz polarization difference for a tropical cyclone in the Indian Ocean. One may note in Fig. 1 good correlation between the near surface reflectivity and the 19 GHz polarization difference.

However, the features of the 19 GHz polarization difference are smoother due to low resolution of the TMI instrument at the 19 GHz frequency. The low resolution of the low frequency channels makes the combined retrieval from satellite observations more challenging that that from airborne observations.

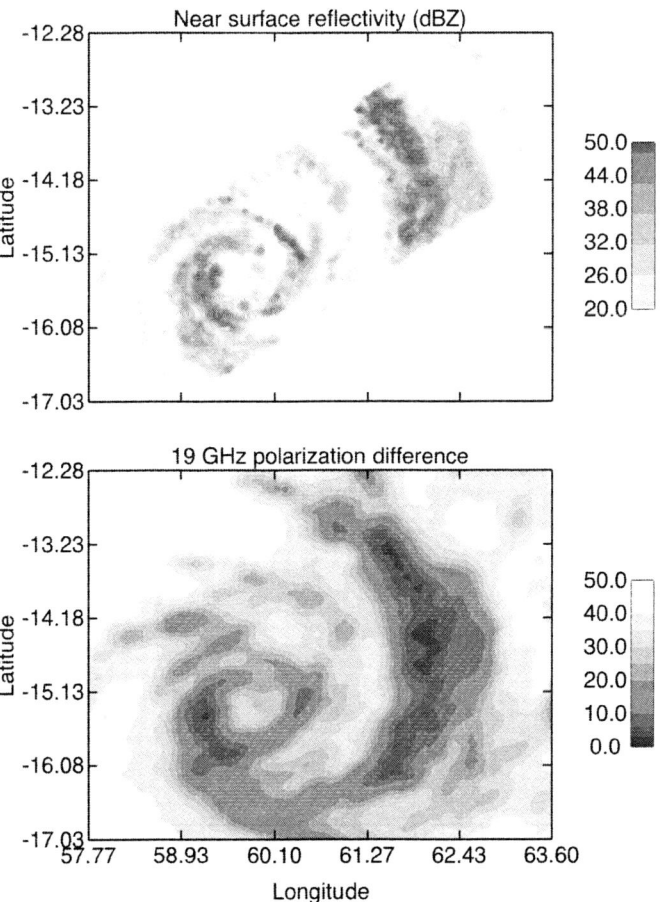

Fig. 1. TRMM near surface attenuation corrected reflectivity and 19 GHz polarization difference for a tropical cyclone in the vicinity of Madagascar on 10 February 1998

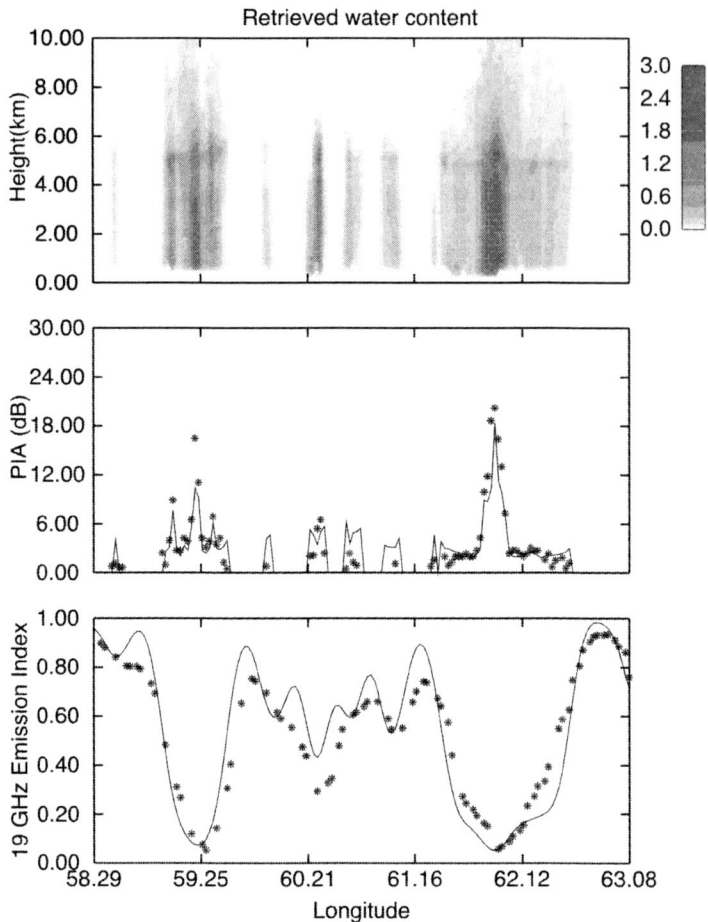

Fig. 2. Combined PR/TMI retrievals. From *top to bottom*: retrieved water contents at nadir (in g/m^{-3}); retrieved PIA (*thin line*) and PIA from a surface reference method (*stars*); calculated 19 GHz emission index (*thin line*) and observed 19 GHz emission index (*stars*)

In addition to reflectivities that are used as strong constraints and the normalized polarization differences at 10 and 19 GHz, the combined retrievals of Grecu et al. (2004) use also estimates of the PIA from a surface reference method by Meneghini et al. (2000). In Fig. 2, a nadir cross-section through retrievals derived from the combined algorithm applied to the data presented in Fig. 1 is shown. One may note in Fig. 2 quite good agreement between the retrieved and the surface return method PIAs. The attenuation correction appears effective even for large attenuation. In regions with little attenuation, the surface reference method PIA cannot be used effectively to adjust the retrievals. These regions are affected most by the radiometer information. Non-negligible differences between the calculated on observed 19 GHz emission indices (which is defined as 1.0 minus the normalized polarization difference) are apparent in Fig. 2. This suggests that important random errors are present in the combined approach. This case is particularly challenging because tropical cyclones are characterized by strong, highly variable surface winds and large amounts of possibly non-precipitating clouds. However, Grecu et al. (2004) found that combined retrievals from PR and TMI observations are in better agreement with estimates from a ground radar in the Kwajalein atoll than retrievals from PR-only and TMI-only observations.

Shown in Fig. 3 is a frequency plot of combined PR/TMI retrievals versus PR-only retrievals. PR-only retrievals are derived using radar algorithm embedded in a combined framework. One may notice in Fig. 3 that the two types of retrievals are different mainly for small water contents.

When the PIA estimate from the surface reference method is large and reliable, its informational content dominates the retrieval. However, when this estimate is unreliable, the adjustment in N_0^* is driven by TMI information. Masunaga and Kummerow (2005) found larger differences between their combined and PR-only retrievals that

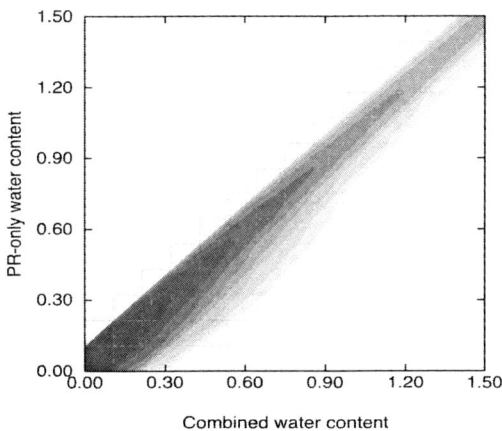

Fig. 3. Frequency plots of PR-only (in gm^{-3}) retrievals versus combined PR-TMI retrievals for the case in Fig. 1

can be attributed to a different assumption in the uncertainty of the surface reference PIA estimates.

The application of the combined algorithm of Grecu et al. (2004) to a large amount of TRMM facilitated the construction of a physically consistent database of precipitation profiles and associated TRMM brightness temperatures. The use of this database with a Bayesian algorithm for precipitation retrieval from TMI-only observations allowed the derivation of TMI-only estimates more consistent with PR-only estimates than the TMI-only official estimates (Grecu and Olson 2006). This application demonstrated that one of the strengths of the combined approach is the development of physically consistent databases of precipitation and radiometer observations that can be used to improve the retrievals from radiometer-only observations.

9.4 Concluding remarks

Combined radar radiometer retrievals are an effective mechanism to mitigate the effect that uncertainties in PSDs have on radar-only retrievals. The roots of combined approaches are in radar profiling algorithms, the passive information serving as basis to adjust assumptions (parameters) in the radar retrievals. Various combined algorithms have been developed. They are in principle similar but are based on different mechanisms to achieve the consistency between the models involved in the retrievals and the actual observations.

Future work should be directed towards a better quantification of uncertainties in the models and the assumptions upon which the retrievals are based. Only a correct quantification of these uncertainties warrants the derivation of unbiased, optimal results. Another very import aspect is that the existing approaches appear to be inclined towards the impact of low frequencies radiometer observations on the liquid phase retrievals. The information existent in the high frequency observations, although challenging, may have a big impact on the retrievals of both liquid and solid phase.

References

Di Michele S, Marzano FS, Mugnai A, Tassa A, Poiares Baptista JPV (2003) Physically based statistical integration of TRMM microwave measurements for precipitation profiling. Radio Sci 38:8072, doi:10.1029/2002RS002636

Ferreira F, Amayenc P, Oury S, Testud J (2001) Study and tests of improved rain estimates from the TRMM precipitation radar. J Appl Meteorol 40:1878–1899

Grecu M, Olson WS (2006) Bayesian estimation of precipitation from satellite passive microwave observations using combined radar-radiometer retrievals. J Appl Meteorol 45:416–433

Grecu M, Olson WS, Anagnostou EN (2004) Retrieval of precipitation profiles from multiresolution, multifrequency, active and passive microwave observations. J Appl Meteorol 43:562–575

Haddad ZS, Smith EA, Kummerow CD, Iguchi T, Farrar MR, Durden SL, Alves M, Olson WS (1997) The TRMM 'day-1' radar/radiometer combined rain-profiling algorithm. J Meteorol Soc Jpn 75:799–809

Hitschfeld W, Bordan J (1954) Errors inherent in the radar measurement of rainfall at attenuating wavelengths. J Meteorol 11:58–67

Iguchi T, Meneghini R (1994) Intercomparison of single-frequency methods for retrieving a vertical rain profile from airborne or spaceborne radar data. J Atmos Ocean Tech 11:1507–1516

Iguchi T, Kozu T, Meneghini R, Awaka J, Okamoto K (2000) Rain-profiling algorithm for the TRMM precipitation radar. J Appl Meteorol 39:2038–2052

Jiang H, Zipser EJ (2006) Retrieval of hydrometeor profiles in tropical cyclones and convection from combined radar and radiometer observations. J Appl Meteorol Clim 45:1096–1115

Kummerow C, Barnes W, Kozu T, Shiue J, Simpson J (1998) The tropical rainfall measuring mission (TRMM) sensor package. J Atmos Ocean Tech 15:808–816

Kummerow C, Simpson J, Thiele O, Barnes W, Chang ATC, Stocker E, Adler RF, Hou A, Kakar R, Wentz F, Ashcroft P, Kozu T, Hong Y, Okamoto K, Iguchi T, Kuroiwa H, Im E, Haddad Z, Huffman G, Krishnamurti T, Ferrier B, Olson WS, Zipser E,. Smith EA, Wilheit TT, North G, Nakamura K (2000) The status of the tropical rainfall measuring mission (TRMM) after two years in orbit. J Appl Meteorol 39:1965–1982

Marzano FS, Mugnai A, Panegrossi G, Pierdicca N, Smith EA, Turk J (1999) Bayesian estimation of precipitating cloud parameters from combined measurements of spaceborne microwave radiometer and radar. IEEE T Geosci Remote 37:596–613

Marzoug M, Amayenc P (1994) A class of single- and dual-frequency algorithms for rain-rate profiling from a spaceborne radar. Part I: Principle and tests from numerical simulations. J Atmos Ocean Techn 11:1480–1506

Masunaga H, Kummerow CD (2005) Combined radar and radiometer analysis of precipitation profiles for a parametric retrieval algorithm. J Atmos Ocean Tech 22:909–929

Meneghini R, Iguchi T, Kozu T, Liao L, Okamoto K, Jones JA, Kwiatkowski J (2000) Use of the surface reference technique for path attenuation estimates from the TRMM radar. J Appl Meteorol 39:2053–2070

Olson WS, Kummerow C, Heymsfield GM, Caylor IJ, Giglio L (1996) A method of combined passive/active microwave retrievals of cloud and precipitation profiles. J Appl Meteorol 35:1763–1789

Petty GW (1994) Physical retrievals of over-ocean rain rate from multichannel microwave imagery Part I. Theoretical characteristics of normalized polarization and scattering indeces. Meteorol Atmos Phys 54:79–99

Smith EA, Turk FJ, Farrar MR, Mugnai A, Xiang XW (1997) Estimating 13.8-GHz path-integrated attenuation from 10.7-GHz brightness temperatures for the TRMM combined PR-TMI precipitation algorithm. J Appl Meteorol 36:365–388

Testud J, Oury S, Amayenc P (2000) The concept of 'normalized' distribution to describe raindrop spectra: A tool for hydrometeor. Remote Sensing. Phys Chem Earth (B) 25:897–902

Weinman JA, Meneghini R, Nakamura K (1990) Retrieval of precipitation profiles from airborne radar and passive radiometer measurements: Comparison with dual-frequency radar measurements. J Appl Meteorol 29:981–993

ii. Ground estimation

10 Rain microstructure from polarimetric radar and advanced disdrometers

Merhala Thurai, V. N. Bringi

Dept. of Electrical and Computer Engineering, Colorado State University, CO, USA

Table of contents

10.1 Introduction 234
    10.1.1 Background 234
    10.1.2 Rain microstructure: relevance 235
    10.1.3 Relating rain microstructure to polarimetric radar measurements 238
10.2 Drop size distributions 242
    10.2.1 Variability 242
    10.2.2 DSD models 243
    10.2.3 DSD estimation from polarimetric radar measurements 248
    10.2.4 DSD estimation from advanced disdrometers 254
    10.2.5 Global DSD characteristics 257
    10.2.6 Seasonal variation 259
10.3 Drop shapes 263
    10.3.1 Axis ratio measurements from an artificial rain experiment 263
    10.3.2 Drop contours 265
    10.3.3 Consistency with polarimetric radar measurements 268
10.4 Drop orientation angles 269
10.5 Fall velocities 274
10.6 Summary 276
References 279

10.1 Introduction

10.1.1 Background

The use of polarimetric techniques in weather radars has in recent years captured a great deal of interest within the operational weather radar community (e.g., Ryzhkov et al. 2005). These systems often operate in three different frequency bands, namely S, C and X-bands, corresponding to approximately 3, 5 and 11 GHz, respectively, and furthermore, make use of dual-linear polarizations at horizontal and vertical states, as proposed back in the late 70's by Seliga and Bringi (1976). Such dual-polarization techniques are particularly useful at low elevation angles because the rain medium consists of highly oriented non-spherical particles (i.e., rain drops), with their axis of symmetry oriented along the vertical, and as such, will give rise to differences in their forward and back scatter amplitudes between vertical and horizontal polarizations.

The retrieval or the estimation of rainfall rates from the back scatter polarization measurements requires fundamentally some knowledge of the rain microstructure. By this, we mean primarily, (a) drop size distribution, (b) drop shapes and their variations due oscillations, (c) drop orientation angles and (d) drop fall velocities. The dual-linear polarization methods make use of the fact that the drops, particularly those larger than 1 mm diameter, are non-spherical in shape and that these shapes are diameter-dependent. By making use of the fact that the forward scatter amplitudes and the back scatter reflectivities will, therefore, show polarization dependence, it becomes possible to retrieve information on the drop size distribution (DSD) from the polarimetric measurements and hence estimate the rainfall rates more accurately than the conventional weather radars which use single polarization (e.g., Doviak and Zrnic 1993).

Rain microstructure can also in parallel be determined from the so-called 'disdrometers'. These are instruments which are designed to measure or infer the DSDs within a given sensor volume. The conventional systems, such as the impact-type Joss-Waldvogel disdrometers (JWD; Joss and Waldvogel 1967), make certain assumptions on the drop diameter dependence on fall velocities and do not allow for non-zero drop 'canting angles'. Other types of disdrometers such as micro-rain radars (MRR) operate on the Doppler principle to derive the DSD from the fall velocity spectra. There are also laser-based optical devices (such as the Parsivel; Löffler-Mang and Joss 2000), which does not image the particle but rather gives the maximum width and velocity of particles 'binned' in a 32 × 32 matrix. The most advanced system of all, at least at present, is the 2-dimensional video disdrometer (abbreviated to 2DVD) whose measurement principles and specifications have been described in detail in Chap. 1. It measures the size, shape, orientation and fall velocity of each individual hydrometeor falling through its sensor area; i.e., it measures all four primary parameters of rain microstructure. Moreover, all four parameters are measured by a direct method through imaging techniques using fast line scan cameras. For this reason, we give here several examples which utilize measurements from this instrument, most of which have only recently been published. The examples are given to highlight certain important properties/features of the rain microstructure. For other disdrometer data and evaluation, readers are referred to publications by other authors, as listed in Table 1.

As for intercomparison between instruments, Krajewski et al. (2006) have given DSD, rainfall and velocity-diameter data taken during intense precipitation which show some discrepancies in certain cases. Here we deal with mostly 2DVD and one other, very promising, instrument, namely POSS (referred to in Table 1). This instrument is similar to the MRR in that it derives the DSD from the Doppler power spectra, but it operates in a bistatic mode and retrieves the DSD within the common volume defined by the transmit and receive antenna patterns.

10.1.2 Rain microstructure: relevance

While rain microstructure is basic to cloud physics of rain formation, here we consider its fundamental importance in three main application areas: (a) in the estimation or retrieval of DSD parameters and rainfall

Table 1. Examples of advanced disdrometers and references for detailed information

Instrument	Reference
Parsivel	Löffler-Mang and Joss (2000)
Dual-beam spectro-pluviometer	Hauser et al. (1984)
2-dimensional video disdrometer	Randeu et al. (2002)
	Kruger and Krajewski (2002)
	Schönhuber et al. (2007)
Micro-rain radar(*)	Peters et al. (2005)
Precipitation Occurrence Sensor System (*)	Sheppard (1990)
	Sheppard and Joe (1994)

(*) These instruments estimate the drop size distribution from the measured Doppler power spectra

rates from polarimetric weather radars, (b) in the formulation of attenuation-correction schemes especially at C- and X-bands, and (c) in evaluating propagation effects in rain for satellite and terrestrial communication systems (Hall et al. 1996; Allnutt 1989). With respect to (a), a large body of literature already exists (e.g., Bringi and Chandrasekar 2001 and the references contained therein) which demonstrate the importance of DSD retrievals for improving the accuracy of rainfall estimations for hydrology and flash flood forecasting. For this reason (and several others such as the ability to identify damaging hail events), the S-band NEXRAD network in the U.S. is being upgraded for dual-polarization capability in the near future (Ryzhkov et al. 2005). In Europe, Météo France already has several new C-band dual-polarized radars for operational applications and are also planning to upgrade existing radars (Gourley et al. 2006). Several other European weather forecasting agencies (e.g., Italy, U.K., Germany, Finland) are in the process of evaluation of dual-polarized radars for operational use.

With respect to (b), reflectivity data when uncorrected for rain attenuation can lead to large errors in rainfall estimation, particularly at C and X-bands (e.g., Park et al. 2004). For S-band this occurs less frequently, but for cases where the propagation path intercepts multiple rain cells of high intensity, significant attenuation has been shown to occur (for example, Ryzhkov and Zrnić 1995). Figure 1 shows a case at X-band to illustrate the need for attenuation correction. It shows comparisons of 1-hourly accumulations from 8 rain gauges with the corresponding (dual-polarized) radar-based estimates, (a) with and, (b) without attenuation correction. Here, the attenuation-correction scheme is based on measurement of differential propagation phase between horizontal (H) and vertical (V) polarizations (the relation between rain microstructure and dual-polarized radar is described in Sect. 10.1.3). Significant reductions in standard error (from 49% to 19%) as well as in bias (from 48% to 2%) are achieved using the attenuation-correction procedures which are based on measured properties of the rain microstructure such as DSD and drop shapes. The authors have further demonstrated that the combined use of the attenuation-corrected reflectivity and the specific differential propagation phase K_{dp} would further reduce the error and bias to 15% and 1.1%, respectively (dual-polarized radar variables are described later in Sect. 10.1.3).

The third area where rain microstructure plays an important role is in the evaluation of propagation effects for wireless communications systems, in particular, those operating at frequencies above 10 GHz. The systems include both line-of-sight terrestrial systems and Earth-space systems. Whereas the DSD governs the relationship between the specific attenuation and rainfall rate required for link availability calculations (Olsen 1981), drop shapes determine the polarization dependence as well as rain-induced depolarization (Oguchi 1983) for systems utilizing orthogonal polarizations. Moreover, drop oscillations and drop orientation angle distributions can give rise to secondary effects such as spread in the cross-polar discrimination.

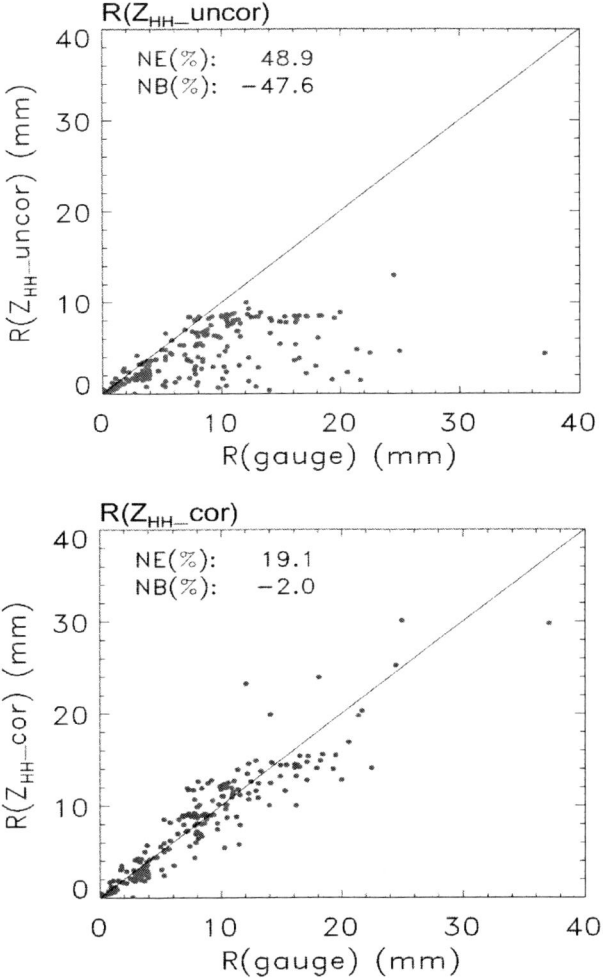

Fig. 1. 1-hour rainfall estimates using Z-R relationships from an X-band polarimetric radar compared with rain gauge data. The upper panel shows the radar estimates derived without attenuation correction while the lower panel shows the estimates derived using an attenuation-correction scheme based on the knowledge of the rain microstructure (from Park et al. 2004)

10.1.3 Relating rain microstructure to polarimetric radar measurements

The polarimetric radar variables most commonly measured are based on linear horizontal (H) and vertical (V) polarization states both for transmit and receive. The principal measurements are (a) reflectivity at H-polarization (Z_h), (b) the differential reflectivity (Z_{DR}) which is the ratio of reflectivities at H and V polarizations, (c) the specific differential phase (K_{dp}) which is proportional to the real part of the difference between the forward scattering amplitudes at H and V polarizations, (d) the linear depolarization ratio (LDR) which is the ratio of the cross-polar reflectivity to the co-polar reflectivity and (e) the correlation coefficient (ρ_{co}) between the received co-polar signals at H and V polarizations. These measurables are generally expressed in terms of the 3 real and 3 complex elements of the polarimetric covariance matrix (e.g., Chap. 3 of Bringi and Chandrasekar 2001) which forms a complete set for randomly varying precipitation media. Here we focus on the relation between the radar measureables and the rain microstructure within the framework of an analytical formulation which is possible for Rayleigh scattering. Such formulations not only enhance understanding of the relationship of the polarimetric measureables to rain microstructure but are also useful in developing the form of the retrieval algorithms needed for estimating DSD parameters as well as rain rates.

We start with a number of definitions and assumptions from both radar and rain microstructure perspectives. As mentioned earlier we assume Rayleigh scattering and further, low radar elevation angles (typically < 15°). Raindrop shapes are assumed to be oblate spheroids defined by their axis ratio ($r = b/a$ = minor axis/major axis) and the equivolumetric spherical diameter (referred to as D_{eq} or simply as D). The drop size distribution is N(D) while the axis ratio distribution is described by the marginal distribution p(r/D). The mean axis ratio is <r/D> while the variance is denoted by var(r/D). For D≥1.5 mm, a linear relation between <r> and D is a good approximation with slope denoted by γ, i.e., <r> = a − γD. Thurai and Bringi (2005) derive a=1.055, γ=0.0653 mm^{-1}. The 2DVD measurements of drop shapes and oscillations will be given in more detail in Sect. 10.3 but suffice here to mention that the <r> is generally governed by the balance between gravitational, surface tension and aerodynamic forces, whereas drop oscillations (i.e., contributing to var(r/D)) are governed by intrinsic mechanisms such as resonance with vortex shedding (e.g., Beard et al. 1989) for small drops (D≈1–1.5 mm) or by aerodynamic feedback for larger drops (e.g., Tokay and Beard 1996).

Finally, we describe drop orientation as the orientation of the symmetry axis with respect to the local vertical direction in terms of zenith angle θ and azimuth angle φ. Further, the canting angle (β) in the polarization plane is defined as the angle between the projection of the drop's symmetry axis on this plane and the projection of the local vertical direction on this same plane (e.g., Holt 1984). The 2DVD measurements of drop orientation angles will be discussed in Sect. 10.4. Here, we assume that the radar elevation angle is $\approx 0°$ and hence $\theta = \beta$ if $\varphi = \pi/2$. The distribution of β is generally due to turbulence or wind shear but Beard and Jameson (1983) show that $\langle\beta\rangle \approx 0$ with $\sigma_\beta < 5°$ if turbulence is the main cause as will be assumed herein (we exclude layers of wind shear which are often localized).

The reflectivity factor is defined as the 6th moment of N(D) and is approximately equal to Z_h (the reflectivity factor at H-polarization). The differential reflectivity ratio ($Z_{DR} = Z_h/Z_v$) assuming that $\beta = 0°$ (perfect orientation) is related to the reflectivity-weighted mean axis ratio $\langle r_z \rangle$ as:

$$(Z_{dr})^{-1} \approx [\langle r_z \rangle]^{7/3} \qquad (1)$$

The above approximation is valid for small values of var(r/D). For details we refer to Jameson (1983) or Chap. 7 of Bringi and Chandrasekar (2001). Note that $\langle r_z \rangle$ is defined via the DSD as:

$$\langle r_z \rangle = \frac{\int D^6 E(r/D) N(D) dD}{\int D^6 N(D) dD} \qquad (2)$$

where E stands for expected value. The above along with Eq. (1) clearly shows that because of the D^6 weighting, the mean axis ratio of the large drops dominate the differential reflectivity ratio (note also that axis ratio r <1). Hence, it is important to characterize the 'tail' of the N(D). As mentioned earlier, the variation between $\langle r/D \rangle$ is linear with D (for D $\geq$ 1.5 mm) and of the form $\langle r/D \rangle = a - \gamma D$ with slope of γ ; it follows that $\langle r_z \rangle = a - \gamma D_z$ where D_z is the reflectivity-weighted mean diameter which is simply the ratio of the 7th to 6th moments of the N(D). This further highlights the importance of the shape of the big drops in determining the Z_{DR} and conversely shows that D_z may be retrieved from measurements of Z_{DR}.

For gamma DSDs (to be defined in Sect. 10.2), the different moments of N(D) can be related to each other via functions of the shape parameter, i.e., the D_z can be related to lower order moments such as the mass-weighted mean diameter (D_m : ratio of 4th to 3rd moments of the DSD) given by:

$$D_m = \frac{\int_0^{D_{max}} N(D) D^4 \, dD}{\int_0^{D_{max}} N(D) D^3 \, dD} \tag{3}$$

As power law relationships are ubiquitous in radar meteorology it is common to relate D_m to Z_{DR} (here the subscript DR in capitals refers to $10\log_{10}(Z_{DR})$, dB), e.g., from Bringi and Chandrasekar (2001):

$$D_m = 1.619 (Z_{DR})^{0.485}, \text{ mm} \tag{4}$$

The above is based on numerical scattering calculations assuming gamma DSDs and the Beard and Chuang (1987) numerical model for equilibrium axis ratios. Note that the power law fit is not based on theoretical grounds but based on a non-linear regression approach; as such it is a mean fit with scatter due to variations in the shape parameter only.

The specific differential phase (K_{dp}) under perfect orientation and zero radar elevation angle can be approximately expressed as (again for details we refer to Bringi and Chandrasekar 2001 and Jameson 1985):

$$K_{dp} = \lambda \int N(D) \int \text{Re}[\hat{h} \cdot \vec{f}(r,D) - \hat{v} \cdot \vec{f}(r,D)] p(r/D) \, dr \, dD \tag{5a}$$

where λ is the radar wavelength and $\vec{f}$ is the vector forward scattering amplitude. Further assuming Rayleigh scattering, K_{dp} can be expressed as:

$$K_{dp} = \left(\frac{180}{\lambda}\right) 10^{-3} CW[1 - \langle r_m \rangle]; \text{ °/km} \tag{5b}$$

where λ (in m) is the radar wavelength, $C \approx 3.75$ (both dimensionless and independent of λ), W is the rain (or, liquid) water content (in g m^{-3}) defined as:

$$W = 10^{-3} \left(\frac{\pi}{6}\right) \rho_w \int D^3 N(D) \, dD \tag{6}$$

where D is in mm, N(D) in mm^{-1} m^{-3} and ρ_w is the water density (1 g cm^{-3}). The $\langle r_m \rangle$ is the mass (or, volume)-weighted mean axis ratio defined as:

$$\langle r_m \rangle = \frac{\int D^3 E(r/D) N(D) dD}{\int D^3 N(D) dD} \tag{7}$$

Once again since $\langle r/D \rangle = a - \gamma D$, it follows that $\langle r_m \rangle = a - \gamma D_m$ and K_{dp} can be related to W, D_m and γ. It also follows that K_{dp} can be related to rain rate (R), D_m and γ. Note that K_{dp} (in contrast to Z_{DR}) depends on the shape and concentration of the 'medium'-sized drops in the distribution as lower order moments (i.e., 4th and 3rd) are involved. If the canting angle distribution is Gaussian with mean=0° and standard deviation σ_β, then Eq. (5b) is modified as:

$$K_{dp}(\sigma_\beta) = K_{dp}(0) \exp(-2\sigma_\beta^2) \tag{8}$$

Since $\sigma_\beta < 5°$, the adjustment factor due to canting is $< 6\%$.

If the transmitted polarization state is H-polarization, then the back scatter return from non-spherical and/or canted particles is composed of both a cross-polar signal (that is V-polarized) and co-polar signal (that is H-polarized). The ratio of the corresponding reflectivities is defined as the linear depolarization ratio (LDR) = Z_{vh}/Z_{hh}. Under Rayleigh scattering the LDR may be expressed as (for details refer to Bringi and Chandrasekar 2001):

$$LDR = \left(\frac{1}{4}\right)\left[1 - \exp(-8\sigma_\beta^2)\right]\left[(1 - \langle r_z \rangle)^2 + \text{var}(r_z)\right] \tag{9}$$

If for simplicity we assume that $\langle r_z \rangle = 1 - \gamma D_z$ (i.e., a=1), we obtain:

$$LDR = \left(\frac{1}{4}\right)\left[1 - \exp(-8\sigma_\beta^2)\right]\left[\gamma^2 D_z^2 + \text{var}(r_z)\right] \tag{10}$$

The above equation clearly shows the dependence of LDR on the standard deviation of the canting angle distribution, the reflectivity – weighted mean diameter and the variance of the axis ratios. Of these three factors, the most important are D_z and σ_β with lesser effect of γ and var(r_z). In contrast to Z_{DR} and K_{dp}, the LDR will depend on both the mean and variance of the reflectivity-weighted axis ratio.

The final radar measurable we consider is the co-polar correlation coefficient (ρ_{co}) which is the correlation coefficient between the co-polar signal returns. Factors that reduce the correlation coefficient from unity (for spheres or for scatterers fixed in space) are related (among other factors) to the shape variation across the DSD. For Mie scattering it is also reduced by changes in the back scatter differential phase across the DSD. However, for Rayleigh scattering and perfect orientation, the magnitude of ρ_{co} may be expressed as:

$$|\rho_{co}| \approx 1 - \left(\frac{1}{2}\right)\frac{[\text{var}(r_z)]}{[\langle r_z \rangle^2]} \qquad (11)$$

For example if var(r_z) = 0.01 and $\langle r_z \rangle$ = 0.8, then $|\rho_{co}|$ = 0.992. If $r_z = a - \gamma D_z$ it follows that var(r_z) = γ^2 var(D_z). Hence, for equilibrium-shaped drops (11) reduces to:

$$|\rho_{co}| \approx 1 - \left(\frac{1}{2}\right)\frac{[\gamma^2 \text{var}(D_z)]}{[\alpha - \gamma D_z]^2} \qquad (12)$$

10.2 Drop size distributions

10.2.1 Variability

The drop size distribution is a highly variable quantity, showing both spatial and temporal variations of the microphysics of rain formation (e.g., Pruppacher and Klett 1997). They also show dependence on location, climate and rain regime as well as seasonal and diurnal variations even for a given location. For example, Kozu et al. (2006) have recently examined the DSD characteristics in three different monsoon regions and found that there were very clear seasonal variations and that the DSD variability also exhibited a diurnal cycle. Figure 2 shows the averaged DSDs (over several years) for 3 and 30 mm h^{-1} rainfall rates, for three different locations, namely, (a) Gadanki, South India, (b) Singapore and (c) Koto Tabang (KT) in Western Sumatra, and for four different time periods, namely, (a) 00–06 hrs, (b) 06–12 hrs, (c) 12–18 hrs and (d) 18–24 hrs. For the higher rainfall rate case (30 mm h^{-1}), the DSD characteristics show considerable amount of diurnal variations, especially for KT. In general, the afternoon DSDs showed broader distributions than the morning ones, implying that the local convection cycle over land which often causes peak afternoon rainfall activity, produces wider distribution of rain drop sizes. Although there are numerous articles on DSD characterization (e.g., Rosenfeld and Ulbrich 2003), individual studies at certain locations over a long period of time such as those in Fig. 2 are valuable in understanding the effect of rain microphysical mechanisms on the rain microstructure. Note, the DSD measurements shown in Fig. 2 were all obtained using measurements from JWD impact-type disdrometers.

10.2.2 DSD models

The physical processes that contribute to the formation of the drop size distribution (DSD) are well-known to be complex from a microphysical viewpoint. In some cases of steady physical processes such as occurring in stratiform rain with bright band or strong convective rain, the DSD can be in either 'size' controlled or 'concentration' controlled domains with most other rain types falling in between (Steiner et al. 2004). Despite these variations, it has been found that they can be generally fitted to three-parameter models such as gamma and log-normal models.

Out of the two, the gamma distribution is the one that is more frequently used, at least in the polarimetric radar field. This distribution can be formulated in the following manner:

$$N(D) = N_0 D^\mu \exp(-\Lambda D) = N_T \frac{\Lambda^{\mu+1} D^\mu}{\Gamma(\mu+1)} \exp(-\Lambda D) \qquad (13)$$

where $N(D)$ is the number of drops with diameter D, and the parameter set (N_0, μ and Λ; Ulbrich 1983) or (N_T, μ and Λ; probability density function form) are the gamma model parameters where Γ represents the gamma function. Note N_T is the zeroth moment of the DSD (total number concentration) in Eq. (13). From a theoretical perspective it is generally preferred as a DSD parameter instead of N_0 whose units depend on μ. The parameter μ is referred to as the shape parameter and is often fixed in order to simplify the DSD retrievals from polarimetric radar measurements. The parameter Λ is related to μ as well as the median volume diameter (D_0) and is given by (Ulbrich 1983):

$$\Lambda = \frac{3.67}{D_0} = \frac{4+\mu}{D_m} \qquad (14)$$

When μ is set to 0, N(D) reduces to an exponential distribution, given by:

$$N(D) = N_0 \exp(-\Lambda D) = N_T \Lambda \exp(-\Lambda D) \qquad (15)$$

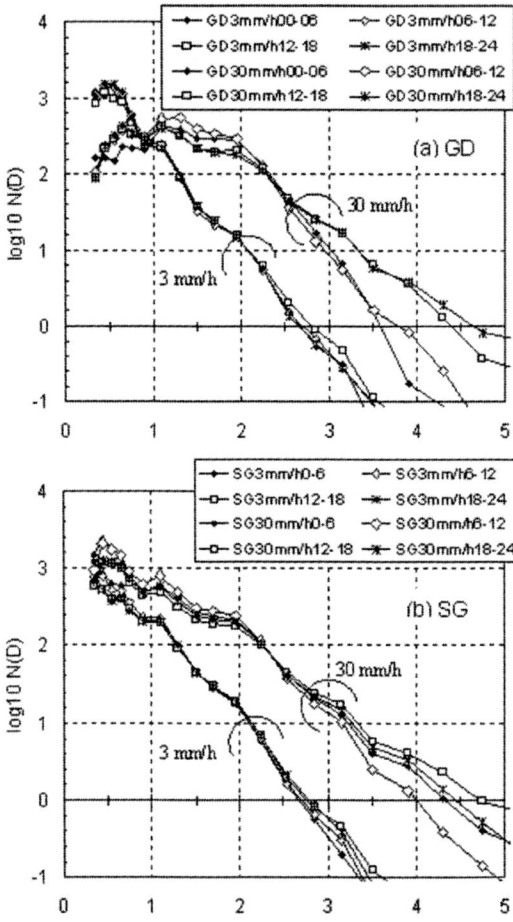

Fig. 2. Averaged DSD characteristics showing diurnal variations in three different Asian monsoon regions, (**a**) Gadanki, India, (**b**) Singapore and (**c**) Koto Tabang, West Sumatra. The averaged DSDs are for 0–6, 6–12, 12–18 and 18–24 local time for rain rates of 3 and 30 mm h^{-1}. For the latter, the afternoon DSDs are seen to have broader distributions (from Kozu et al. 2006)

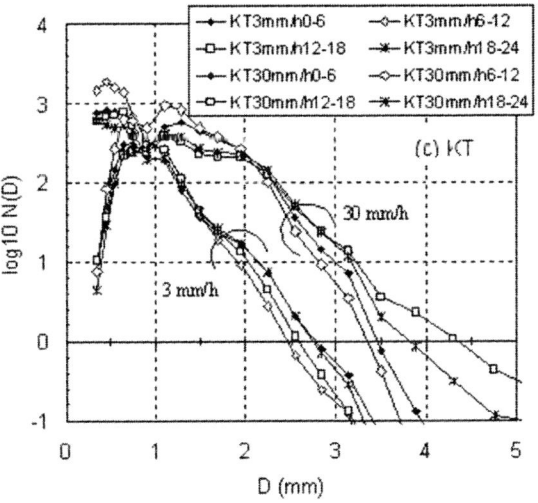

Fig. 2. (Continued)

with only two parameters (N_0 and Λ or N_T and Λ) characterizing the DSD. The well-known Marshall-Palmer (MP) DSD model is a special case of the exponential distribution, with N_0 set to 8000 mm^{-1} m^{-3} and Λ (mm^{-1}) = 4.1 $R^{-0.21}$, R being the rainfall rate in mm h^{-1}. Other special cases include Joss-thunderstorm, Joss-widespread and Joss-drizzle (all given in Joss et al. 1968).

The comparison of DSD shapes with widely varying liquid water contents (W) is made possible by normalizing and scaling the N(D) in different ways depending on the application (e.g., Illingworth and Blackman 2002). Recall from the earlier section that one important radar variable is K_{dp} which is proportional to the product of WD_m. Thus, normalizing the D by D_m is the first step. Scaling the N(D) by the factor N_w is often done based on the water content and D_m as follows (for details refer to Bringi and Chandrasekar 2001):

$$N_W = \frac{4^4}{(\pi \rho_W)} \left[\frac{10^3 W}{D_m^4} \right] \quad \text{mm}^{-1} \text{ m}^{-3} \quad (16)$$

where ρ_w is the water density (1 g cm^{-3}), W is the water content in g m^{-3} (see Eq. (6)) and D_m (see Eq. (3)) is the mass-weighted mean diameter (in mm). The physical interpretation of N_w is that it is the same as the N_0 parameter of an exponential DSD (see Eq. (15)) which has the same W and D_m as the actual N(D). From measured N(D) it possible to compare the shapes of DSDs with widely varying water contents by simply plotting the (scaled) $N(D)/N_w$ versus (normalized) D/D_m (for example, see Fig. 4 of Bringi et al. 2003). Assuming Rayleigh scattering and a power law form for the terminal fall speed of drops, $V(D)=3.78D^{0.67}$ (Atlas and Ulbrich 1977), it is easy to show that the reflectivity factor Z is related to rain rate R as:

$$Z = \frac{\alpha'(\mu)}{\left(\sqrt{N_W}\right)} R^{1.5} \tag{17}$$

where $a'(\mu)$ depends on μ only and other constants. Testud et al. (2001) further generalized (17) to be applicable to any general DSD, not necessarily of the gamma form. Note that there is no assumption in (17) that N_w be constant, rather it can vary with rain rate. In one extreme case N_w may vary linearly with R (the so-called equilibrium-type or 'concentration-controlled' DSD) in which case the exponent of R in (17) becomes 1 (linear Z-R relation). In another extreme, N_w may be constant and the exponent of R in (17) is fixed at 1.5 ('size-controlled').

Another, alternate way of describing the DSD without explicitly invoking a model (e.g., gamma model) is in terms of the rain water content (or, alternately total concentration N_T), the mass-weighted mean diameter (D_m; see Eq. (3)) and the normalized standard deviation of the mass spectrum σ_M (Ulbrich and Atlas 1998). For an arbitrarily-shaped DSD, the σ_M is defined as:

$$\frac{\sigma_M}{D_m} = \left\{ \frac{\int (D-D_m)^2 D^3 N(D) dD}{D_m^2 \int D^3 N(D) dD} \right\}^{1/2} \tag{18}$$

The ratio σ_M/D_m in Eq. (18) can also be thought of as another shape parameter, similar to μ in Eq. (13). This ratio for a gamma DSD reduces to:

$$\frac{\sigma_M}{D_m} = \left\{ \frac{1}{(4+\mu)} \right\}^{1/2} \tag{19}$$

The σ_M is a useful parameter; for example, referring back to Fig. 2, one would expect higher σ_M for the afternoon convective rain compared with the morning events, for the same rainfall rates. The degree to which measured DSDs fall within the gamma model may be checked by using Eq. (19) (see also Bringi et al. 2003).

Based on the DSD, the 'still-air' rainfall rate (R) is defined as:

$$R = \left(\frac{\pi}{6}\right)\left[\int v(D)D^3 N(D)\,dD\right] \qquad (20)$$

where v(D) is the drop terminal velocity which at sea-level is given to a very good approximation by the Atlas et al. (1973) fit to the data of Gunn and Kinzer (1949):

$$v(D) = 9.65 - 10.3\exp(-0.6D);\ \mathrm{m\,s^{-1}} \qquad (21)$$

In the above, D is in units of mm. An altitude adjustment factor has been given by Beard (1985). Measurements of v(D) using the 2DVD will be given in Sect. 10.5 from different locations.

It is well-known that attenuation of the signal due to rain along the propagation path is related to the extinction cross-section of the drops weighted by N(D). For simplicity if the drops are assumed spherical in shape then the specific attenuation is defined as:

$$k = 4.343 \times 10^3 \left[\int \sigma_{ext}(D)N(D)\,dD\right];\ \mathrm{dB\,km^{-1}} \qquad (22)$$

where σ_{ext} is the extinction cross-section in m^2, and N(D) dD in m^{-3}. Atlas and Ulbrich (1977) have approximated k as:

$$k = 4.343 \times 10^3 C_\lambda \left[\int D^n N(D)\,dD\right] \qquad (23)$$

where C_λ and n depend on both wavelength and temperature. As a first approximation, n ≈ 4 at C and X-band frequencies so that k is proportional to the 4th moment of the DSD. Recall from the discussion following Eq. (7) that K_{dp} is proportional to the product WD_m which by definition is related also to the 4th moment of the DSD. If we further approximate k ≈ k_h (the specific attenuation at H-polarization due to oblate, perfectly oriented drops) then k_h can be linearly related to K_{dp} as $k_h ≈ \alpha K_{dp}$ which forms the basis of attenuation-correction of the measured Z_h for dual-polarized radars. Similarly, the specific differential attenuation between H and V polarized signals ($k_{hv} = k_h - k_v$) can also be linearly related to K_{dp} which forms the basis of correcting the measured Z_{DR} for differential attenuation. More generally, the k_h–K_{dp} and k_{hv}–K_{dp} relations may be expressed as power laws of the form $k_h = \alpha\, K_{dp}^{b}$ and $k_{hv} = \beta\, K_{dp}^{b}$ where the exponent (b) is close to 1 whereas the coefficients depend on wavelength, temperature (mainly for α), the mean drop shapes and the D_m if it exceeds around 2.5 mm. Thus, the attenuation-correction procedures also depend on some aspects of the rain microstructure.

10.2.3 DSD estimation from polarimetric radar measurements

D_m estimation from Z_{DR}

Equation (4) in Sect. 10.1.3 gives a relatively simple formula to obtain a direct estimate of D_m from attenuation-corrected Z_{DR}. However, due to the nature of DSD and shape fluctuations, there will be some uncertainties associated with the D_m estimates. Figure 3 shows examples of calculations made at (a) S- (b) C- and (c) X-band, respectively. The calculations use disdrometer (2DVD and Joss-Waldvogel) measurements of N(D) from many different climatic regimes. The N(D) has been fitted to the normalized gamma model with parameters (N_w, D_m and μ: the fitting procedure has been described in Bringi et al. 2003). A standard model for the drop axis ratios was used, based on the oblate approximation to the equilibrium drop shapes of the numerical model of Beard and Chuang (1987). The intensity contours represent the frequency of occurrence on a $\log_{10}$ scale. The scatter in the data indicates the amount of uncertainty associated with the D_m retrieval (mainly due to μ variations). Note the effect of non-Rayleigh scattering at C-band for $D_m > 2$ mm. Another point to note is that the presence of large drops (4–8 mm) can introduce additional errors due to their non-oblate shapes, particularly at C-band where the resonance effects have their greatest impact. These effects are not included in Fig. 3(b) but will be discussed briefly in Sect. 10.3.

Estimating N_w

Assuming D_m is retrieved using Z_{DR} (e.g., by mean fits to data as illustrated in Fig. 3), one can further assume a most probable value for the shape parameter μ (though variable, generally assumed to be 3 for convective rain and 0 for stratiform rain) in the gamma DSD. The N_w parameter can be derived in two possible ways, first by recognizing that for Rayleigh scattering Z_h can be expressed as:

$$Z_h = F_Z(\mu)N_w D_m^7 \qquad (24)$$

where $F_z(\mu)$ is weakly dependent on μ (see Chap. 7 of Bringi and Chandrasekar 2001). Based on Eq (24) a power law (using the rain microstructure data) can be fitted of the form:

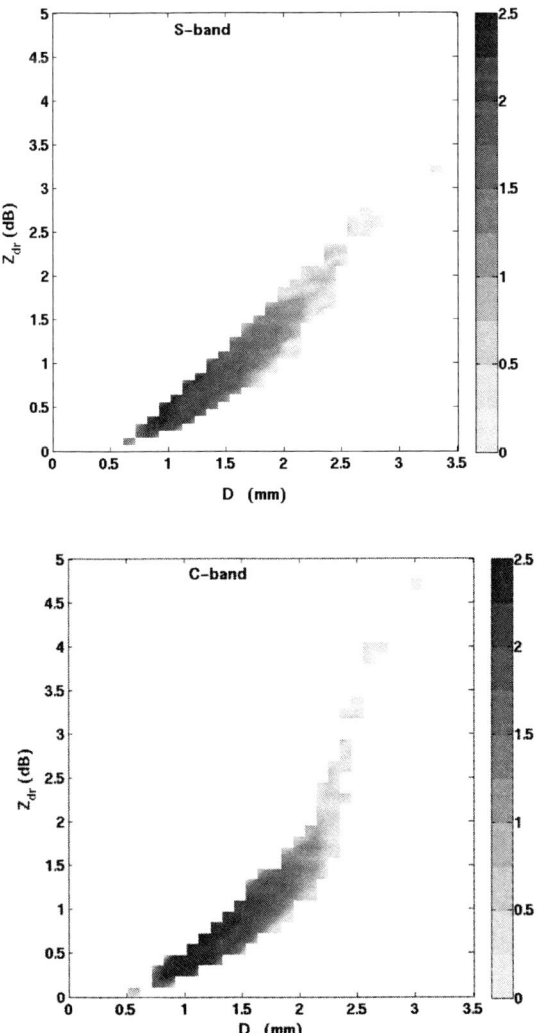

Fig. 3. Calculations of Z_{DR} versus the mass-weighted mean diameter (D_m) at S-, C- and X-bands, based on measured drop size distributions from a variety of climatic regimes. The grey scale contours the frequency of occurrence of ZDR – Dm pairs on a log scale

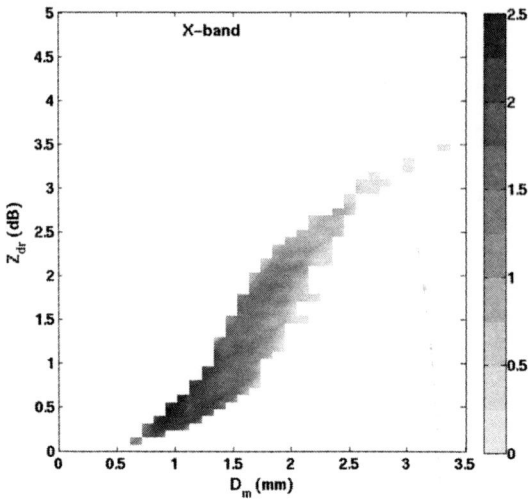

Fig. 3. (Continued)

$$N_w = aZ_h Z_{DR}^{-b} \tag{25}$$

The second formulation is based on Eq. (16) and the discussion following Eq. (7) where K_{dp} was shown under Rayleigh scattering to be proportional to the product WD_m. It follows that N_w is proportional to $K_{dp} D_m^{-5}$. Using rain microstructure data, the N_w estimator can be expressed as (e.g., Gorgucci et al. 2002):

$$N_w = \alpha K_{dp}^b Z_{DR}^{-c} \tag{26}$$

The above procedures (which are based on Sects. 10.1.3 and 10.2.2) formulate the gamma parameter estimators based on dual-polarized radar data. However, the μ parameter is assumed to be known a priori since it is difficult to estimate it using Z_h, Z_{DR} and K_{dp} alone. Brandes et al. (2002) have tried to overcome this problem by deriving a relation between μ and Λ (see Eq. (13)) from 2DVD measurements. In this case, the parameters N_T and D_m can be estimated from Z_h and Z_{DR} data alone. However, a universal μ- Λ relation valid for all rain types does not appear to be valid (Ulbrich and Atlas 2007).

Once the DSD is characterized from radar measurements, rainfall rate (R) and W can be estimated using Eqs. (20) and (6), respectively. Alternately, the rain rate (or W) can be formulated directly in terms of radar observations (assuming Z_h and Z_{DR} have been corrected for rain attenuation, especially at C- and X-bands) as power laws in three forms (see Chap. 8 of Bringi and Chandrasekar 2001):

$$R = a\, Z_h Z_{DR}^b \tag{27a}$$

or,

$$R = a\, K_{dp}^b \tag{27b}$$

or,

$$R = a\, K_{dp} Z_{DR}^{-c} \tag{27c}$$

The coefficients and exponents in the above equations depend on the rain microstructure, a theme that pervades this entire Chapter.

Caveats

Prior to any DSD retrievals from polarimetric measurements, two factors need to be established. Firstly, that the radar calibration is sufficiently accurate, particularly for Z_h (within ± 1 dB) and Z_{DR} (within ± 0.2 dB). There are several techniques (e.g., Joe and Smith 2001) to determine the absolute calibration of the radar, e.g., (a) solar flux, (b) metal sphere, (c) external source, (c) comparison with disdrometers, and (d) using the rain medium itself for calibration. The latter makes use of the fact that the range profile of the measured differential phase Φ_{dp} can be reconstructed from the corresponding range profiles of Z_H and Z_{DR}. This has been demonstrated at S-band by Goddard et al. (1994a,b), at C-band by Tan et al. (1995) and more recently by Thurai and Hanado (2005), but these techniques assume that Z_{DR} accuracy is within ± 0.2 dB. Techniques developed for Z_{DR} calibration include, (a) vertically-pointing in rain with the azimuth scanning from 0–360° (Gourley et al. 2006), or (b) comparison with disdrometers, or (c) a combination of solar flux and regular receiver calibration via injection of test signals at the receiver input (Zrnic et al. 2006).

The second factor that needs to be established is that the radar echoes are due to rain and not due to non-meteorological echoes (such as ground clutter, biological scatterers, etc.) nor due to hydrometeors other than rain, such as hail, graupel, melting or wet snow or even dry snow. There are several ways to identify and remove non-meteorological echoes (see for example, Gourley et al. 2007). In addition, there are several different hydrometeor classification schemes (for example, Ryzhkov et al. 2005) to identify rain from non-rain hydrometeors using distinguishing characteristics of the various

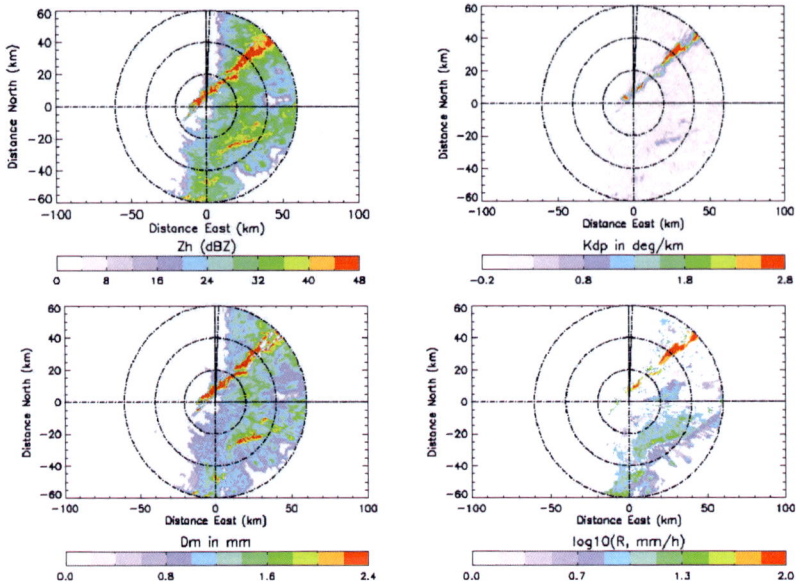

Fig. 4. PPI scan data taken at 1.5 deg elevation angle showing a 'snap-shot' of the 'Baiu' front event at 13:43 UTC on 8 June 2004. *Top left* shows the attenuation-corrected reflectivity (Z_h) field and the *top right* shows the specific differential propagation phase (K_{dp}) field. The two lower panels show the retrieved median diameter D_m and $\log_{10}$ of rainfall rate. The convective line of cells along the north-east direction shows high reflectivity (up to 55 dBZ) and high K_{dp} (up to around 2.4° km^{-1}), caused by high D_m (up to 2.4 mm) and rainfall rates as much as 70 mm h^{-1}. Details of the field campaign can be seen (from Nakagawa et al. 2005)

polarimetric signatures. These use fuzzy logic schemes along with 'class membership functions' to identify different types of hydrometeors and are now being adapted for operational use.

An example at C-band

An example of retrieving D_m and R in an 'all-rain' event is shown in Fig. 4. The data were taken from the C-band radar located in Okinawa, Japan, named COBRA (Nakagawa et al. 2003). The top left panel shows a low elevation angle PPI sweep of attenuation-corrected reflectivity while the top right panel shows the K_{dp} field. These data were collected during a long-duration 'Baiu' frontal event, the analysis of which is described in detail by Bringi et al. (2006).

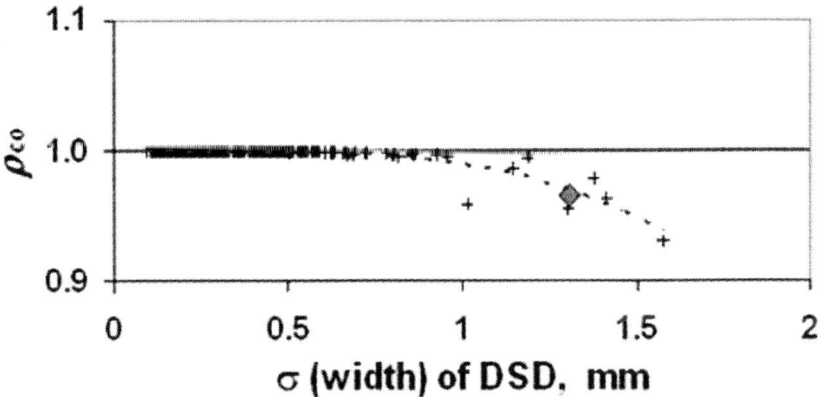

Fig. 5. Calculations (+ signs) using the DSD and axis ratios measured by the 2DVD for a localized convective event in Ontario, Canada, showing that ρ_{co} at C-band could be used to estimate the width of the DSD in heavy rainfall. Also marked on the plot (as shaded diamond) is one measurement point captured by the King City radar during this event. The width here refers to the standard deviation of the mass spectrum (σ_M; see Eq. (18))

Convection along the 'Baiu' front is clearly visible; the retrieved D_m shows values as high as 2.4 mm, with rain rates reaching nearly 100 mm h^{-1}. South of this line of convection, the D_m values are generally lower (< 1.5 mm) and lower rain rates (< 15 mm h^{-1}).

Use of ρ_{co}

The use of Z_h, Z_{DR} and K_{dp} does not enable a direct estimation of the DSD spectral width. Recent analysis using C-band radar data suggests that the co-polar correlation coefficient, ρ_{co}, (referred to in Sect. 10.1.3) could be useful for estimating the width of the mass spectrum (σ_M), defined in Eq. (18). Figure 5 shows calculations made using 2DVD measurements of rain microstructure from Ontario, Canada, which show that ρ_{co} at C-band begins to reduce in magnitude (from ~1) for $\sigma_M > 1$ mm. This localized convective event was simultaneously observed with the C-band King City radar (Hudak et al. 2006), located 30 km away from the 2DVD site. The measured ρ_{co}, averaged over the 2DVD site is marked in the figure. This is the first evidence of the lowering of ρ_{co} which has been quantitatively linked with the increase in DSD spectral width. This reduction arises for rain events with broad DSDs normally associated with high D_m (>2 mm) and quite often with heavy rainfall rate. As mentioned near the end of Sect. 10.1.3, non-Rayleigh scattering effects cause the back scatter differential phase between H and V polarizations to vary across the size range 4–7 mm, leading to de-correlation of the H and V received signals. Further investigations in the near future will address this possibility of using ρ_{co} to characterize the DSD width since such information cannot be obtained accurately from Z_h, Z_{DR} and K_{dp} alone.

10.2.4 DSD estimation from advanced disdrometers

There are several instruments which are designed to infer or measure in-situ DSDs. As mentioned in Sect. 10.1.1, the 2DVD is one such system. It is also the most advanced disdrometers currently available. Its operating principles, specifications and rain microstructure measurements are fully described in Chap. 2 of this book and hence will not be repeated here.

There are also Doppler radars which are capable of inferring the DSDs from the power spectral density using a known relationship between drop diameter and terminal fall-speed. One such example is the micro-rain radar operating in the 20 GHz band (Peters et al. 2005). Another example is the Precipitation Occurrence Sensor System (POSS) described in detail by Sheppard (1990). This is a bistatic, continuous wave, horizontally polarized, X-band Doppler radar. The transmitter and receiver antenna are angled from the vertical by 20 degrees to define a measurement volume extending to about 3 m above the radomes. Precipitation falling through the measurement volume generates a continuous Doppler velocity signal that is sampled to produce a 1 minute average of 960 Doppler velocity spectral measurements. Using a relationship between the terminal velocity of raindrops and their size, the velocity spectrum is mathematically inverted to estimate raindrop size distribution (DSD). Moments of this DSD are then calculated to estimate several rainfall parameters.

DSD measurements from various (co-located) disdrometers have been compared in the past by several investigators (e.g., Krajewski et al. 2006; Tokay et al. 2001; Peters et al. 2005; Sheppard and Joe 1994). The last reference (Sheppard and Joe 1994) shows comparisons of 1-minute averaged DSDs from POSS and 2 other types of disdrometers, namely JWD and 2D grey scale spectrometer. Similar comparisons between POSS and 2DVD have been made more recently by Thurai et al. (2007a). The multi-panel Fig. 6 shows a sub-set of one example. These measurements were taken during a stratiform precipitation event in

Ontario, Canada, which lasted over 10 hours. The top panel (a) shows the time series of the DSD (in grey scale) from the POSS measurements.

As seen from the plot, the drops were mostly smaller than 3 mm. The D_m parameter of the DSD is superimposed on the plot. The second panel (b) compares the D_m from POSS with those from the co-located 2DVD and the third panel compares the rainfall rates. The agreement is excellent throughout the event, both for D_m and for rainfall rates. The fourth panel compares the reflectivity computed using the DSDs from

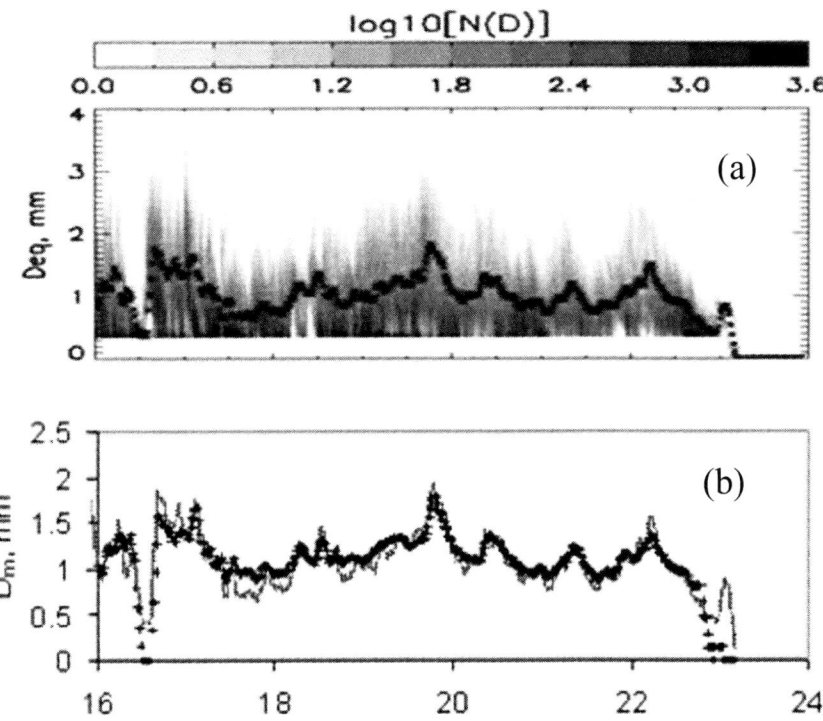

Fig. 6. DSDs for the 30 November 2006 'cold rain' event in Ontario, Canada. (a) DSD time series from POSS, together with the D_m values superimposed as black stars; (b) D_m comparisons between POSS and 2DVD; (c) rain rate comparisons between POSS and 2DVD; (d) Z_h comparisons computed from 2DVD and POSS, together with those extracted from the King City radar scans; (e) & (f) DSD comparisons at 16:44 and 19:49 UTC from 2DVD and POSS. In (b), (c), (d), (e) and (f), the + marks are from 2DVD and the grey lines are from POSS; further, the circles in (d) show the radar measured reflectivity values over the 2DVD and POSS site, taken from RHI and low elevation PPI scans

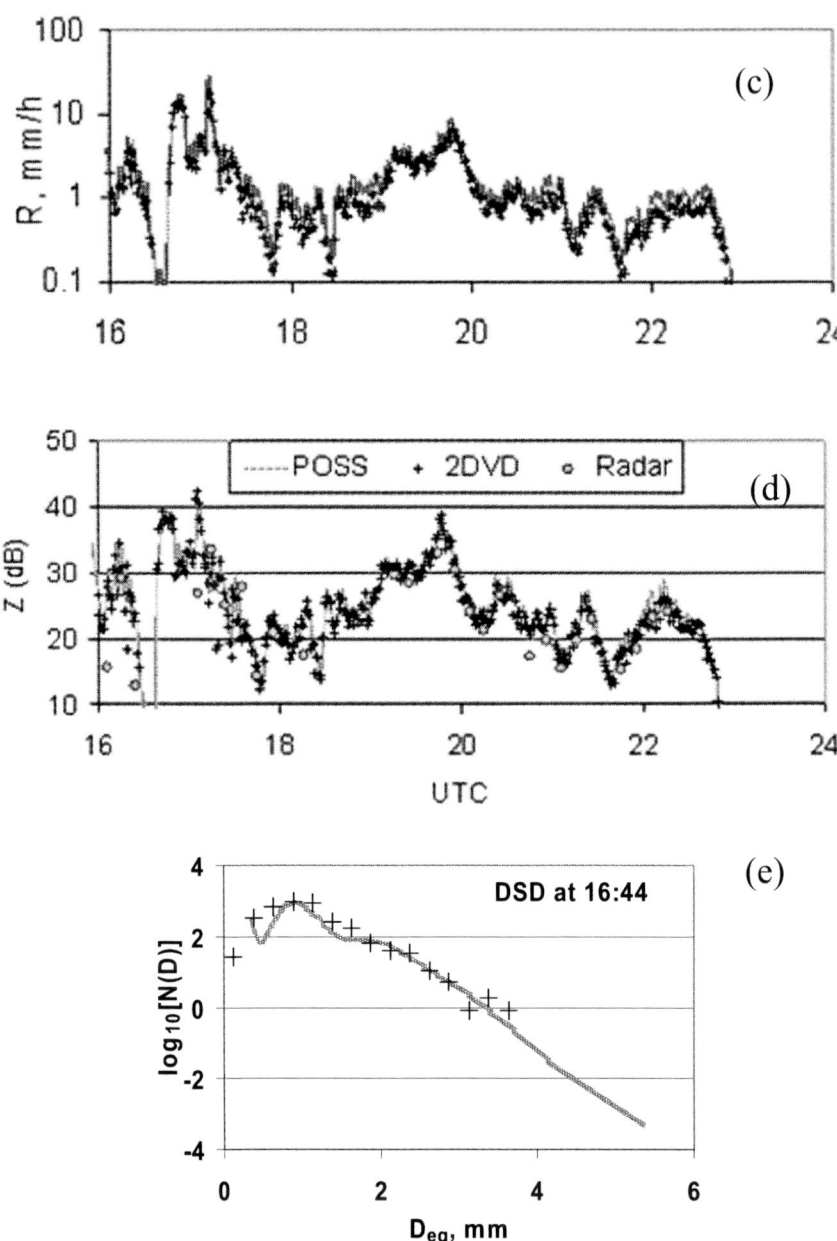

Fig. 6. (Continued)

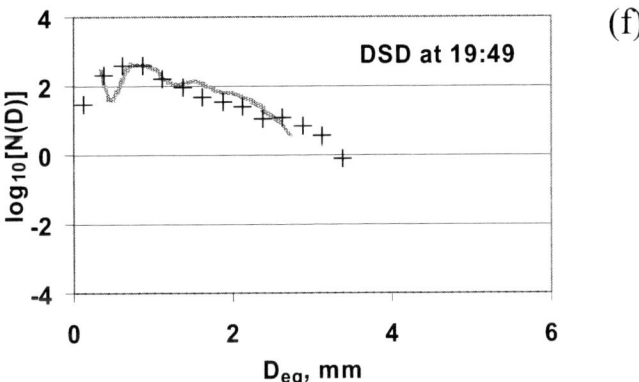

Fig. 6. (Continued)

the two instruments. In addition, the panel shows reflectivity measurements over the disdrometer site from the King City radar PPI and RHI sweeps with time samples spaced at around 10 minutes apart. These data demonstrate the accuracy of the King City radar reflectivity calibration, since there was no need for any systematic offsets to be applied to the radar data. Later in Sect. 10.3, we will show the consistency between the Z_h-Z_{DR} from the radar and the computations using the 2DVD-based rain microstructure measurements for this event.

The DSDs themselves are compared for two different time periods in the last panel. They represent rainfall rates in the 5–11 mm h^{-1} range and D_m values in the 1.5–1.8 mm range. Agreement is once again very close, despite the significant differences in measurement principle, sampling volumes, and space/time averaging between the two instruments. Such comparisons of both integral parameters as well as the actual DSDs demonstrate that systematic and random measurement errors for both instruments are low, a condition necessary for accurate measurements of rainfall characteristics as well as for accurate characterization of rain microstructure.

10.2.5 Global DSD characteristics

The variability of the DSD across different climatic regimes can be demonstrated by examining a plot of $\log_{10}(<N_w>)$ versus $<D_m>$ where angle brackets denote averages. For example, Fig. 7a shows such data retrieved from disdrometer measurements (2DVD or Joss-Waldvogel types) as well as from polarimetric radar data for stratiform rain. The radar retrievals of N_w and D_m follow the general principles elucidated in

Sect. 10.2.3 and are spatial averages from a small number of PPI sweeps, the averaging regions being selected by expert examination of the data (details are in Bringi et al. 2003). Most of the radar data are from the trailing stratiform regions of convective squall lines or mesoscale convective complexes. The disdrometer data used in the calculations of N_w and D_m (according to Eqs. (16) and (3), respectively) represent 2-minute integration of the DSD. The shape parameter μ and σ_M/D_m were also calculated and found to be consistent with Eq. (19). A simple classification of stratiform rain was based on the standard deviation (σ_R) of R over 5 consecutive samples being < 1.5 mm h^{-1}. This threshold is based on the fact that stratiform rain is 'steady' and was checked against several datasets where radar data confirmed that a 'bright band' was present. Only rain rates > 0.5 mm h^{-1} were considered.

For stratiform rain there appears to be a clear inverse relation between $\log_{10}(<N_w>)$ and $<D_m>$; in fact, it is quite remarkable that a straight line fit results from the composite disdrometer and radar retrievals, these data encompassing a number of regimes from near equatorial to the U.S. High Plains. From a microphysical perspective, stratiform rain results via the melting of snowflakes and/or tiny graupel or rimed particles. If the bright band is 'strong', then it likely reflects melting of larger, low density and dry snowflakes into relatively larger raindrops, whereas if the bright band is 'weak' then it may reflect the melting of tiny, compact graupel or rimed snow particles (Fabry and Zawadski 1995). In essence, the large, low density snowflakes lead to DSDs that have smaller $<N_w>$ and larger $<D_m>$ relative to the tiny, compact graupel or rimed snow particles. In Fig. 7a the two horizontal lines represent two Z-R relations on this plot based on Eq. (17). It is clear that the coefficient of the Z-R relation varies with the microstructure of stratiform rain.

Figure 7b shows similar results for convective rain. Again, the classification criteria from disdrometer data was based on R > 5 mm h^{-1} and σ_R > 1.5 mm h^{-1}. For the radar data it was relatively easy to select convective rain cells by examination of the PPI sweeps. For reference the Marshall-Palmer N_0=8000 mm^{-1} m^{-3} is drawn as a horizontal line in Fig. 7b. Note a cluster of data points with $<D_m>$=1.5–1.75 mm and $\log_{10}<N_w>$=4–4.5, the regime varying from near equatorial (Papua New Guinea) to sub-tropics (Florida, Brazil) to oceanic (TOGA-COARE, Kwajalein, SCSMEX). This cluster may be referred to as a 'maritime'-like cluster where rain DSDs are characterized by a higher concentration of smaller–sized drops. The Fort Collins flash-flood event is unusual for Colorado as the data fall in the 'maritime'-like cluster. The vertical structure of reflectivity in this event was highly unusual for summer time Colorado storms resembling instead the vertical profile of Z in oceanic convection (Petersen et al. 1999).

The second 'cluster' is characterized by $<D_m>=2-2.75$mm and $\log_{10}<N_w>=3-3.5$, the regime varying from the U.S. High Plains (Colorado) to continental (Graz, Austria) to sub tropics (Sydney, Australia; and Arecibo, Puerto Rico). The 'continental'-like cluster may be defined which reflects rain DSDs characterized by a lower concentration of larger-sized drops as compared with the previously-defined 'maritime'-like cluster. One of the main implications of Fig. 7 is that dual-polarized radar data can be used to estimate a 'pol-based' Z-R relation according to Eq. (17) as described by Bringi et al. (2004) without the need for classification of rain types. Another implication is that for conventional non-polarimetric radars, a climatological Z-R relation may be derived from Fig. 7 using an appropriate N_w=constant value along with Eq. (17).

10.2.6 Seasonal variation

Even for a given location, the DSD characteristics can differ greatly, depending on the rain production mechanisms. An example from Okinawa is shown in Fig. 8. It compares histograms of (a) D_0 and (b) N_w for warm shallow rain and typhoon, the former occurring in December-February period and the latter occurring in the August-October period (D_0 is closely related to D_m via Eq. (14)). The main microphysical difference between the two rain types is warm rain formation at altitudes less than 3 km for the shallow rain case and ice phase microphysics for the typhoon case. For the shallow rain case, additional measurements from the 2DVD had been used to derive N_w, D_0 and μ in order to validate the mode of the histograms derived from the C-band polarimetric radar data (COBRA) from Okinawa. The typhoon case, on average, has mean N_w and mean D_0 values which lie in the tropical maritime 'cluster' identified earlier in Fig. 7b, whereas for the warm shallow rain, the mean N_w and the mean D_0 are distinctly different from this 'cluster' with smaller $<D_0>$ values and significantly larger $<N_w>$ values. Both were included in Fig. 7b. Further analysis (not given here) has shown that even for the same rain rate interval (say 20–40 mm h^{-1}) the range of N_w and D_0 values from the shallow rain events were 'shifted' away from the typhoon range, that is, the shallow rain is dominated by a larger concentration of smaller drops compared with the typhoon case for the same rain rate interval.

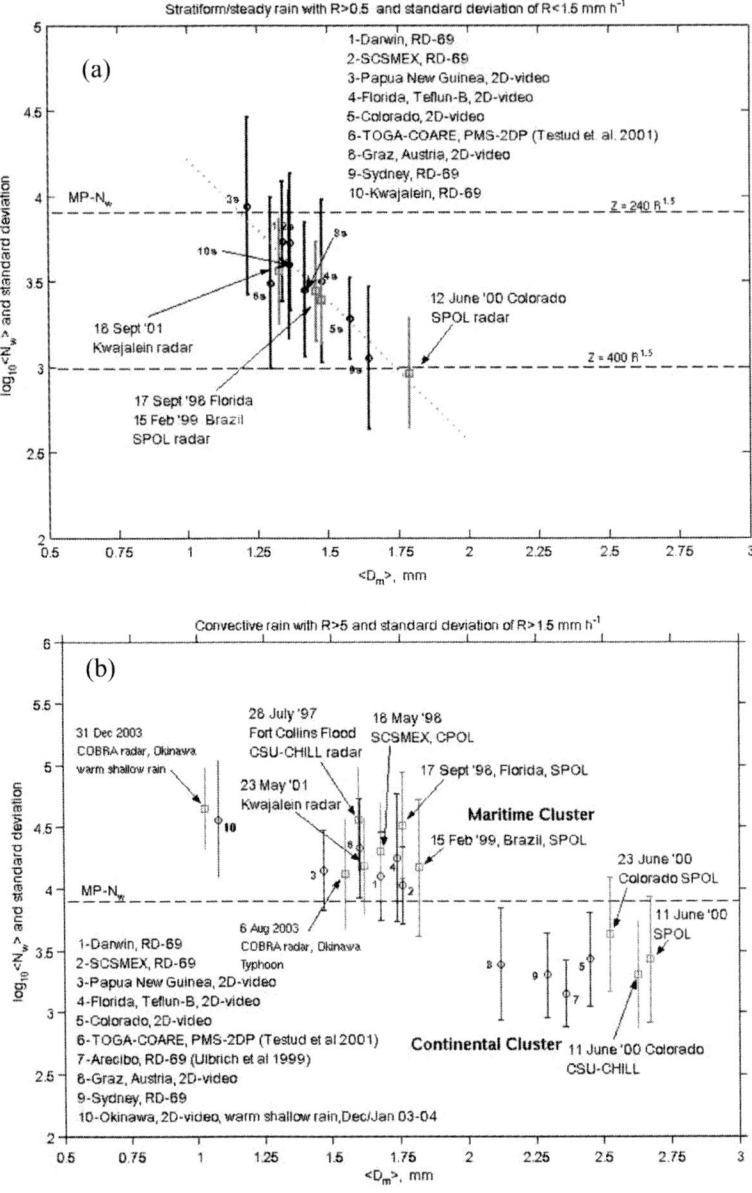

Fig. 7. (a) The average value of $\log_{10} N_w$ (with $\pm 1\sigma$ std dev bars) versus average D_m from disdrometer data (numbered open circles) and dual-polarization radar retrievals (open squares as marked) for stratiform rain. Dotted line is the least squares fit. (b) As in (a) except data for convective rain. Note that N_w is the 'normalized' intercept parameter (see Eq. (16)) and D_m is the mass-weighted mean diameter of a 'normalized' gamma DSD

Another set of histogram examples from Okinawa, this time from a long-duration event which occurred during the 'Baiu' season (May-June), is shown in Fig. 8c. (For details of field campaign, see Nakagawa et al. 2005). This event composed of intense cells along the convection line embedded in large regions of stratiform rain (as shown in Fig. 4). Note that the Fig. 8c histograms were derived from 2DVD data over the entire period of the event most of which was dominated by stratiform rain. Hence, the overall average values for N_w and D_0 estimated to be 6,000 mm^{-1} and 1.42 mm, respectively, lie on the tropical part of the stratiform 'line' in Fig. 7a.

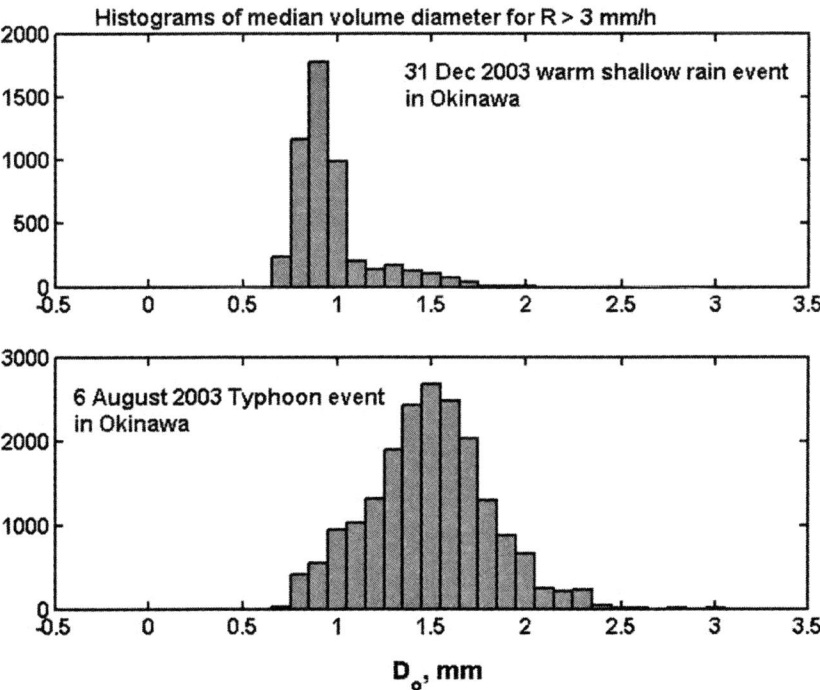

Fig. 8a. Histograms of D_0 for warm shallow rain (*top*: from 2DVD)) on 31 December 2003 compared with the typhoon case (*bottom*: from COBRA radar) on 6 August 2003 in Okinawa, Japan

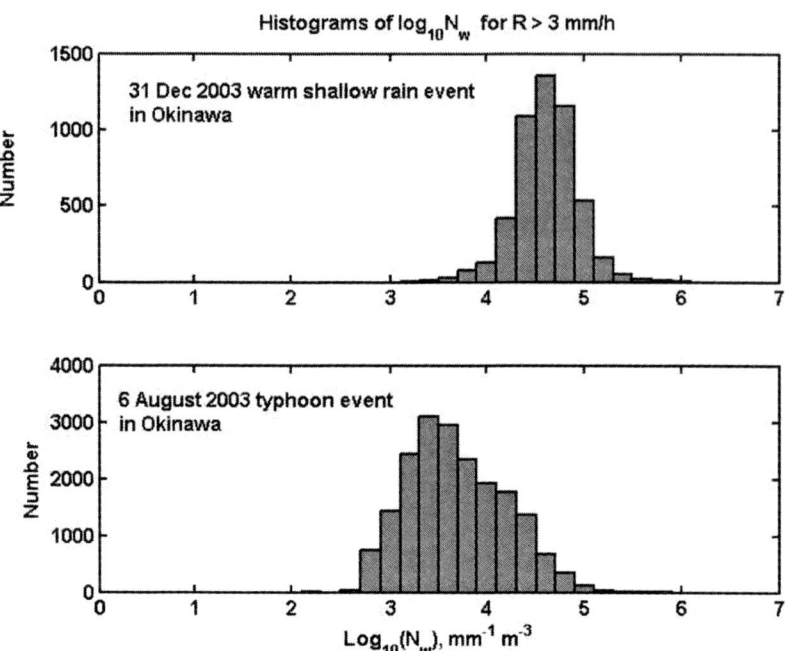

Fig. 8b. As in Fig. 8a, except histograms of $\log_{10}N_w$ are shown

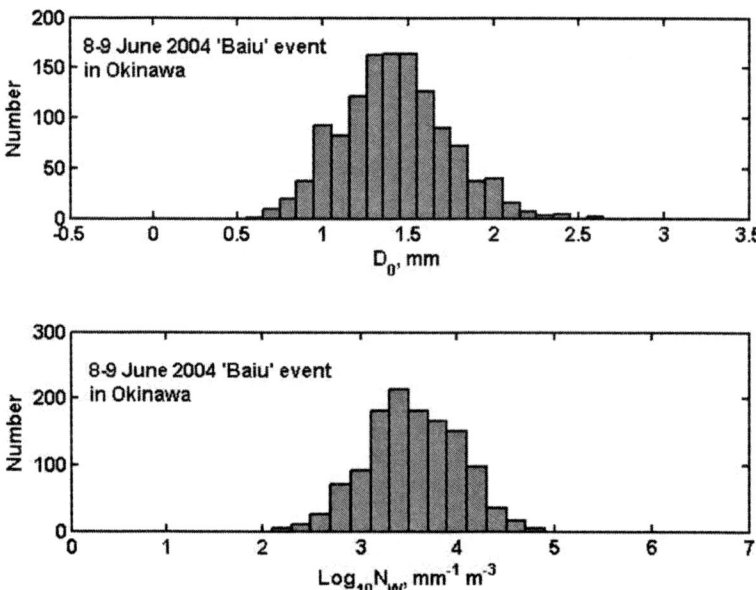

Fig. 8c. Histograms of D_0 (*top* panel) and $\log_{10}N_w$ (*bottom*) for the long duration 'Baiu' event on 8–9 June 2004, again in Okinawa, Japan

10.3 Drop shapes

Raindrop shapes have been studied for the last several decades (for example, McDonald 1954) but only in the last few years it has become possible to characterize it using experimental measurements and contouring techniques. Back in 1970, Pruppacher and Beard (1970) published their well known wind-tunnel measurements which showed clearly that the drop axis ratios (defined by the ratio between maximum vertical chord and the maximum horizontal chord) decreased monotonically with the drop equivalent diameter, D_{eq}. The implications for polarimetric radar were subsequently examined by Seliga and Bringi (1976), followed by experimental demonstration with the Chilbolton dual-polarization radar (Goddard et al. 1982) which showed that in order to explain the measured Z_{DR}, drop oscillations needed to be taken into account. Since then, several different types of methods have been used to characterize the drop shapes in their 'equilibrium' state as well as the oscillation modes (e.g., Beard 1984).

In this section, we summarize the latest set of results using the 2DVD instrument and give examples to show that the 2DVD-based shape contours are consistent with the radar observations. Accurate information on drop shapes is of course needed for effectively using the radar-measured polarimetric parameters Z_{DR}, K_{dp} and ρ_{co} in DSD parameter estimation as discussed in Sects. 10.1.3 and 10.2.3.

10.3.1 Axis ratio measurements from an artificial rain experiment

The latest measurements of drop axis ratios have come from an artificial rain experiment (Thurai and Bringi 2005) where drops generated from a hose were allowed to fall freely through the slats of a rail-road bridge 80 m above ground level, this height being sufficient for drops to reach terminal velocity as well as steady-state oscillations. A precisely calibrated 2DVD unit was placed on the ground in order to let drops fall through the 10 by 10 cm sensor area of the instrument. This experiment captured images of more than 115000 drops with D_{eq} ranging up to 9 mm, for which the effective mean axis ratios (denoted as <r/D> in Sect. 10.1.3) and their standard deviations ($\sigma(r/D)$) were derived. These are given in Table 2 (from Thurai and Bringi 2005). For 1.5< D_{eq}< 4 mm, the results closely follow the empirically-derived formula based on the Chilbolton radar polarimetric measurements (Goddard et al. 1995). The results also closely followed the formula given in Brandes et al. (2002)

but note that this formula represents a fitted equation to a number of different experimental measurements (e.g., laboratory, wind-tunnel and airborne probes) with different levels of measurement and sampling errors.

A revised equation (with more precision) was later given by Thurai et al. (2007b) to represent the axis ratio measurements from the 80 m bridge experiment. The formula is valid for drops larger than 1.5 mm since the resolution of the 2DVD instrument (see Chap. 2) did not enable the smaller drops to be measured accurately. The revised equation is as follows:

$$\frac{b}{a} = 1.065 - 6.25 \times 10^{-2} (D_{eq}) - 3.99 \times 10^{-3} (D_{eq}^2)$$

$$+ 7.66 \times 10^{-4} (D_{eq}^3) - 4.095 \times 10^{-5} (D_{eq}^4) \text{ for } D_{eq} > 1.5 \text{mm} \quad (28)$$

For smaller drops, the laboratory data by Beard and Kubesh (1991) can be considered more accurate for $0.7 \leq D_{eq} \leq 1.5$ mm. A fit to these data is given by:

$$\frac{b}{a} = 1.17 - 0.516(D_{eq}) + 0.47(D_{eq}^2) - 0.132(D_{eq}^3)$$

$$- 8.5 \times 10^{-3} (D_{eq}^4) \text{ for } 0.7 \leq D_{eq} \leq 1.5 \text{mm} \quad (29)$$

Below 0.7 mm, drops can be assumed spherical, i.e.:

$$\frac{b}{a} = 1 \text{ for } D_{eq} < 0.7 \text{mm} \quad (30)$$

Recently, Huang et al. (2007) have investigated drop axis ratios in natural rain using 2DVD measurements from different locations and have shown that under calm wind conditions, the mean axis ratios fit well to the above Eq. (28). Moreover, axis ratio distributions obtained from 2DVD measurements in an equatorial climate (western Sumatra) have shown to closely agree with those from the 80 m bridge experiment, for D_{eq} up to 3 mm (beyond that there were insufficient number of drops to derive statistically meaningful mean and standard deviations).

Table 2. Mean and standard deviation of axis ratio distributions derived from the 2DVD data taken during the artificial rain experiment

Diameter interval	Mean axis ratio	Standard Deviation
1.0–1.5	0.980	0.036
1.5–2.0	0.948	0.037
2.0–2.5	0.911	0.028
2.5–3.0	0.881	0.031
3.0–3.5	0.844	0.037
3.5–4.0	0.808	0.050
4.0–4.5	0.771	0.073
4.5–5.0	0.732	0.081
5.0–5.5	0.704	0.077
5.5–6.0	0.671	0.071
6.0–6.5	0.645	0.072
6.5–7.0	0.617	0.071
7.0–7.5	0.589	0.075
7.5–8.0	0.553	0.068
8.0–8.5	0.520	0.070
8.5–9.0	0.474	0.065
9.0–9.5	0.446	0.067
9.5–10.0	0.424	–

10.3.2 Drop contours

The 2DVD line scan camera images can also be processed to derive the contoured shapes of each individual drop, by filtering out the quantization noise of the instrument and subsequently using interpolation to obtain a 'smooth' contour. The images of all 115000 drops from the 80 m bridge experiment were subsequently processed in such manner and reported in Thurai et al. (2007b). Figure 9 shows the contours on a $\log_{10}$ probability scale for 3, 4, 5 and 6 mm drops. Superimposed on these contours in light blue are the equivalent oblate version given by Eq. (28) above. Whilst the 3 mm and to some extent the 4 mm drops show agreement with oblate shapes, the larger drops show increasingly flatter bases. The contours in black in Fig. 9 represent the most-probable shapes (also close to the mean shapes) and are over-plotted to show the extent of the deviation from their corresponding oblate approximations.

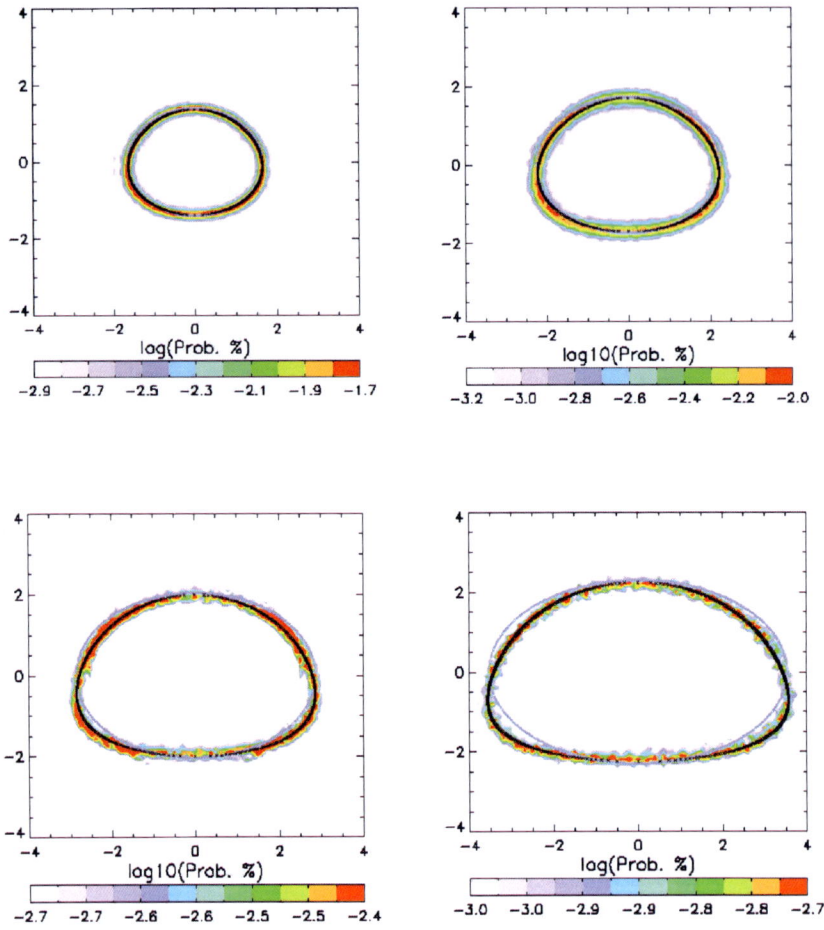

Fig. 9. Drop shape 'probability contours' for 3 mm (*top left*), 4 mm (*top right*), 5 mm (*bottom left*) and 6 mm (*bottom right*) drops. Superimposed in black and light blue are the curves derived using Eq. (31) and their oblate approximations given by Eq. (28), respectively. The larger drops show increasingly flatter base and deviate more and more from oblate shapes

The most probable contour for each drop diameter interval has been fitted to a smoothed conical equation whose x and y Cartesian coordinates are given by:

$$x = c_1 \sqrt{1-\left(\frac{y}{c_2}\right)^2} \left[\cos^{-1}\left(\frac{y}{c_3 c_2}\right)\right]\left[c_4\left(\frac{y}{c_2}\right)^2 + 1\right] \quad (31)$$

where the coefficients c_1, c_2, c_3 and c_4 have been fitted to D_{eq} (in mm) – dependent polynomials, in the following manner:

$$c_1 = -\frac{1}{\pi}\left(0.02914\,D_{eq}^2 + 0.9263 D_{eq} + 0.07791\right)$$

$$c_2 = -0.01938\,D_{eq}^2 + 0.4698\,D_{eq} + 0.09538$$

$$c_3 = -0.06123\,D_{eq}^3 + 1.3880\,D_{eq}^2 - 10.41\,D_{eq} + 28.34$$

$$c_4 = -0.01352\,D_{eq}^3 + 0.2014\,D_{eq}^2 - 0.8964\,D_{eq} + 1.226 \text{ for } D_{eq} > 4\text{mm}$$

$$c_4 = 0 \text{ for } 1.5 \leq D_{eq} \leq 4\text{mm}$$

Equation (31) is an enhanced version of the equation given in Wang (1982). The parameters c_1, c_2 and c_3 are the same as a, c and λ, respectively in Wang's shape model. The term containing c_4 is an additional term required to represent the mean shapes for drops larger than 4 mm whose base becomes increasingly flatter with larger size. Figure 10 shows the mean contours derived from Eq. (31) for D_{eq} up to 6 mm. Remarkably, they are consistent with the shapes predicted by the full numerical model of Beard and Chuang (1987), which is also included for comparison.

Huang et al. (2007) have examined the 2DVD data in natural rain in Okinawa and west Sumatra (Indonesia) and found that Eq. (31) can also represent the most probable contours, at least for D_{eq} up to 4 mm. (Beyond that, there have not been sufficient number of drops in natural rain to derive statistically meaningful probability contours.) In addition to the mean or the most probable contours, the shape variations were also investigated, these variations occurring as a result of drop oscillations. The inner and outer contour limits corresponding to 95% probability level for the 4 mm drops have shown that the shape variation occurs more in the vertical than in the horizontal. Moreover, the vertical variation appears to be somewhat higher at the top than at the bottom.

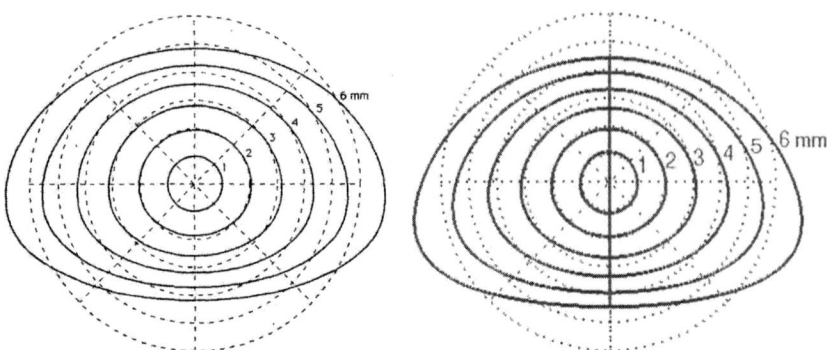

Fig. 10. Computed shapes from the Beard-Chuang (1987) model (*left*) compared with the shapes derived using Eq. (31) (*right*) for 1, 2, 3, 4, 5 and 6 mm drops. The origin is at the center of mass. The dashed circles of each diameter are divided into 45 degree sectors

To quantify (approximately) the variations for the 4 mm drop probability dimensions, if we let the drop center of gravity to be at the origin (0, 0), the vertical dimensions lie in the range 1.7 ± 0.5 mm at the top and –1.65 ± 0.35 mm at the bottom. In the horizontal plane, the corresponding limits are 2.2 ± 0.3 mm and –2.2 ± 0.3 mm. In other words, in the horizontal plane, the variations are very similar whereas in the vertical plane, the variations are slightly higher at the top. These results indicate that the drop oscillation mode should be treated as 'oblate-conical' instead of 'oblate-prolate' mode which is often used in many theoretical drop oscillation studies.

10.3.3 Consistency with polarimetric radar measurements

As explained in Sect. 10.1.3, the mean drop shapes affect the radar polarimetric parameters such as Z_{DR}, K_{dp} and ρ_{co}. Consistency between calculations based on the drop shape measurements from the disdrometer and the probable variation from polarimetric data should, therefore, confirm the validity of the mean shapes. Bringi et al. (2006) have shown this for a long duration 'Baiu' event in Okinawa. Having established the calibration accuracy of the C-band (COBRA) radar, the variation of Z_{DR} and K_{dp} with Z_h were examined (after applying an 'optimal' attenuation-correction procedure). Figure 11 shows one example from the 'Baiu' event. The top panel shows the grey-scale intensity plot shows K_{dp} versus Z_h on a $\log_{10}$ scale, derived from a full PPI scan taken during the event. Overlaid on the plot is the expected trend (i.e., mean ± standard deviation) based only on the 2DVD (located 15 km away) based DSDs and axis ratios. K_{dp} values extend from near 0 up to nearly 3° km^{-1}, the latter corresponding to ~75 mm h^{-1} rainfall rate. The close agreement between the radar-derived and the 2DVD-based variations gives credibility to the statistical 'representativeness' of the 2DVD measured DSDs and drop shapes. If either the reflectivity calibration of the radar or of the two rain microstructure parameters had been wrong, the agreement in Fig. 11a would have been much poorer.

Another example, this time taken in Ontario, Canada, with the King City C-band radar and a 2DVD (located 30 km away, as was mentioned in Sect. 10.3.4) is shown in the bottom panel of Fig. 11. The event corresponding to this figure is the same as the one in Fig. 6 which did not require any significant attenuation corrections to be applied. Once again, the grey-scale intensity represents the Z_{DR} versus Z_h variation on a log$_{10}$ scale, derived from a full PPI scan taken during the event. The black marks denote the individual calculations using the 2DVD-based axis ratios and 1-minute integrated DSDs' for that event. The points lie along the radar-based variation. Rainfall rates up to ~25 mm h^{-1} are represented in the comparisons (as was shown earlier in Fig. 6). The event was a stratiform (but not steady state) rain event, with the melting layer at around 2 km.

In both panels of Fig. 11, the 2DVD-based calculations used the full shapes described by Eq. (31), but the number of drops above 4 mm were limited and hence the effects due to non-oblate shapes of the (larger) drops could not be quantified. However, calculations using model-based DSDs have shown that Z_{DR} could be affected for cases with unusual DSDs with large D_0 (> 3 mm), for example, as can occur along the leading edge of severe convective storms or aloft due to localized 'big drop' zones, especially at C-band, where the resonance effects have greater impact. For such cases, Thurai et al. (2007b) have recommended the use of the mean contoured shapes rather than their oblate approximations for retrieving rain rates and rain microstructure.

10.4 Drop orientation angles

We now report on the preliminary results of orientation angles (θ, Φ: zenith and azimuth angles, respectively) derived from the canting angles obtained from the two camera images. The data set is from the 80 m fall bridge experiment referred to earlier. Since these data were obtained

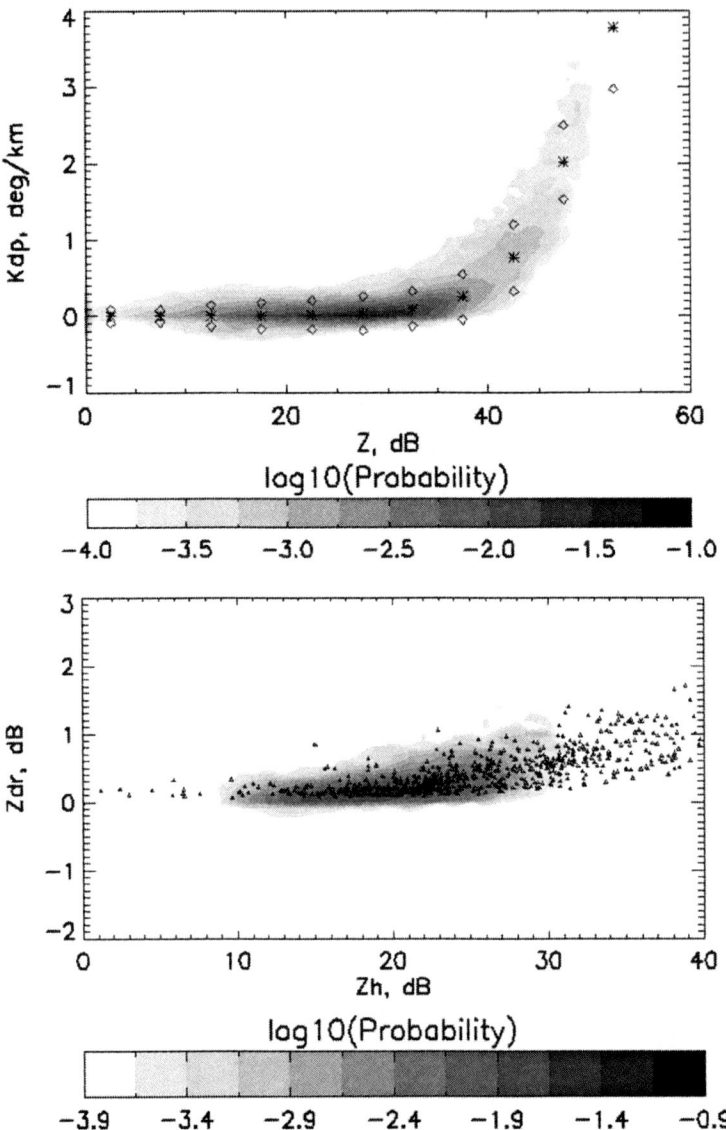

Fig. 11. Variation of K_{dp} (*top*) and Z_{DR} (*bottom*) versus Z_h for the 'Baiu' event in Okinawa and the 'cold rain' event in Ontario, respectively. For the former, the mean variation and the $\pm 1\sigma$ (standard deviation) computed using the 2DVD measurements during the 'Baiu' event are overplotted; for the latter, the individual comparisons from the 1-minute DSD from the 2DVD data taken during the cold rain event are overplotted

under calm wind conditions, the orientation angle distributions form a 'baseline' which may be compared with natural rain. Previously, the algorithm for deriving (θ, Φ) has been described in (Schönhuber et al. 2000; Schauer 1998) and applied to drops > 3.5 mm under (a) artificial rain conditions (35 m fall) and (b) for a (low wind) natural rain event from Papua-New Guinea. They found that the mean zenith angle under calm conditions was close to 5°. Later, the algorithm was further developed by Joanneum Research to allow orientation angles to be derived for drops > 2 mm (later 2DVD units were equipped with higher speed cameras which reduced the quantization noise).

When developing dual-polarized radar algorithms (see, also, Sect. 10.1.3) for rain rate or DSD parameter retrievals using Z_{DR} and/or K_{dp}, the assumption is often made that the canting angle (β) distribution in the plane of polarization is Gaussian with zero mean and standard deviation (σ_β) of 5-10°. Beard and Jameson (1983) argue that σ_β should be < 5° due to turbulence effects. The canting angle is the angle between the projection of the drop's symmetry axis on the polarization plane and the projection of the local vertical direction on this same plane (e.g., Holt 1984).

The orientation of the symmetry axis of a spheroid in 3D is defined by its zenith or polar angle (θ) and its azimuth angle (Φ). As such the orientation distribution of the symmetry axis is described on a spherical surface, i.e., p(Ω) dΩ gives the probability that the symmetry axis lies within the solid angle interval (Ω, Ω+ dΩ) and the Fisher distributions (Mardia 1972) are appropriate on a spherical surface as opposed to assuming a priori the Gaussian shape (see Sect. 2.3.6 of Bringi and Chandrasekar 2001). The previously defined canting angle (β) can be derived from (θ, Φ) and the radar elevation angle (usually assumed to be 0). It is also common to assume that the probability density function (pdf) of Φ is uniform in the interval (0,2π). In general, the marginal pdf of θ, or $p_\Omega(\theta) = p(\theta) \sin\theta$ is not Gaussian (see Fig. 2.9a of Bringi and Chandrasekar 2001). However, p(θ) may be assumed to be Gaussian (mean θ = 0; σ_θ) in which case p(β) will also be Gaussian with mean=0 and $\sigma_\beta \approx \sigma_\theta$. Simulations have shown this to be valid for σ_θ at least up to 25° (Huang 2003).

The 2D video disdrometer has two, orthogonally placed line scan cameras which give two 'views' of the raindrop as it passes through the sensor area. If the drops fall vertically through the two light planes (typical plane separation is around 6 mm), the canting angle is 0°. This is true even if the drop has a horizontal velocity component. However, if the drop is canted as it enters the sensor area then the 'distorted' image is more difficult to 'correct' for which is a precursor step to determining the 'true' canting angle (the details are given in Schönhuber et al. 2000; Schauer 1998). Here, the term canting angle is used (even though it is defined for radar applications) since each camera image can be thought of as being in the 'polarization' plane of a radar beam at zero elevation angle. As such, two canting angles are derived for each drop (the angle being defined from the vertical line which is perpendicular to the light planes).

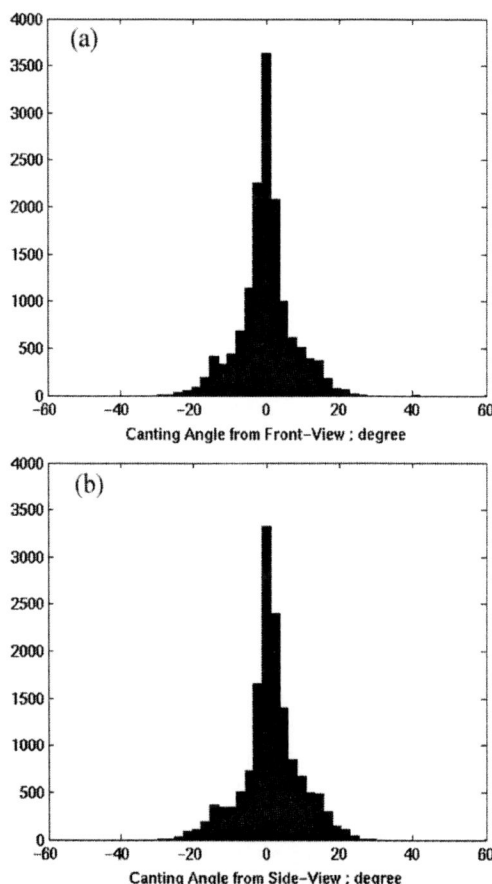

Fig. 12a, b. The histograms of canting angle from Front (Camera A) and Side (Camera B) views. Note that they were almost symmetric with mean of 0° and standard deviation of 7.2° and 7.8°, respectively

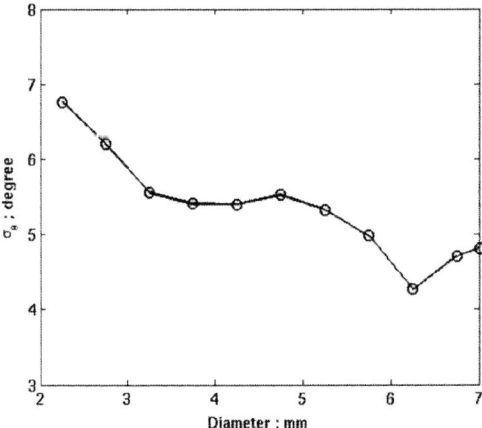

Fig. 13a. The standard deviation of θ versus drop size (D_{eq}) from the 80 m fall bridge experiment. The size intervals are from 2 mm to 7 mm with 0.5 mm step. The last data point represents those drops great than 7 mm. Note that in calm conditions prevalent during the experiment, the large drops are more stably oriented (smaller σ_θ) than small drops (larger σ_θ)

For all drops with $D_{eq} \geq 2$ mm from the 80 m fall bridge experiment, the histogram of canting angles derived from Camera A and Camera B are shown in Figs. 12a,b. Note that the shape of the canting angle histograms are approximately Gaussian with mean $\beta \approx 0$ and σ_β of around 7°. The marginal pdf, $p_\Omega(\theta)$, derived from the two canting angles has been investigated by Huang et al. (2007). The shape was not Gaussian, rather it was skewed with mode of $\theta \approx 3°$. The corresponding marginal pdf, $p(\Phi)$, was fairly uniform in the range 0–2π.

Further, for each drop class diameter interval from 2–7 mm (with bin width of 0.5 mm), the σ_θ has been calculated as a function of the mid-point of the diameter class and is shown in Fig. 13a. As seen from the graph, σ_θ falls with increase in D_{eq}, the inference being that the larger drops are more stably oriented than the small ones. From these data σ_θ reduces from 6.8° at 2 mm to 4.8° at 7 mm. These results support the dual-polarized radar observations made by Huang et al. (2003) who derived the mean variation of σ_β as a function of Z_{DR} in a summer-time convective rain storm in Colorado using the CSU-CHILL radar. Their Fig. 2 from that conference paper is reproduced in Fig. 13b (for details of the methodology please refer to Huang et al. 2003). They comment ' …that σ_β decreases with Z_{DR} and reflects the fact that larger drops are more stably oriented as compared to small-sized drops'. The 2DVD

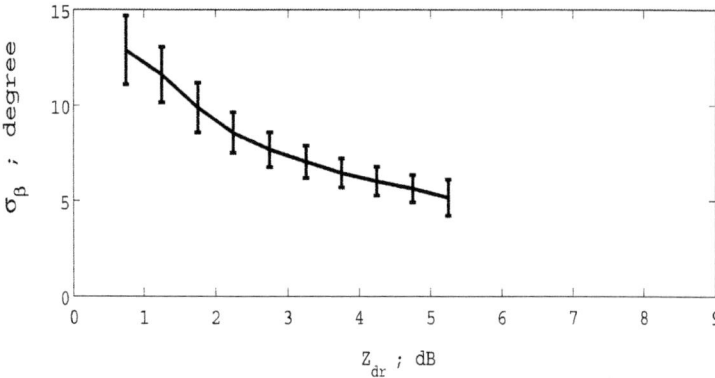

Fig. 13b. The mean σ_β versus Z_{DR} from 11 June 2000 convective rain event from STEPS project in eastern Colorado. The vertical bars are mean ± 1σ. The data are from a number of PPI sweeps using the CHILL radar operated by Colorado State University

estimation of orientation angle distributions from the 80 m fall bridge experiments under calm conditions is consistent with the radar-based results of Huang et al. (2003).

10.5 Fall velocities

For terminal velocities of rain drops, the Gunn-Kinzer (1949) data measured at sea level are often used. As mentioned earlier, Atlas et al. (1973) have fitted these data to a diameter-dependent equation, given previously in Eq. (21). Brandes et al. (2002) have proposed a fit based on combining GK data with Beard and Pruppacher (1969) measurements up to 8 mm, given by:

$$v = -0.1021 + 4.932\,(D) - 0.9551\,D^2 + 0.07934\,D^3 - 0.002362\,D^4 \tag{33}$$

where again D is in mm. The above two formulas have been compared with data from the artificial rain (80 m bridge) experiment which generated drops as large as 9 mm. A small correction factor had to be applied in order to take into account the terrain height of the location at which the 80 m bridge experiment was conducted. Assuming a US standard atmosphere, the air density was calculated for the terrain height of 480 meters above mean sea level (msl) which was then used to calculate the corresponding correction factor for the fall speed using:

$$v(h) = \left(\frac{\rho_0}{\rho(h)}\right)^m v_0 \qquad (34)$$

where ρ_0 and $\rho(h)$ are the air density at ground level and at height h, respectively, v_0 is the drop terminal velocity at sea level and m is typically 0.4 (Foote and Du Toit 1969). For better accuracy, m is set to a diameter-dependent expression given by (Beard 1985):

$$m(D) = 0.375 + 0.025D \qquad (35)$$

which results in a 1.8% increase in fall speed for 1 mm drop (an increase because of the more rarified atmosphere).

Comparisons between the 80 m bridge experimental data and the two formulas quoted above are shown in the first three columns of Table 3. The experimental values also have numbers below them in brackets to represent the equivalent mean velocities at sea level. Excellent agreement is found between all three for D_{eq} up to 6 mm. Beyond that the formula of Atlas et al. (1973) gives rise to a plateau whereas the experimental data and the fit by Brandes et al. (2002) shows a slight tendency towards decreasing velocity with increasing diameter for the large drops. It is not clear why such a decrease should occur, but one possible reason is that for drops larger than 6.5 mm, the drag increases due to increased distortion when compared with the increase in weight. Some observations in the past have also reported a similar trend. Laws (1941) used previously published data on fall velocity measured in natural rain conditions showing that drops for diameters larger than 5.5 mm, the terminal velocity decreases with increasing size. The decrease in velocity has also been evident in the adjusted velocities aloft, as seen in Fig. 6 of Beard (1976).

The last three columns of Table 3 show the 2DVD measured mean velocities in three different locations with varying altitudes. These are data from Colorado, western Sumatra and San Juan (Puerto Rico) at altitudes of 1500, 900 and 0 m, respectively above mean sea level. Once again, the equivalent velocities at sea level are included within brackets. For the San Juan case, there was no need for altitude correction since the measurements were taken almost at sea level, whereas the data from western Sumatra and Colorado required an increase of around 4% and 7%, respectively, for example for the 2 mm drop. The agreement with the Atlas et al. (1973) fit is quite close after the reference sea level adjustment, at least for drops up to 4 mm diameter.

Implications from Table 3 are two fold: (a) that the Foote and Du Toit (1969) adjustment for the altitude results in accurate estimates of terminal velocities, and (b) if rainfall estimates are to be derived from polarimetric radar measurements at significant altitudes, a correction factor needs to be applied for the drop terminal velocities based on, for example, Eq. (34). This need arises because the D_{eq}-dependence of fall velocity appears inside the integral equation for rainfall rate given in Eq. (20) in Sect. 10.2.2. If direct rainfall rate estimates are to be made using Z_h, Z_{DR} and K_{dp} for high altitude regions, then an appropriate correction factor for the rain rate algorithm needs to be incorporated. Matrosov et al. (2002) have considered this for X-band cases and specifically mention that noticeably different rainfall rates can occur if the correction factor is not included in the rain rate algorithm.

10.6 Summary

The last few years have seen considerable advances in the knowledge of rain micro-structure in different climates. This has been made possible because of the variety of measurements, both from dual-polarization radars and from advanced disdrometers such as 2DVD. Studies in several different locations have highlighted the following aspects of rain micro-structure:

The average DSD parameters ($<N_w>$ and $<D_m>$) derived from both disdrometers and dual-polarized S- and C-band radars from different locations ranging from near equatorial to mid-latitudes have shown the well-known fact that a single Z-R relation cannot capture the wide variability in the DSD. For stratiform rain an inverse, nearly linear relation between $\log_{10}<N_w>$ and $<D_m>$ has been observed. For convective rain, 'maritime' and 'continental' clusters were observed in this plane, the 'maritime' cluster showing an $<N_w> \approx$ 10000–30,000 $mm^{-1} m^{-3}$ and $<D_m> \approx$ 1.5–1.75 mm, whereas the continental cluster showing corresponding values 1000–3000 $mm^{-1}m^{-3}$ and 2–2.75 mm. Dual-polarized radars can be used to estimate the N_w and D_m parameters of the DSD as well as to derive 'pol-based' Z-R relations without rain type classification.

DSDs also show distinct seasonal variations as well as diurnal variations. Calculations using the 2DVD measurements in several climatic zones have shown that Z_{DR} can be used to estimate the mass-weighted mean diameter (which is close to the median volume diameter, D_0) of the DSD. More recent calculations have indicated that for heavy rainfall

Table 3. Mean fall velocities in rain and the standard deviations from various locations, compared with two published formulas and the artificial rain data. Values given in brackets are mean velocity data at sea level, after altitude corrections using Eqs. (34) and (35)

D, mm ↓	Atlas et al. (1973)	Brandes et al. (2002)	80 m bridge experimental data (0.5 km a.s.l)	Colorado (1.5 km a.s.l)	San Juan, Puerto Rico (at sea level)	Sumatra, Indonesia (0.9 km a.s.l)
1.5 ± 0.1	5.46	5.530	5.61 ± 0.25 (5.51)	5.70 ± 0.17 (5.34)	5.39 ± 0.23	5.46 ± 0.24 (5.26)
2.0 ± 0.1	6.55	6.637	6.56 ± 0.24 (6.44)	6.85 ± 0.16 (6.41)	6.58 ± 0.23	6.79 ± 0.25 (6.53)
2.5 ± 0.1	7.35	7.480	7.38 ± 0.25 (7.24)	7.78 ± 0.16 (6.88)	7.37 ± 0.26	7.73 ± 0.25 (7.43)
3.0 ± 0.1	7.95	8.102	7.95 ± 0.21 (7.80)		7.91 ± 0.30	8.39 ± 0.26 (8.05)
3.5 ± 0.1	8.39	8.544	8.41 ± 0.21 (8.24)		8.28 ± 0.29	8.89 ± 0.29 (8.52)
4.0 ± 0.1	8.72	8.842	8.68 ± 0.20 (8.50)		8.56 ± 0.32	9.24 ± 0.28 (8.85)
4.5 ± 0.1	8.96	9.027	8.87 ± 0.23 (8.68)		8.76 ± 0.26	
5.0 ± 0.1	9.14	9.129	8.99 ± 0.22 (8.80)			
5.5 ± 0.1	9.27	9.173	9.10 ± 0.21 (8.90)			
6.0 ± 0.1	9.37	9.182	9.13 ± 0.22 (8.92)			
6.5 ± 0.1	9.44	9.174	9.19 ± 0.23 (8.98)			
7.0 ± 0.1	9.50	9.164	9.22 ± 0.24 (9.00)			
7.5 ± 0.1	9.54	9.162	9.18 ± 0.22 (8.96)			
8.0 ± 0.1	9.57	9.177	9.10 ± 0.20 (8.88)			
8.5 ± 0.1	9.59	9.213	8.99 ± 0.25 (8.76)			

rates, the co-polar correlation coefficient could be used to estimate the standard deviation of the DSD mass spectrum. When combined with estimates of N_w and D_m, these three parameters are sufficient to characterize the DSD for most rain types.

Regarding drop shapes, contouring algorithms have been utilized to derive the 'probable shapes' for drop diameters ranging up to 6 mm. The larger drops deviate more and more from the oblate shapes, with increasingly flattened bases. More detailed measurements have indicated that the drops oscillate between oblate and conical shapes rather than between oblate and prolate shapes.

There has been consistency between radar measured Z_{DR} (and K_{dp}) versus Z_h with calculations based on 2DVD measurements, using the drop shape information as well as measured DSD. Two examples in very different climate regions have been used to illustrate this. However, more examples are needed, especially for cases with unusual DSDs with large drop diameters, in order to verify the shape of large drops and their implications for radar parameters, especially Z_{DR} at C-band.

Under calm conditions, the 2DVD data show that large drops are more stably oriented than the small ones (the standard deviation of the zenith angle varying from 7° at 2 mm to 5° at 6–7 mm diameters). These data are consistent with prior dual-polarized radar based measurements of canting angle variations which were found to decrease with increasing Z_{DR}. It is also in agreement with canting angle variations due to turbulence deduced theoretically (<5°). Thus, when deriving dual-polarized radar algorithms, the assumption can be made that the standard deviation of the canting angle varies between 5–7° in natural rain except in regions of very large wind shear.

Drop fall velocities can vary depending on the altitude. At sea level, the approximated formulas for Gunn-Kinzer data agree very well with the measurements. At higher altitudes, correction factors are required depending on the drop diameter and the reduced air density.

This has implications for the various rain rate estimation algorithms using Z_h, Z_{DR} and K_{dp} normally derived assuming sea level fall speeds in that a corresponding adjustment to the estimation formula is necessary at higher altitudes.

Acknowledgements

Much of the work reported in this paper was supported by the National Science Foundation via grants ATM-0140350 and ATM-0603720. The deployment of the 2DVD near Toronto was supported by the GPM ground validation program via NASA grant NNX06AG89G. The authors would like to thank Dr. D. Hudak, Dr. B. Sheppard and several other colleagues at Environment Canada as well as Dr K. Nakagawa and colleagues at the National Institute for Information and Communications Technology of Japan for their cooperation. We also wish to thank Dr. M. Schönhuber and Günter Lammar of Joanneum Research, Austria, for their close cooperation on the various 2DVD related work and to Dr. G.J. Huang of Colorado State University, for assistance with data analysis.

References

Allnutt J (1989) Satellite-to-Ground Radiowave Propagation. Peter Peregrinus Ltd., on behalf of the Institute of Electrical Engineers, Chap. 4 and 5

Atlas D, Ulbrich CW (1977) Path and area integrated rainfall measurement by microwave attenuation in the 1–3 cm band. J Appl Meteorol 16:1322–1331

Atlas D, Srivastava RC, Sekkon RS (1973) Doppler radar characteristics of precipitation at vertical incidence. Rev Geophys Space GE 2:1–35

Beard KV (1976) Terminal velocity and shape of cloud and precipitation drops aloft. J Atmos Sci 33:851–864

Beard KV (1984) Oscillation modes for predicting raindrop axis and back scatter ratios. Radio Sci 19:67–74

Beard KV (1985) Simple altitude adjustments for raindrop velocities for doppler radar analysis. J Atmos Ocean Tech 2:468–486

Beard KV, Chuang C (1987) A new model for the equilibrium shape of raindrops. J Atmos Sci 44:1509–1524

Beard KV, Jameson AR (1983) Raindrop canting. J Atmos Sci 40:448–454

Beard KV, Kubesh RJ (1991) Laboratory measurements of small raindrop distortion. Part 2: Oscillation Frequencies and Modes. J Atmos Sci 48:2245–2264

Beard KV, Pruppacher HR (1969) A determination of the terminal velocity and drag of small water drops by means of a wind tunnel. J Atmos Sci 26:1066–1072

Beard KV, Ochs HT, Kubesh RJ (1989) Natural oscillations of small rain drops. Nature 342:408–410

Brandes EA, Zhang G, Vivekanandan J (2002) Experiments in rainfall estimation with a polarimetric radar in a sub-tropical environment. J Appl Meteorol 41:674–684

Bringi VN, Chandrasekar V (2001) Polarimetric Doppler weather radar. Cambridge University Press, 636 pp

Bringi VN, Chandrasekar V, Hubbert J, Gorgucci E, Randeu WL, Schönhuber M (2003) Raindrop size distribution in different climatic regimes from disdrometer and dual-polarized radar analysis. J Atmos Sci 60:354–365

Bringi VN, Tang T, Chandrasekar V (2004) Evaluation of a new polarimetrically based Z–R relation. J Atmos Ocean Tech 21:612–623

Bringi VN, Thurai M, Nakagawa K, Huang GJ, Kobayashi T, Adachi A, Hanado H, Sekizawa S (2006) Rainfall estimation from C-band polarimetric radar in Okinawa, Japan: comparisons with 2D-video disdrometer and 400 MHz wind profiler. J Meteorol Soc Jpn 84:705–724

Doviak RJ, Zrnic DS (1993) Doppler radar and weather observations. Academic Press, New York, 562 pp

Fabry F, Zawadski I (1995) Long term radar observations of the melting layer of precipitation and their interpretations. J Atmos Sci 52:838–851

Foote GB, Du Toit PS (1969) Terminal velocity of raindrops aloft. J Appl Meteorol 8:245–253

Goddard JWF, Cherry SM, Bringi VN (1982) Comparison of dual-polarisation radar measurements of rain with ground-based disdrometer measurements. J Appl Meteorol 21:252–256

Goddard JWF, Tan J, Thurai M (1994a) A technique for the calibration of meteorological radars using differential phase. Electron Lett 30:166–167

Goddard JWF, Eastment JD, Tan J (1994b) Self-consistent measurements of differential phase and differential reflectivity in rain. In: Proceedings International Geoscience and Remote Sensing Symposium, August 1994, Pasadena, vol 1, pp 369–371

Goddard JWF, Morgan KL, Illingworth AJ, Sauvageot H (1995) Dual-wavelength polarization measurements in precipitation using the CAMRA and Rabelais radars. In: Preprints 27th International Conference on Radar Meteorology. American Meteorological Society, Vail, CO., pp 196–198

Gorgucci E, Chandrasekar V, Bringi VN, Scarchilli G (2002) Estimation of raindrop size distribution parametric radar measurements. J Atmos Sci 59:2373–2384

Gourley JJ, Tabary P, Parent du Chatelet J (2006) Data quality of the Meteo-France C-Band polarimetric radar. J Atmos Ocean Tech 23:340–1356

Gourley JJ, Tabary P, Parent du Chatelet J (2007) A fuzzy logic algorithm for the separation of precipitation from non-precipitating echoes using polarimetric radar observations. J Atmos Ocean Tech 24:1439–1451

Gunn R, Kinzer GD (1949) The terminal velocity of fall for water droplets in stagnant air. J Meteorol 6:243–248

Hall MPM, Barclay LW, Hewitt MT (eds) (1996) Propagation of radiowaves. Institution of Electrical Engineers, Chap. 12

Hauser D, Amayenc P, Nutten B, Waldteufel P (1984) A new optical instrument for simultaneous measurements of raindrop diameter and fall speed distributions. J Atmos Ocean Tech 1:256–269

Holt AR (1984) Some factors affecting the remote sensing of rain by polarization diversity radar in the 3-to-35 GHz range. Radio Sci 19:1399–1412

Huang GJ (2003) Evaluation and application of polarimetric radar data for the measurement of rainfall. PhD dissertation, Colorado State University

Huang GJ, Bringi VN, Hubbert J (2003) An algorithm for estimating the variance of the canting angle distribution using polarimetric covariance matrix data. In: Proceedings 31st Conference on Radar Meteorology. 6-12 August 2003, Seattle, Washington

Huang GJ, Bringi VN, Schönhuber M, Thurai M, Shimomai T, Kozu T, Marzuki M, Harujpa W (2007) Drop shape and canting angle distributions in rain from 2 D video disdrometer. In: Proceedings 33rd Conference on Radar Meteorology. August 2007, Cairns, Australia, p. 8A.8

Hudak DP, Rodriguez GW, Lee A, Ryzhkov, Fabry F, Donaldson N (2006) Winter precipitation studies with a dual polarized C-band radar. In: Proceedings 4th European Conference on Radar in Meteorology and Hydrology, 18-22 September 2006, Barcelona, Spain. http://www.erad2006.org

Illingworth AJ, Blackman TM (2002) The need to represent raindrop size spectra as normalized gamma distributions for the interpretation of polarization radar observations. J Appl Meteorol 41:286–297

Jameson AR (1983) Microphysical interpretation of multi-parameter radar measurements in rain. Part I: Interpretation of polarization measurements and estimation of raindrop shapes. J Atmos Sci 40:1792–1802

Jameson AR (1985) Microphysical interpretation of multi-parameter radar measurements in rain. Part III: Interpretation and measurements of propagation differential phase shift between orthogonal linear polarisations. J Atmos Sci 42:607–614

Joe P, Smith PL (2001) Summary of the radar calibration workshop. In: Proceedings 30th Conference on Radar Meteorology. American Meteorological Society, paper 3.1, Munich, Germany

Joss J, Waldvogel A (1967) A raindrop spectrograph with automatic analysis. Pure Appl Geophys 68:240–246

Joss J, Thams JC, Waldvogel A (1968) The accuracy of daily rainfall measurements by radar. In: Preprints 13th Conference on Radar Meteorology. Montreal, Canada, pp 448–451

Kozu T, Reddy KK, Mori S, Thurai M, Ong JT, Rao DN, Shimomai T (2006) Seasonal and diurnal variations of raindrop size distribution in Asian Monsoon region. J Meteorol Soc Jpn 84A:195–209

Krajewski WF, Kruger A, Caracciolo C, Golé P, Barthes L, Creutin J-D, Delahaye J-Y, Nikolopoulos EI, Ogden F, Vinson J-P (2006) DEVEX-disdrometer evaluation experiment: basic results and implications for hydrologic studies. Adv Water Resour 29:311–325

Kruger A, Krajewski W (2002) Two-dimensional video disdrometer: a description. J Atmos Ocean Tech 19:602–617

Laws JO (1941) Measurements of the fall velocity of water drops and rain drops. J Hydrol 22:709–721

Löffler-Mang M, Joss J (2000) An optical disdrometer for measuring size and velocity of hydrometeors. J Atmos Ocean Tech 17:130–139

Mardia KV (1972) Statistics of directional data. Academic press, NewYork

Matrosov SY, Clark KA, Martner BE, Tokay A (2002) X-band polarimetric radar measurement of rainfall. J Appl Meteorol 41:941–952

McDonald JE (1954) The shape and aerodynamics of large raindrops. J Meteorol 11:478-494 also: The shape of raindrops. Sci Am 190:64–68

Nakagawa K, Hanado H, Satoh S, Takahashi N, Iguchi T, Fukutani K (2003) Development of a new C-band bistatic polarimetric radar and observation of Typhoon Events. In: Proceedings 31st Conference Radar Meteorology. American Meteorological Society, 6–12 August, pp 863–866

Nakagawa K, Kitamura Y, Iwanami K, Hanado H, Okamoto K (2005) Field campaign of observing precipitation in the 2004 rainy season of Okinawa, Japan. In: Proceedings IEEE International Geoscience and Remote Sensing Symposium. 25–29 July, Seoul, Korea, vol 7 pp 5088–5091

Oguchi T (1983) Electromagnetic wave propagation and scattering in rain and other hydrometeors. In: Proceedings IEEE.1983, 71(9), pp 1029–1079

Olsen RL (1981) Cross-polarization during precipitation on terrestrial links: a review. Radio Sci 16:781–812

Park SG, Maki M, Iwanami K, Bringi VN (2004) Correction of radar reflectivity and differential reflectivity for rain attenuation and estimation of rainfall at X-band wavelength. In: Proceedings 6th International Symposium on Hydrological Applications of Weather Radar, 2–4 February, Melbourne, Australia

Peters G, Fischer B, Münster H, Clemens M, Wagner A (2005) Profiles of raindrop size distributions as retrieved by microrain radars. J Appl Meteorol 44:1930–1949

Petersen WA, Carey LD, Rutledge SA, Knievel JC, Johnson RH, Doesken NJ, McKee TB, Vonder Haar T, Weaver JF (1999) Mesoscale and radar observations of the Fort Collins flash flood of 28 July 1997. B Am Meteorol Soc 80:191–216

Pruppacher HR, Beard KV (1970) A wind tunnel investigation of the internal circulation and shape of water drops fall at terminal velocity in air. Q J Roy Meteor Soc 96:247–256

Pruppacher HR, Klett KV (1997) Microphysics of Clouds and Precipitation. 2nd edn. Kluwer Academic Publishers, 954 pp

Randeu WL, Schönhuber M, Lammer G (2002) Real-time measurements and analyses of precipitation micro-structure and dynamics. In: Proceedings 2nd European Conference on Radar Meteorology. 18-22 November 2002, Delft, The Netherlands, pp 78–83. http://www.copernicus.org/erad/index2002.html

Rosenfeld D, Ulbrich CW (2003) Cloud microphysical properties, processes, and rainfall estimation opportunities. Meteorol Monogr 30:237–258

Ryzhkov A, Zrnić DS (1995) Precipitation and attenuation measurements at 10 cm wavelength. J Appl Meteorol 34:2121–2134

Ryzhkov A, Schuur TJ, Burgess DW, Heinselman PL, Giangrande SE, Zrnić DS (2005) The joint polarization experiment: polarimetric rainfall measurements and hydrometeor classification. B Am Meteorol Soc 86:809–824

Schauer G (1998) Distrometer-based determination of precipitation parameters for wave propagation research. Diploma Thesis, Technical Univ. of Graz (Supervisors: Randeu WL and Schönhuber M), 98 pp

Schönhuber M, Randeu WL, Urban HE, Poiares Baptista JPV (2000) Field measurements of raindrop orientation angles. In: ESA SP-444 Proceedings, Millennium Conference on Antennas and Propagation. 9–16 April 2000, Davos, Switzerland

Schönhuber M, Lammer G, Randeu WL (2007) One decade of imaging precipitation measurement by 2D-video-distrometer. Adv Geosci 10:85–90

Seliga TA, Bringi VN (1976) Potential use of radar differential reflectivity measurements at orthogonal polarizations for measuring precipitation. J Appl Meteorol 15:69–76

Sheppard BE (1990) The measurement of raindrop size distributions using a small Doppler radar. J Atmos Ocean Tech 7:255–268

Sheppard BE, Joe PI (1994) Comparison of raindrop size distribution measurements by a Joss-Waldvogel disdrometer, 1 PMS 2DG spectrometer and a POSS Doppler radar. J Atmos Ocean Tech 11:874–887

Steiner M, Smith JA, Uijlenhoet R (2004) A microphysical interpretation of radar reflectivity–rain rate relationships. J Atmos Sci 61:1114–1131

Tan J, Goddard JWF, Thurai M (1995) Applications of differential propagation phase in polarisation-diversity radars at S- and C-band. In: Proceedings International Conference on Antennas and Propagation. IEE Conference Publication Number 407, April, Eindhoven, Netherlands

Testud J, Oury S, Amayenc P, Black RA (2001) The concept of 'normalized' distributions to describe raindrop spectra: a tool for cloud physics and cloud remote sensing. J Appl Meteorol 40:1118–1140

Tokay A, Beard KL (1996) A field study of raindrop oscillations. Part I: Observation of size spectra and evaluation of oscillation causes. J Appl Meteorol 35:1671–1687

Tokay A, Kruger A, Krajewski W (2001) Comparison of drop size distribution measurements by impact and optical disdrometers. J Appl Meteorol 40:2083–2097

Thurai M, Bringi VN (2005) Drop axis ratios from a 2D video disdrometer. J Atmos Ocean Tech 22:963–975

Thurai M, Hanado H (2005) Absolute calibration of C-band weather radars using differential propagation phase in rain. Electron Lett 41:1405–1406

Thurai M, Hudak D, Bringi VN, Lee GW, Sheppard B (2007a) Cold rain event analysis using 2-D video disdrometer, C-band polarimetric radar, X-band vertically-pointing radar and POSS, Paper 10.7A (also P. 10.5). In: Proceedings 33rd Conference on Radar Meteorology. August 2007, Cairns, Australia

Thurai M, Huang GJ, Bringi VN, Randeu WL, Schönhuber M (2007b) Drop shapes, model comparisons, and calculations of polarimetric radar parameters in rain. J Atmos Ocean Tech 24:1019–1032

Ulbrich CW (1983) Natural variations in the analytical form of the raindrop size distribution. J Appl Meteorol 22:1764–1775

Ulbrich CW, Atlas D (1998) Rainfall microphysics and radar properties: analysis, methods for drop size spectra. J Appl Meteorol 37:912–923

Ulbrich CW, Atlas D (2007) Microphysics of rain drop size spectra: tropical continental and maritime storms. J Clim Appl Meteorol 46:1777–1791

Wang PK (1982) Mathematical description of the shape of conical hydrometeors. J Atmos Sci 39:2615–2622

Zrnic DS, Melnikov VM, Carter JK (2006) Calibrating differential reflectivity on the WSR-88D. J Atmos Ocean Tech 23:944–951

11 On the use of spectral polarimetry to observe ice cloud microphysics with radar

Herman Russchenberg[1], Lennert Spek[1], Dmitri Moisseev[2], Christine Unal[1], Yann Dufournet[1], Chandrasekhar Venkatachalam[2]

[1]Delft University of Technology, IRCTR, The Netherlands
[2]Colorado State University, Fort Collins, CO, USA

Table of contents

11.1	Introduction	286
11.2	The concept of spectral polarimetry	287
11.3	Microphysical model of ice particles	288
	11.3.1 The shape of ice crystals	289
	11.3.2 Canting angles of ice crystals	290
	11.3.3 Mass density of ice crystals	290
	11.3.4 Velocity of ice crystals	291
	11.3.5 Bulk parameters	292
11.4	Radar observables of ice particles	293
11.5	Retrieval of microphysical parameters	296
	11.5.1 Dependence on DSD parameters of plates and aggregates	296
	11.5.2 The curve fitting procedure	297
	11.5.3 Quality of retrieval technique	302
11.6	Application to radar data	303
	11.6.1 Retrieval algorithm results	304
	11.6.2 Comparison of IWC with LWC	304
	11.6.3 Relation between IWC and reflectivity	307
	11.6.4 Influence of the shape parameter of the DSD	308
11.7	Summary and conclusions	310
References		311

11.1 Introduction

The formation of rainfall is a complex process. Warm rain is formed under conditions where the temperature is above the freezing point of water. Raindrops then result from collision and coalescence. In cold rain also ice particles and super-cooled water droplets at temperatures below the freezing point come into play. The microphysical transition of ice particles into raindrops is understood in common terms, but details of this and the impact on rainfall rates are still largely unknown. Aircraft measurements of ice particles are difficult and mostly lead to 'snapshot' information, whereas continuous observations of the entire transition process are needed. This is where remote sensing enters the scene. Weather radars are very useful tools to study the entire rain cell and advanced systems, combining Doppler and polarization capabilities, can deliver a wealth of information. The interpretation of radar observations of ice precipitation is, however, quite complex. A typical ice precipitation event can consist of a mixture of different particle types, such as pristine ice particles, aggregates, or graupel, and therefore one has to take into account the differences in scattering of the radar waves due to this variety. This has led to a number of reported relationships between the radar reflectivity and ice water content; see for example Sekhon and Srivastava (1970), Smith (1984), Matrosov (1992). There is, however, not a unique one. Radar techniques to discriminate ice crystal types have been developed. These techniques are based on a conceptual idea of stratiform rain as a layered structure, where each layer is homogeneous in terms of type of hydrometeor. In other words: mixtures of ice crystal types do not occur at a given latitude. Matrosov (1998) has investigated the use of dual-wavelength radar for estimation of snow parameters. They have shown that using measurements taken at two wavelengths, where at least one of them is located in a non-Rayleigh region, one can estimate parameters of a particle size distribution. In these studies, it was assumed that there is only one type of particles present in the observation volume. In a similar vain, Matrosov et al. (1996) have shown that by using dual-polarization radar measurements taken at several elevation angles, it is possible to discriminate various types of ice particles, such as planar crystals, columnar crystals and aggregates in a homogeneous cloud. The use of VHF profiler measurements for the retrieval of the size distribution of ice particles above the freezing level in the stratiform region of a tropical squall line was demonstrated by Rajopadhyaya et al. (1994). This technique is based on a velocity-diameter relationship (e.g., Langleben 1954 or

Locatelli and Hobbs 1974) and vertically pointing Doppler observations. These methods deliver information about the ice crystals averaged over the radar volume in a cloud layer at a given altitude. However, since the radar volume is very large and given the fact that it is not uncommon that different types of ice crystals occur *within* the cloud layer, the radar signal may be due to different categories of ice crystals in the resolution volume. To alleviate this problem, the spectral dual-polarization method has been developed (Moisseev et al. 2004). It was shown that a combination of Doppler measurements and dual-polarization observations can be used to distinguish between different types of ice hydrometeors within a radar volume. In this chapter, we will expand this concept, using observations of the *spectral differential reflectivity*.

11.2 The concept of spectral polarimetry

Spectral polarimetry is based on combined Doppler and polarization measurements, with a close look on the polarization dependence of the radar signal per velocity bin of the Doppler spectrum. In addition to the more traditional approach of expressing the Doppler spectrum in its statistical moments, spectral polarimetry can give a wealth of detailed microphysical information of precipitation. Central to the methods are the spectrally resolved polarization parameters. Figure 1 gives the principle of one of those: the spectral differential reflectivity. One measures the Doppler velocity spectrum of the radar signal at different polarizations, in this example horizontal and vertical, and determines the difference between the signal strengths at these polarizations for every separate velocity bin of the spectrum. This difference is called the spectral differential reflectivity.

What is the physical information contained in the spectral differential reflectivity? Figure 1 shows it conceptually in the case of rainfall. It is well known that small raindrops are more or less spherical, whereas larger droplets are oblate because the air resistance flattens their base while they are falling. Also, small raindrops have a smaller fall speed than large ones. This means that the spectral differential reflectivity will increase with increasing fall speed, as depicted in Fig. 1. Inversely, this means that measurements of the spectral polarization properties can be used to estimate the raindrop shape. A complicating factor in this scheme is formed by turbulence. The variation of wind speeds at scales smaller than the radar volume will mix up particles of different shapes and velocities. Effectively, this leads to a flattening of the spectral

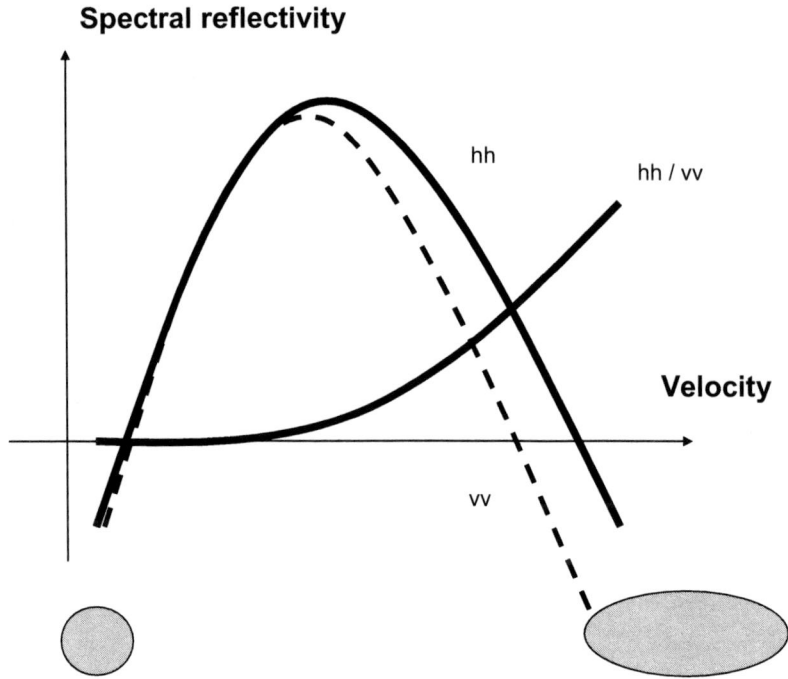

Fig. 1. The principle of the spectral differential reflectivity

differential reflectivity curve: at a given velocity one may have spherical as well as oblate particles. Yanovsky et al. (2005) have developed a method to use this property to estimate the intensity of turbulence.

In this chapter, we will discuss how the spectral differential reflectivity can be used to estimate microphysical properties of ice particles. It will be illustrated with examples of radar observations of a cloud layer in light precipitation, a few hundred meters above the melting layer where the falling ice particles turn into raindrops. The method is based on a microphysical model of ice particles, its link to expected radar observables and a curve fitting technique to relate them to observations.

11.3 Microphysical model of ice particles

The scattering of radar waves by ice particles depends on the mass density, size, shape and orientation of each individual particle, and the statistical distributions of those parameters. Furthermore, the velocity of the particles contributes to the radar signal in variations of the phase distribution: Doppler processing can resolve this. This leads to a considerable complexity as there are more than 60 different types of ice crystals (Magono and Lee 1966). The occurrence of the different ice particles depends on temperature, pressure and humidity of the atmosphere. Aggregates are formed as a combination of pristine ice particles, see for example (Rajopadhyaya et al. 1994) and (Szymer and Zawadzki 1999). A summary of the ice particle types that may be present above the melting layer in stratiform precipitation is given in Table 1. The given size ranges are approximate.

11.3.1 The shape of ice crystals

Although ice crystals may have complex shapes, for cm-wavelength radar studies details of the shape are not so important, and approximate shapes suffice. To this end, hydrometeors are generally modeled as spheroids (Bringi and Chandrasekar 2001). The relation between the smallest and biggest particle dimension for snow crystals is given by a power law,

$$w(D) = \xi D^\zeta \tag{1}$$

where w is the smallest dimension of the spheroid and D the largest dimension. Values for ξ and ζ can be found in (Matrosov et al. 1996; Auer and Veal 1970) for plates, dendrites and aggregates (see also Table 2).

Table 1. Types and typical sizes of snow particles. The given diameter denotes the maximum particle dimension of the ice particles

Type	Diameter (mm)	Reference
Plates	$0.015 \leq D \leq 3$	Mitchell 1996; Pruppacher and Klett 1978
Dendrites	$0.3 \leq D \leq 4$	Mitchell 1996
Aggregates	$0.5 \leq D \leq 8$	Mitchell et al. 1990; Mitchell 1996
Graupel	$2 \leq D \leq 8$	Pruppacher and Klett 1978
Hail	$5 \leq D \leq 25$	Mitchell 1996; Pruppacher and Klett 1978

Table 2. Parameters of the shape-diameter relations for different ice crystals given by Eq. (1). Data are taken from Matrosov et al. (1996), Auer and Veal (1970) and Bringi and Chandrasekar (2001)

Type	Size (cm) from	to	Shape ξ	ζ
Plates	0.0015	0.3	0.0141	0.449
Dendrites	0.03	0.4	0.00902	0.377
Aggregates	0.05	0.8	0.8	1
Hail	0.5	2.5	0.8	1

11.3.2 Canting angles of ice crystals

Falling hydrometeors will be canted due to external forces like wind, turbulence and draft. The canting angle is described by two orientation angles δ and α, where δ is the angle between the zenith and one symmetry axis of the ice particle and α is the azimuth angle. Under normal conditions the azimuth angle α is uniformly distributed between 0 and 2π. For plates, dendrites and aggregates, the orientation angle δ follows a Fisher probability function with $\kappa=30$ and a mean equals to zero according to Bringi and Chandrasekar (2001). Static electricity can orient ice crystals into a preferred direction, but in this chapter we assume that such is not the case.

11.3.3 Mass density of ice crystals

The density of ice particles is important, because it determines their electromagnetic properties: they are mixtures of air and ice (occasionally water also) and to calculate the radar scattering the refractive index of the composite has to be known. The mass density of ice particles can be modeled as function of maximum particle dimension D (Pruppacher and Klett 1978),

$$\rho_e = kD^l \qquad (2)$$

where ρ_e denotes the density of the ice crystals. Pruppacher and Klett (1978) give values for the variables k and l for plates and dendrites (Table 3). Because aggregates occur in many different appearances, their density is difficult to model. However, in Fabry and Szymer (1999), different density-diameter relations are examined. Based on the comparison of radar observations of reflectivity and velocity obtained at

Table 3. Parameters of the density-diameter relations for different ice crystals given by Eq. (2) Data are taken from Pruppacher and Klett (1978), El-Magd et al. (2000) and Fabry and Szymer (1999)

Type	Size (cm) from	to	Density (gcc^{-1}) k	l
Plates	0.0015	0.3	0.9	0
Dendrites	0.03	0.4	0.2468	−0.377
Aggregates	0.05	0.8	0.015	−1
Graupel	0.2	0.8	0.55	0
Hail	0.5	2.5	0.9	0

different radar frequencies, they derived a relation $\rho_e = 0.015D^{-1}$; this is used in this chapter.

11.3.4 Velocity of ice crystals

The fall velocity of ice crystals is due to the balance between gravity, air viscosity and the microphysical particle properties. The terminal fall velocity v_t of hydrometeors is given by Mitchell (1996).

$$v_t = \frac{a\nu}{D}\left(\frac{2mD^2g}{\rho_a \nu^2 A}\right)^b \quad (3)$$

where A is the area projected to the normal flow of the ice particle, ρ_a is the density of air, m is the mass of the particle, g is the gravitational constant, D is the largest dimension of the particle and ν is the kinematic viscosity of air.

To obtain velocity-size relations of ice particles dependent on their diameter, mass and area of the particles need to be parameterized as a function of diameter. For these parameterizations power laws are used (Mitchell 1996). They can be expressed as a function of maximum particle dimension D,

$$m(D) = \alpha D^\beta \quad (4)$$

$$A(D) = \gamma D^\sigma \quad (5)$$

Combining Eqs. (3)–(5), the terminal fall velocity of ice particles (cm s^{-1}) is expressed as a power law of the maximum particle dimension D (cm),

$$v_t(D) = a\, \nu \left(\frac{2\alpha g}{\rho_a \nu^2 \gamma} \right)^b D^{b(\beta+2-\sigma)-1} = CD^B \qquad (6)$$

The values of α, β γ and σ, for the different types of ice particles can be found in (Mitchell 1996); see Tables 4 and 5. The values of a and b are derived in Khvorostyanov and Curry (2002).

Table 4. Parameters of the area-diameter relations for different ice crystals given by Eq. (5). Data are taken from Mitchell (1996) and Heymsfield and Kajikawa (1987)

Type	Size (cm)		Area (cm^2)	
	from	to	γ	σ
Plates	0.0015	0.01	0.24	1.85
Plates	0.01	0.3	0.65	2
Dendrites	0.03	0.4	0.21	1.76
Aggregates	0.05	0.8	0.2285	1.88
Graupel	0.2	0.8	0.5	2
Hail	0.5	2.5	0.625	2

Table 5. Parameters of the mass-diameter relations for different ice crystals given by Eq. (4). Data are taken from Mitchell (1996) and Heymsfield and Kajikawa (1987)

Type	Size (cm)		Mass (g)	
	from	to	α	β
Plates	0.0015	0.3	0.00739	2.45
Dendrites	0.03	0.4	0.003	2.3
Aggregates	0.05	0.8	0.003	2.1
Graupel	0.2	0.8	0.049	3.06
Hail	0.5	2.5	0.466	3

11.3.5 Bulk parameters

Several 'bulk' parameters need to be defined. Their expression is simple and assumes a spherical shape of the ice particles, based on the equivolumetric diameter. Central to the definition of bulk parameters is the drop size distribution. The gamma distribution and its special case, the exponential distribution, are commonly used in the literature. In this work, we use he exponential distribution to simplify matters: two parameters in the exponential distribution have to be retrieved, instead of three parameters for the gamma distribution. The gamma distribution is given by

$$N(D) = N_w D^\mu \exp\left(-(3.67 + \mu)\frac{D}{D_0}\right) \quad (7)$$

with N_w the intercept parameter (mm^{-1} m^{-3}) and D_0 the median volume diameter (mm). The exponential distribution is obtained when $\mu = 0$.

Ice water content (IWC) or liquid water content (LWC)

The water content, liquid or ice, is derived by integrating over all particle masses present in the volume.

$$IWC, LWC = \int_{D=0}^{\infty} m_{ice\,liquid}(D) N_{ice\,liquid}(D) dD \quad (8)$$

with IWC and LWC as the ice or liquid water content (g m^{-3}) and m the mass of the particle (g). The mass-diameter relation is given by Eq. (4). To obtain the total ice water content from the retrieved drop size distributions a summation is done over the ice water content obtained for plates and aggregates.

Number of particles

The particle concentration (m^{-3}), or total number of particles, is

$$N_T = \int_{D=0}^{\infty} N(D) dD \quad (9)$$

11.4 Radar observables of ice particles

Backscattering of radar waves is described by the radar cross-section. At S-band radar frequencies, the radar cross-section of spheroidal hydrometeors can be determined using the Rayleigh scattering theory (Russchenberg 1992; Bringi and Chandrasekar 2001). The Maxwell-Garnett equation gives the effective dielectric constant for mixtures depending on their volume fractions of ice in air. Ray (1972) gives a way to calculate the permittivity of ice over a broad spectral range and temperature. With the given description of the properties of ice particles, the radar cross-section of the different types of particles can be calculated. It is plotted versus the velocity in Fig. 2. Common radar observables for precipitation are the horizontal equivalent reflectivity and the differential reflectivity. Their spectral representations are given in Eqs. (10) and (11), respectively. Precipitation above the melting layer consists of multiple particle types and the radar observables are, therefore, given by a summation over the n types present in the radar volume.

$$sZ_{HH}(v)dv = \frac{\lambda^4}{\pi^5 |K|^2} \sum_{i=1}^{n} N_i(D_i\{v\}) \sigma_{HHi}(D_i\{v\}) \left|\frac{dD_i}{dv}\right| dv \qquad (10)$$

$$sZ_{DR}(v)dv = \frac{\sum_{i=1}^{n} N_i(D_i\{v\}) \sigma_{HH,i}(D_i\{v\}) \left|\frac{dD_i}{dv}\right| dv}{\sum_{i=1}^{n} N_i(D_i\{v\}) \sigma_{VV,i}(D_i\{v\}) \left|\frac{dD_i}{dv}\right| dv} \qquad (11)$$

where the subscripts HH and VV denote, respectively, horizontal and vertical transmitting and receiving polarization modes of the radar, i represents the particle type, N(D) is the drop size distribution, v is related to the terminal fall velocity and σ is the radar cross-section; K is representative for the refractive index of the particles. Integration over the entire drop size distribution leads to the equivalent reflectivity factor:

$$Z_e = \frac{|K|^2}{|K_r|^2} \int_{D=0}^{\infty} D^6 N(D) dD \qquad (12)$$

where K is related to the relative permittivity of ice and K_r corresponds to the relative permittivity of water. Z_e, D and the drop size distribution, N(D), are expressed in $mm^6 \, m^{-3}$, mm and $mm^{-1} \, m^{-3}$, respectively.

In this chapter, we only discuss measurements of stratiform precipitation where large fall velocities are not present; hail and graupel are discarded as possible scatterers. The remaining categories of possible types of particles are aggregates, the plates and the dendrites (see Fig. 2). If the concentrations of different particle types are comparable and one of the respective radar cross-sections is significantly smaller, the scattering by this particle type is not seen. In case of a population of plates and dendrites, we assume that the radar backscattering is dominated by plates, since the radar cross-section of dendrites is much smaller than the one of plates. When dendrites and plates have similar radar cross-sections, other microphysical properties (like the axial ratio) are similar and then there is no possibility to differentiate them. This is, however, different for aggregates and plates: their shapes will be significantly different to be able to dissociate them.

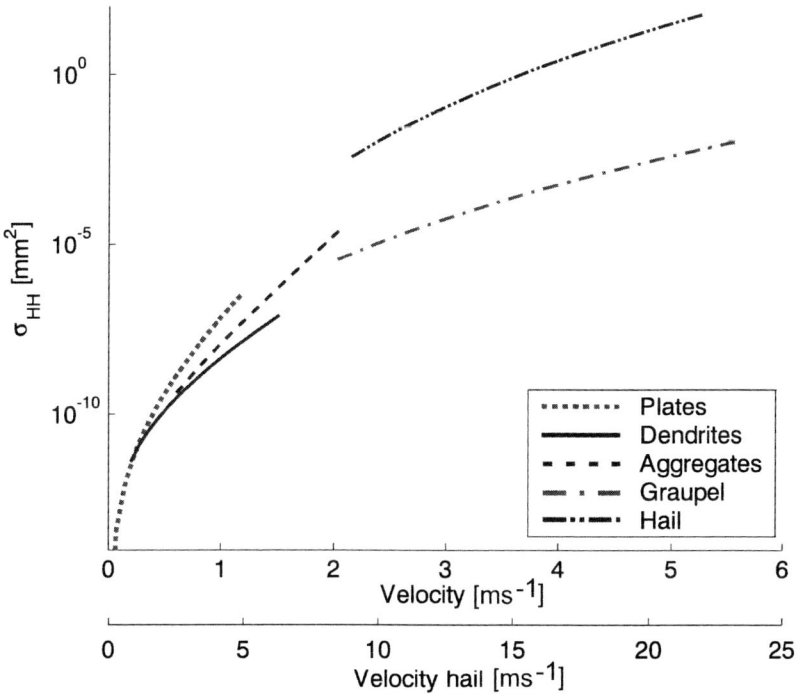

Fig. 2. Radar cross-section for different ice particles types depending on their terminal fall velocity. The radar cross-section is calculated with a drop size distribution equal to one for all diameters and a frequency of 3 GHz. The elevation angle of the radar is 45 degrees

11.5 Retrieval of microphysical parameters

The microphysical model described above is used to simulate the spectral equivalent reflectivity and spectral differential reflectivity. These simulated spectra can then be fitted to the measurements using a non-linear least squares optimization. With this approach, microphysical properties of plates and aggregates from spectral radar measurements above the melting layer in stratiform precipitation can be extracted. Doppler spectra broadening is due to several factors, such as turbulence, wind, rocking of hydrometeors, etc. (Doviak and Zrnic 1993). On top of that, ambient wind will cause a shift on the fall velocities. According to Doviak and Zrnic (1993), spectral broadening can be modeled as a convolution of the spectral radar observables with a Gaussian convolution kernel with mean velocity v_0 and width σ_0. Finally, assuming two types of ice crystals (aggregates and plates) in the radar volume, the model of the spectral radar observables depends on 6 parameters:

$$sZ_{HH}^{mod}(v) = sZ_{HH}(v, N_w^{agg}, D_0^{agg}, N_w^{pla}, D_0^{pla}, v_0, \sigma_0) \qquad (13)$$

$$sZ_{DR}^{mod}(v) = sZ_{DR}(v, N_w^{agg}, D_0^{agg}, N_w^{pla}, D_0^{pla}, v_0, \sigma_0) \qquad (14)$$

The curve fitting technique will lead to the retrieval of these model parameters, but before this is applied, the sensitivity of the radar observables to changes in the microphysical properties of the ice crystals is described.

11.5.1 Dependence on DSD parameters of plates and aggregates

The model of horizontal and differential spectral reflectivity is dependent on six parameters: the two particle size distribution parameters of plates and aggregates, the spectral broadening and the ambient wind velocity. Before the parameters can be derived from spectral radar measurements, it is necessary to verify if the parameters have a significant effect on the two spectral radar observables. If a change in one parameter has no effect on the spectral observables, the parameter cannot be determined correctly. By changing the six parameters of the model one by one, whilst keeping the other five parameters constant, a good insight is provided on the dependence of the spectral radar observables on the different parameters of the model.

In Figs. 3–5, the plots are shown for changing the different parameters of the spectral model over a realistic range. The following conclusions can be drawn:

- An increase in N_w for aggregates leads to an increase in sZ_{HH} and a decrease in sZ_{DR}. The more aggregates there are, the more the total sZ_{DR} tends to the spectral differential reflectivity of aggregates, close to zero dB because the, on average, spherical shape of aggregates.
- An increase in D_0 for aggregates leads to an increase and a wider spectrum for sZ_{HH}. sZ_{DR} decreases for the same reason as explained for an increase in N_w, the contribution of aggregates to the total spectrum becomes dominant.
- An increase of N_w for plates hardly affects the observed sZ_{HH}, due to the fact that the radar cross-section for plates is significantly smaller than those of aggregates. On the other hand sZ_{DR} increases with increasing N_w, due to their oblate shape of plates. With an increasing number of plates, the observed spectral differential reflectivity tends more to the spectral differential reflectivity of plates.
- An increase in D_0 generates a similar effect as an increase of N_w for plates.
- The effect of spectral broadening on the horizontal reflectivity is that the maximum of the spectrum becomes lower and the spectrum becomes wider and more symmetric as well. Next to that, an increase of spectral broadening flattens out the spectral differential reflectivity.

In summary, the spectral horizontal reflectivity is mostly dependent on the drop size distribution of aggregates, the spectral broadening factor and the ambient wind velocity. The spectral differential reflectivity depends on all the six parameters. All the six parameters have a significant effect on the spectral radar observables, which makes their retrieval possible.

11.5.2 The curve fitting procedure

The retrieval algorithm is described here to obtain the six parameters by fitting modeled spectra to measured spectra. An optimization procedure minimizes the difference (or error) between the fitted spectrum and the measured spectrum by varying the six input parameters. The minimization is carried out on the spectral differential reflectivity:

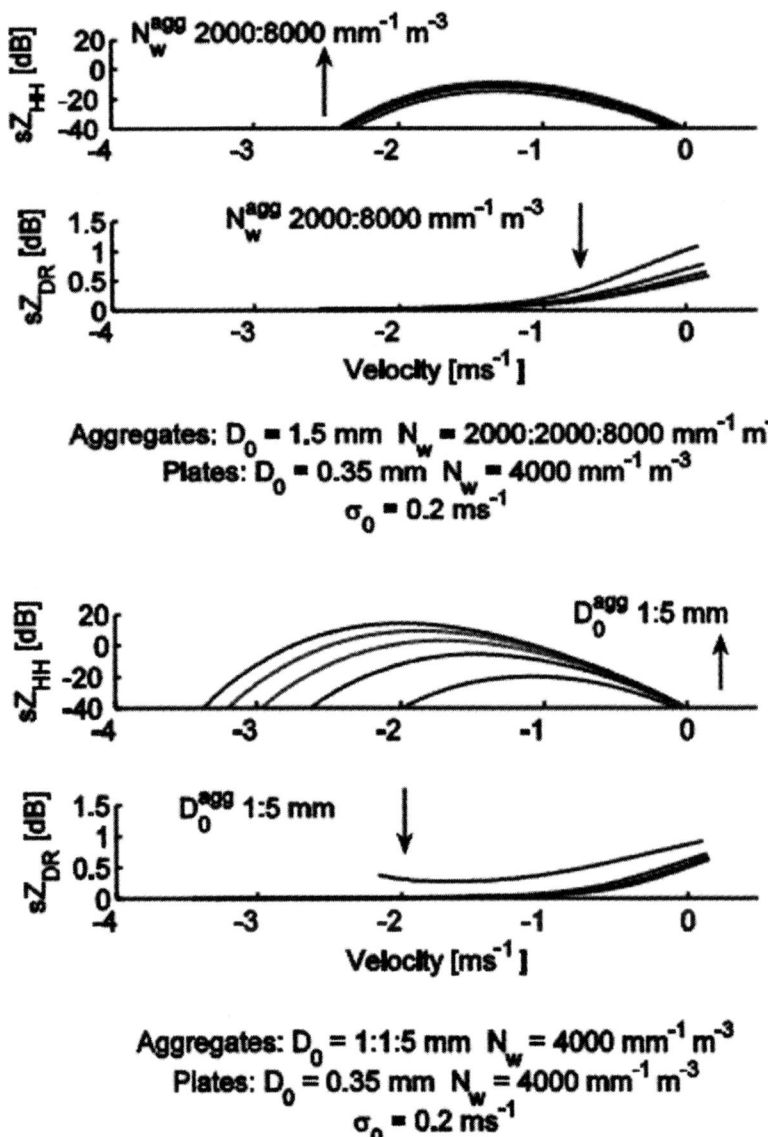

Fig. 3. Dependence of sZ_{HH} and sZ_{DR} on the parameters of the drop size distribution of aggregates

Chapter 11 - Spectral polarimetry to observe ice cloud microphysics   299

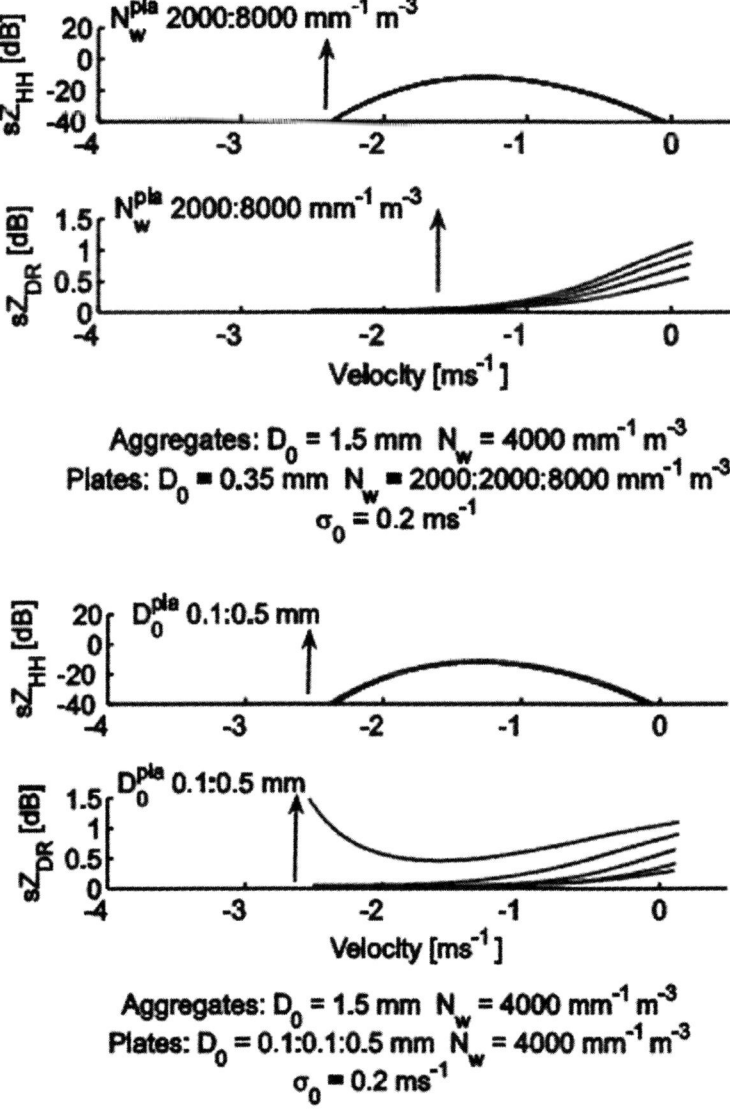

Fig. 4. Dependence of sZ_{HH} and sZ_{DR} on the parameters of the drop size distribution of plates

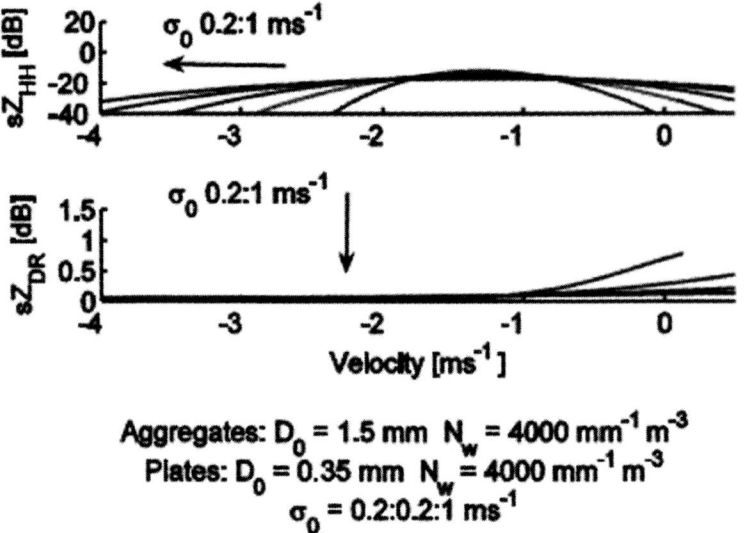

Fig. 5. Dependence of sZ_{HH} and sZ_{DR} on the spectral broadening factor

$$\min_{\Psi} \sum_{v=v_{min}}^{v_{max}} \left[sZ_{DR}^{meas}(v) - sZ_{DR}^{mod}(v,\Psi) \right]^2 \quad (15)$$

where Ψ contains all four drop size distribution parameters, the spectral broadening and the ambient wind velocity. The error as function of the six parameters is called the cost function or objective function of the minimization problem. For the implementation, a non linear least squares algorithm is used. Six parameter non linear least squares optimizations are usually difficult to solve and time consuming.

The six parameters minimization problem can be simplified by separating the retrieval of the intercept parameters of plates and aggregates, using:

$$sZ_{DR}^{meas}(v) = \quad (16)$$

$$\frac{sZ_{HH}^{mod,\alpha gg}(v,1,D_0^{\alpha gg},v_0,\sigma_0)N_w^{\alpha gg} + sZ_{HH}^{mod,pla}(v,1,D_0^{pla},v_0,\sigma_0)N_w^{pla}}{sZ_{VV}^{mod,\alpha gg}(v,1,D_0^{\alpha gg},v_0,\sigma_0)N_w^{\alpha gg} + sZ_{VV}^{mod,pla}(v,1,D_0^{pla},v_0,\sigma_0)N_w^{pla}}$$

$$sZ_{HH}^{meas}(v) = sZ_{HH}^{mod,\alpha gg}(v,1,D_0^{\alpha gg},v_0,\sigma_0)N_w^{\alpha gg} +$$
$$sZ_{HH}^{mod,pla}(v,1,D_0^{pla},v_0,\sigma_0)N_w^{pla} \qquad (17)$$

Such linear combinations of non linear functions can be solved using the variable projection method (Rust 2003). It allows the derivation of estimates of the intercept parameters without non linear fitting. A second simplification of Eq. (15) is done on the estimation of the ambient wind velocity v_0. The ambient wind velocity creates a shift of the measured spectral horizontal reflectivity with respect to the modeled one. Assuming the other five parameters of the model are known, the shift between the modeled and the measured spectrum can be obtained by determining the lag of the cross-correlation of the measured $sZ_{HH}^{meas}(v)$ and the modeled spectra

$$sZ_{HH}^{mod}(v,N_w^{\alpha gg},D_0^{\alpha gg},N_w^{pla},D_0^{pla},\sigma_0)$$

The separation of the intercept parameters and the ambient wind velocity results in a three parameter non linear least squares problem given by

$$\min_{D_0^{\alpha gg},D_0^{pla},\sigma_0} \min_{N_w^{\alpha gg},N_w^{pla}} \min_{v_0} \sum_{v=v_{min}}^{v_{max}}[sZ_{DR}^{meas}(v)-sZ_{DR}^{mod}(v,\Psi)]^2 \qquad (18)$$

The retrieval of spectral broadening is separated from the retrieval of the median volume diameters of plates and aggregates by optimizing the spectral broadening factor based on the spectral horizontal reflectivity. The cost function for the optimization of the spectral broadening factor is given by

$$L(\sigma_0) = \sum_{v=v_{min}}^{v_{max}}[sZ_{HH}^{meas}(v)-sZ_{HH}^{mod}(v,\sigma_0)|_{D_0^{\alpha gg},D_0^{pla}}] \qquad (19)$$

The conditional values for the median volume diameters of plates and aggregates imply that the cost function for the spectral broadening is calculated every time the cost function for the median volume diameter for plates and aggregates is calculated.

The total derived optimization procedure is given by

$$\min_{D_0^{agg}, D_0^{pla}} \min_{\sigma_0} \min_{N_w^{agg}, N_w^{pla}} \min_{v_0} \sum_{v=v_{min}}^{v_{max}} \left[sZ_{DR}^{meas}(v) - sZ_{DR}^{mod}(v, \Psi) \right]^2$$

(20)

where the optimization for the spectral broadening factor and the ambient wind velocity is based on the spectral horizontal reflectivity.

11.5.3 Quality of retrieval technique

To get insight in the quality of the optimization procedure, the optimization is applied on simulated Doppler spectra. By comparing the input parameters used to create a simulated spectrum with the parameters obtained with the retrieval algorithm, conclusions can be drawn on their errors.

The simulated spectra are created using Eqs. (13) and (14). To generate signals with real statistical properties, noise is added according to Chandrasekar et al. (1986). The values of the parameters are selected randomly from the depicted intervals, given in Table 6. In addition to the constraints on the input parameters of the model, the retrieval algorithm is only applied on the simulated spectra when the spectral horizontal reflectivity $sZ_{HH}(v)$ dv exceeds -10 dB with respect to the maximum of the spectrum and the maximum spectral differential reflectivity exceeds 0.5 dB. The first threshold is to ensure that the spectrum has a sufficient signal-to-noise ratio to perform the optimization and the second threshold is to ensure that the amount of plates is detectable.

The root mean square error of each parameter is given in Table 7. The error on both intercept parameters is large. Due to the layered structure of the retrieval algorithm, an error on the median volume

Table 6. Regions of variables

Parameter	Region	
D_0^{agg}	0.5–5	mm
N_w^{agg}	0–8000	$mm^{-1} m^{-3}$
D_0^{pla}	0.02–0.5	mm
N_w^{pla}	0–8000	$mm^{-1} m^{-3}$
σ_0	0.1–0.7	$m\,s^{-1}$
v_0	0–1	$m\,s^{-1}$

Table 7. The root mean square (RMS) errors of the six retrieved parameters

Parameter	RMS error		Relative RMS error
D_0^{agg}	0.60	mm	17%
D_0^{pla}	0.067	mm	15%
N_w^{agg}	2662	$mm^{-1} m^{-3}$	87%
N_w^{pla}	8038	$mm^{-1} m^{-3}$	136%
σ_0	0.0052	$m\, s^{-1}$	2%
v_0	0.11	$m\, s^{-1}$	16%

Table 8. The root mean square (RMS) errors of the integral parameters

Parameter	RMS error		Relative RMS error
Z_e	0.030	dB	0.15%
IWC	0.041	$g\, m^{-3}$	28%
N_t	1940	m^{-3}	64%

diameter will be corrected by the estimated value of the intercept parameter to obtain the correct spectral reflectivity. Because the reflectivity is related to the sixth moment of the diameter and proportional to the intercept parameter, the error on the median volume diameter will have a large effect on the intercept parameter.

The same exercise is carried out on integral parameters, the equivalent reflectivity, the ice water content (IWC) and the number of particles. The same dataset is used to obtain the errors on the drop size distributions and on the integral parameters. The results are given in Table 8. The root mean square error of the equivalent reflectivity is very small because the errors on the median volume diameter and intercept parameter cancel out in the estimate of the reflectivity. Because the error of the intercept parameter and the media particle size are not independent, the final error of the number concentration is not so large.

11.6 Application to radar data

The developed retrieval technique is applied to real radar measurements. The data is collected by the radar TARA (Heijnen et al. 2000) during a moderate stratiform rain event in Cabauw, The Netherlands (Russchenberg et al. 2005). The reflectivity values of rain vary between 20 and 35 dBZ. The elevation angle of the radar is set to 45 degrees to ensure a significant Doppler-polarization signature. The measurements are carried out in alternating polarization and wind mode (Unal et al.

2005) where VV, HV, HH and two offset beams measurements are collected in a data block of 5 ms. The Doppler spectrum is calculated from a time series of 512 samples (2.56 s). Ten Doppler spectra are averaged to obtain the final Doppler spectrum which will be the input for the inversion algorithm (25.6 s). The range and the Doppler resolution are 15 m and 1.8 cm s^{-1}, respectively.

The spectral horizontal reflectivity and the spectral differential reflectivity are shown in Fig. 6. They consist of Doppler spectra of the radar observables for every height. A target approaching the radar has a Doppler velocity negative (convention). The variability of the mean Doppler velocity versus height is mainly related to the ambient wind velocity. The spectral reflectivity is calibrated and its sum over all the Doppler velocities gives the commonly used reflectivity factor. The melting layer is located between 1280 m and 2000 m. Below the melting layer, there is rain and above the melting layer, there is a precipitating cloud.

11.6.1 Retrieval algorithm results

The Doppler spectra used as input for the retrieval algorithm are selected at least 200 m above the top of the melting layer (2000 m). Next, the constraints on the values of the spectral reflectivity and spectral differential reflectivity, which are considered in the simulation, are applied.

For spectral reflectivity values under the −10 dB clipping level, the spectral differential reflectivity is very affected by noise. An example of the spectral horizontal and differential reflectivity data with their obtained fits as well as the obtained six parameters is given in Fig. 7.

Regarding the obtained values of the drop size distributions as well as the retrieved values of spectral broadening and ambient wind velocity, it is concluded that the outputs of the inversion algorithm for plates and aggregates are consistent for small variations in height. The consistency of the six retrieved parameters versus the time is also successfully verified (500 s were considered). Figures 8 and 9 illustrate the time consistency of, the median volume diameter and the particle concentration, respectively, for both aggregates and plates.

11.6.2 Comparison of IWC with LWC

The ice water content is estimated from the obtained drop size distribution parameters of plates and aggregates using the inversion algorithm. The liquid water content is estimated from the drop size distribution parameters of rain, which are retrieved with the algorithm

Chapter 11 - Spectral polarimetry to observe ice cloud microphysics

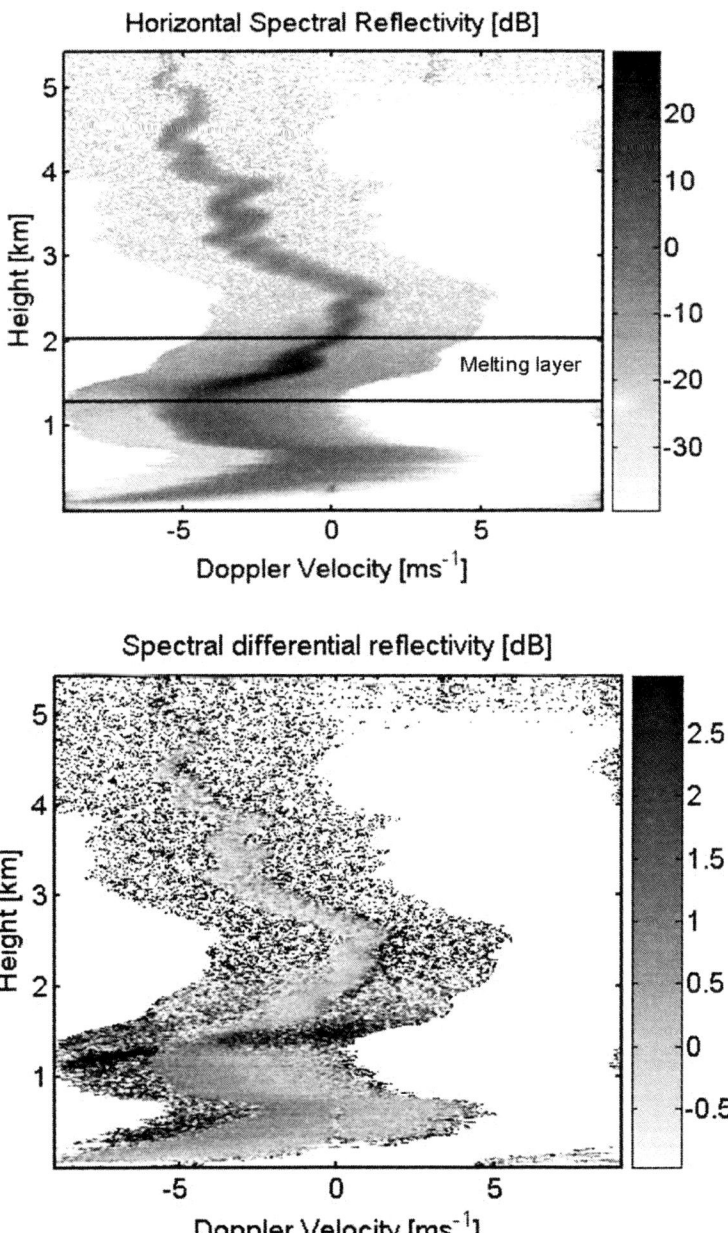

Fig. 6. Example of spectral radar observables (Doppler spectra of radar observables for every height) obtained with TARA: spectral reflectivity and spectral differential reflectivity

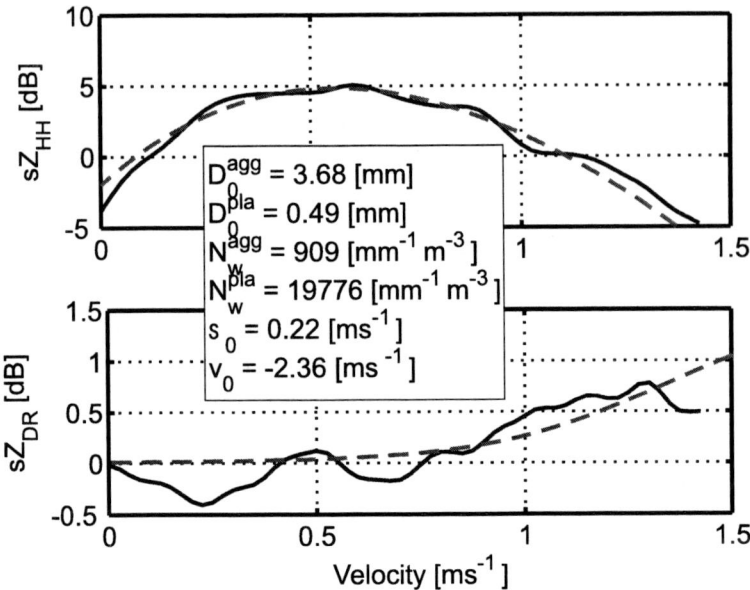

Fig. 7. Measured spectral horizontal and differential reflectivity (*solid line*) with the obtained fits and six parameter values (*dashed line*) for the height 2227 m

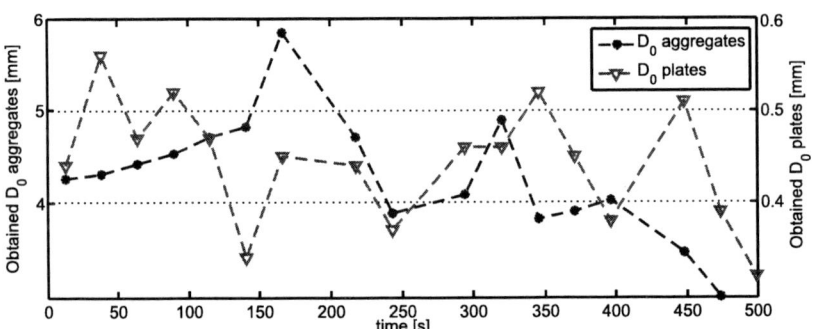

Fig. 8. Time series of retrieved median volume diameter for both aggregates and plates

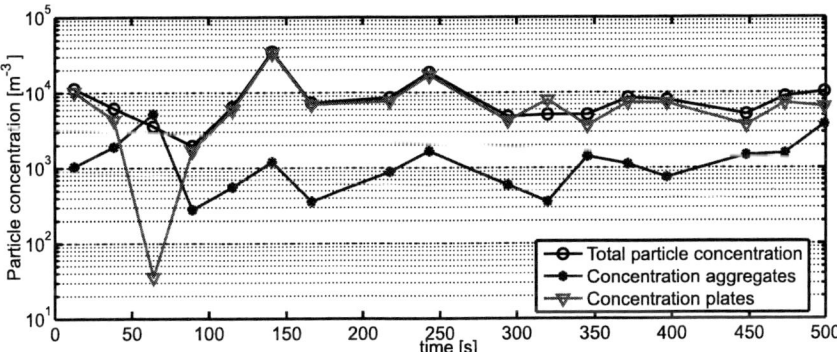

Fig. 9. Time series of retrieved particle concentration for both aggregates and plates

developed by Moisseev et al. (2006). In Fig. 10, the estimated ice water content and the estimated liquid water content are given as function of time. They show a good agreement, particularly after 250 s. From 0 to 250 s, the ice water content is larger compared to the liquid water content values. It may indicate vaporization of precipitation, which results in a decrease of liquid water content at 950 m where the Doppler spectra, used as input for the inversion algorithm of rain, are selected.

11.6.3 Relation between IWC and reflectivity

There is little knowledge on the microphysical properties of ice crystals above the melting layer of precipitation. In literature, several relations are discussed to obtain the ice water content from measured reflectivity values. The goal of the relations is to estimate the vertical structure of the ice water content in clouds. The obtained knowledge is used in climate research and weather forecast (Liu and Illingworth 2000).

The relation between the ice water content and the equivalent reflectivity at radar frequency of 3 GHz is be derived from the expressions given in Hogan et al. (2006) for temperatures ranging from 0 to $-10°C$, resulting in

$$IWC = 0.02 Z_e^{0.6} \qquad (21)$$

with IWC as the ice water content in g m^{-3} and Z_e the equivalent reflectivity in mm^6 m^{-3}. This relation is based on measurements of ice

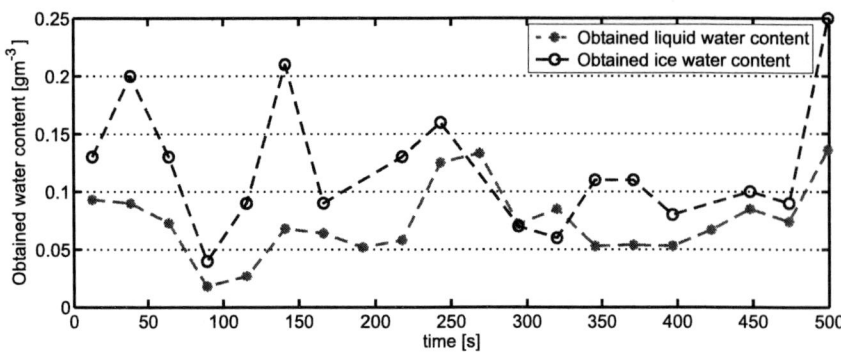

Fig. 10. Retrieved ice water content (above melting layer) and liquid water content (rain) versus time

particles in non-precipitating ice clouds. Using the outputs of the retrieval algorithm, which provides drop size distribution parameters of both plates and aggregates, it is possible to separate the ice water content and equivalent reflectivity values for aggregates and plates. In Fig. 11, the estimated ice water contents are plotted versus the estimated equivalent reflectivities for both plates and aggregates, together with Hogan's relation. In case of plates, there is a good agreement between the relation provided by Hogan et al. (2006) and the results obtained with the inversion algorithm. In case of aggregates the relation between the estimated ice water content and the horizontal reflectivity is obtained by a curve fit:

$$IWC = 0.0023 Z_e^{0.68} \tag{22}$$

The exponent in the estimated ice water content-reflectivity relation (26), is in good agreement with the exponents used in ice water content-reflectivity relations in literature. They are generally between 0.55 and 0.74 (Liu and Illingworth 2000).

11.6.4 Influence of the shape parameter of the DSD

The most common drop size distribution used in the literature is the gamma distribution, which consists of 3 parameters, N_w, D_0 and the shape parameter μ. The exponential distribution is a simplification of the gamma distribution ($\mu=0$). There is no good reason to fix the shape

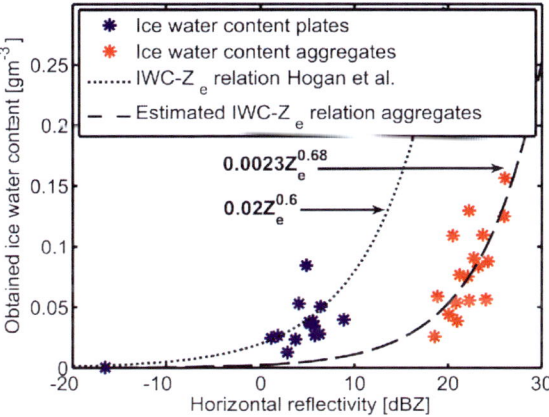

Fig. 11. Estimated ice water content versus estimated reflectivity values for both plates and aggregates

parameter to zero, other than to reduce the number of free parameters in the model. To investigate the dependence of the output of the inversion algorithm on the value of the shape parameter, the drop size distribution parameters of plates and aggregates are also obtained from the radar dataset with values for the shape parameter of two, four and six. The DSD follows now a gamma distribution and the same shape parameter is used for plates and aggregates.

The obtained median volume diameters for both aggregates and plates vary with the chosen value of the shape parameter. The maximum variation in the obtained value of the median volume diameter is 1.4 mm and 0.09 mm for aggregates and plates, respectively. The obtained differences are significantly larger than the root mean square errors: 0.6 mm for aggregates and 0.067 mm for plates. The intercept parameter will also vary with the value of the shape parameter. Therefore, the drop size distributions retrievals depend on the value of the shape parameter.

The same exercise is carried out on the ice water content and the number of particles. The maximum difference in the retrieved ice water content as function of the shape parameter is close to 0.04 g m^{-3}, which is the root mean square error on the ice water content. Concerning the number of particles, the maximum difference, 4000 m^{-3}, is larger than the Root Mean Squared error given in Table 8 (1940 m^{-3}). Using the root mean square error on the ice water content for comparison, it is concluded that the IWC is not significantly dependent on the chosen value of the shape parameter of the DSD of plates and aggregates.

11.7 Summary and conclusions

In this chapter, a methodology is described to discriminate between different types of ice particles present in a radar observation volume using spectral polarimetry. When two dominating types of particles are present, parameters of particle size distribution for each category of ice particles can be estimated. The retrieval procedure is based on a microphysical model of a Doppler power spectrum and the spectral differential reflectivity. This model of spectral radar observables depends on six parameters: the drop size distribution parameters of the two types of ice particles, the spectral broadening and the ambient wind velocity. The algorithm uses non linear least squares approach to fit the modeled spectral radar observables to spectral radar measurements, by varying the input parameters. We have shown that the six parameters minimization problem can be reduced to a three parameters minimization problem. The proposed methodology can be used for other applications, where two or more types of ice particles have to be separated. The error analysis of the proposed method was carried out on radar signal simulations. It was shown that the relative RMS of the median volume diameter for plates and aggregates is 15% and 17%, respectively. The retrieval of the intercept parameters of the DSD is less accurate. Aggregation of these parameters into lump parameters like ice water content and the number concentration leads to realistic and fairly accurate estimates. The final accuracy of the number concentration is around 60–65 % and around 30 % for the ice water content.

The obtained particle size distribution parameters are consistent over space and time. Integral parameters, like the ice water content and the number of particles, were calculated from the retrievals. The time series of ice water content and rain water content show a good agreement with each other. Based on the retrieved size distributions, ice water content-radar reflectivity relation can be established for different categories of ice crystals.

Acknowledgments

The authors from the Delft University of Technology acknowledge the support of 'Climate changes spatial planning' and 'Earth Research Centre Delft'. D. Moisseev and V. Chandrasekar were supported by the National Science Foundation.

References

Auer AH, Veal DL (1970) The dimension of ice particles in natural clouds. J Atmos Sci 27:919–926

Bringi VN, Chandrasekar V (2001) Polarimetric Doppler Weather Radar: Principles and Applications. Cambridge University Press, 636 pp

Chandrasekar V, Bringi VN, Brockwell PJ (1986) Statistical properties of dual-polarized radar signals. In: Preprints 23rd Conference on Radar Meteorology. American Meteorological Society, Snowmass, pp 193–196

Doviak RJ, Zrnic DS (1993) Doppler radar and weather observations. Academic Press, London, 562 pp

El-Magd A, Chandrasekar V, Bringi VN, Strapp W (2000) Multiparameter radar and in situ aircraft observation of graupel and hail. IEEE T Geosci Remote 38:570–578

Fabry F, Szymer W (1999) Modeling of the melting layer. Part II: Electromagnetic. J Atmos Sci 56:3593–3600

Heijnen SH, Lighthart LP, Russchenberg HWJ (2000) First measurements with TARA: a S-band transportable atmospheric radar. Phys Chem Earth (B) 25:995–998

Heymsfield AJ, Kajikawa (1987) An improved approach to calculating terminal velocities of plate-like crystals and graupel. J Atmos Sci 44:1088–1099

Hogan RJ, Mittermaier MP, Illingworth AJ (2006) The retrieval of ice water content from radar reflectivity factor and temperature and its use in evaluating a mesoscale model. J Appl Meteorol 44:860–875 (Table 2)

Khvorostyanov VI, Curry JA (2002) Terminal velocities of droplets and crystals: power laws with continuous parameters over the size spectrum. J Atmos Sci 59:1872–1844

Langleben MP (1954) The terminal velocity of snowflakes. Q J Roy Meteor Soc 80:174–181

Liu C-L, Illingworth AJ (2000) Toward more accurate retrievals of ice water content from radar measurements of clouds. J Atmos Sci 39:1130–1146

Locatelli JD, Hobbs PV (1974) Fall speeds and masses of solid precipitation particles. J Geophys Res 79:2185–2197

Magono C, Lee CW (1966) Meteorological classification of natural snow crystals. J Fac Sci, Hokkaido Univ 2:321–335

Matrosov SY (1992) Radar reflectivity in snowfall. IEEE T Geosci Remote 30:454–461

Matrosov SY (1998) A dual-wavelength radar method to measure snowfall rate. J Appl Meteorol 7:1510–1521

Matrosov SY, Reinking RF, Knopfli RA, Bartram BW (1996) Estimation of hydrometeor types and shapes from radar polarization measurements. J Atmos Ocean Tech 13:85–96

Mitchell DL (1996) Use of mass- and area-dimensional power laws for determining precipitation particle terminal velocities. J Atmos Sci 53:1710–1723

Mitchell DL, Zhang R, Pitter RL (1990) Mass dimensional relationships for ice particles and the influence of riming on snow flakes. J Appl Meteorol 29:153–163

Moisseev DN, Chandrasekar V, Unal CMH, Russchenberg HWJ (2006) Dual-polarization spectral analysis for retrieval of effective raindrop shapes. J Atmos Ocean Tech (accepted)

Moisseev DN, Unal CMH, Russchenberg HWJ, Chandrasekar V (2004) Radar observations of snow above the melting layer. In: Proceedings 3rd European Conference on Radar in Meteorology and Hydrology. 6–10 September, Visby, Sweden, pp 407–411. http://www.copernicus.org/erad/2004/

Pruppacher HR, Klett JD (1978) Microphysics of clouds and precipitation. Reidel Publishing Company, Dordrecht, 954 pp

Rajopadhyaya DK, May PT, Vincent RA (1994) The retrieval of ice particle size information from VHF wind profiler Doppler spectra. J Atmos Sci 11:1559–1568

Ray PS (1972) Broadband complex refractive indices of ice and water. Appl Optics 11:1836–1844

Russchenberg HWJ (1992) Ground based remote sensing of precipitation using multi-polarized FM-CW Doppler radar. Delft University Press, 206 pp

Russchenberg HWJ, Bosveld F, Swart D, ten Brink H, de Leeuw G, Uijlenhoet R, Arbesser-Rastburg B, van der Marel H, Ligthart L, Boers R, Apituley A (2005) Ground-based atmospheric remote sensing in The Netherlands: European outlook. IEICE T Commun 6:2252–2258

Rust BW (2003) Fitting nature's basic functions part IV: the variable projection algorithm. Comput Sci Eng 5:74–79

Sekhon RS, Srivastava RC (1970) Snow size spectra and radar reflectivity. J Atmos Sci 27:299–307

Smith PL (1984) Equivalent radar reflectivity factors for snow and ice particles. J Clim Appl Meteorol 23:1258–1260

Szymer W, Zawadzki I (1999) Modeling of the melting layer. Part I: Dynamics and microphysics. J Atmos Sci 56:3573–3592

Unal CMH, Russchenberg HWJ, Moisseev DN (2005) Simultaneous wind velocity estimation and dual-polarization measurements of precipitation and clouds by an S-band profiler. In: Proceedings 32nd Conference on Radar Meteorology. American Meteorological Society, Albuquerque, CD-ROM

Yanovsky FJ, Russchenberg HWJ, Unal CMH (2005) Retrieval of information about turbulence in rain by using Doppler-polarimetric radar. IEEE Transactions on microwave theory and techniques 53:444–450

12 Performance of algorithms for rainfall retrieval from dual-polarization X-band radar measurements

Marios N. Anagnostou[1], Emmanouil N. Anagnostou[1,2]

[1]Hellenic Center for Marine Research, Institute of Inland Waters, Anavissos-Attikis, Greece
[2]Civil and Environmental Engineering, University of Connecticut, Storrs, CT, USA

Table of contents

12.1 Introduction...313
12.2 X-band dual-polarization systems 316
12.3 Attenuation correction schemes for X-band dual-polarization radar observations...318
12.4 Rainfall estimation algorithms..319
    12.4.1 Review of microphysical retrieval algorithms.............319
    12.4.2 Rainfall retrieval algorithms 324
    12.4.3 Data..325
12.5 Algorithm evaluation 328
    12.5.1 Evaluation of the DSD retrieval techniques 329
    12.5.2 Evaluation of rainfall retrieval techniques...................333
12.6 Closing remarks 337
References 337

12.1 Introduction

Quantitative precipitation measurement remains a key topic in radar meteorology. On the one hand, in urban hydrology at the small catchment scale, it is required the repetitive high sampling and space resolution that also required over a wide area that can only be achieved with weather radar. On the other hand, the study of the hydrological cycle at global scale requires large scale observations such as those provided by satellites. Satellites use onboard microwave active or passive sensors for measuring rainfall or Infrared radiometers for measuring cloud top temperature. However, there are open questions on the accuracy of precipitation retrievals from satellites. Ground validation on the basis of in situ measurements is impaired since in different climate regimes there are different cloud and precipitation microphysical processes that can affect satellite rain retrievals.

Weather radars' capability to monitor precipitation at high spatial and temporal scales has stimulated great interest and support within the hydrologic community. The US National Weather Service (NWS) is using an extensive network of weather surveillance Doppler radar (WSR-88D) systems (Heiss et al. 1990), which can bring dramatic advancements to the precipitation monitoring with direct implications on the improvement of real-time forecasting of river floods and flash floods. Precipitation, though, may originate from varying meteorological systems, ranging from cold frontal systems to thunderstorms and tropical systems, where rainfall estimates based on these classical single polarization radar observations have quantitative limitations (e.g., Smith et al. 1996; Fulton et al. 1998; Anagnostou et al. 1999). These limitations arise from uncertainties associated with the lack of uniqueness in reflectivity to rainfall intensity transformation, radar system calibration and contamination by ground returns problems, as well as precipitation profile and complex terrain effects. Recent considerations concern the upgrade of WSR-88D systems to include dual-polarization capability, expected to moderate the effect of Z-R variability and radar calibration (Ryzhkov et al. 2005; Bringi et al. 2004; Zrnic and Ryzhkov 1999), while deploying local radar units is an option to fill up critical gaps in the WSR-88D network. Use of small and cost effective X-band radar units for this purpose (e.g., CASA NSF Engineering Research Center) is particularly stressed in cases of regions prone to localized severe weather phenomena, like tornados and flash floods, and over mountainous basins not well covered (due to terrain blockage) by operational weather radar networks.

Advances in weather radar technology have led to the development of polarimetric systems that are becoming more suitable to hydrological and hydrometeorological applications. First, Seliga and Bringi (1976, 1978) used the anisotropy information arising from the oblateness of raindrops to estimate rainfall. This information was exploited by producing new parameters such as the differential reflectivity (Z_{DR}) and the differential propagation phase shift (Φ_{DP}). The Φ_{DP} is a powerful tool for the quantification of rain-path attenuation in short wavelength radar observations (Anagnostou et al. 2006a, b; Park et al. 2005; Matrosov et al. 2005) and for the estimation of precipitation parameters including hydrometeor size distributions (Brandes et al. 2004; Zhang et al. 2001; Vivekanandan et al. 2004; Gorgucci et al. 2000). Due to rain-path attenuation, such shorter wavelengths (X- and C-band) undergo makes longer wavelengths (S-band) more attractive in the quantification of rainfall. Note that even at C-band significant attenuation issues associated with convective storms can occur. Several polarimetric relations for rain rate estimation have been suggested during the last two decades, using Z_H, Z_{DR} and the specific propagation differential phase shift K_{DP} (Ryzhkov et al. 2001; Brandes et al. 2001; May et al. 1999; Anagnostou et al. 2004; Matrosov et al. 2002). All these studies have shown that (a) there is an improvement in rainfall estimation if polarimetric radar is used, and (b) polarimetric rainfall estimation techniques are more robust with respect to Drop Size Distribution (DSD) variation than conventional Z-R relations. However, there is still no definitive compromise on the degree of improvement and the choice of the optimal polarimetric relationship (Ryzhkov et al. 2005). Furthermore, improving local flood and flash flood forecasting requires accurate quantitative rainfall measurements at small temporal (minutes) and spatial (hundred of meters to few kilometers) scales. The ability of short wavelength radar (X-band) to monitor precipitation at high spatio-temporal scales has stimulated great interest and support within the hydrologic community. As mentioned earlier in this Section, the use of small size X-band radar units is sought as an approach to fill up critical gaps in operational weather `radar networks (consisting primarily of S-band, e.g., WSR-88D network in US and C-band radars, e.g., radar networks in Europe). This would be particularly significant for providing high resolution rainfall observations over small scale watersheds, urban areas and mountainous basins not well covered by operational radar networks.

A primary disadvantage of X-band frequency is the enhanced rain path attenuation in Z_H and Z_{DR} measurements, compared to S-band (and moderately to C-band), including the potential for complete signal loss in cases of signal propagation through more than 10 km paths of high rainfall intensity, on the one hand. On the other hand, power independent parameters such as Φ_{DP} exhibit greater phase change per unit rainfall rate at shorter wavelengths. As a result, the sensitivity of Φ_{DP} to rainfall intensity at X-band can be about three times that of S-band observations. Consequently, X-band frequency offers an increased sensitivity on differential phase-based estimation of weak targets (such as stratiform rain rates) compared to S-band and C-band systems. Furthermore, a radar beam at X-band is associated with greater resolution than the lower frequencies (S-/C-band) for the same antenna size and is less susceptible to side lobe effects. As a result, X-band systems offer mobility and therefore cost efficiency, since they require low power units and small antenna sizes.

This Chapter is based on the following three testable hypotheses: (a) The use of differential propagation phase shift can provide accurate estimates of rain-path specific and differential attenuation provided there is no total lose of the transmitted power; (b) attenuation-corrected X-band dual polarization radar measurements offer higher sensitivity compared to lower frequency radar in the estimation of low rain rates, and a comparable accuracy in the estimation of moderate-to-high rainfall rates and DSD parameters; and (c) X-band can provide high spatial and temporal resolution estimates but is limited by range to less than 50 km and up to 120 km in heavy and low-to-moderate rain rates, respectively.

This Chapter explores the synergy of rainfall observations from multiple sensors needed to test the above hypotheses. The second part of the Chapter presents the state-of-the-art technology of today's research X-band polarimetric radar systems. Most of these systems are mobile and some are static systems. However, as we discussed earlier, the major drawback in systems is the atmospheric attenuation effect. Anagnostou et al. (2006a), Matrosov et al. (2005) and Park et al. (2005) have shown that Φ_{DP} can provide stable estimates of the path specific attenuation at horizontal polarization A_H and specific differential attenuation A_{DP} along a radar ray. The last two parts of the Chapter focuses on methods to estimate rainfall from X-band dual-polarization radar systems based on the microphysical properties of rain. Rainfall estimation techniques can be broadly quantified to physical based and empirical techniques. Physical based techniques are mainly power law algorithms where the coefficients have been calculated based on simulations. These algorithms are evaluated based on in-situ disdrometer spectra observations.

12.2 X-band dual-polarization systems

Throughout the years, radars have been used for operational and research purposes in a variety of applications (aviation, military, meteorology, etc.) Short wavelength radar systems (3 cm wavelength) became more attractive for research purposes (see Fig. 1 for a list of

Chapter 12 - Rainfall retrieval from dual-polarization X-band radar 317

Fig. 1. Sample photos of current X-band Polarimetric radars: *Left panel* from *top* to *bottom*—the National Research Institute for Earth Science and Disaster Prevention 'MP-X' multi-parameter radar system, the NOAA Environmental Technology Laboratory 'HYDRO' and one of the two radar systems from the University of Massachusetts and the University of Puerto Rico Mayaguez. In the *right panels* from *top* to *bottom* – the second radar system from the University of Massachusetts and the University of Puerto Rico, the Centre National de la Recherché Scientifque 'HYDRIX' and the National Observatory of Athens 'XPOL' radars

current X-band dual-polarization research radars) and lately for operational use (CASA http://www.casa.umass.edu) due to their small size and low cost as they are designed to require smaller size dish and very low power signal source to attain the requisite resolution and signal measurement of precipitation. These systems can either be mobile (trailer mounted, containerized or airborne) or static. Either way, they can constitute a low-cost solution to the problem of hydrologic forecasting for urban and small-scale flood-prone basins and coastal areas, but have some significant limitations. Since, they are low power systems, they have range limitations. Otherwise, they lose their main advantage which is sensitivity and spatial resolution. Another major limitation is that measurements at X-band undergo severe co-polar (A_H) and differential (A_{DP}) attenuation that can cause significant reduction of the horizontal reflectivity (Z_H) and differential reflectivity (Z_{DR}) signal, which must be corrected because it introduces errors in the rainfall estimation. Therefore, uses of X-band Polarimetric systems are mainly for gap filling (used for filling up blockages of large operation weather radar systems) purposes, or to monitor storms over small scale watersheds prone to flash floods. Another important aspect is that due to the small antenna size the systems can be deployed in today's cell phone network installments, or on small mobile platforms suitable for research studies.

12.3 Attenuation correction schemes for X-band dual-polarization radar observations

As mentioned in the introduction, the major issue in X-band rainfall estimation studies is the atmospheric attenuation effect. This subject can be a standalone chapter if one wishes to elaborate, thus here we will only give a brief discussion of the available attenuation correction algorithms and references were the reader could refer to.

The fundamental aspect that brought X-band back to the interest of hydrometeorologists for rainfall estimation is that the horizontal versus vertical polarization differential phase shift Φ_{DP} measurement can be used as a constraint parameter for the effective estimation of specific co-polar, AH, and differential, ADP, attenuation profiles (e.g., Testud et al. 2000; Vulpiani et al. 2005; Matrosov et al. 2002; Anagnostou et al. 2006b; Park et al. 2005). As shown by a recent elaborate study by Anagnostou et al. (2006b), this aspect minimizes the uncertainty due to rain path attenuation at X-band due to the fact that Φ_{DP} is not affected by attenuation (provided that backscattering signals are above the minimum detectable level) and it is almost linearly related with the range integrated co-polar attenuation, expressed in dB. Once the A_H range profile is estimated by means of a rain path attenuation Φ_{DP} constrained technique, A_{DP} can then be retrieved directly from A_H given that A_H and A_{DP} are almost linearly related (i.e., $A_H \approx \gamma A_{DP}$).

12.4 Rainfall estimation algorithms

Rainfall is a stochastic process that varies both in space and time thus making its accurate estimation an extremely difficult task. Since radar does not measure rainfall directly but rather we relate the measured variable (e.g., radar reflectivity) with rainfall properties, accurate estimation of rainfall requires a model that would best describe those physical properties of rain. For decades, rainfall estimates were derived from single radar measurements - radar reflectivity factor (Z). The conventional single-polarization Doppler radar has been broadly used to estimate rainfall (e.g., Atlas and Ulbrich 1990; Joss and Waldvogel 1990).

In order to better understand the characteristics of rainfall, research in the advances of radar technology has turned to dual-polarization systems which provide additional information on the physical parameters of rainfall. These radar observations are the differential reflectivity (Z_{DR}), which is the difference of horizontal to vertical polarization reflectivity in logarithmic scale, the horizontal to vertical polarization differential propagation phase shift (Φ_{DP}) and its range derivative (K_{DP}). In this Section, we discuss rainfall estimation techniques developed for X-band dual-polarization radar measurements. We first describe two microphysical retrieval algorithms and continue with the description of currently published rainfall algorithms. The Section closes with an evaluation of the various rainfall and microphysical retrieval techniques on the basis of coincident radar and disdrometer observations of rainfall from two climatic regimes.

12.4.1 Review of microphysical retrieval algorithms

A more accurate model of the distribution of raindrop sizes and shapes forms better derivation for precipitation estimation algorithms. It has been shown by a number of studies (Bringi and Chandrasekar 2001; Brandes et al. 2004; Anagnostou et al. 2007) that the various polarimetric radar observables (Z_H, Z_{DR}, K_{DP}) depend on raindrop shape, which is directly related to drop size. Hence, these parameters contain information about DSD that should allow more accurate estimation of rain rates. In this Section, we provide brief review of two microphysical retrieval algorithms used to estimate N_W and D_0 on the basis of dual-polarization radar observations (Z_H, Z_{DR} and K_{DP}) at X-band.

Polarimetric radar variables

The polarimetric radar parameters which are the most important for quantitative rain estimation are the horizontal polarization reflectivity, Z_H (mm^{6}m^{-3}), vertical polarization reflectivity (Z_V, mm^{6}m^{-3}) and differential reflectivity, Z_{DR} (dB) and the specific differential phase shift, K_{DP} (° km^{-1}). These variables depend on the raindrop size distribution, DSD, and the drop scattering amplitudes as follows:

$$Z_{H,V} = \frac{4\lambda^4}{\pi^4 |K_w|^2} \int_{D_{min}}^{D_{max}} |f_{HH,VV}(D)|^2 N(D) dD \qquad (1)$$

$$Z_{DR} = 10 \log\left(\frac{Z_H}{Z_V}\right) \qquad (2)$$

$$K_{DP} = \frac{180\lambda}{\pi} \int_{D_{min}}^{D_{max}} \text{Re}[f_{HH}(0,D) - f_{VV}(0,D)] N(D) dD \qquad (3)$$

where D (mm) is the raindrop equivalent volume diameter and D_{min} and D_{max} are the diameters of smallest and largest drops in the distribution. The $f_{HH,VV}(D)$ and $f_{HH,VV}(0,D)$ are the backscattering and the forward scattering amplitudes of a drop at horizontal and vertical polarization, K_W is the dielectric factor of water, λ (cm) is the radar wavelength and N(D) (mm^{-1}m^{-3}) is the count of raindrop of size D. The $f_{HH,VV}(D)$ and $f_{HH,VV}(0,D)$ parameters depend on the assumed raindrop shape-size relationship as discussed in a subsequent Section.

A gamma distribution model (or a similar model such as log-normal distribution) can adequately describe many of the natural variations in the shape of raindrop size distribution. The raindrop size distribution model used here is the 'normalized gamma distribution' function as presented in recent polarimetric radar rainfall studies (e.g., Testud et al. 2000; Bringi and Chandrasekar 2001):

$$N(D) = N_w f(\mu) \left(\frac{D}{D_0}\right)^{\mu} e^{-(4+\mu)\left(\frac{D}{D_0}\right)} \quad (mm^{-1} m^{-3}) \qquad (4)$$

with

$$f(\mu) = \frac{6}{4^4} \frac{(4+\mu)^{(\mu+4)}}{\Gamma(\mu+4)} \qquad (5)$$

where N_W (in $mm^{-1} m^{-3}$) is called 'normalized intercept parameter' and is the same as N_0 of an equivalent exponential DSD that has the same liquid water content (in gm^{-3}) and median raindrop diameter D_0 (mm) as the gamma DSD, while the μ is the shape parameter. Values for the three parameters are typically obtained on the basis of disdrometer measured raindrop spectra (e.g., Bringi et al. 2002). In short, the water content (W, in gr/m^3) and D_0 are calculated directly from the measured spectra, based on which we obtain N_W. The μ value is then estimated by minimizing in a least squares sense the divergence of modeled and measured raindrop size frequency distributions.

Simulation of radar parameters from DSD spectra

As shown by the integral Eqs. (1)–(3), information on the DSD, as well as hydrometeor phase (liquid, solid, mixed) and shape are needed to relate polarimetric radar measurements to precipitation and other radar parameters. As indicated by past investigations based on models and observations, the shape of raindrops can be well approximated by oblate spheroids (e.g., Pruppacher and Beard 1970; Beard and Chuang 1987; Bringi et al. 1998). The spheroid minor-to-major axis ratio (r) can be approximately related to the equivolumetric spherical diameter (D). Here, two raindrop shape-size relationships will be used. The first relationship is given by Brandes et al. (2002):

$$r = 0.9951 + 0.0251D - 0.03644D^2 + 0.00503D^3 - 0.00025D^4 \qquad (6)$$

The second is a linear relationship between r and D (D is in mm) originally presented by Pruppacher and Beard (1970):

$$r = 1.03 - \beta D \qquad (7)$$

A point to note about this relationship is that β may vary and that this variability can be determined on the basis of polarimetric radar parameters (Matrosov et al. 2002; Gorgucci et al. 2000, 2001; Bringi et al. 2002).

On the basis of DSD parameter sets determined from raindrop spectra the radar variables (Z_H, Z_{DR}, K_{DP}, A_H, A_{DP}, etc.) at X-band frequency were computed from T-matrix scattering calculations (Barber and Yeh 1975), assuming (1) the axis ratio models of Eqs. (6) and (7), (2) a Gaussian canting angle distribution with zero mean and standard deviation 10°, (3) a 8 (mm) maximum drop diameter, and (4) a dielectric constant of the water evaluated for an assumed atmospheric temperature.

There are, however, different approaches as to how to implement the gamma DSD model and incorporate the raindrop shapes. The inverse problem is how to determine the governing parameters of the DSD from radar observations to produce accurate estimates of rain rate and its DSD properties. Two methods originally developed for S-band dual-polarization measurements, named constrained-method and (Beta) β-method, have been parameterized and evaluated for the X-band frequency as described in Anagnostou et al. (2007). The methods are described next and evaluated in subsequent Sections.

Constrained-method

Many studies have shown that the governing parameters of the gamma DSD model are not mutually independent. This aspect can be of great significance because it can help to reduce the number of unknowns, thus enable the retrieval of the DSD parameters from a pair of more independent radar measurements, i.e., the horizontal-polarization reflectivity (Z_H) and differential reflectivity (Z_{DR}).

Analysis of DSD spectra revealed a high correlation between the shape size relation μ of the gamma distribution and the parameter μ which is defined as ($1/D_0$) and led to the derivation of an empirical μ-Λ relation,

$$\Lambda\left(\frac{1}{mm}\right) = 0.036\mu^2 + 0.735\mu + 1.935 \qquad (8)$$

Therefore, with the constrained and the fixed axis ratio relation we can calculate the radar parameters (Z_H and Z_{DR}) for D_0 and N_W.

Hence, the method starts with the estimation of D_0 and liquid water content (W) parameters based on relationships derived from scattering calculations using raindrop spectra and the Brandes et al. (2002) axial ratio model. There is a relation between the non-attenuated X-band radar parameters (Z_H in $mm^6 m^{-3}$ and Z_{DR} in $10^{0.1 Z_{DR}}$ units) and the median-drop diameter (D_0 in mm) and liquid water content (W in gm^{-3}) through the following best-fit relations:

$$D_0 = 0.25 Z_{DR}^3 - 1.23 Z_{DR}^2 + 2.35 Z_{DR} + 0.8 \tag{9}$$

$$W = (4.2 \cdot 10^{-4}) Z_H \cdot 10^{0.108 Z_{DR}^2 - 0.867 Z_{DR}} \tag{10}$$

Having determined D_0 and W, N_W can then be determined from the following equation:

$$N_W = \log_{10}\left[\left(\frac{3.67^4}{\pi}\right) 10^3 \frac{W}{D_0^4}\right] \tag{11}$$

The β (Beta) method

The major advantage of this method is that it treats the raindrop shape-size relations as a variable according to the raindrop axis ratio of Eq. (7). The method starts with estimating the β parameter using a non-linear regression approach described by Gorgucci et al. (2000) for scattering simulations performed for S-band frequency and modified by Park et al. (2005) for X-band frequency.

$$\beta = 0.94 \left(\frac{K_{DP}}{Z_H}\right)^{0.26} (\xi_{DR} - 1)^{0.37} \tag{12}$$

where $\xi_{DR} = 10^{0.1 Z_{DR}}$ is the differential reflectivity in linear units (a ratio), the Z_H in $mm^6 m^{-3}$. Incorporating the β term, an expression can be derived for the N_W and D_0 were

$$D_0 = 0.627 Z_H^{0.057} Z_{DR}^{(-0.03 \beta^{-1.22})} \tag{13}$$

$$\log_{10} N_W = 2.48 Z_H^{0.099} Z_{DR}^{(-0.067 \beta^{-1.15})} \tag{14}$$

In order to avoid any noise contamination from K_{DP} in rain retrieval, we set as lower K_{DP} and Z_H thresholds for applying this method the 0.10 (° km^{-1}) and 10 (dBZ), respectively.

12.4.2 Rainfall retrieval algorithms

This Section explores a large variety of published rainfall polarimetric algorithms. The algorithms are listed below:

(i) Standard Z_H-R relation for X-band (Kalogiros et al. 2006 here called STD):

$$R = 0.15 Z_H^{0.39} \tag{15}$$

(ii) The (Z_H, N_W)-R relation (Testud et al. 2000 here called TE00):

$$\left(\frac{R}{N_W}\right) = 0.0017 \left(\frac{Z_H}{N_W}\right)^{0.59} \tag{16}$$

(iii) The (Z_H, Z_{DR})-R relation (Bringi and Chandrasekar 2001 hereafter called BC01_1):

$$R = 0.0093 \, Z_H^{1.07} \cdot \left(10^{0.1 Z_{DR}}\right)^{-5.97} \tag{17}$$

(iv) The (K_{DP}, Z_{DR})-R relation (Bringi and Chandrasekar 2001 here called BC01_2):

$$R = 28.6 K_{DP}^{0.95} \cdot \left(10^{0.1 Z_{DR}}\right)^{-1.37} \tag{18}$$

(v) The (Z_H, Z_{DR})-R relation (Brandes et al. 2003 here called BA03):

$$\log\left(\frac{R}{Z_H}\right) = 0.15 Z_{DR}^2 - 0.89 Z_{DR} - 1.96 \tag{19}$$

(vi) The K_{DP}-R relations (Park et al. 2005 called here PR05):

$$R = \begin{cases} R = 0.007 Z_H^{0.82} \text{ for} \left(Z_H \leq 35 \text{dBZ and } K_{DP} \leq 0.3° \text{km}^{-1} \right) \\ 19.63 K_{DP}^{0.823} \text{ (otherwise)} \end{cases} \qquad (20)$$

(vii) The (Z_H, Z_{DR} and N_W)-R relation (Anagnostou et al. 2004 here after called AN04):

$$R = 63.7 N_W \left(\frac{Z_H}{N_W} \right)^{-0.16} Z_{DR}^{-0.07} \left(\frac{K_{DP}}{N_W} \right)^{1.12} \text{ (for } \beta = 0.062 \text{)} \qquad (21)$$

(viii) The (Z_H, Z_{DR}, β)-R relation (Bringi et al. 2004 here called BR04):

$$a = 0.54 \beta^{0.43}, c = 0.48 \beta^{-0.7}$$
$$R = a Z_H \xi_{DR}^c \qquad (22)$$

(ix) The (β, K_{DP})-R relations (Matrosov et al. 2002, 2005 here called MA02):

$$R = 1.1 Z_H^{0.3} K_{DP}^{0.32} Z_{DR}^{-0.82} \qquad (23)$$

12.4.3 Data

We used data from two different climatologic regimes to demonstrate the performance of the above rainfall and DSD retrieval algorithms. The first dataset is based on measurements from a mid latitude widespread storm with embedded convection in the urban area of Athens Greece. The data consist of radar observations from the National Observatory of Athens mobile X-band dual-polarization Doppler weather radar (XPOL) and coincident measurements from a 2D-video disdrometer at 10-km range from the radar. Figure 2 shows frequency and sample time series plots of Z_H and Z_{DR} parameters measured by NOA's XPOL radar and derived from the 2-D video disdrometer spectra. As noted from the figure the measured data range from low to moderate intensity stratiform rainfall to a few convective cell rain rates of higher intensity.

The second data set is associated with a maritime convective regime in Japan. It includes radar and disdrometer data from Ebina (35.4°N, 139.4°E), Japan where the National Research Institute for Earth Science and Disaster Prevention (NIED) is operating a dual-polarization and Doppler X-band radar (named MP-X) (Maki et al.

2005). For the validation of MP-X there is a network of in-situ stations that consists of four rain gauges and three, Joss-Waldvogel type (JW), disdrometers at approximate 10 km intervals along an azimuth of about

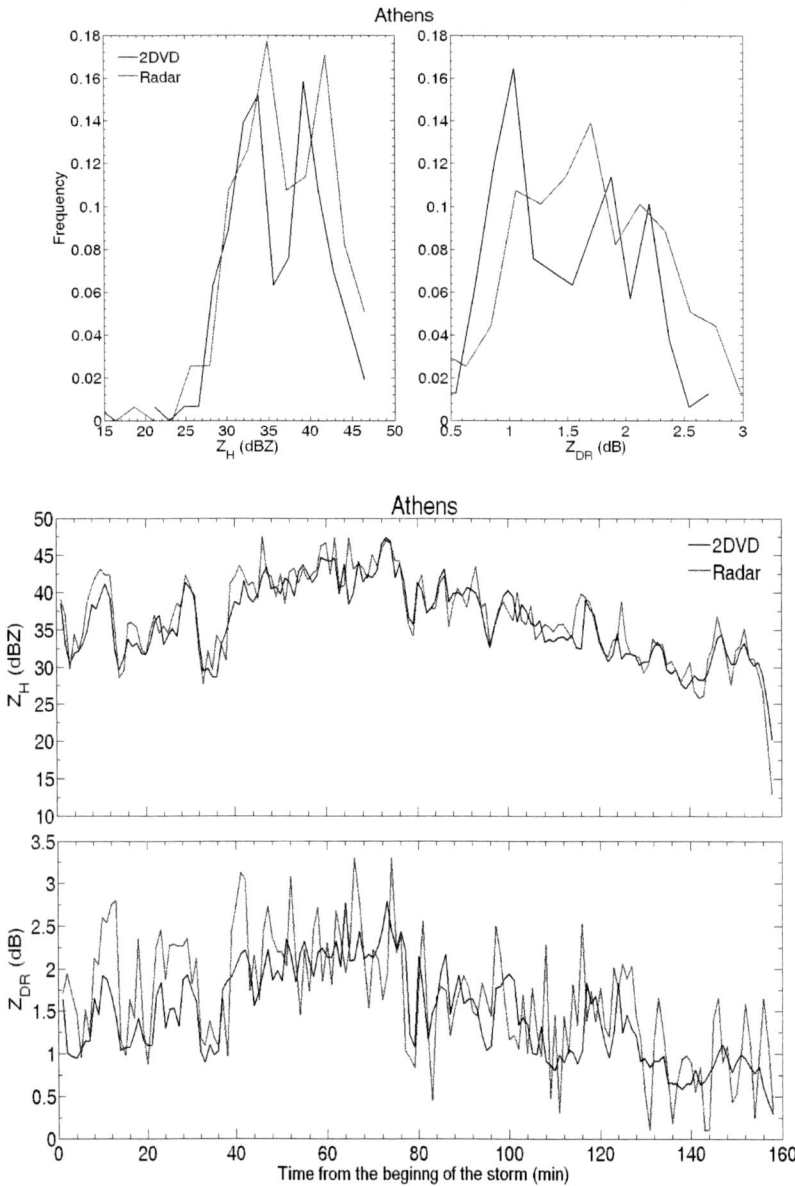

Fig. 2. Relative frequency (*upper panel*) and collocated sample time series (*lower panel*) plots of Z_H and Z_{DR} from XPOL and the 2D-video disdrometer in Athens

Chapter 12 - Rainfall retrieval from dual-polarization X-band radar

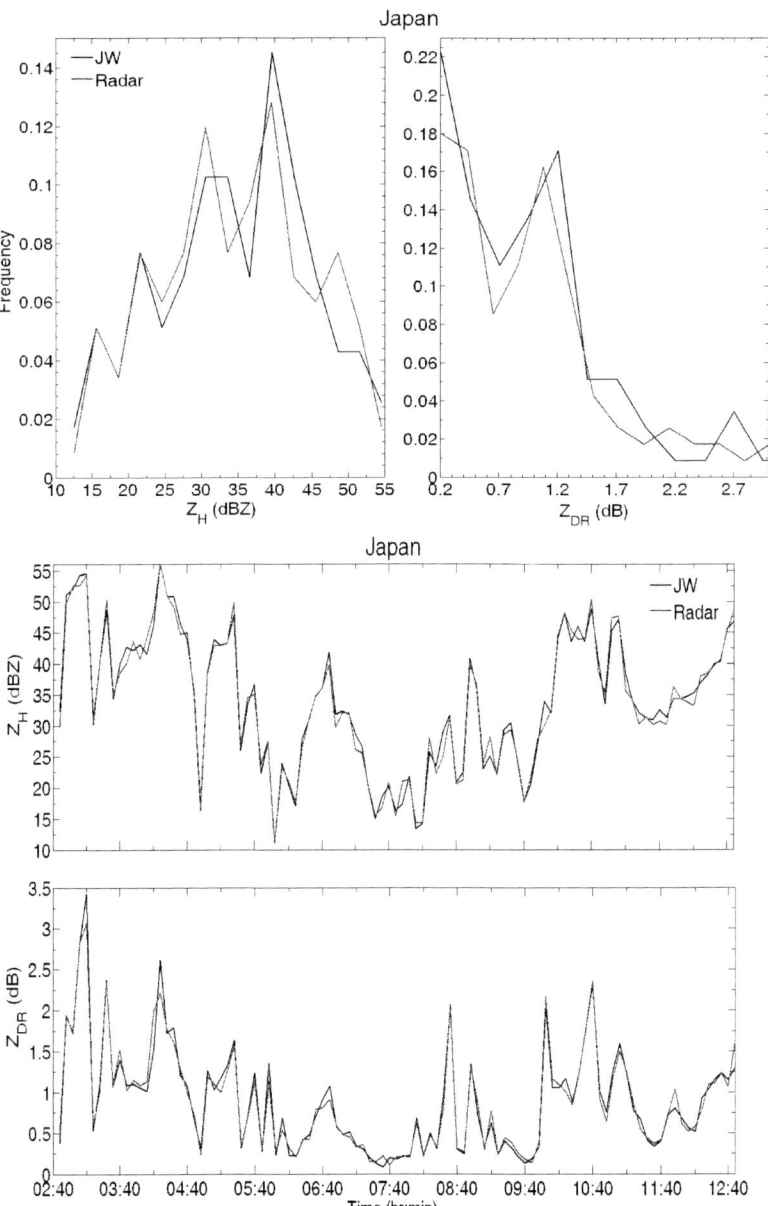

Fig. 3. Same as in Fig. 2, but for the Japan site. The radar there is the MP-X and the disdrometer is of JW type

257°. In this study, we used about 17 h of coincident MP-X and JW measurements from one of the in situ sites (~18 km range from the radar) during the passage of a Typhoon on 9 August 2003. As shown in

Fig. 3 both the disdrometer and radar measured high (40–55 dBZ) to moderate (30–40 dBZ) reflectivities during this storm passage.

A point to note from Figs. 2 and 3 is that radar measurements from both sites gave good agreement with the radar parameters derived from the in situ disdrometer measured raindrop spectra. In the Athens site, though, the XPOL radar exhibits higher noise in the Z_{DR} compared to the MP-X radar in Japan, which may be due to the storm structure, interference activity from marine type X-band radar and intensities in the two sites.

12.5 Algorithm evaluation

The two radar/disdrometer coincident datasets (from Athens and Japan) described above are used here to evaluate the different microphysical and rainfall retrievals algorithms. As described in Sect. 12.4.1 for microphysical retrievals and 12.4.2 for rainfall retrievals, simulated Z_H, Z_{DR} and K_{DP} from observe disdrometer spectra are used to calculate β, N_W and D_0 using Eqs. (9)–(14), and statistically compared with N_W and D_0 estimated from the same source of measured spectra. This approach, addresses the numerical stability of each algorithm. The same approach is devised here to evaluate the rainfall Eqs. (15)–(23). Furthermore, the performance of the algorithms is evaluated based on actual radar data, using as reference the disdrometer derived variables.

Evaluation is performed based on visual and statistical comparison methods. Visual include time series and scatter plots used to show the co-variation of the two technique estimates in comparison to the corresponding parameters derived from spectra. Statistical methods include mean relative error (MRE), correlation coefficient and relative root mean square error (RRMS) of the retrieved from radar parameters (hereafter named estimated) versus the disdrometer derived (hereafter named reference) variable. MRE is defined as the ratio of mean difference between the estimates and the reference versus the reference:

$$\text{MRE} = \frac{\sum \text{reference} - \sum \text{estimated}}{\sum \text{reference}} \quad (24)$$

while the relative root mean square error is given as the following relation:

$$\text{RRMS} = \frac{\sqrt{\frac{1}{N}\sum(\text{reference} - \text{estimated})^2}}{\frac{1}{N}\sum \text{reference}} \qquad (25)$$

12.5.1 Evaluation of the DSD retrieval techniques

To evaluate the two DSD retrieval techniques we use as input simulated (or actual X-band Polarimetric) radar parameters and compare estimates from the algorithms to corresponding DSD parameters (N_W and D_0) derived from the measured raindrop spectra. Visual and statistical comparisons are discussed next.

Time series of the retrieved DSD parameters using the two techniques are shown in Figs. 4 (for the Athens region) and 5 (for Japan).

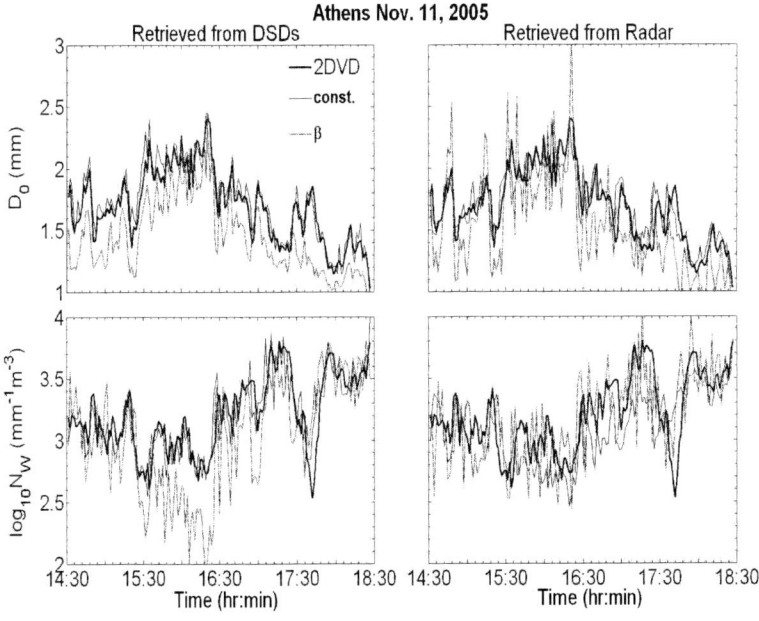

Fig. 4. Time series plot of the N_W and D_0 comparing the two methods using simulated radar and actual X-band radar data with the disdrometer observations

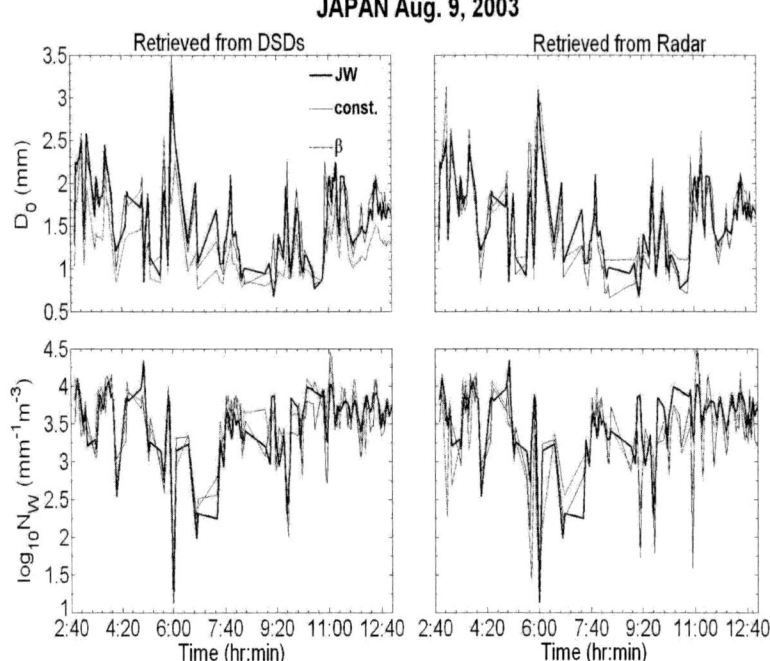

Fig. 5. As in Fig. 4 but using the data from Japan

The plots show time series comparisons of DSD parameters estimated from simulated (*left panels*) and actually measured (*right panels*) radar parameters. The plots demonstrate close coincidence for all time series with a greater agreement exhibited for the Japan storm case, which is consistent to the closer agreement in the radar parameters shown in Fig. 3. In Tables 1 and 2 we show the associated bulk statistics from Athens and Japan, respectively, using only disdrometer observations and simulated radar products from the observed spectra. In Tables 3 and 4 we compare the two algorithms on the basis of actual X-band radar observations.

A point to note is that the disdrometer data were three minute averages; this was done to reduce the noise in the measurements and that the bulk statistics are calculated for Z_H and K_{DP} parameters greater than 10 dBZ and $0.1°$ km^{-1}, respectively.

Table 1. Bulk statistics comparing the two methods with only disdrometer observations from Athens and retrievals from simulated radar parameters

β/constrained	Correlation	MRE	RRMS
N_W	0.62/0.84	0.14/0.004	0.12/0.05
D_0	0.84/0.94	0.09/0.005	0.10/0.06

Table 2. Bulk statistics comparing the two methods with only disdrometer observations from Japan and retrievals from simulated radar parameters

β/constrained	Correlation	MRE	RRMS
N_W	0.80/0.81	0.007/0.009	0.15/0.12
D_0	0.84/0.88	0.004/0.002	0.12/0.08

Table 3. Bulk statistics comparing the two methods with disdrometer observations from Athens and retrievals from XPOL radar observations

β/constrained	Correlation	MRE	RRMS
N_W	0.59/0.67	0.15/0.03	0.18/0.09
D_0	0.76/0.84	0.10/0.02	0.14/0.09

Table 4. Bulk statistics comparing the two methods with disdrometer observations from Japan and retrievals from MP-X radar observations

β/constrained	Correlation	MRE	RRMS
N_W	0.63/0.69	0.04/0.04	0.16/0.15
D_0	0.86/0.87	0.010/0.007	0.13/0.14

In Table 5 we show the relative effect of measurement error in the retrieval of DSD parameters. The table summarizes the difference of RRMS evaluated based on actual radar data to that of the simulated radar retrievals normalized by the simulated radar retrieval RRMS (presented in %). Results in this table indicate that measurement error can have significant effect in the retrieval of DSD parameters. This was particularly apparent in the Athens data where XPOL exhibited larger random deviations from the disdrometer observations. Another point to note is that the constrained method is most susceptible to the radar measurement error. This is apparent in both Athens and Japan datasets.

Table 5. Relative RRMS increase (in %) due to radar measurement error

β/constrained	Athens	Japan
N_W	50/80	6/25
D_0	40/50	8/75

In Figs. 6 and 7 we show the corresponding scatter plots of D_0 and N_W from the two datasets. The figures have four panels, were on the y-axis we plot the retrieved DSD parameters from either using simulated radar products or actual radar observations, while on the x-axis are the estimated DSD parameters derived from the disdrometer measured spectra. The upper two panels are from simulated radar parameters and the lower two panels are based on actual radar data. Points to note from the scatter plots is the enhanced scatter doing from upper to lower panel plots indicating the effect of radar measurement error on the retrieval uncertainty and the enhanced scatter in the retrieval of N_W relative to D_0.

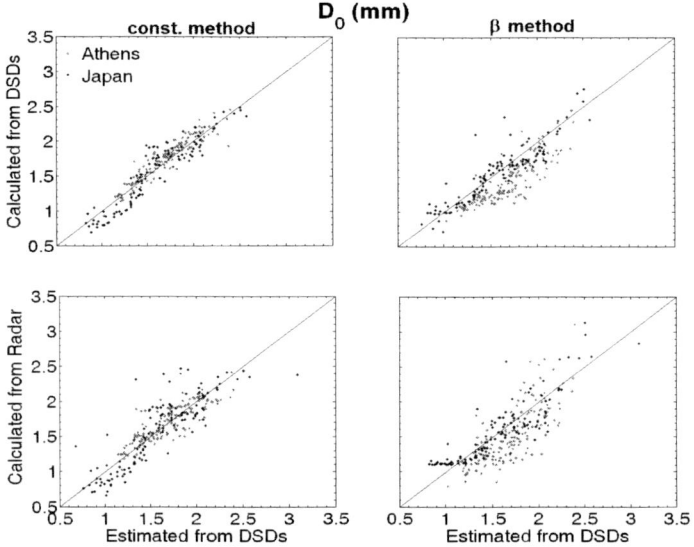

Fig. 6. Scatter plots of measured versus estimated D_0 using the two methods. The *black dots* are the data from Japan and the *light gray* are from Athens

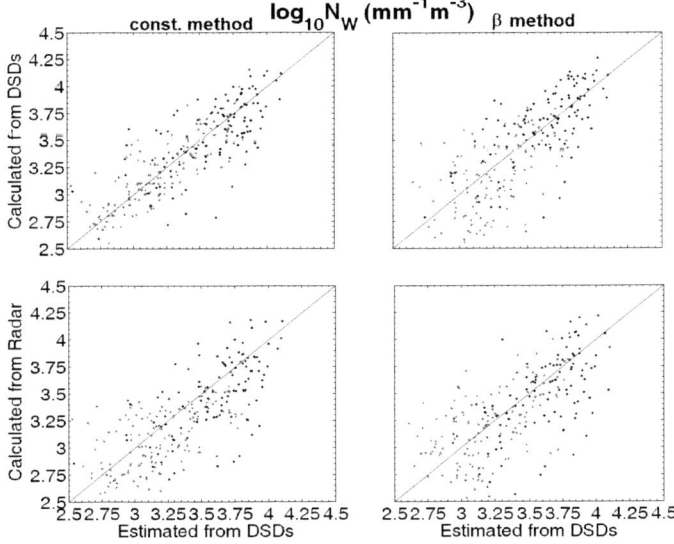

Fig. 7. Same as in Fig. 6 but for N_W

12.5.2 Evaluation of rainfall retrieval techniques

Similar statistics and visual comparisons are performed for all the nine rainfall algorithms discussed earlier in this Chapter. A point to note is that N_W used by Eqs. (16) and (21) is derived by the constrained method discussed above. Again, evaluation of the rainfall algorithms is performed for both simulated and actual radar measurements and reference is rainfall rates derived from the disdrometer measured raindrop spectra. First, in Tables 6 and 7 we show the bulk error statistics of the algorithms applied on simulated and actual radar data, accordingly. Several points are noted from the presented statistics. First, the STD algorithm exhibits lower correlation and increased RRMS relative to the other polarimetric techniques. A second observation is that among the different polarimetric techniques we cannot draw a clear winner in terms of significant improvements on the various statistics at both climatic regimes. The techniques that exhibit the most consistent improvement on both simulated and actual radar data in the Athens dataset are the AN04, MA02, BC01_2, BR04 and BA03. The corresponding techniques in the Japan datasets are BA03, PA05, AN04, MA02 and BR04. The order of best performing techniques is different at the two climatic regimes indicating a regional dependence of the radar algorithms.

Table 6. Bulk error statistics of the rainfall algorithms applied on simulated radar data and compared against rainfall rates derived from the disdrometer observations

Athens/Japan	Correlation	MRE	RRMS
(STD)	0.83/0.89	0.08/0.13	0.28/0.39
(TE00)	0.93/0.92	0.03/0.02	0.20/0.26
(AN04)	0.99/0.97	0.05/0.01	0.10/0.25
(BC01_1)	0.97/0.92	0.08/0.05	0.11/0.21
(BC01_2)	0.98/0.93	0.02/0.05	0.10/0.27
(BA03)	0.95/0.92	0.03/0.02	0.16/0.22
(BR04)	0.97/0.91	0.02/0.06	0.10/0.28
(MA02)	0.98/0.93	0.01/0.03	0.09/0.27
(PA05)	0.90/0.93	0.10/0.06	0.32/0.27

Table 7. Same as in Table 6 but using actual X-band radar observations

Athens/Japan	Correlation	MRE	RRMS
(STD)	0.73/0.71	0.09/0.14	0.36/0.68
(TE00)	0.89/0.80	0.04/0.15	0.23/0.52
(AN04)	0.90/0.86	0.06/0.10	0.12/0.53
(BC01_1)	0.83/0.85	0.09/0.18	0.33/0.61
(BC01_2)	0.90/0.86	0.06/0.19	0.20/0.54
(BA03)	0.82/0.88	0.04/0.10	0.33/0.38
(BR04)	0.88/0.81	0.05/0.25	0.24/0.55
(MA02)	0.90/0.86	0.07/0.11	0.23/0.53
(PA05)	0.84/0.87	0.10/0.14	0.36/0.46

Scatter plots of rain rates from the nine algorithms applied to actual radar measurements versus disdrometer derived rainfall rates are shown in Fig. 8. The vertical axis is rainfall observed from the 2-D video disdrometer in Athens and the JW disdrometer in Japan, while the horizontal axis is rainfall rates estimated by the XPOL in Athens or the MP-X radar in Japan. Notice that both datasets, from Japan (light gray diamonds) and Athens (black solid circles), are on the same subplots. Clearly, Japan data are associated with higher rain rates reaching 120 mm/h, while the Athens data are up to 17 mm/h. The scatter plots visually confirm the observations drawn earlier from the bulk statistics.

Specifically, STD exhibits enhanced scatter relative to most of the polarimetric techniques. The AN04, MA05, BA03, BC01_2 and PR05 seem to have the least scatter.

Another qualitative comparison based on time series plots of three selected polarimetric (algorithms that use Z_H, Z_{DR} and K_{DP} radar parameters) rainfall algorithms (Anagnostou et al. 2004; Brandes et al. 2004; Matrosov et al. 2005) and the standard Z-R relation (that uses a single polarization observation, Z_H) is shown in Fig. 9 (for the Athens dataset) and 10 (for the Japan dataset). Our observations derived from these plots are similar to what discussed earlier in this Section. A point to note is the larger deviations exhibited by the STD method relative to the polarimetric techniques. This is apparent in both datasets where STD under and overestimates segments of the storm relative to the disdrometer rainfall measurements. The other techniques exhibit remarkable coincidence with the disdrometer measurements particularly over the high rain rates in the Japan dataset.

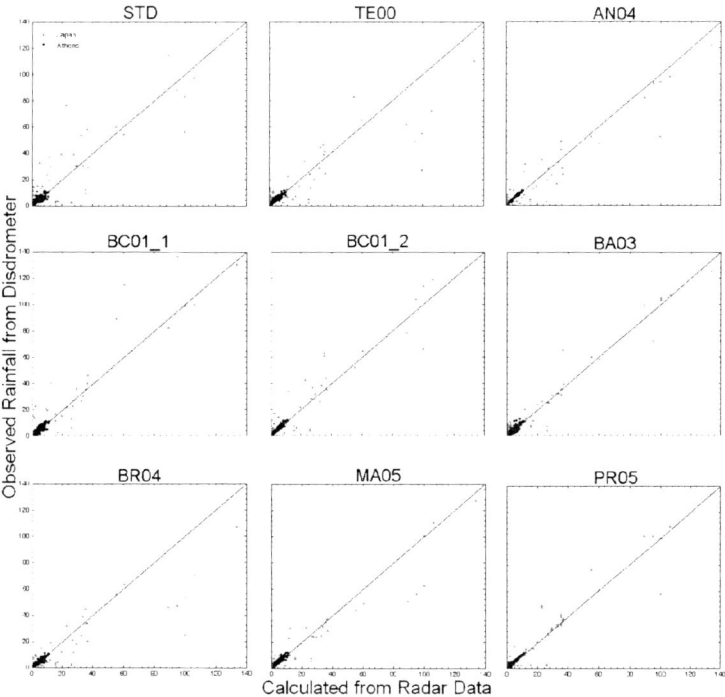

Fig. 8. Scatter plots of radar estimated from the different techniques versus disdrometer derived rainfall rates (in mm/h)

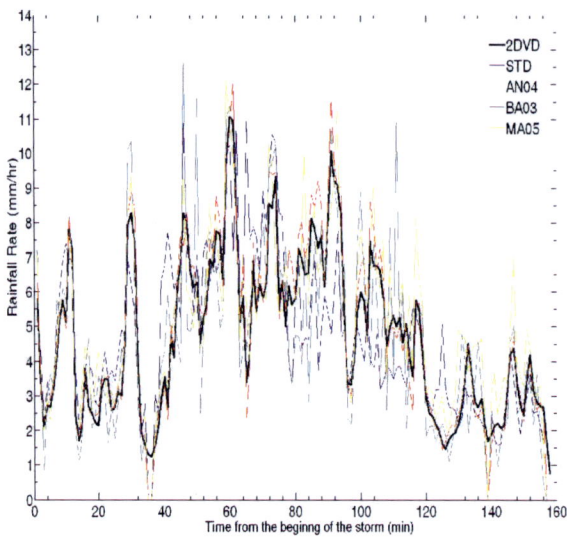

Fig. 9. Rainfall time series from Athens of the standard Z-R relation and three of the polarimetric algorithms compared to observed rainfall from a 2-D video disdrometer

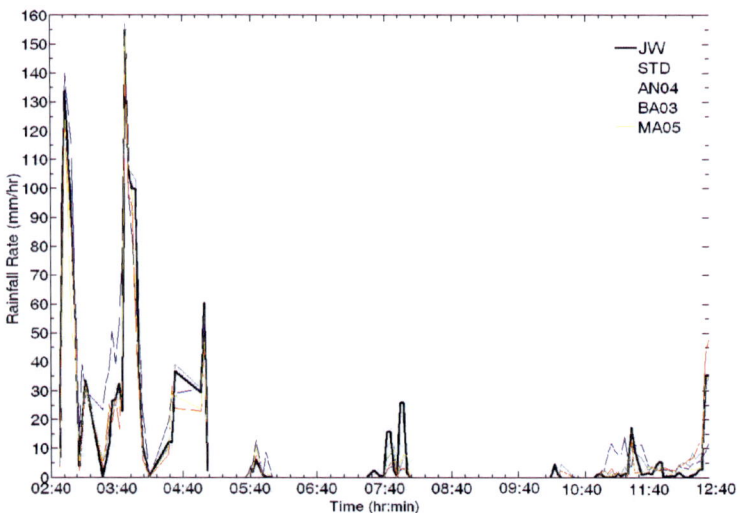

Fig. 10. Rainfall time series from Japan of the standard Z-R relation and three of the polarimetric algorithms compared to observed rainfall from a JW disdrometer

12.6 Closing remarks

The use of X-band weather radar systems have advanced to the level that can have direct implications to meteorological and hydrological forecasting. This is due to advancements in dual-polarization techniques, which dramatically improved the rainfall estimation from X-band observations through minimizing the uncertainty in attenuation correction and enhanced sampling resolutions and sensitivity to rainfall. These are generally low power short-range (<60 km) systems and primarily suitable for hydrometeorological applications and as 'gap'-filler radars for operational radar networks.

This Chapter focused on the estimation of microphysical and rainfall retrievals for X-band dual-polarizations weather systems. The rain-path attenuation correction that can be a significant uncertainty source at X-band was discussed and recent approaches based on dual-polarization information were presented.

We reviewed and evaluated two microphysical techniques for estimating raindrop size distribution and several rainfall retrieval algorithms. X-band dual-polarization radar and in situ disdrometer observations from two climatologically different regimes were used to evaluate the DSD and rain retrieval techniques. Results from these comparisons demonstrated close coincidence between radar and disdrometer measurements of DSD with the tropical storm case from Japan exhibiting greater agreement. The study demonstrated marginal differences in the performance of the rainfall retrieval techniques. The order of best performing techniques is different at the two climatic regimes indicating a regional dependence of the radar algorithms. The study shows that measurement error can have a error propagation effects in the DSD and rainfall retrievals.

Acknowledgments

Dr. Masayuki Maki and his research group contributed the MP-X radar and JW disdrometer data from Japan. Dr. Yiannis Kalogiros of the National Observatory of Athens contributed the XPOL radar and 2DVD disdrometer observations from Athens.

References

Anagnostou EN, Krajewski WF, Smith J (1999) Uncertainty quantification of mean-areal radar-rainfall estimates. J Atmos Ocean Tech 16:206–215

Anagnostou EN, Anagnostou MN, Krajewski WF, Kruger A, Miriovsky BJ (2004) High-resolution rainfall estimation from X-band polarimetric radar measurements. J Hydrometeorol 5:110–128

Anagnostou EN, Grecu M, Anagnostou MN (2006a) X-band polarimetric radar rainfall measurements in Keys Area Microphysical Project. J Atmos Sci 63:187–203

Anagnostou MN, Anagnostou EN, Vivekanandan J (2006b) Correction for rain path specific and differential attenuation of X-band dual-polarization observations. IEEE T Geosci Remote 44:2470–2480

Anagnostou MN, Anagnostou EN, Vivekanandan J, Ogden FL (2007) Comparison of raindrop size distribution estimates from X-band and S-band polarimetric observations. IEEE Geosci Remote Sensing Letters 4: 601–605

Atlas D, Ulbrich CW (1990) Early foundations of the measurements of rainfall by radar. In: Atlas D (ed) Radar in Meteorology. American Meteorological Society, Boston, USA, pp 86–97

Barber P, Yeh C (1975) Scattering of electromagnetic waves by arbitrarily shaped dielectric bodies. Appl Optics 14:2864–2872

Beard KV, Chuang D (1987) A new model for the equilibrium shape of raindrops. J Atmos Sci 44:1509–1524

Brandes EA, Ryzhkov AV, Zrnić DS (2001) An evaluation of radar rainfall estimates from specific differential phase. J Atmos Ocean Tech 18:363–375

Brandes EA, Zhang G, Vivekanandan J (2002) Experiments in rainfall estimation with polarimetric radar in a subtropical environment. J Appl Meteorol 41:674–685

Brandes EA, Zhang G, Vivekanandan J (2003) An evaluation of a drop distribution–based polarimetric radar rainfall estimator. J Appl Meteorol 42:652–660

Brandes EA, Zhang G, Vivekanandan J (2004) Drop size distribution retrieval with polarimetric radar: Model and Application. J Appl Meteorol 43:461–475

Bringi VN, Chandrasekar V (2001) Polarimetric Doppler weather radar: principles and applications. Cambridge University Press, Cambridge, UK, p 662

Bringi VN, Chandrasekar V, Xiao R (1998) Raindrop axis ratios and size distributions in Florida rainshafts: An Assessment of Multiparameter Radar Algorithms. IEEE T Geosci Remote 36:703–715

Bringi VN, Huang G-J, Chandrasekar V, Gorgucci E (2002) A methodology for estimating the parameters of a gamma raindrop size distribution model from polarimetric radar data: Application to a Squall-Line Event from the TRMM/Brazil Campaign. J Atmos Ocean Tech 19:633–645

Bringi VN, Tang T, Chandrasekar V (2004) Evaluation of a new polarimetrically based Z–R Relation. J Atmos Ocean Tech 21:612–623

Fulton RA, Breidenbach JP, Seo D-J, Miller DA, O'Bannon T (1998) The WSE-88D rainfall algorithm. Weather Forecast 13:377–395

Gorgucci E, Scarchilli G, Chandrasekar V, Bringi VN (2000) Measurement of mean raindrop shape from polarimetric radar observations. J Atmos Sci 57:3406–3413

Gorgucci E, Scarchilli G, Chandrasekar V, Bringi VN (2001) Rainfall estimation from polarimetric radar measurements: Composite algorithms immune to variability in raindrop shape–size relation. J Atmos Ocean Tech 18:1773–1786

Heiss WH, McGrew DL, Sirmans D (1990) NEXRAD: Next generation weather radar (WSR-88D) Microwave J 33:79–98

Joss J, Waldvogel A (1990) Precipitation measurements and hydrology. in radar in meteorology. In: Atlas D (ed) Radar in Meteorology. American Meteorological Society, Boston, MA, USA, pp 577–606

Kalogiros J, Anagnostou MN, Anagnostou EN (2006) Rainfall retrieval from polarimetric X-band radar measurements. In: Proceedings 4th European Conference on Radar in Meteorology and Hydrology, 18–22 September 2006, Barcelona, Spain. http://www.erad2006.org

Maki M, Iwanami K, Misumi R, Park S-G, Moriwaki H, Maruyama K, Watabe I, Lee D-I, Jang M, Kim H-K, Bringi VN, Uyeda H (2005) Semi-operational rainfall observations with X-band multi-parameter radar. Atmos Sci Lett 6:12–18

Matrosov SY, Clark KA, Martner BE, Tokay A (2002) X-band polarimetric radar measurements of rainfall. J Appl Meteorol 41:941–952

Matrosov SY, Kingsmill DE, Martner BE, Ralph FM (2005) The utility of X-band polarimetric radar for continuous quantitative estimates of rainfall parameters. J Hydrometeorol 6:248–262

May PT, Keenan TD, Zrnic DS, Carey LD, Rutledge SA (1999) Polarimetric radar measurements of tropical rain at a 5-cm wavelength. J Appl Meteorol 38:750–765

Park S-G, Bringi VN, Chandrasekar V, Maki M, Iwanami K (2005) Correction of radar reflectivity and differential reflectivity for rain attenuation at X band. Part I: Theoretical and empirical basis. J Atmos Ocean Tech 22:1621–1632

Pruppacher HR, Beard KV (1970) A wind tunnel investigation of the internal circulation and shape of water drops falling at terminal velocity in air. Q J Roy Meteor Soc 96:247–256

Ryzhkov AV, Schuur TJ, Zrnic DS (2001) Radar rainfall estimation using different polarimetric algorithms. In: Preprints 30th International Conference on Radar Meteorology. American Meteorological Society. Boston, Munich, Germany, pp 641–643

Ryzhkov AV, Giangrande SE, Schuur TJ (2005) Rainfall estimation with a polarimetric prototype of WSR-88D. J Appl Meteorol 44:502–515

Seliga TA, Bringi VN (1976) Potential use of radar differential reflectivity measurements at orthogonal polarizations for measuring precipitation. J Appl Meteorol 15:69–76

Seliga TA, Bringi VN (1978) Differential reflectivity and differential phase shift: Applications in radar meteorology. Radio Sci 13:271–275

Smith JA, Seo D-J, Baeck ML, Hudlow MD (1996) An intercomparison study of NEXRAD precipitation estimates. Water Resour Res 32:2035–2045

Testud J, Le Bouar E, Obligis E, Ali-Mehenni M (2000) The rain profiling algorithm applied to polarimetric weather radar. J Atmos Ocean Tech 17:332–356

Vivekanandan J, Zhang G, Brandes E (2004) Polarimetric radar estimators based on a constrained gamma drop size distribution model. J Appl Meteorol 43:217–230

Vulpiani G, Marzano FS, Chandrasekar V, Lim S (2005) Constrained iterative technique with embedded neural network for dual-polarization radar correction of rain path attenuation. IEEE T Geosci Remote 43:2305–2314

Zhang G, Vivekanandan J, Brandes E (2001) A method for estimating rain rate and drop size distribution from polarimetric radar measurements. IEEE T Geosci Remote 39:830–841

Zrnic DS, Ryzhkov AV (1999) Polarimetry for weather surveillance radars. B Am Meteorol Soc 80:389–406

iii. Underwater estimation

13 Underwater acoustic measurements of rainfall

Eyal Amitai[1,2], Jeffrey A. Nystuen[3]

[1]NASA Goddard Space Flight Center, Greenbelt, MD, USA
[2]George Mason University, Fairfax, VA, USA
[3]Applied Physics Laboratory, University of Washington, Seattle, WA, USA

Table of contents

13.1 Introduction..343
    13.1.1 Why measure rainfall at sea? 343
    13.1.2 Why listen to rainfall underwater?..............344
    13.1.3 What instrumentation is used to measure rainfall at sea? 344
    13.1.4 Using sound to measure drop size distribution and rain rate 345
13.2 Listening to rainfall in a shallow water pond 348
13.3 Oceanic field studies of the acoustic measurement of rainfall . 349
13.4 Listening to rainfall 2000 meters underwater – the Ionian Sea Rainfall Experiment 350
    13.4.1 Rain type classification and wind speed estimates.......358
13.5 Conclusions and outlook...360
References 361

13.1 Introduction

13.1.1 Why measure rainfall at sea?

A better understanding of the global water cycle is important to the field of climate research. Rainfall is a significant component of the dynamics of the atmosphere and the oceans. It is part of the heat, momentum and water budgets. During the process of rainfall formation a large amount of heat is being released into the atmosphere, driving atmospheric circulation. Thus, changes in rainfall patterns are an indication of climate change. Rainfall is an important part of upper ocean hydrology, affecting both the temperature and salinity structure of the ocean. On the time scale of months and longer, low salinity, high temperature barrier layers can form (e.g., Lukas and Lindstrom 1991). These layers block vertical mixing and permit the possibility of lateral movements of these water masses (Cronin and McPhaden 1998), thus affecting large-scale ocean circulation.

13.1.2 Why listen to rainfall underwater?

Rainfall is very difficult to measure at sea: Ships are unstable platforms, surface moorings are also unstable and subject to vandalism, satellites have poor temporal coverage and large spatial averaging, while rain is highly variable in both time and space. Improving measurements of rainfall and rainfall monitoring methods at sea is, therefore, a major challenge. The ambient sound field in the ocean contains a lot of information about the physical, biological and anthropogenic processes in the ocean. It is a combination of natural and manmade sounds. Interpretation of the ambient sound field can be used to quantify these processes (e.g., Nystuen and Selsor 1997; Nystuen et al. 2000; Ma and Nystuen 2005). In the frequency range from 1 to 50 kHz, the general character of ocean ambient sound is a slowly changing background that is closely associated with local wind speed, interspersed with shorter time scale events such as rain storms, ships and animal calls. In particular, the underwater ambient sound generated by raindrops striking the ocean surface and trapping bubbles has a unique signal in this frequency band. In fact, when rain is present, the sound from rain dominates the underwater sound field (Fig. 1).

13.1.3 What instrumentation is used to measure rainfall at sea?

Acoustic data are collected on hydrophones. Hydrophones are simple, robust sensors that can be deployed on most ocean instrumentation systems including surface or sub-surface moorings, bottom mounted systems, drifters and autonomous underwater vehicles. A dedicated oceanic underwater recorder called an Acoustic Rain Gauge (ARG) will be described in Sect. 13.3. As with other remote sensing techniques, the acoustic measurement does not interfere with the process being

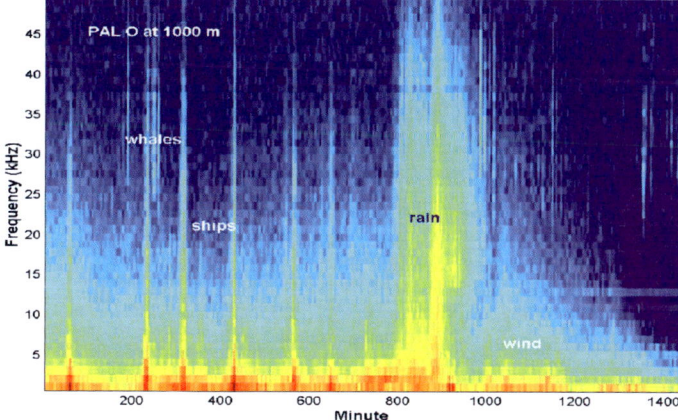

Fig. 1. A day-long example (12 March 2004) of ambient sound in the sea collected by a 1000 m depth hydrophone in the Ionian Sea (Greece). Signals are present from wind, rain, ships and whales

monitored. The sensor is away from the harsh (on instruments) environment of the air-sea interface. The likelihood of vandalism, a surprisingly big problem for surface instrumentation even in remote ocean locations, is reduced. Furthermore, because the measurement is passive, no potentially harmful sound is introduced into the marine environment.

13.1.4 Using sound to measure drop size distribution and rain rate

Laboratory studies of drop splashes (Franz 1959; Pumphrey et al. 1989; Medwin et al. 1992; Nystuen and Medwin 1995) have identified two components to the sound generated by a raindrop. These are the splat (impact) of the drop onto the water surface and then the subsequent formation of a bubble underwater during the splash. The relative importance of these two components of sound depends on the raindrop size. For most raindrops, it is the bubble that is, by far, the loudest sound source. Initially, a newly formed bubble radiates sound and then it absorbs, mostly at its resonant frequency. The important observation is that the size of the bubble is inversely proportional to its resonance (ringing) frequency. Larger bubbles ring at lower frequencies. The sound radiated is often loud and narrowly tuned in frequency (a pure tone). But quickly, after just tens of milliseconds, a bubble in water becomes a quiet adult bubble and changes roles. It absorbs sound and is especially efficient absorbing sound at its resonance frequency.

Naturally occurring raindrops range in size from about 300 microns diameter (a drizzle droplet) to over 5 mm diameter (often at the beginning of a heavy downpour). As the drop size changes, the shape of the splash changes and so does the subsequent sound production. Laboratory and field studies (Medwin et al. 1992; Nystuen 2001) have been used to identify five acoustic raindrop sizes (Table 1). For tiny drops (diameter <0.8 mm), the splash is gentle and no sound is detected. On the other hand, small raindrops (0.8–1.2 mm diameter) are remarkably loud. The impact component of their splash is still very quiet, but the geometry of the splash is such that a bubble is generated by every splash in a very predictable manner (Pumphrey et al. 1989).

These bubbles are relatively uniform in size and therefore relatively uniform in ringing frequency and are very loud underwater. Small raindrops are present in almost all types of rainfall, including light drizzle and are, therefore, responsible for the remarkably loud and unique underwater 'sound of drizzle' heard between 13 and 25 kHz, the

Table 1. Acoustic raindrop sizes. The raindrop sizes are identified by different physical mechanisms associated with the drop splashes (Medwin et al. 1992; Nystuen 2001)

Drop size	Diameter	Sound source	Frequency range	Splash character
Tiny	< 0.8 mm	Silent		Gentle
Small	0.8–1.2 mm	Loud bubble	13–25 kHz	Gentle, with bubble in every splash
Medium	1.2–2.0 mm	Weak impact	1–30 kHz	Gentle, no bubbles
Large	2.0–3.5 mm	Impact Loud bubbles	1–35 kHz 2–35 kHz	Turbulent Irregular bubble entrainment
Very large	>3.5 mm	Loud impact Loud bubbles	1–50 kHz 1–50 kHz	Turbulent Irregular bubble entrainment penetrating jet

resonance frequency for these bubbles. Interestingly, the splash of the next larger raindrop size, medium (1.2–2.0 mm diameter), does not trap bubbles underwater and, consequently, medium raindrops are relatively quiet, much quieter than the small raindrops. The only acoustic signal from these drops is a weak impact sound spread over a wide frequency band. For large (2.0–3.5 mm diameter) and very large (>3.5 mm) raindrops, the splash becomes energetic enough that a wide range of bubble sizes are trapped underwater during the splash, producing a loud sound that includes relatively low frequencies (1–10 kHz) from the larger bubbles. For very large raindrops, the splat of the impact is also very loud with the sound spread over a wide frequency range (1–50 kHz). Thus, each drop produces sound underwater with unique spectral features that can be used to acoustically identify the presence of that drop size within the rain (Fig. 2).

Raindrops of different sizes produce different sounds allowing for the inversion of the sound field to measure drop size distributions (DSD) within the rain (Nystuen 1996, 2001). While the DSD are obtained only for four size categories, in many situations this is still a good measure of rainfall rate, rain accumulation and other interesting features of rainfall such as radar reflectivity factor. The full inversion of the sound field to measure DSD, however, depends on a strong signal from small raindrops. Unfortunately, the sound production mechanism

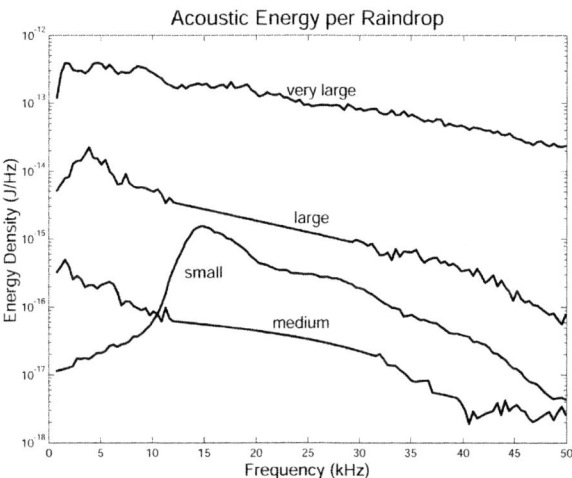

Fig. 2. Radiated acoustic energy densities for very large (>3.5 mm), large (2.0–3.5 mm), medium (1.2–2.0 mm) and small (0.8–1.2 mm) raindrop sizes. This forms the mathematical basis for the inversion of the sound field to obtain drop size distribution (Nystuen 2001)

for this signal is suppressed by wind (e.g., Nystuen and Farmer 1989; Nystuen 1993). Consequently, there will be situations where the full inversion of the sound field is not possible. In these situations, an acoustic rain rate (R) estimation algorithm based on empirical relations might be more practical. A simple regression between the sound pressure level (S) measured by a hydrophone in a given frequency band and R measured by a rain gauge or a radar can be used to derive S-R relations in a form of a power low (e.g., Nystuen et al. 1993; Nystuen et al. 2000; Ma and Nystuen 2005). However, distinctive features of the sound, associated with different components of the DSD, allow a classification according to rain type (Black et al. 1997; Nystuen and Amitai 2003). By first classifying the rain, improved R measurements can be achieved (Nystuen and Amitai 2003). The acoustic features of rainfall that are useful for classification include the sound levels, the ratio of sound levels measured at different frequency bands and the temporal variances of the sound levels within selected frequency bands.

13.2 Listening to rainfall in a shallow water pond

As a transition stage from laboratory to the ocean, the Atlantic Oceanographic and Meteorological Laboratory (AOML) Rain Gauge Facility was used to compare underwater acoustic rainfall measurements to a variety of land-based automatic rainfall instruments (Nystuen et al. 1996, Nystuen 1999). The data collected at this facility allowed simulated radar reflectivity factor (Z) as calculated from a nearby Joss-Waldvogel disdrometer (an instrument measuring DSD; Joss and Waldvogel 1969) to be compared to underwater ambient sound measurements from a hydrophone mounted in a sheltered shallow water pond (Amitai et al. 2004). The sheltering of the pond reduced the effect of wind on the measurements.

Figure 3 shows an example of a time series of S at 4–10 KHz band recorded from the hydrophone and compared to the simulated radar reflectivity factor during a 5-h rainstorm. Due to the strong dependency of both S and Z on higher moments of the DSD, the correlation between 1 min averaged S and Z is higher than the correlation between S and R calculated from the disdrometer observations. This suggests that underwater sound measurements could be used to evaluate radar-based Z_e measurements (Z_e for *observed reflectivity* as opposed to Z for *calculated reflectivity*) without the need to generate intermediate R-based products. In fact, S-Z_e relations have never been demonstrated

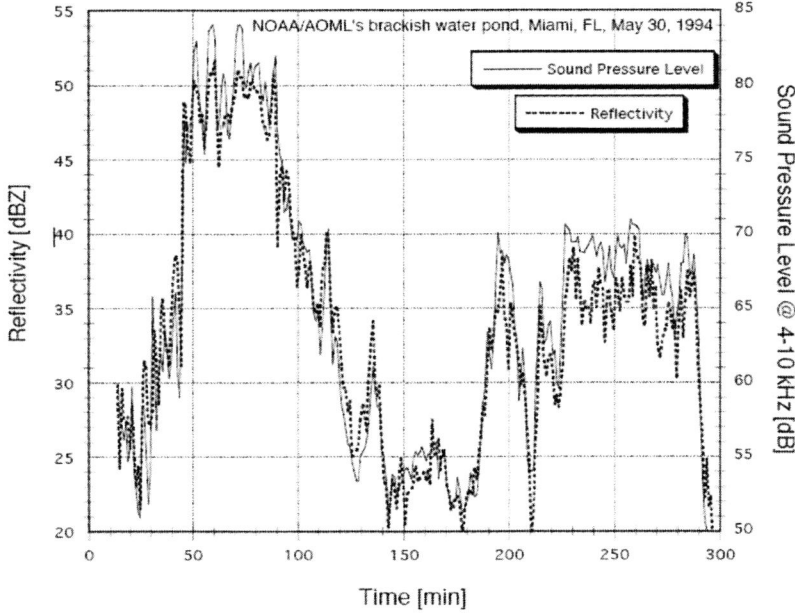

Fig. 3. Sound pressure level as measured by a hydrophone mounted 1.5 m below water surface and the radar reflectivity factor based on disdrometer observations taken 50 m from the hydrophone, during a 5-h rainstorm in Miami, Florida (Amitai et al. 2004)

using actual weather radar observations in general and in a deep-sea situation in particular. This will be shown in Sect. 13.4.

13.3 Oceanic field studies of the acoustic measurement of rainfall

Rainfall has long been recognized as a producer of underwater sound (Heindsman et al. 1955; Bom 1968) and can be detected acoustically at sea even in high sea states (Lemon et al. 1984; Nystuen and Farmer 1989). Shaw et al. (1978) first suggested that the sound produced by rainfall is a signal that could be used to quantify rainfall rate. Evans et al. (1984) noted the detection of rainfall from a hydrophone mounted on the sea floor at full ocean depth (5 km); Wilkerson and Proni (2000, 2001) monitored rainfall over the ocean using the deep-water (1.5 km) hydrophone array at the U.S. Navy Atlantic Undersea Test Evaluation Center (AUTEC) range in the Bahamas.

Acoustic classification of underwater sound is required before quantification of the signal. This depends on identifying unique features of the sound spectrum of rainfall including the spectral peak signal from small drops (Figs. 1 and 2) and relatively more high frequency components compared with the sound from wind waves and shipping. The Canadian Atlantic Storms Program (CASP) (Nystuen and Farmer 1989), the South China Sea Experiment (SCSMEX) (Nystuen et al. 2000) and the Ionian Sea Rainfall Experiment (Nystuen et al. 2008) compared coastal radar measurements of rainfall to verify the co-located acoustic rainfall detections. In deep water ocean locations radar is generally not available and so studies have used co-located satellite rainfall measurements or co-located surface mounted rain gauges (Ma and Nystuen 2005). Weather classification using passive acoustic drifters is compared with satellite data in Nystuen and Selsor (1997).

The basis for the acoustic quantification of rainfall rate depends on a relationship between sound level and rainfall rate. The sound produced from drizzle and small raindrops in the frequency band from 13 to 25 kHz is wind speed dependent (e.g., Laville et al. 1991; Nystuen 1993; Ma and Nystuen 2005; Ma et al. 2005) and, consequently, quantification of rainfall rate at sea has focused on the frequency band from 4 to 10 kHz. In this frequency band the signal from rain is mostly from larger raindrops and is highly correlated with rainfall rate. Empirical algorithms for rainfall rate using this frequency band have been developed using co-located surface mounted rain gauges from coastal locations in the Gulf of Mexico (Nystuen et al. 1993) and the South China Sea (Nystuen et al. 2000). Ma and Nystuen (2005) use long-time series from deep ocean surface moorings to develop a relationship for open ocean measurements. Coefficients in these algorithms different slightly and attempts to unify these relationships is ongoing.

13.4 Listening to rainfall 2000 meters underwater – the Ionian Sea Rainfall Experiment

Because of its inherent temporal and spatial variability, rainfall is a relatively difficult geophysical quantity to measure. Useful measurements often require both spatial and temporal averaging. One interesting feature of the acoustical measurement is that the listening area for hydrophone, its effective 'catchment', is proportional to its depth and yet the signal should be almost independent of depth if the sound source is uniformly distributed on the sea surface. Theoretical calculations suggest that the listening area on the water surface is defined by a radius that is roughly three times the depth of the hydrophone (Nystuen 2001). These calculations consider the radiation pattern for a surface source (vertically oriented dipole) and the attenuation along the acoustic path. The attenuation is due to geometric spreading, scattering and absorption, although scattering is assumed to be minimal and refraction for steep sound paths is also minimal. This feature allows acoustic measurements of rainfall to have sampling with high temporal resolution (order of seconds) (Nystuen and Amitai 2003) and spatial averaging coverage that can be orders of magnitude greater than of in situ rain gauges and comparable to a radar sampling volume. It was time to explore the underwater measurements of rainfall and to test these theoretical calculations in the deep sea.

In early 2004, a field experiment was carried out in the Ionian Sea to investigate the accuracy of acoustic rainfall measurements in the open sea (the experiment was part of a collaborative research effort between the University of Washington, the University of Connecticut, the George Mason University and the National Observatory of Athens, Greece sponsored by the United States National Science Foundation. The authors of this Chapter and Professor Emmanouil Anagnostou were the lead scientists for this experiment). The main objective was to explore the spatial averaging of the acoustic signal by deploying several ARGs vertically separated on a single mooring and comparing those measurements to simultaneous radar observations. The experiment provided for the first time synergistic measurements from a high resolution dual polarization X-band radar (see Chap. 12 of this book), a dense rain gauge network of 15 gauges within 2×2 km^2, a 2-D video distrometer (Chaps. 1 and 12 of this book) and underwater ARGs. The location of the experiment (Fig. 4) offered deep water (over 3 km deep) within the coverage area of the coastal radar. Four ARGs were deployed at 60, 200, 1000 and 2000 m depths (hereafter referred to as ARG60, ARG200, ARG1000 and ARG2000, respectively).

The radar was deployed at Methoni, Greece, 17 km east of the mooring, about 300 m from shore 20 m above sea level. It was operated on a continuous single elevation surveillance scan mode (Plan Position Indication, PPI). The antenna elevation angle was set at 2°, which is associated with a 600 m beam elevation at the mooring location. The beam azimuth and range sampling resolution was set to 0.5° and 150 m and the radar beamwidth is 0.95°. The beam rotation speed was set to 6°/sec resulting in one full swath scan every 1 min. The radar measured

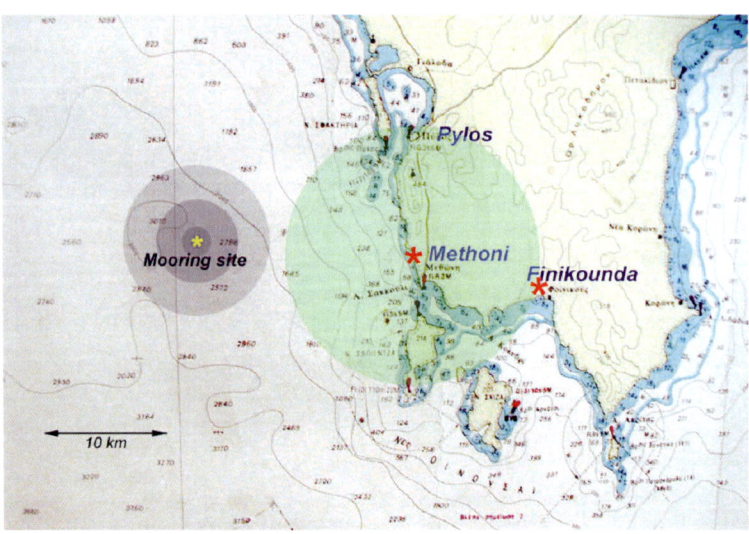

Fig. 4. The Ionian Sea Rainfall Experiment area: The X-band polarimetric radar is in Methoni (36.829°N, 21.705°E). The mooring is 17 km west of the radar site (36.845°N, 21.516°E, yellow star) and consists of four ARGs at 60, 200, 1000 and 2000 m depths. The gray circles represent theoretical listening areas for each ARG. The dense rain gauge network and a 2-D video distrometer are in Finikounda. The green circle represents the area within 10 km of the radar

parameters include reflectivity (Z_e, in mm^{6}m^{-3}) at two polarizations (horizontal—Z_h and vertical—Z_v), differential phase (f_{DP}) and Doppler wind. The polarimetric measurements Z_h and Z_{DR} (Z_h/Z_v) were assessed for calibration biases and noise through comparison with in situ measurements from the 2-D video disdrometer (located 10 km the radar site) and corrected for rain-path attenuation on the basis of the algorithm of Anagnostou et al. (2005, 2006). Chapter 12 in this book describes the X-band radar performance. The results presented here used a corrected Z_h mapped onto a Cartesian grid with a 300 m × 300 m pixel resolution (referred to as Z_e).

The ARGs consist of a low-noise wideband hydrophone, signal pre-amplifiers and a recording computer. The nominal sensitivity of these instruments is –160 dB relative to 1 V/μPa with an instrument noise equal to an equivalent oceanic background noise level of about 28 dB relative to 1 μPa^{2}Hz^{-1}. Band-pass filters are present to reduce saturation from low frequency sound (high pass at 300 Hz) and aliasing from above 50 kHz (low pass at 40 kHz). The ARG sensitivity also rolls off above its resonance frequency, about 40 kHz. A data collection sequence takes about 20 s and consists of four 10.24 ms time series each separated by 5 s. Each of these time series is fast Fourier transformed (FFT) to obtain a 512-point (0–50 kHz) power spectrum. These four spectra are spectrally compressed to 64 frequency bins, with frequency resolution of 200 Hz from 100 to 3000 Hz and 1 kHz from 3 to 50 kHz. The sound spectra are automatically evaluated individually to objectively identify the sound source, e.g., wind waves, rainfall, drizzle, shipping, etc. The data are stored in memory and the time to the next data collection sequence is set based on the assumed source. For example, if rainfall is detected the next data collection sequence occurs 30 s later; if wind is detected the next data collection sequence is 5 min later. This allows the ARG to conserve energy during non-rainy periods and to record data for up to 1 year without servicing. However, it also means that up to 5 min of rainfall can be missed from the beginning of a rain event.

The acoustical measurements and rain gauge network measurements were continuous from mid-January to mid-April. The radar was operating during 7 weeks (10 February through 1 April 2004), however, it needed to have an operator present and so its temporal coverage was not continuous. Nevertheless, roughly ten precipitation events were captured by all three systems (i.e., radar, rain gauges and ARGs). For these events, 1099 radar scans were recorded with Z_e>20 dBZ at the 300 × 300 m^2 pixel over the mooring. The total gauge accumulation at times the radar was operating was about 90 mm, with only four events of more than 10 mm. All rain events were recorded except for a few hours during which there were technical problems associated with the radar. These problems occurred mainly on 1 April. However, during these few hours on 1 April the gauges recorded about 20 mm out of the 110 mm that they recorded during the whole 7-week period. The climatologic rain amount for 7 weeks during this time period for Methoni is 120 mm. Most of the rain events recorded by the radar were characterized by weak rain intensities (stratiform rain). Only 60 radar scans had Z_e>40 dBZ at the 300 × 300 m^2 pixel over the mooring. We chose four events that were characterized by a large dynamic range of rain intensities to allow comparison with a large dynamic range of sound levels.

Perhaps the most serendipitous rain event was an intense isolated squall line that passed over the mooring very rapidly on 8 March 2004. The S-Z_e comparison for this event is presented in Fig. 5. The curves represent time series of the radar Z_e and the four ARGs S at 5 kHz during the 60 min (14:35–15:35 UTC) centered at the time of the squall line overpass. The sound frequency of 5 kHz is chosen as this is frequency that has a high correlation between S and Z_e according to observations in shallow water (Amitai et al. 2004). It is also the frequency band that has been used to quantify rainfall rate using underwater sound (Ma and Nystuen 2005). As shown in the figure, all four ARGs detected rain and the S-Z_e correlation was found to be very high (Amitai et al. 2007).

A curious feature of this rain event is a 'moment of silence' at Min 909 (Nystuen et al. 2008). The ARGs are totally independent of one another and so this is a real feature. Note that the sound level at 200 m is actually quieter than the sound levels before or after the squall line. This suggests that the sound production mechanism for wind (breaking waves) has been disrupted by the intense rain. It is widely reported, but undocumented, that 'rain calms the seas'. This is acoustical evidence that this is true. The radar reports no echo return at Min 909 (Fig. 5) in the pixel directly over the mooring and the radar scan at Min 915 shows an abrupt end to the backside of the squall. Note, however, that severe attenuation of the X-band radar signal through the rain cell may also cause 'no echo' return from the radar. The acoustical evidence suggests that this is not the case here and that there really is no rain on the backside of the squall line.

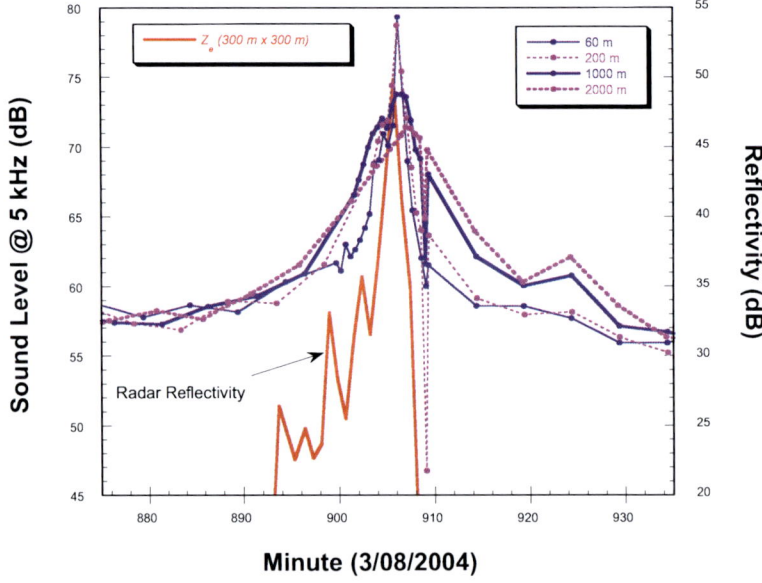

Fig. 5. A 60 min time series of underwater sound pressure level in the 5-kHz band as recorded by ARGs mounted in the Ionian Sea at 60, 200, 1000 and 2000 m below sea level and the 300 m × 300 m radar corrected reflectivity (Z_e) over the mooring. The dots on each sound curve represent the actual sample time (Amitai et al. 2007)

During the passage of a squall line directly above the ARGs the sound levels at the ARGs should decrease with increasing depth because of (1) attenuation by seawater and (2) incomplete filling of the listening area. Separation of these two factors from each other requires analyzing the sound field at times when the rain field is uniform. A perfectly uniform rain field over 100 km^2 (corresponding roughly to the size of the listening area by the deepest ARG) is difficult to find, although events of wide spread stratiform rainfall might satisfy this requirement. Alternatively, examining sound fields at non-rainy periods can be used provided that the sound source is uniform at the surface. Moderate wind conditions meet this requirement, as the time and length scales of wind are larger than the expected listening area of the ARGs. This approach assumes that attenuation of wind-generated sound (i.e., sound from white caps) and rain-generated sound is the same. The correction for attenuation (absorption) as a function of frequency can also be estimated from theoretical calculations (Nystuen et al. 2008). At 5 kHz the absorption is 0.5 dB for ARG1000 and 1 dB for ARG2000. Analyzing the underwater sound from wind (Fig. 4 in Amitai et al. 2007) suggests an additional 0.5, 3 and 6 dB correction for ARG200, ARG1000 and ARG2000, respectively. This is assumed to be an unexpected instrument sensitivity change with depth and was confirmed by post deployment testing. Figure 5 represents the sound levels after correction for instrument sensitivity.

Figure 5 shows the radar reflectivity at the pixel over the mooring, however, upon averaging the radar reflectivity over a larger area centered over the mooring (see Fig. 5 in Amitai et al. 2007), the peak reflectivity (max Z_e) is reduced (i.e., averaged down) compared to the centered pixel Z_e, while the low Z_e are increased (i.e., averaged up). This smoothing effect, a result of incomplete filling of the observed area, has the same effect as having a wider radar beam. The same effect is found with the ARGs, as seen in the figure, due to an increase of the listening area with increasing depth: the shallow ARGs detects (1) higher sound levels than the deep ARGs at the peak and (2) lower sound levels before and after the squall line passes.

Based on theoretical calculations of the effective listening area at the sea surface, the listening radius at the sea surface for the different ARGs were calculated. If the listening radius defines the area receiving 90% of the sound, then the listening radii for the ARGs at 2 kHz are 172, 620, 2800 and 5000 m, roughly three times the depth of the sensor. At 20 kHz the radii are 165, 550, 2000 and 3400 m. At the principal rainfall signal of 5 kHz the radii are 170, 610, 2735 and 4800 m. Most of the energy is arriving from a much smaller area centered over the mooring. For example, at 5 kHz 50% of the energy is arriving from a surface area with radii 58, 200, 970 and 1860 m, which is roughly the depth of the ARG.

Estimation for the actual listening area was first assessed based on the radar echo motion (Amitai et al. 2007). For ARG1000 and ARG2000 the theoretical listening radius, as mentioned above, is defined as 2.7 and 4.8 km, respectively. Analyzing the radar images reveals that the squall line propagated at a speed of about 30 km/h. For a storm that moves toward the mooring at a speed of 0.5 km/min such a listening radius is equivalent to 5.5 (9.5) min of rain detection by the ARG1000 (ARG2000) before it starts to rain directly over the mooring, or about 5 (9) min before the uppermost ARG (ARG60) detects rain. In other words, the total time period ARG1000 (ARG2000) will detect rain is expected to be about 10 (18) min longer than that of ARG60. This is in agreement with the observations presented in Fig. 5, in which ARG1000 detected rain for about 10 more minutes than ARG60.

The change in the effective listening area at the sea surface with increase depth of the ARG was evaluated also by comparing the acoustic rain rate estimates with the radar rain rate estimates (Nystuen et al. 2008). The acoustic estimates were derived by applying an empirical algorithm for rainfall rate based on the Ma and Nystuen (2005) algorithm and is given by $R_{acoustic} = 0.5 \cdot 10^{((S-44.2)/15.4)}$ where S is the sound level at 5 kHz in dB relative to 1 $\mu Pa^2\ Hz^{-1}$. The radar estimates were derived by applying a power law Z-R relations based on the 2-D video disdrometer observations, in which the exponent is derived from the dBZ-dBR scatter-plot regression slop and the coefficient is tuned to the total disdrometer rain amount (Anagnostou et al. 2008). Figure 6 presents for the 8 March event the correlation coefficients for the agreement between the acoustic rain rate estimates at 60, 200, 1000 and 2000 m depths and the radar rain rate estimates averaged over a circle of different radii centered over the mooring. ARG rainfall rates were interpolated to match the time of each radar scan and comparisons were performed over periods where both ARG and XPOL rainfall values are greater than zero. The linear correlation coefficient between the radar estimates for the pixel over the mooring and the acoustic estimates drops sharply with the ARG depth. However, the highest correlation between any ARG and the radar estimates over a larger area is about the same. The averaging radii producing the highest correlation between the radar and the ARGs increases from 450, 750, 1500 to 3000 m for the measurements at 60, 200, 1000 and 2000 m, respectively. Same trend in the correlation was found when the sound levels were compared directly with the radar reflectivities for times of $Z_e>20$ dB (Amitai et al. 2007).

These comparisons show, in general, an increase in effective listening area with increasing ARG depth as shown in Fig. 6. However, for some situations of inhomogeneous filling, averaged sound level recorded by deep ARGs might be below signal to noise value (S/N) for rainfall detection, similar to radar observations of rainfall and the listening area may even decrease for the deep ARG (Amitai et al. 2007).

Comparison of rainfall accumulation estimates from the radar and the ARGs were performed for several rain events. The rain rates were estimated using the algorithm described in Anagnostou et al. (2008). This is a disdrometer derived relationship that combines reflectivity (Z_h) and differential reflectivity (Z_{DR}) measurements when Z_h is greater than 20 dBZ and Z_{DR} greater than 0.1 dB; otherwise, the standard Z_h-rainfall rate relationship described above is used. Table 2 shows the Radar/ARG rain accumulation ratio for spatial averaging radius of the radar data at the maximum correlation. While the application of the un-normalized S-R relations resulted with acoustic rainfall estimates of almost twice as large as the radar estimates, the ratio remained stable form event-to

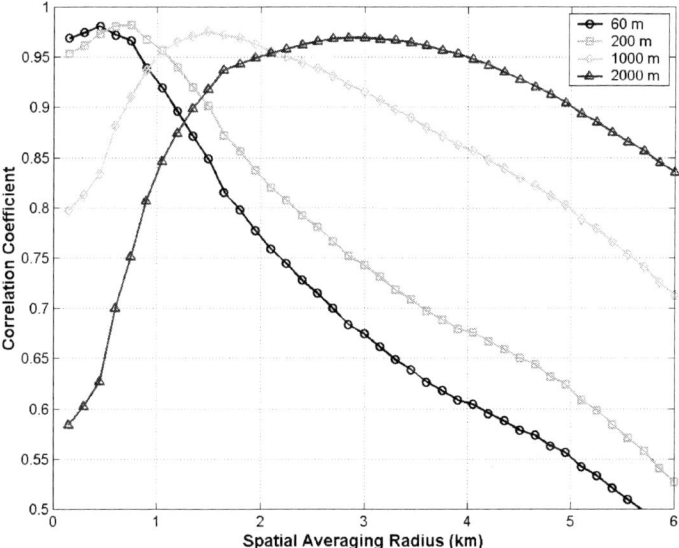

Fig. 6. Correlation coefficients between the acoustic rain rate estimates at 60, 200, 1000 and 2000 m depths and the radar rain rate estimates averaged over a circle of different radii centered over the mooring (Nystuen et al. 2008). The averaging radii producing the highest correlation between the radar and the ARGs increases from 450, 750, 1500 to 3000 m for the measurements at 60, 200, 1000 and 2000 m, respectively

event and was found to be independent of the ARG depth. This result verifies the stability of the acoustic rainfall rate algorithm, but suggests that S-R relationship coefficient need to be modified.

13.4.1 Rain type classification and wind speed estimates

The dataset from the AOML Rain Gauge Facility was also used to develop and improve the acoustical rain type classification algorithm and rain rate estimation algorithm (Black et al. 1997; Nystuen and Amitai 2003). During the convective phase the sound signature drop sizes are relatively large and this is reflected in the sound by relatively more low frequency sound. In contrast, in many types of stratiform rainfall, the drop size distribution is dominated by small drops that produce relatively more high frequency sound. Therefore, the ratio of high to low frequency sound levels often change as the rain goes from a mostly convective situation to a mostly stratiform one. Black et al. (1997) proposed an acoustic classification of convective and stratiform rainfall types based on the ration of the sound intensity in a high frequency band (10–30 kHz) to a lower frequency band (4–10 kHz). Nystuen and Amitai (2003) proposed an objective acoustic classification algorithm in which the rain is classified into several more types such that within each class the relationship between sound intensity and rain rate is nearly linear.

The squall line event in the Ionian Sea Experiment on 8 March was characterized by a fast moving isolated convective front absent of stratiform rain behind. In contrast, on 12 March, the rain event was much longer. It started as a relatively intense (convective rain) and then tapers off as it changed to more wide spread and light rain (stratiform

Table 2. Radar/ARG rain accumulation ratio (correlation and relative RMS) for the different rain events (Anagnostou et al. 2008)

Storm	60 m	200 m	1000 m	2000 m
12 February	0.60	0.60	0.57	0.60
	(0.73, 0.64)	(0.82, 0.43)	(0.70, 0.59)	(0.64, 0.39)
8 March	0.58	0.54	0.47	0.43
	(0.99, 0.50)	(0.98, 0.34)	(0.97, 0.53)	(0.97, 0.42)
9 March	0.66	0.67	0.67	0.70
	(0.60, 0.52)	(0.80, 0.48)	(0.69, 0.27)	(0.68, 0.58)
12 March	0.54	0.55	0.56	0.50
	(0.86, 0.33)	(0.9, 0.28)	(0.90, 0.24)	(0.89, 0.26)

rain). Co-detection of rainfall by the radar and the ARGs remained excellent and the comparison of rainfall rates from the radar and the ARGs remained highly correlated with correlation coefficients of order 0.9 (Table 2). The change in rain type is evident in the classification diagrams for the 12 March event (Fig. 7). The advance of time is indicated with arrows. Under 'wind only' conditions the sound by breaking waves can be used to quantify wind speed (Vagle et al. 1990). By comparing the sound intensity at a lower frequency (8 kHz) to the sound intensity at a higher frequency (20 kHz), a well defined locus of points identifies 'wind only' conditions. These wind only points are shown on the classification diagrams to contrast with the detected rain signal. The diagram shows that at the start of the event that the classification is for high wind of about 10 m/s then heavy rainfall,

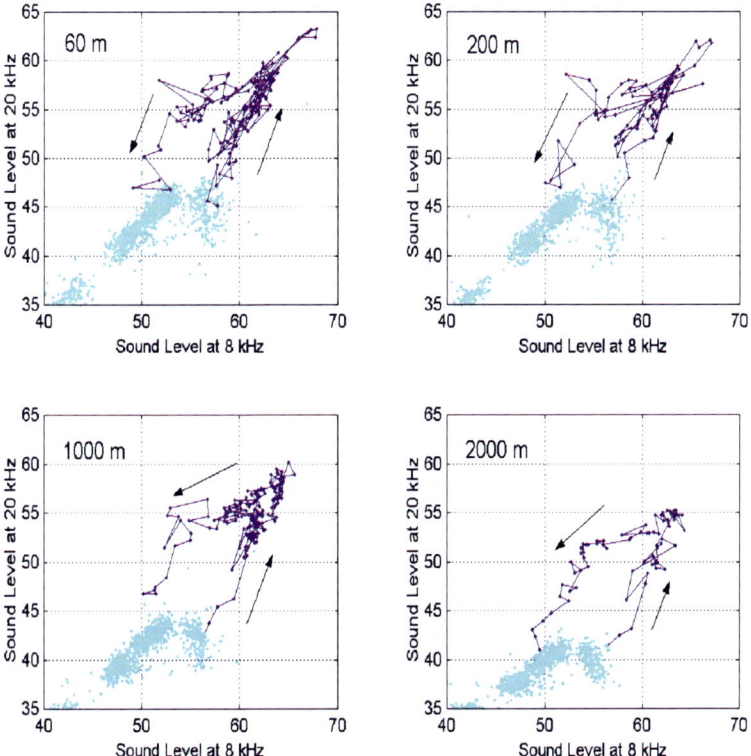

Fig. 7. The sound level at 20 kHz and 8 kHz for the 12 March rainfall event. The *light points* are reference 'wind only' points. The *dark points* are during the rainfall event on 12 March. The *arrows* indicate the advance of time. Initially the rainfall is 'heavy' (convective) and then it is tapers off into 'drizzle' rain (stratiform) (Nystuen et al. 2008) [units: decibels relative to 1 $\mu Pa^2 Hz^{-1}$]

presumably convective, follows with the highest sound levels until they drops at 8 kHz, but remain relatively high at 20 kHz. This indicates that the character of the rain has changed, containing relatively more small raindrops and that the rainfall has become more stratiform in nature. The attenuation of the sound is stronger as the sound frequency increases. This is evident by comparing the diagrams for the different ARGs. As the depth of the ARG increases the sound levels decrease more sharply at 20 kHz than at 8 kHz.

13.5 Conclusions and outlook

For several years we have known that a high correlation exists between radar reflectivity and underwater sound of rainfall based on observation in shallow water, but recently, for the first time such correlation is found in the deep sea. Comparisons between the acoustical measurements at 60, 200, 1000 and 2000 m depths and simultaneous ground-based polarimetric X-band radar observations over the acoustic mooring show acoustic detection of rain events and storm structure at all depths and are in agreement with the radar observations. The effective listening area at the ocean surface increases with the depth of the ARG and can be in the same order of the footprints of a typical ground- or space-based radar pixel -a few square kilometers. Application of independent empirical sound level-rain rate relations yielded high correlation of rain accumulation estimates with the radar estimates. The new results demonstrate the potential for evaluating or perhaps even calibrating ground and space-based radar rainfall observations using underwater sound. The reader is referred to Amitai et al. (2007), Nystuen et al. (2008) and Anagnostou et al. (2007, 2008) for detailed information and more results obtained in the 2004 Ionian Sea Rainfall Experiment.

Precipitation can be monitored from a variety of in situ and remote sensing platforms including spaceborne, ground- and ocean-based platforms. Each type of platform, instrument and measurement technique has its own advantage and disadvantage. Underwater sound measurements of rainfall have the potential to make a major contribution to the field of rainfall measurement as they allow detection, classification and quantification of rainfall in oceanic locations where other surface techniques for rainfall measurement are mostly unavailable. Further studies with more field experiments are required to better assess this potential. These studies should attempt to integrate in a unified measurement methodology observations derived from platforms

–the multi-sensor approach. Comparing and merging measurements made by several instruments and reconciling these different measurement techniques will provide understanding and confidence for all of the methods on the way to achieve improved rainfall estimates.

References

Amitai E, Nystuen JA, Anagnostou EN, Anagnostou MN (2007) Comparison of deep underwater measurements and radar observations of rainfall. IEEE Geosci Remote S 4(3):406–410, DOI: 10.1109/LGRS.2007.895681

Amitai E, Nystuen JA, Liao L, Meneghini R, Morin E (2004) Uniting space, ground and underwater measurements for improved estimates of rain rate. IEEE Geosci Remote S 1:35–38, DOI: 10.1109/LGRS.2004.824767

Anagnostou MN, Anagnostou EN, Vivekananda J (2006) Correction for rain-path specific and differential attenuation of X-band dual-polarization observations. IEEE T Geosci Remote 44:2470–2480

Anagnostou EN, Grecu M, Anagnostou MN (2005) X-band polarimetric radar rainfall measurements in keys area microphysical project. J Atmos Sci 63:187–203

Anagnostou MN, Nystuen JA, Anagnostou EN, Nikolopoulos EI, Amitai E (2007) Evaluation of underwater rainfall measurements during the Ionian Sea Rainfall Experiment (ISREX). In: Proceedings IEEE International Geoscience and Remote Sensing Symposium (IGARSS 2007). Barcelona, Spain

Anagnostou MN, Nystuen JA, Anagnostou EN, Nikolopoulos EI, Amitai E (2008) Evaluation of underwater rainfall measurements during the Ionian Sea Rainfall Experiment. IEEE Transactions on Geoscience and Remote Sensing (under review)

Black PG, Proni JR, Wilkerson JC, Samsury CE (1997) Oceanic rainfall detection and classification in tropical and subtropical mesoscale convective systems using underwater acoustic methods. Mon Weather Rev 125:2014–2042

Bom N (1968) Effect of rain on underwater noise level. J Acoust Soc Am 45:150–156

Cronin MF, McPhaden MJ (1998) The upper ocean salinity balance in the western equatorial Pacific. J Geophys Res 103:27567–27587

Evans DL, Watts DR, Halpern D, Bourassa S (1984) Oceanic winds measured from the seafloor. J Geophys Res 89:3457–3461

Franz G (1959) Splashes as sources of sound in liquids. J Acoust Soc Am 31:1080–1096

Heindsman TE, Smith RH, Arneson AD (1955) Effect of rain upon underwater noise levels. J Acoust Soc Am 27:378–379

Joss J, Waldvogel A (1969) Raindrop size distribution and sampling size errors. J Atmos Sci 26:566–569

Laville P, Abbott GD, Miller MJ (1991) Underwater sound generation by rainfall. J Acoust Soc Am 89:715–721

Lemon DD, Farmer DM, Watts DR (1984) Acoustic measurements of wind speed and precipitation over a continental shelf. J Geophys Res 89: 3462–3472

Lukas R, Lindstrom E (1991) The mixed layer of the western equatorial Pacific Ocean. J Geophys Res 96:3343–3357

Ma BB, Nystuen JA (2005) Passive acoustic detection and measurement of rainfall at sea. J Atmos Ocean Tech 22:1225–1248

Ma BB, Nystuen JA, Lien R-C (2005) Prediction of underwater sound levels from rain and wind. J Acoust Soc Am 117:3555–3565

Medwin H, Nystuen JA, Jacobus PW, Ostwald LH, Synder DE (1992) The anatomy of underwater rain noise. J Acoust Soc Am 92:1613–1623

Nystuen JA (1993) An explanation of the sound generated by light rain in the presence of wind. In: Kerman BR (ed) Natural Physical Sources of Underwater Sound. Kluwer Academic Press, Dordrecht, The Netherlands, pp 659–668

Nystuen JA (1996) Acoustical rainfall analysis: Rainfall drop size distribution using underwater sound field. J Acoust Soc Am 13:74–84

Nystuen JA (1999) Performance of automatic rain gauges under different rainfall conditions. J Atmos Ocean Tech 16:1025–1043

Nystuen JA (2001) Listening to raindrops from underwater: An acoustic disdrometer. J Atmos Ocean Tech 18:1640–1657

Nystuen J, Amitai E (2003) High temporal resolution of extreme rainfall rate variability and the acoustic classification of rainfall. J Geophys Res-Atmos 108(D8):8378–8388

Nystuen JA, Amitai E, Anagnostou EN, Anagnostou MN (2008) Spatial averaging of oceanic rainfall variability using underwater sound: Ionian Sea Rainfall Experiment 2004. J Acoust Soc Am (to appear) 2007

Nystuen JA, Farmer DM (1989) Precipitation in the Canadian Atlantic Storms Program: Measurements of the acoustic signature. Atmos Ocean 27:237–257

Nystuen JA, McGlothin CC, Cook MS (1993) The underwater sound generated by heavy precipitation. J Acoust Soc Am 93:3169–3177

Nystuen JA, McPhaden MJ, Freitag HP (2000) Surface measurements of precipitation from an ocean mooring: The acoustic log from the South China Sea. J Appl Meteorol 39:2182–2197

Nystuen JA, Medwin H (1995) Underwater sound generated by rainfall: Secondary splashes of aerosols. J Acoust Soc Am 97:1606–1613

Nystuen JA, Proni JR, Black PG, Wilkerson JC (1996) A comparison of automatic rain gauges. J Atmos Ocean Tech 13:62–73

Nystuen JA, Selsor HD (1997) Weather classification using passive acoustic drifters. J Atmos Ocean Tech 14:656–666

Pumphrey HC, Crum LA, Bjorno L (1989) Underwater sound produced by individual drop impacts and rainfall. J Acoust Soc Am 85:1518–1526

Shaw PT, Watts DR, Rossby HT (1978) On the estimation of oceanic wind speed and stress from ambient noise measurements. Deep Sea Res 25:1225 1233

Vagle S, Large WG, Farmer DM (1990) An evaluation of the WOTAN technique for inferring oceanic wind from underwater sound. J Atmos Ocean Tech 7:576–595

Wilkerson JC, Proni JR (2000) Monitoring tropical and subtropical rainfall over the ocean using underwater acoustic techniques. In: Proceedings 5th European Conference on Underwater Acoustics. ECUA 2000, Lyon, France, pp 741–746

Wilkerson JC, Proni JR (2001) Underwater acoustical measurement of wind and rainfall in the Bahamas during Hurricane Irene. In: Proceedings 17th International Congress on Acoustics (ICA). International Commission for Acoustics, Rome, Italy, CD-ROM

Part III. Prediction of precipitation

14 Probabilistic evaluation of ensemble precipitation forecasts

Bodo Ahrens[1], Simon Jaun[2]

[1]Institute for Atmosphere and Environment, Goethe-University Frankfurt a.M., Germany
[2]Institute for Atmospheric and Climate Science, ETH Zurich, Switzerland

Table of contents

14.1 Introduction 367
14.2 Rain station precipitation data 370
14.3 Forecast data by the limited-area prediction system COSMO-LEPS 371
14.4 Observational references 373
14.5 Skill scores 376
14.6 Results and discussion 379
14.7 Conclusions 384
References 386

14.1 Introduction

Weather forecast systems have to be evaluated and evaluation errors have to be quantified. Nowadays, limited-area numerical weather prediction systems provide meteorological forecasts with kilometer-scale horizontal grid spacing. High resolution precipitation forecasts are of primary interest. For example, in flood forecasting systems the precipitation details are a crucial input parameter. Here, as an illustrative example, daily area-mean precipitation forecasts in Switzerland with a total area of 41,300 km^2 and in Swiss mountainous catchments with a typical area as small as about 1,500 km^2 shall be evaluated (cf. Fig. 1).

Recently, ensemble prediction systems (EPS) became operational which predict forecast probabilities by integration of an ensemble of numerical weather prediction models from slightly different initial states and model parameters (Ehrendorfer 1997; Palmer 2000). The motivation for the EPS is that the spread in the ensemble forecasts indicates forecast uncertainty and the interpretation of the forecast probabilities provides better results than interpretation of one single deterministic forecast that is initiated by the best known but nonetheless uncertain atmospheric state. Zhu et al. (2002) showed with a simple cost-loss model that for most users the ensemble forecasts offer a higher economic value than the deterministic forecast.

Here, EPS precipitation forecasts of the limited-area EPS COSMO-LEPS (Montani et al. 2003) with grid-spacing of 10 km are evaluated. The evaluation period covers the years 2005 and 2006 and the evaluation areas are Switzerland and three Swiss catchments (cf. Fig. 1). These three catchments are one pre-alpine catchment, the Thur and two alpine catchments, the Aare (part of an elongated wet anomaly extending along the northern rim of the Alps) and the Hinterrhein (relatively dry inner-alpine area).

The most important ingredient of the evaluation of meteorological forecasts is the comparison with meteorological observations. But what is the best observational reference? Rain station data is commonly

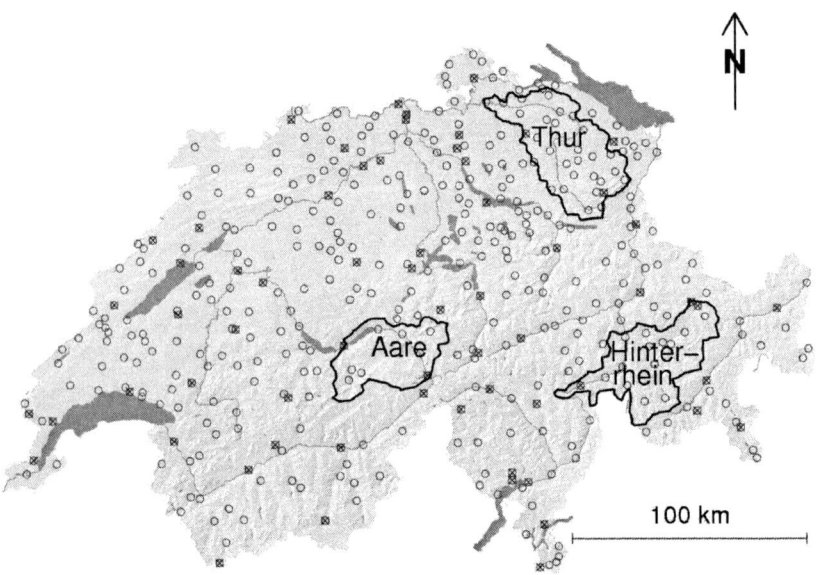

Fig. 1. Switzerland (total area: 41,300 km^2) and three catchments named Thur (1,700 km^2), Aare (1,200 km^2) and Hinterrhein (1,500 km^2). The circles show the locations of the rain station network ALL and the subset indicated by the crosses show the locations of the network SUB

preferred to remote sensing data, in particular radar data, because of the relatively large measurement uncertainties especially in mountainous area (e.g., Young et al. 1999; Ciach et al. 2000; Adler et al. 2001).

A typical distance between precipitation observation sites with daily observation frequency in the European Alps is 10 km and substantially more if near-real-time data is considered (cf. Fig.1 for the distribution of precipitation stations in Switzerland). This is a comparatively dense observation network but precipitation is a quantity with high spatial variability. Therefore, it is a valid question to ask if such a density of observations allows for evaluation of daily catchment precipitation forecasts. What is the uncertainty in observational estimates of catchment-mean precipitation and is the resulting evaluation uncertainty small enough to compare different versions of the EPS over reasonably short (e.g., three months) evaluation periods?

Observational estimates of catchment-mean precipitation can be determined by a various set of methods. The simplest method is the approximation of the catchment-mean precipitation by the arithmetic mean of the in-catchment rain-station observations. More elaborate methods regionalize the observations and average the resulting precipitation field in the catchment. Regionalization can be made by some fitting approach yielding a precipitation analysis. For example, a recent analysis of precipitation for the European Alps by Frei and Schär (1998) has a time resolution of 24 h and a spatial grid of about 25 km with regionally even lower effective resolution depending on the available surface station network. This type of analysis is useful for model validation at the 100 km scale (see, e.g., Ahrens et al. 1998; Ferretti et al. 2000; Frei et al. 2003), but probably yields substantial evaluation uncertainties at smaller scales. The fitting analysis is a smoothing regionalization. This deteriorates the application in higher-moment evaluation statistics if the network is not dense enough. The statement 'dense enough' critically depends on the pixel support of the observations (what is the area an observation is representative for?) and the analysis scheme.

Another regionalization approach is stochastic simulation of precipitation fields with conditioning on the available station data. The idea of this is that the data is respected and the spatial variability is represented more realistically than in the analysis. Additionally, an ensemble of observation based fields (i.e., observational references) can be simulated. Then the forecast can be compared with an ensemble of references which are equally valid realizations of precipitation fields given the available measurements. This allows for easy quantification of the evaluation uncertainty that is caused by the averaging uncertainty as will be shown below.

A set of useful evaluation statistics has to be chosen. There are many of them discussed in the literature and the interested reader is referred to, for example, Murphy and Winkler (1987), Wilks (2006), Wilson (2001). For illustration, we apply a small set of skill scores only: the commonly used Brier Skill Score (BSS) and the recently developed Mutual Informations Skill Scores (MISs) (Ahrens and Walser 2007). Both skill scores assess the probability forecasts of dichotomous events (e.g., the probability of more than 10 mm precipitation in the period and area of interest). The observational reference is typically assumed to be certain: the observed event probability is either zero or one; and the uncertainty in the observed catchment precipitation is often neglected. Here, the uncertainty of rain station averaging to the catchment scale will be considered explicitly.

14.2 Rain station precipitation data

This contribution investigates precipitation in Switzerland and in Swiss catchments in the years 2005 and 2006. The considered temporal resolution of the evaluation is daily. The observational references are based on precipitation data as observed by the Swiss conventional precipitation station network available through the national weather service MeteoSwiss with more than 300 stations and a mean next-neighbor distance of about 7 km. The data from this dense network is named ALL here. Also considered in the evaluation is a coarser data subset with 65 stations, which are located close to stations of the automatic measurement network ANETZ of MeteoSwiss with mean next-neighbor station distance of about 17 km. This subset resembles the data availability in case of near real-time evaluation or in less densely observed regions and is named SUB. ANETZ data itself is not applied to avoid problems with mixing of different station types in the evaluation. Figure 2 shows the time series of daily Swiss-mean precipitation as estimated by the arithmetic mean of the values observed by the network SUB.

The spatial distributions of the two station sets are illustrated in Fig. 1. Within the catchments considered, the numbers of stations are of the order of ten in case of ALL but only of two in case of SUB. Therefore, differences in the evaluation with the different data sets are to be expected.

Chapter 14 - Probabilistic evaluation of ensemble precipitation forecasts 371

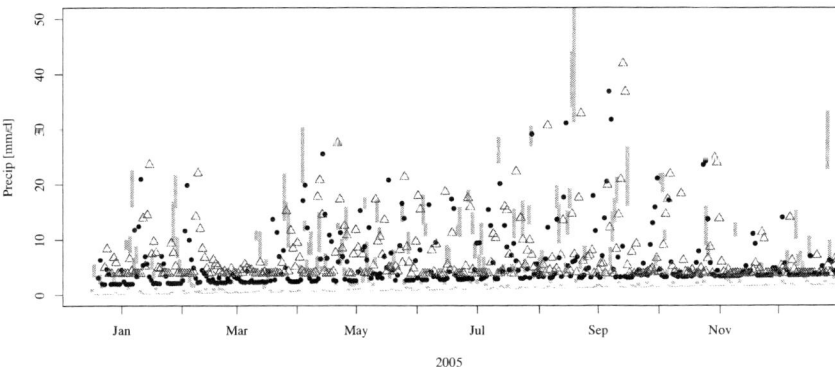

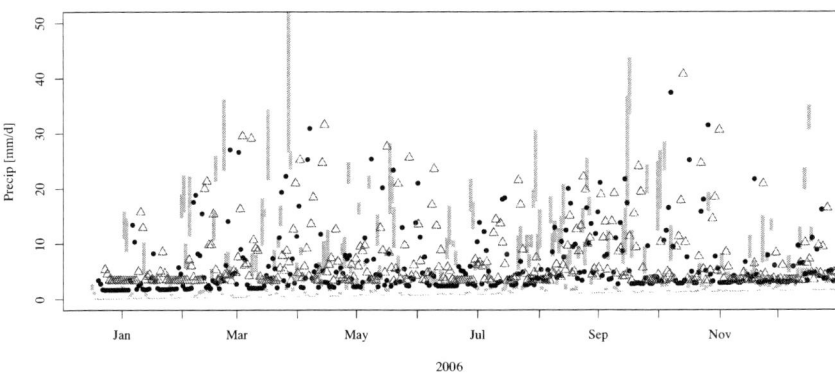

Fig. 2. Swiss-mean time series of daily precipitation as derived from the observations by the network SUB with arithmetic observations averaging (*bullets*), Kriging analysis (*triangles*) and stochastic interpolation (90% confidence intervals are illustrated by bars)

14.3 Forecast data by the limited-area prediction system COSMO-LEPS

The experimentally evaluated forecast ensemble data are supplied by the consortium for small-scale modeling limited-area ensemble prediction system COSMO-LEPS (Montani et al. 2003; http://www.cosmo-model.org). The COSMO-LEPS implementation is formally validated in

Marsigli et al. (2005). We selected the years 2005 and 2006 as our evaluation period. In this period, the ensemble size was set to ten until January 2006 and to 16 afterwards (with small changes in the physics of the prediction model) and each ensemble member's forecast with grid-spacing of 10 km was initiated each day at 12:00 UTC. Here, precipitation simulations for the forecast hours 18 to 42 h, 42 to 66 h and 66 to 90 h (the one-, two- and three-day forecasts, respectively) are assessed.

Each LEPS member is nested into a different representative forecast of a coarser-grid global EPS (the operational ensemble forecast of the European Centre for Medium Range Weather Forecasts, Reading). These representative global members are selected by grouping the global members into ten (16 from February 2006) clusters based on the analysis of wind and vorticity fields over a domain covering most of Europe (Molteni et al. 2001). From each cluster the central member (with minimum distance to all cluster members) is chosen to host a limited-area forecast. In the evaluation presented below, we consider limited-area EPS members not weighted with the cluster size. This is unlike the operational approach but the differences in skill are small (cf. Ahrens and Jaun 2007). Additionally, we want to keep the ensemble size constant in the evaluation period to ease the discussions. Therefore, we consider the ten heaviest global clusters only.

Figure 3 shows the one-day forecast of the LEPS member that is driven by the most representative member (the central member from the cluster with about 25% of the global members) for 21 August 2005. This precipitation event led to major flooding in the northern European Alps. Also given in Fig. 3 are interpolated precipitation observations (cf. next section). The forecast depicts the coarse-scale features of the precipitation pattern but also over-estimates precipitation substantially in the central region of the event.

The direct model output at grid-box scale should not be applied and some temporal and spatial smoothing of the output is recommended for being numerically representative (e.g., Grasso 2000; Ahrens 2003). Here, daily catchment means of precipitation are evaluated with averages over at least 15 model grid-boxes and thus the forecasts can be assumed numerically representative.

Chapter 14 - Probabilistic evaluation of ensemble precipitation forecasts 373

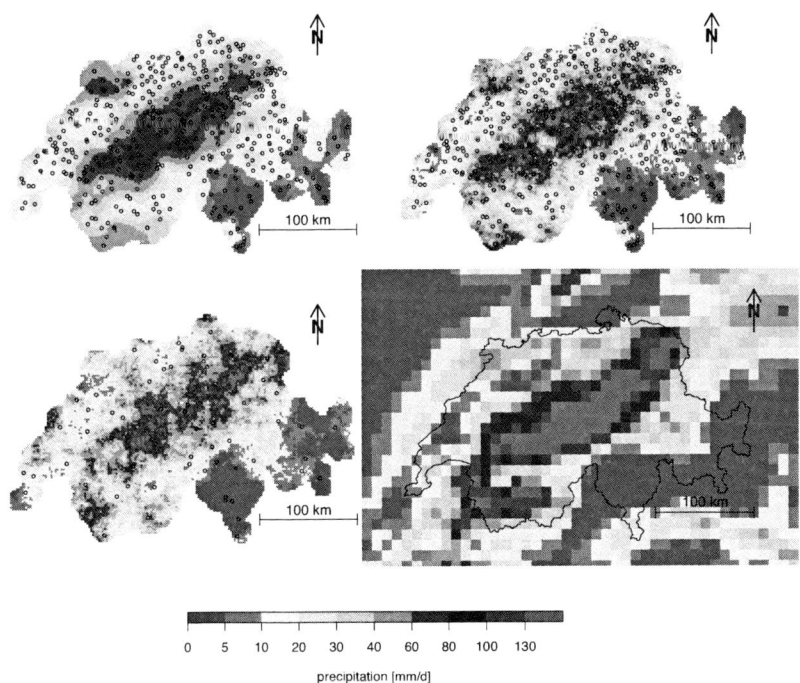

Fig. 3. Precipitation of 21 August 2005 in Switzerland as interpolated by Kriging (*upper left panel*), by stochastic simulation with ALL stations (*upper right*) or SUB stations (*lower left*) and as predicted by a 1-day forecast of the most representative COSMO-LEPS member (*lower right*). The locations of the considered stations are indicated by small circles

14.4 Observational references

The precipitation forecasts shall be evaluated at catchment scale. But how to estimate representative observational references from the limited number of rain gauge stations available? This has to be done by interpolation and averaging to the catchment scale. Table 1 summarizes the observational references applied here in the evaluation.

The simplest applied method for estimation of daily catchment-mean precipitation is arithmetic averaging of values observed within the catchment. This method results in a single precipitation value per day

Table 1. Averaging methods and abbreviations of derived references

Abbreviation	Method	Type of reference
DOR1	Arithmetic averaging of in-catchment observations	Deterministic
DOR2	Ordinary Kriging	Deterministic
MRR	Conditioned stochastic simulation	Multiple deterministic
POR	Conditioned stochastic simulation	Probabilistic
DOR3	Ensemble average of MRR	Deterministic

and a specific precipitation event ($x_t \leq x_0$ with x_t the precipitation at some day t with x_0 a chosen event threshold) is observed or is not. In this sense the method delivers a Deterministic Observational Reference (abbrev. DOR1).

Another method is ordinary Kriging with a spherical variogram model as interpolation method followed by catchment averaging (named DOR2). Kriging variants are often proposed and applied in precipitation analysis (Creutin and Obled 1982; Atkinson and Lloyd 1998; Beck and Ahrens 2004). For the necessary variogram estimation we adopted a sub-optimal but robust approach (Ahrens and Beck 2007). From the daily data of the year 2005 and 2006 we estimated from standardized observations a climatological variogram range to about 40 km with a sill of 1 $(mm/d)^2$ (by construction). For daily analyses the sill is rescaled with the data variance. For either data set, ALL and SUB, a local neighborhood of 8 stations is considered in interpolation. Figure 3 shows the Kriging analysis for the day 21 August 2005 with ALL data. The estimation of the Kriging interpolation errors is extremely difficult in case of precipitation since the stationarity and normality assumptions of Kriging are not very well fulfilled. Here, the areal precipitation estimate through ordinary Kriging is considered a deterministic observational reference (DOR2) because no uncertainty in interpolation is considered.

An alternative interpolation approach is based on stochastic simulation of an ensemble of precipitation fields with conditioning on the available station data. The idea is to simulate stochastically an ensemble of field realizations that 'honor' the observed data (their point values, their areal mean and their covariance structure) (Journel 1974; Chilès 1999; Ahrens and Beck 2007). Therefore, the spatial variability is represented more realistically in the stochastic realizations than in Kriging. For the evaluation exercise, an ensemble of 1,000 observation-based realizations is generated per day and catchment. Each ensemble member can be dealt with as an observational reference. Thus, there are 1,000 references and subsequently 1,000 evaluation results per day. The spread in the evaluation results is an uncertainty measure for the evaluation without troublesome estimation and interpretation of the Kriging variance. Using the reference ensemble this way is named multiple deterministic realization references (abbrev. by MRR). The observational ensemble can also be averaged yielding daily ensemble averages. This resembles the Kriging estimate in case of very large ensembles and of identical selection of variograms and additional interpolation parameters. The resulting reference is deterministic and named DOR3.

Additionally, the ensemble of realizations can be interpreted probabilistically. This means that the observational ensemble is used to determine event occurrence probabilities in the interval [0, 1] that an precipitation event is observed or not. This yields a Probabilistic Observational Reference (POR) and allows the comparison of probabilistic EPS forecasts against the probabilistic reference POR by comparison of probability distributions.

Stochastic interpolation is done by conditioned sequential Gaussian simulation (e.g., Johnson 1987; Chilès 1999) as it is implemented in the geostatistical software package gstat (Pebesma 2004). Sequential simulation involves the generation of a Gaussian random field, conditioned to the observed data that honors the variogram of the random field. Since daily precipitation is a non-Gaussian, non-negative process, the data has been normalized by a logarithmic transformation and the appliance of variogram estimates for the transformed data based on rescaling of the climatological variogram with an estimated climatological range of about 100 km. For each day and data set an ensemble of realizations with one thousand members is generated and applied in the following comparisons. Each ensemble member is less accurate than the Kriging analysis in a squared-error sense by construction, but respects the covariance structure given by the observations and is a possible realization given the observational information at hand.

Figure 3 shows two realizations of stochastic interpolation: one is conditioned on ALL and the other on SUB observations. As expected the stochastic interpolation is rougher than Kriging. Additionally, it can be seen that the conditioning by ALL is more restrictive than by SUB by comparison with the Kriging interpolation of the dense ALL network data. Figure 4 illustrates that in case of the SUB network there is even for daily and Swiss averages substantial scatter in the observational reference.

Optimal interpolation of precipitation fields is an active field of research. The remaining deficiencies of the Kriging analysis and stochastic simulation upscaling motivate the discussion of the advantages of PORs over DORs. Nevertheless, the applied methods are state-of-the-art for daily high resolution precipitation interpolation.

14.5 Skill scores

An often applied performance measure for the evaluation of probabilistic forecasts, which is also applied here, is the Brier Skill Score, BSS (cf. Stanski et al.1989; Wilks 2006). The BSS compares probability forecasts $Y_t = P(y_t \leq y_0)$ at dates $t = 1, 2,..., T$ of forecast events $y_t \leq y_0$ (y_0 is a chosen event threshold: e.g., 10 mm/d in case of precipitation forecasts y_t) with the observed probabilities $O_t = P(x_t \leq x_0)$ of some observational quantity x_t with related threshold x_0. Commonly, the observations are assumed perfect and thus O_t is in the set $\{0, 1\}$ - the event occurred or did not. This is the assumption made in the evaluation with the DORs. Figure 4 shows that our knowledge about observed event occurrence is uncertain: for several precipitation days the 90th percentile threshold is within the confidence interval of the reference values and thus the event occurrence probability is in between 0 and 1. Therefore, the POR is useful to be applied and the O_t's co-domain is the interval $[0, 1]$. The BSS with the probabilistic reference POR allows for a consequent probabilistic evaluation of the LEPS forecasts.

The BSS is defined by

$$BSS = 1 - \frac{BS(Y,O)}{BS(C,O)} \tag{1}$$

with the Brier score

$$BS(Y,O) = 1/T \sum_{t=1}^{T}(Y_t - O_t)^2 \tag{2}$$

The BS is essentially the mean squared error of the probabilistic forecast. The BS(C,O) of some climatological forecast C is introduced

Chapter 14 - Probabilistic evaluation of ensemble precipitation forecasts 377

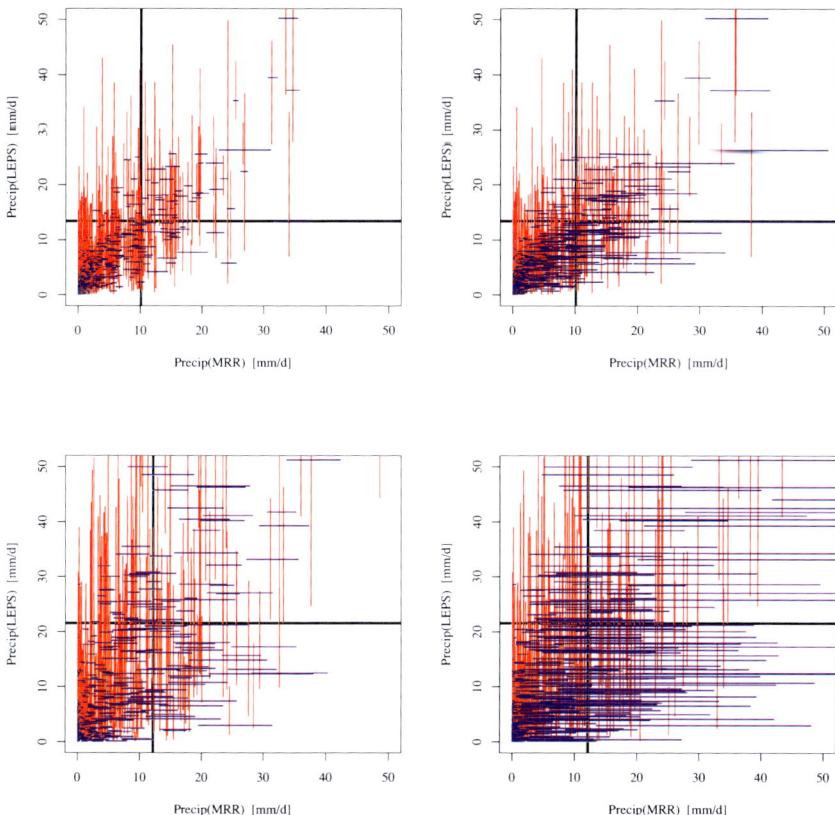

Fig. 4. The median 1-day forecasts LEPS versus the median observation-based stochastic averages MRR of the ALL (*left panels*) or SUB (*right panel*) networks. The precipitation values are daily Swiss (*upper*) or daily Aare (*lower panels*) averages. The bars indicate the 90% confidence intervals of the forecasts (vertical bars) and of the interpolations (horizontal bars). The thick black lines give the 90th percentiles of the interpolations and the forecasts

as a reference forecast in the BSS for normalization. The skill score equals one in case of perfect forecasts (a perfect forecast of an uncertain observational reference is uncertain itself) and zero if the evaluated forecast skill compares to the skill of the climatology.

The estimation of forecast probabilities from small EPS leads to biased BSS values (Müller et al. 2005). The COSMO-LEPS ensemble size is ten only. Therefore, we de-biased the BSS following Weigel et al. (2007). Another issue is the estimation of the climatological probability of some precipitation forecast threshold. It can not be estimated reliably because of the short period of available COSMO-LEPS data. We applied instead the 90th percentiles in 2005 and 2006 depending on the data set (forecast or observation-based, selected catchment) as thresholds. For example, the threshold for the MRR reference with ALL is 10.1 mm/d in Switzerland and 12.2 mm/d in the Aare catchment. The thresholds for the LEPS forecasts are 14.4 and 21.6 mm/d, respectively, for the first forecasts day and thus event precipitation forecasts are much larger than observed in the Aare catchment. This is consistent with larger total means and ensemble spread in the forecast ensembles in comparison with the reference ensembles. For the third forecast day the thresholds are slightly smaller (12.8 and 20.2 mm/d, respectively). The data set dependent selection of thresholds is equivalent to some forecast post-processing and improves the BSSs slightly.

Here, a new set of skill score is applied also (Ahrens and Walser 2007): the Mutual Information skill Score MIS_Y that quantifies the fraction of useful information in the forecasts and the MIS_O that quantifies the fraction of information in the observational reference that is explained by the forecasts. Here, the single threshold version is applied, but an extension to multiple thresholds exists. The scores are based on the information entropy available in the time series of forecasts probabilities and of probabilities given by the observational reference:

$$H(Y) = -\sum_{k=1}^{K} P(Y_k) \log P(Y_k) \tag{3}$$

$$H(O) = -\sum_{k=1}^{K} P(O_k) \log P(O_k) \tag{4}$$

with K classes of probabilities values. The number of classes K is set to 11 since in case of the LEPS forecasts with M = 10 members the maximum number of populated probability classes is K = M + 1 = 11. The Mutual Information MI between the time series of probabilities Y_t and O_t is given by

$$MI(Y,O) = H(Y) + H(O) - H(Y,O) \tag{5}$$

and after normalization the skill scores are

$$MIS_Y = \frac{MI(Y,O)}{H(Y)} \quad \text{and} \quad MIS_O = \frac{MI(Y,O)}{H(O)} \tag{6}$$

Chapter 14 - Probabilistic evaluation of ensemble precipitation forecasts

The information entropy is a measure of the variability in the time series and of the uncertainty in the forecasts or observational references. For example, if all forecast members deliver always the same forecasts then the a priori knowledge about a future forecast is larger. The forecast uncertainty is smaller than in case of spread in the forecasts. This does not imply that the forecasts are better. In contrary, the spread in the forecasts is too small in case of differences between forecasts and observations, which leads to smaller mutual information between forecasts and observation than in case of reasonable forecast spread. The concept of mutual information is illustrated in Fig. 5 (cf. Cover and Thomas 1991). Obviously, the mutual information increases with an increase of uncertainty in the observational reference (e.g., because of decreasing station density). This implies that the relative amount of useful information as quantified by MIS_Y increases. At the same time the amount of explained information in the observational reference as quantified by MIS_O decreases.

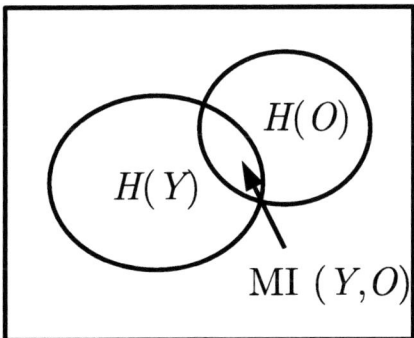

Fig. 5. Venn diagram indicating the mutual information common to the time series Y_t and O_t. The mutual information MI(Y,O) is defined by the intersection of the two sets, whereas the joint entropy H(Y,O) (not shown) is defined by the union of the two sets

14.6 Results and discussion

Table 2 summarizes the skill of the COSMO-LEPS one-day forecasts in Switzerland and for the evaluation period 2005 and 2006. The results show that with ALL stations the skill scores vary only slightly if the differently estimated deterministic references DOR1, DOR2, or DOR3 are applied in the evaluation. Also the scatter in the ensemble of evaluations against the ensemble of stochastic realizations conditioned

Table 2. Skill scores of the LEPS forecasts in Switzerland for the evaluation period 2006 and 2006. The one-day forecast evaluation is shown. The skill scores are estimated against different observational references (DOR1, DOR2, DOR3, MRR and POR) introduced in Sect. 14.4 based on either the ALL or the SUB station network. For the reference ensemble MRR the 90% confidence intervals of the evaluation results are given

	DOR1	DOR2	DOR3	MRR	POR
			ALL		
BSS	0.50	0.50	0.51	0.50–0.53	0.54
MIS_Y	0.15	0.15	0.15	0.15–0.16	0.19
MIS_O	0.48	0.47	0.48	0.47–0.52	0.44
			SUB		
BSS	0.53	0.51	0.49	0.47–0.55	0.56
MIS_Y	0.16	0.15	0.15	0.14–0.17	0.23
MIS_O	0.50	0.50	0.49	0.46–0.54	0.41

to ALL stations' observations, MRR, is reasonably small. But, even for the total area of Switzerland there is some uncertainty in the observational reference because of the errors made in rain station data interpolation. This uncertainty is even more evident in case of application of the SUB network in the evaluation. The impact of the analysis scheme used in the estimation of the DORs is smaller than the impact of the data sample size.

The skill score values of BSS and MIS_Y are largest in case of the reference POR that applies event observation probabilities derived from the ensemble of stochastic realizations. Here, the evaluation objective is evaluation of area-mean precipitation forecasts. The areal reference is uncertain and this reference uncertainty should not lead to a decrease of the forecast skill scores. The increase in the skill scores using POR is interpreted as an increase of forecast performance relative to our knowledge about the observational truth.

The values for MIS_O using the deterministic references DOR1, DOR2 and DOR3 are larger in case SUB than in case of the denser station network ALL. This is against intuition. The forecasts should not explain more of the observational reference if this reference gets more and more uncertain. This over-explanation effect is because of the fitting characteristics of the deterministic references: less observation yields smoother time series and thus smaller values of H(O). Over-explanation does not occur if evaluation against single observation

Table 3. Same as Table 2, but in the Aare catchment

	DOR1	DOR2	DOR3	MRR	POR
			ALL		
BSS	0.31	0.29	0.27	0.24–0.33	0.32
MIS_Y	0.11	0.10	0.10	0.09–0.11	0.19
MIS_O	0.32	0.32	0.32	0.28–0.34	0.31
			SUB		
BSS	0.39	0.38	0.28	0.19–0.34	0.37
MIS_Y	0.12	0.12	0.10	0.08–0.11	0.21
MIS_O	0.36	0.36	0.29	0.24–0.34	0.24

realizations (as with MMR) or against the probabilistic reference POR is performed.

Table 3 summarizes the skill of the COSMO-LEPS one-day forecasts for the Aare catchment. As to be expected the forecast skill of the COSMO-LEPS is smaller in the smaller evaluation domain Aare than in the Swiss domain. Additionally, the observational references are more uncertain yielding an increase in the evaluation uncertainty expressed in the larger confidence interval in the evaluation against MMR. This increase in the evaluation uncertainty leads to an amplified increase of the MIS_Y with POR and it can be stated that the fraction of useful forecast information is almost as large in the Aare catchment as in Switzerland given the larger reference uncertainty in the Aare catchment.

The Figs. 6, 7 and 8 show the COSMO-LEPS forecasts in Switzerland and the Aare domain for the one-, two- and three-day forecasts. Generally, the skill scores decrease with lead time of the forecast and are better in the larger evaluation domain. But, there are exceptions. Given the evaluation uncertainty in the Aare catchment there is no significant difference in the BSS (Fig. 6) between the one- and two-day forecasts for the 10% heaviest rain events which are evaluated. The fraction of useful forecast information is comparable relative to the reference uncertainty in the Swiss and Aare domain. And it is interesting to note that in terms of BSS the three-day forecast for Switzerland performs better than the one-day forecast in the Aare domain. But, not in terms of MIS_Y that obviously considers better the forecast (smaller domain and thus higher temporal variability and more information in the time series) and evaluation difficulties. The main evaluation conclusions are independent of the chosen rain station network (ALL or SUB), but it is to be emphasized that the evaluation uncertainty with ALL stations is already substantial in case of the smaller evaluation domain.

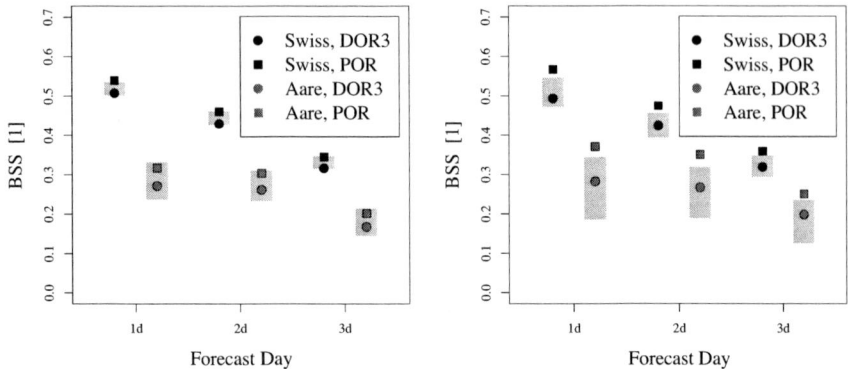

Fig. 6. BSSs of ensemble forecasts in the Swiss and Aare areas with different observational references (DOR3, MRR, POR). The references are derived from a dense (ALL, *left panel*) and a coarse (SUB, *right panel*) station network. The evaluation results against MRR are given by the 90% confidence intervals (greybars)

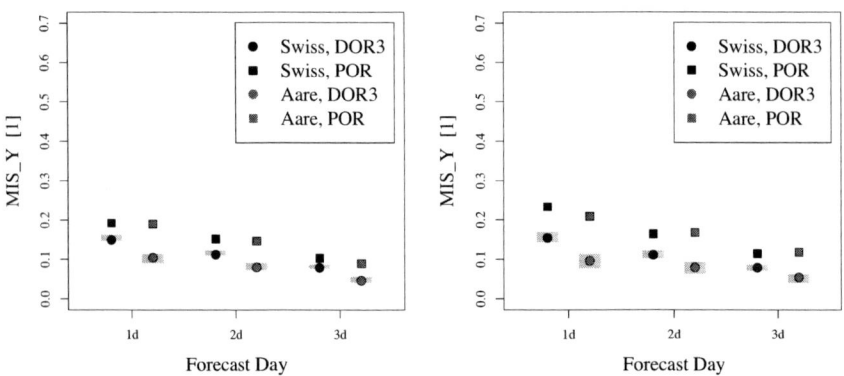

Fig. 7. Same as Fig. 6, but for the skill score MIS_Y

Chapter 14 - Probabilistic evaluation of ensemble precipitation forecasts

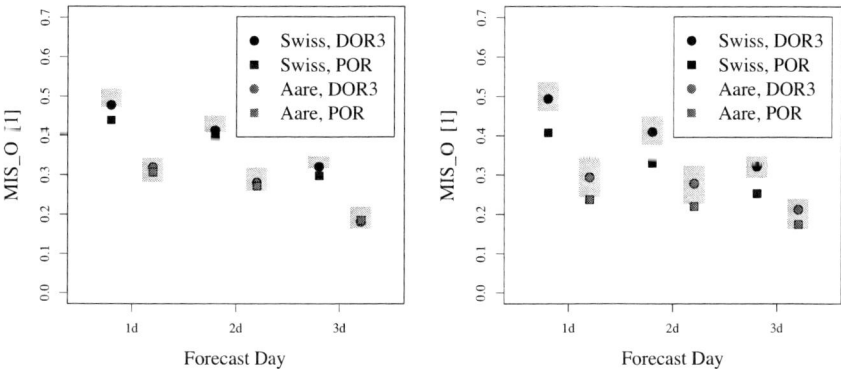

Fig. 8. Same as Fig. 6, but for the skill score MIS_O

Here, as the event threshold usually the 90th percentile value is chosen. Figure 9 shows the evaluation results for thresholds determined as the 80th and 99th percentiles. In case of one-day forecasts the MIS_Y value increases with the extremeness of the evaluated events: the skill score values in the Aare catchment with reference POR and ALL stations are 0.17, 0.19 and 0.22 for the 80, 90 and 99th percentile events, respectively. But the forecast skill decreases slower with forecast lead time for the less extreme event: the skill scores are 0.10, 0.09, 0.08, respectively, for the three-day forecasts.

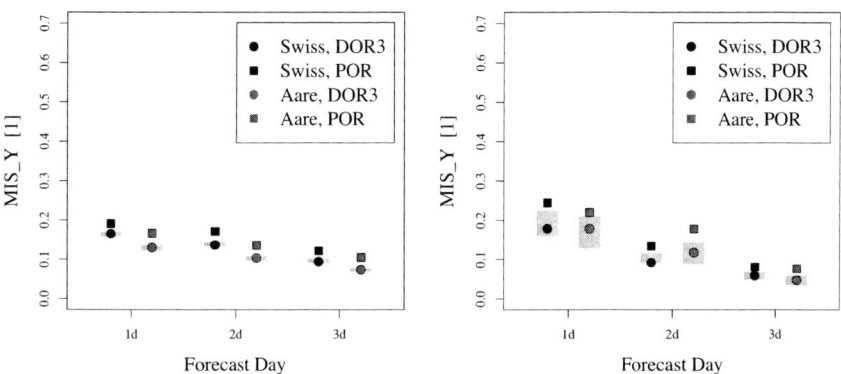

Fig. 9. Same as Fig. 7 left panel, but with the event thresholds for the 80th (*left panel*) and 99th (*right panel*) percentiles of the precipitation time series instead of the (usually chosen here) 90th percentile

Figure 9 shows also – through the confidence intervals for the evaluation against MRR – that the evaluation uncertainty increases with the event extremeness. More extreme events are less frequent (e.g., in the two-year evaluation period there are only 7 events with more than the 99th percentile precipitation) and the evaluation period less representative. This can also be illustrated by partitioning of the two-year period in three-month evaluation periods and estimation of the skill scores. In the Aare catchment the BSS is 0.32 (cf. Fig. 6 and Table 4) applying the two-year period and POR. Applying the three-month partition the mean BSS is 0.31, but with large scatter in the sample of three-monthly evaluation results with the smallest lower boundary of the MRR 90% confidence interval as small as –0.07 and the upper boundary as large as 0.75. Therefore, it is extremely important to consider confidence intervals seriously. Here, we discuss only evaluation uncertainty because of spatial interpolation uncertainty. In an operational EPS evaluation additional uncertainties through precipitation measurements and natural variability in the precipitation series have to be considered too.

Besides Switzerland and the Aare catchment, the Thur and Hinterrhein catchments have also been used as evaluation domains. Table 4 summarizes the results. The most difficult forecast target area is the relatively dry inner-alpine area Hinterrhein. The forecast performances in the Aare and Thur catchments are similar with a small advantage in the pre-alpine catchment Thur. But these conclusions have to be considered carefully by looking at the confidence intervals (not shown) since the uncertainty of the observational references vary between the evaluation domains and it has been shown above that this influences the evaluation results substantially.

14.7 Conclusions

Precipitation forecasts of EPS have to be evaluated. Here, the performance evaluation of area mean forecasts is discussed. It is common practice in the evaluation of probabilistic areal precipitation forecasts that the observational reference is assumed perfect, i.e., neglecting errors because of spatial interpolation of rain station data, for example. This Chapter shows that generating reference ensembles of stochastically interpolated fields conditioned on the available data is a simple method for the consideration of interpolation uncertainty in the evaluation. These ensembles allow the determination of ensembles of

Table 4. Evaluation results for different evaluation domains (Switzerland, Thur, Hinterrhein, Aare) and forecast lead times using the probabilistic observational reference POR based on ALL stations

	1 day	2 days	3 days
		Switzerland	
BSS	0.54	0.46	0.34
MIS_Y	0.19	0.15	0.10
MIS_O	0.44	0.40	0.30
		Thur	
BSS	0.39	0.34	0.25
MIS_Y	0.20	0.13	0.10
MIS_O	0.39	0.31	0.26
		Hinterrhein	
BSS	0.21	0.20	0.17
MIS_Y	0.11	0.08	0.06
MIS_O	0.23	0.22	0.15
		Aare	
BSS	0.32	0.31	0.20
MIS_Y	0.19	0.14	0.09
MIS_O	0.31	0.26	0.18

comparisons, if the forecast is compared against every single reference ensemble member. The spread in the comparison ensemble easily delivers an evaluation uncertainty. This is demonstrated by estimation of ensembles of the Brier skill score, BSS, and the mutual information skill scores, MIS_Y and MIS_O. Additionally, the observational ensembles can be considered as a probabilistic reference in a probabilistic evaluation. This compares forecast probabilities against observed probabilities of events (the reference POR) with appropriate skill scores in a fully probabilistic evaluation.

Evaluation experiments with different rain station densities illustrate the estimated skill scores' dependence on the quality of observational reference. In case of increasing reference uncertainty, the values of BSS and MIS_Y with probabilistic reference are increasingly higher than with a single, deterministic reference. This is fair, since in doubt it should be assumed that the forecasts perform well. Therefore, we suggest the application of an ensemble of references for (a) the estimation of evaluation uncertainty and (b) the estimation of the potentially best performance of the forecast given the reference's uncertainty.

The experimental evaluation performed in this Chapter mainly gives the expected results. The performance of the limited-area ensemble prediction system COSMO-LEPS decreases with forecast lead time in the large evaluation area of Switzerland ($\sim$41,300 km^2). In the large Swiss area the performance is better than in small mountainous catchments ($\sim$1,500 km^2). But, as shown by the application of the reference ensembles, the regional performance differences of the EPS forecasts are difficult to quantify because of regional differences in the quality of the observational reference. This reference uncertainty even prevents that the intuitively expected decrease of the LEPS performance with different forecast lead times can be proven with significance in the smaller catchments.

Besides horizontal interpolation errors, additional sources of evaluation uncertainty have to be considered. In mountainous areas the vertically inhomogeneous distribution of stations can lead to systematic errors (e.g., Sevruk 1997). Further, wind and evaporation loss of the rain gauges yields precipitation under-catch up to several ten percent (e.g., Rubel and Hantel 1999). These error sources additionally illustrate how challenging the evaluation of precipitation forecasts in small areas is–and will be–with ever increasing resolution of forecasts and forecast applications.

Acknowledgements

Data are provided by MeteoSwiss, Zurich. S.J. acknowledges support through the National Centre for Competence in Research (NCCR-Climate) and the State Secretariat for Education and Research SER (COST 731). We thank the maintainers and developers of the statistical software system R and especially E. Pebesma for the development of the software package gstat.

References

Adler RF, Kidd C, Petty G, Morissey M, Goodman HM (2001) Intercomparison of global precipitation products: the third precipitation intercomparison project (PIP-3). B Am Meteorol Soc 82:1377–1396

Ahrens B (2003) Evaluation of precipitation forecasting with the limited area model ALADIN in an alpine watershed. Meteorol Z 12(5):245–255

Ahrens B, Beck A (2008) On upscaling of rain-gauge data for evaluating numerical weather forecasts. Meteorol Atmos Phys (in print)

Ahrens B, Jaun S (2007) On evaluation of ensemble precipitation forecasts with observation-based ensembles. Adv Geosci 10:139–144

Ahrens B, Walser A (2008) Information based skill scores for probabilistic forecasts. Mon Weather Rev 136:352–363

Ahrens B, Karstens U, Rockel B, Stuhlmann R (1998) On the validation of the atmospheric model REMO with ISCCP data and precipitation measurements using simple statistics. Meteorol Atmos Phys 68:127–142

Atkinson PM, Lloyd CD (1998) Mapping precipitation in Switzerland with ordinary and indicator Kriging. J Geogr Information and Decis Anal 2: 65–76

Beck A, Ahrens B (2004) Multiresolution evaluation of precipitation forecasts over the European Alps. Meteorol Z 13:55–62

Chilès J-P (1999) Geostatistics: modeling spatial uncertainty. John Wiley & Sons, New York

Ciach G, Morrissey ML, Krajewski WF (2000) Conditional bias in radar rainfall estimation. J Appl Meteorol 39:1941–1946

Cover TM, Thomas JA (1991) Elements of information theory. John Wiley & Sons, New York

Creutin J, Obled C (1982) Objective analyses and mapping techniques for rainfall fields: an objective comparison. Water Resour Res 18:413–431

Ehrendorfer M (1997) Predicting the uncertainty of numerical weather forecasts: a review. Meteorol Z 6:147–183

Ferretti R, Paolucci T, Zheng W, Visconti G, Bonelli P (2000) Analyses of the precipitation pattern in the Alpine region using different cumulus convection parameterizations. J Appl Meteorol 39:182–200

Frei C, Schär C (1998) A precipitation climatology of the Alps from high-resolution rain-gauge observations. Int J Climatol 18:873–900

Frei C, Christensen JH, Déqué M, Jacob D, PL Vidale (2003) Daily precipitation statistics in regional climate models: evaluation and intercomparison for the European Alps. J Geophys Res 108:4124–4142

Grasso LD (2000) The differentiation between grid spacing and resolution and their application to numerical modeling. B Am Meteorol Soc 81:579–580

Johnson ME (1987) Multivariate statistical simulation. Wiley, New York

Journel AG (1974) Geostatistics for conditional simulation of ore bodies. Econ Geol 69:673–687

Marsigli C, Boccanera F, Montani A, Paccagnella T (2005) The COSMO-LEPS mesoscale ensemble system: validation of the methodology and verification. Nonlinear Proc Geoph 12:527–536 SRef-ID: 1607-7946/npg/2005-12–527

Molteni F, Buizza R, Marsigli C, Montani A, Nerozzi F, Paccagnella T (2001) A strategy for high-resolution ensemble prediction. I: Definition of representative members and global-model experiments. Q J R Meteor Soc 127:2069–2094

Montani A, Capaldo M, Cesari D, Marsigli C, Modigliani U, Nerozzi F, Paccagnella T, Patruno P, Tibaldi S (2003) Operational limited-area ensemble forecasts based on the Lokal Modell. ECMWF Newsletter 98:2–7

Müller WA, Appenzeller C, Doblas-Reyes FJ, Liniger MA (2005) A debiased ranked probability skill score to evaluate probabilistic ensemble forecasts with small ensemble sizes. J Climate 18:1513–1523

Murphy AH, Winkler RL (1987) A general framework for forecast verification. Mon Weather Rev 115:1330–1338

Palmer TN (2000) Predicting uncertainty in forecasts of weather and climate. Rep Prog Phys 63:71–116

Pebesma EJ (2004) Multivariable geostatistics in S: the gstat package. Comput Geosci 30:683–691

Rubel F, Hantel M (1999) Correction of daily rain gauge measurements in the Baltic Sea drainage basin. Nord Hydrol 30:191–208

Sevruk B (1997) Regional dependency of precipitation-altitude relationship in the Swiss Alps. Climatic Change 36:355–369

Stanski HR, Wilson LJ, Burrows WR (1989) Survey of common verification methods in meteorology. World Weather Watch Tech. Rept. No.8, WMO/TD No.358. World Meteorological Organization, Geneva, Switzerland, 114 pp

Weigel AP, Liniger MA, Appenzeller C (2007) The discrete Brier and ranked probability skill scores. Mon Weather Rev 135:118–124

Wilks DS (2006) Statistical methods in the atmospheric sciences. Academic Press, San Diego

Wilson C (2001) Review of current methods and tools for verification of numerical forecasts of precipitation. COST717-Working Group Report WDF 02 200109 1. http://www.smhi.se/cost717/

Young CB, Nelson BR, Bradley AA, Smith JA, Peters-Lidard CD, Kruger A, Baeck ML (1999) An evaluation of NEXRAD precipitation estimates in complex terrain. J Geophys Res 104:19691–19703

Zhu Y, Toth Z, Wobus R, Richardson D, Mylne K (2002) The economic value of ensemble-based weather forecasts. B Am Meteorol Soc 83:73–83

15 Improved nowcasting of precipitation based on convective analysis fields

Thomas Haiden, Martin Steinheimer

Central Institute for Meteorology and Geodynamics (ZAMG), Vienna, Austria

Table of contents

15.1 Introduction 389
15.2 The INCA system 393
15.3 Advection forecast 397
15.4 Convective analysis fields 401
15.5 Cell evolution algorithm 403
15.6 Verification and parameter sensitivity 407
15.7 Orographic effects in convective initiation 412
15.8 Conclusions 415
References 416

15.1 Introduction

In quantitative prediction of precipitation the term 'nowcasting' usually refers to forecasts of up to 2–3 h ahead. Within this time range the precipitation field is in many cases closely related to its initial state. Thus there is prognostic information contained in the precipitation pattern observed at analysis time. Nowcasting methods are designed to make use of this information based on various extrapolation techniques.

The local time evolution of a two-dimensional meteorological field $\psi(x,y)$ can formally be written

$$\frac{\partial \psi}{\partial t} = \frac{d\psi}{dt} - \vec{V} \cdot \nabla \psi , \qquad (1)$$

where $\vec{V} = (u, v)$ is not necessarily a horizontal wind vector but a more general motion vector. In the case of convective cells, for example, $\vec{V}$ can be different from the wind at any atmospheric level. Purely advection-based nowcasting methods are based on the assumption that the first term on the right hand side of Eq. (1) can be neglected, and

$$\psi(t) = \psi(t_0) \qquad (2)$$

following the motion. The local rate of change is then solely determined by the advection of existing patterns. Precipitation systems often exhibit a high Lagrangian – low Eulerian persistence, which justifies this assumption. Expressed in terms of scales, advection dominates if

$$\tau_{EVOL} \gg \frac{L}{U}, \qquad (3)$$

where τ_{EVOL} is the time scale of Lagrangian evolution of the system, L is the length scale over which significant along-flow variations of precipitation occur, and U is the translation speed. The condition of Eq. (3) is fulfilled best for quasi-stationary (in the Lagrangian sense), fast-moving systems with sharp along-flow precipitation gradients such as fronts or squall lines. Of course such systems undergo a life-cycle of formation, intensification, weakening, and dissipation. They are typically composed of smaller-scale sub-structures which have a shorter evolution time τ_{EVOL}. Nevertheless, as the inequality expressed by Eq. (3) illustrates, if precipitation gradients are sufficiently sharp (L sufficiently small) advection algorithms will give useful nowcasts even if the evolution timescale is short.

Convective cells can develop in synoptic environments with weak mid-tropospheric winds, which allows them to remain more or less stationary with respect to the ground. Often the topography plays a role both in the triggering and the anchoring of convection to a certain area (Banta 1990). In such cases the translation speed U may become arbitrarily small so that inequality expressed by Eq. (3) is no longer fulfilled. A purely advection-based algorithm then degenerates to an Eulerian persistence forecast. Unfortunately, such quasi-stationary systems can lead to disastrous hydrological consequences. Due to the near-zero translation, extreme precipitation amounts may accumulate in a given area (Caracena et al. 1979).

In general, precipitation systems move and evolve. Mathematically, this is equivalent to taking into account higher-order terms in the Lagrangian time evolution Eq. (2)

$$\psi(t) = \psi(t_0) + (t-t_0)\frac{d\psi}{dt} + \frac{(t-t_0)^2}{2}\frac{d^2\psi}{dt^2} + \ldots \qquad (4)$$

Attempts to improve the advection forecast by taking into account the linear term while neglecting higher-order terms have been of limited success due to the life-cycle behavior of precipitating convection. In order to mathematically represent an evolution $\psi(t)$ that undergoes a life-cycle of accelerated growth, decelerated growth, and weakening, a polynomial of at least third order would be necessary. Fitting such a polynomial requires observations at four consecutive times, with at least three of them being non-zero. Such a method can be used if a convective cell has already existed in the radar data for at least three time-steps. However, it is unlikely that the time and magnitude of the maximum intensity can be reliably predicted in this way unless the latest observation is already well within the stage of decelerating growth and thus itself close to the maximum. Consequently, algorithms designed to predict the formation of new convective cells, or convective initiation (CI), cannot be based solely on statistical methods but must include some physical considerations. The object-oriented cell evolution algorithm in the GANDOLF nowcasting system, for example, employs a mixed methodology of extrapolation, cell developmental stage classification, and physical relationships (Hand 1996; Pierce et al. 2000).

One of the most important predictors of CI appears to be boundary-layer mass convergence. As shown in an experimental nowcasting study by Wilson and Schreiber (1986), CI in the High Plains area around Denver is closely associated with boundary-layer convergence zones ('boundaries'). Human forecasters at the Denver Stapleton Airport were able to predict thunderstorm initiation by monitoring radar-detected boundaries and associated cumulus cloudiness. This resulted in improved forecast skill compared to automated advection forecasts. The main problem was that the boundaries indicated the location area of new cell formation but not their precise position and initiation time. According to Wilson and Mueller (1993), small-scale (a few km) structures in the temperature, humidity, and wind field appear to determine where and when cells will form. The methodology of using boundaries to identify areas of incipient CI was later incorporated into the Autonowcaster (ANC) system (Mueller et al. 2003).

A related approach is used in the GANDOLF system (Hand 1996). Convective cells are classified into different stages of development. Based on radar data and on a conceptual model of storm evolution, the current state of a cell is diagnosed and future states predicted. New ('daughter') cells close to existing ones are initiated if the near-surface mass-convergence predicted by a numerical weather prediction (NWP) model is sufficiently strong.

During the 2000 Sydney Olympics Forecast Demonstration Project (FDP) various nowcasting systems were tested and evaluated (Pierce et al. 2004). These systems use different methods of determining precipitation motion vectors such as area tracking, individual cell tracking, and NWP model winds. Two of the systems, namely GANDOLF and ANC, have convective evolution and initiation capability. The main findings with regard to convective cell prediction in the Sydney FDP can be summarized as follows (Wilson et al. 2004). (1) Predictive skill above pure translation occurs when boundaries can be identified and used to nowcast cell evolution. (2) For nowcasts beyond 60 min, boundary characteristics are more important for storm initiation than early detection of cumulus clouds. (3) The accuracy of nowcasts even for periods ≤ 60 min is generally quite low.

For the development of a cell initiation and evolution module within the INCA system these results served as a guideline. They indicate the importance of a high resolution analysis of wind, temperature, and humidity in the boundary layer. In Austria's alpine terrain, boundary layer mass-convergence is often related to the topography, which adds a deterministic component to cell initiation (Haiden 2004). With the current version of the INCA wind field analysis, the ability to correctly detect these convergence lines depends critically on the skill of the NWP model, and on the density of the surface station network. Only if a convergence line is either correctly predicted by the NWP model or captured by the surface station data it will be present in the INCA analysis and give a signal for CI or intensification.

Section 15.2 gives a brief overview of the INCA analysis and nowcasting system used operationally at ZAMG. The advection nowcast which serves as a reference for the cell evolution forecasting experiments is described in Sect. 15.3. A discussion of the convective analysis fields used for the cell evolution algorithm is given in Sect. 15.4, and the algorithm itself is described in Sect. 15.5. Verification and parameter sensitivity are presented in Sect. 15.6, followed by some thoughts about orographic effects in convection initiation (Sect. 15.7). The concluding section summarizes the main problems and proposes priorities for possible further research in the area.

15.2 The INCA system

The basic structure of the INCA analysis and nowcasting system is shown in Fig. 1. The forecast of an NWP model is trilinearly interpolated to the high resolution ($\Delta x = \Delta y = 1$ km, $\Delta z = 200$ m) INCA grid, and serves as a first guess $\theta^{NWP}(i, j, m)$.

Differences between surface observations and the first guess are determined and spatially interpolated using inverse distance weighting (IDW) both in geometric and in physical space. In the case of temperature, humidity, and wind, geometrical distance weighting is used in the horizontal, while in the vertical the distance weighting is performed in potential temperature space. The three-dimensional

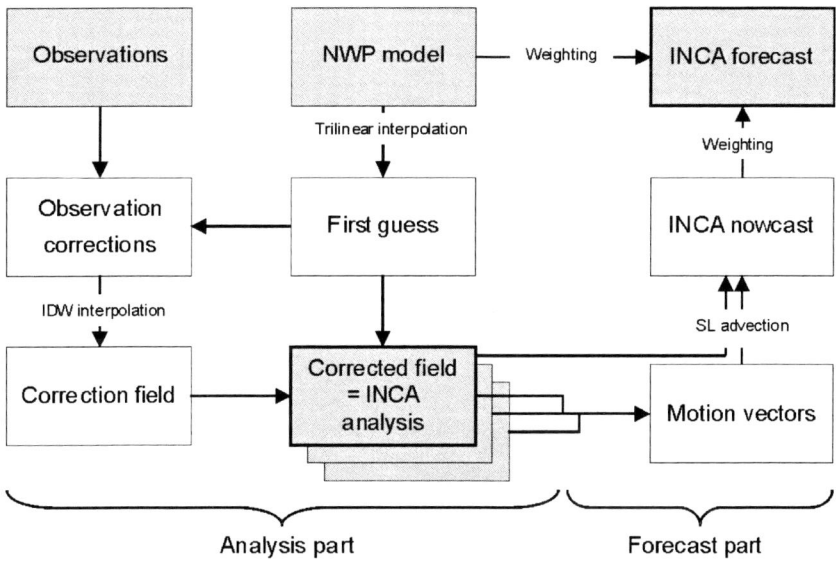

Fig. 1. Basic structure of the INCA analysis and forecasting system. The analysis part consists in correcting and downscaling NWP model fields using observations. In the case of precipitation, consecutive analyses are used to generate a nowcast which is then combined with the NWP forecast (Steinheimer and Haiden 2007)

'distance' between INCA grid point (i, j, m) and the k-th surface station is given by:

$$r_{ijmk} = \sqrt{(x_k - x_i)^2 + (y_k - y_j)^2 + c^2(\theta_k^{NWP} - \theta_{ijm}^{NWP})^2}, \quad (5)$$

where the parameter c has the dimension of an inverse temperature gradient. Based on cross-validation its optimum value was found to be close to $3 \cdot 10^4$ m/K. It means that a distance of 1 K in potential temperature space is equivalent to a horizontal distance of ≈30 km. The three-dimensional temperature difference field obtained from the interpolation

$$\Delta\theta(i,j,m) = \frac{\sum_k \frac{\theta_k^{OBS} - \theta_k^{NWP}}{r_{ijmk}^2}}{\sum_k \frac{1}{r_{ijmk}^2}} \quad (6)$$

is then added to the NWP first guess, giving

$$\theta^{INCA}(i,j,m) = \theta^{NWP}(i,j,m) + \Delta\theta(i,j,m). \quad (7)$$

In the case of wind, an additional step has to be performed, because the trilinear interpolation of the NWP wind and the interpolation Eqs. (5) – (7) do not generate a wind-field that is mass-consistent with respect to the INCA topography. A relaxation algorithm is run to adapt the wind-field to the high resolution terrain. During the relaxation, horizontal wind vectors at the station locations are fixed at their observed values. The result is a wind field that is kinematically consistent with the topography and reproduces, within grid resolution, the surface wind observations.

Meteorological fields routinely analyzed with INCA are listed in Table 1. In addition to temperature, humidity, and wind, some two-dimensional fields are generated. They do not use NWP output as a first guess but combine remote sensing (satellite, radar) and surface station data. For precipitation, a short description of the combination algorithm is given in Sect. 15.3. Convective analysis fields derived from the temperature, humidity, and wind analyses include classical convective diagnostics such as lifted condensation level (LCL), convective

Table 1. Meteorological fields operationally analyzed and forecast by the INCA system. SFC = surface station observations, SAT = MSG satellite data (cloud types), RAD = precipitation radar data. Convective analysis fields are derived from the temperature, humidity, and wind analysis and therefore use surface observations only indirectly

Field	Dimensionality	Update interval	Observations used	Nowcasting method
Temperature	3	1 h	SFC	local weighting
Humidity	3	1 h	SFC	local weighting
Wind	3	1 h	SFC	local weighting
Cloudiness	2	15 min	SFC, SAT	motion vectors
Precipitation	2	15 min	SFC, RAD	motion vectors
Global radiation	2	1 h	SFC, SAT	local weighting + motion vectors
Convective analysis fields	2	1 h	(SFC)	–

available potential energy (CAPE), equivalent potential temperature, and various instability indices.

The INCA system merges observation-based nowcasts and NWP forecasts with a weighting function that varies with lead time, similar to the NIMROD system (Golding 1998). For precipitation this function decreases linearly from 1 to 0 in the forecast range interval between +2 and +6 h. For the first 2 h, the weight is 1 (pure nowcast), beyond +6 h it is 0 (pure NWP forecast). One problem of this method in the case of precipitation is that, at times shortly before CI, if no convective cells are yet observed but CI is correctly predicted by the NWP model, it will be suppressed and delayed in the combined forecast because the advection algorithm yields zero precipitation. Experiments were carried out using a variable weighting function that starts to give weight to the NWP forecast at an earlier forecast time if it differs less from the analysis. However, it turned out that the skill of the NWP model in predicting CI is not yet high enough to gain an advantage from giving it more weight within the first 2 h of the forecast.

The high horizontal grid resolution of 1 km is an essential property of the INCA system. It enables the analysis scheme to better assimilate locally influenced station observations because at this resolution the true elevation and exposition of most surface stations coincides reasonably well with the corresponding values on the numerical grid. Although in steep terrain a 1 km resolution is still not entirely adequate, it was chosen as an operationally feasible compromise for a domain size encompassing the entire area of Austria (600 × 350 km², cf. Fig. 2). Another reason for using a 1 km grid is that it corresponds to the resolution of the radar data used in INCA.

In the vertical, a z-system is used where z is the height above the 'valley-floor surface'. In mountainous or hilly terrain, the valley-floors of adjacent valleys are generally found at comparable heights. Thus one may define a hypothetical surface that is smooth compared to the actual topography and connects major valley-floors (Haiden 1998). This surface represents a useful local reference height for the z-System. The vertical resolution of INCA is currently equidistant at $\Delta z = 200$ m. The system has 21 levels (surface included) parallel to the valley-floor surface, covering the lowest 4000 m above this surface. For the wind analysis a z-coordinate with horizontal levels and $\Delta z = 125$ m is used.

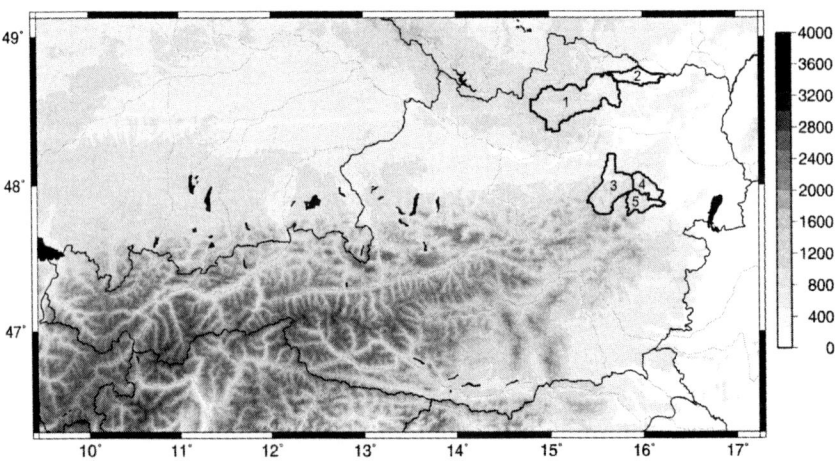

Fig. 2. INCA domain and topography. Domain size is 600 × 350 km, the horizontal resolution is 1 km. River catchments referred to in this study are Kamp (1), Pulkau (2), Traisen (3), Triesting (4), Piesting (5) (Steinheimer and Haiden 2007)

The irregular shape and reduced volume of grid elements intersecting the terrain is taken into account using the shaved element approach (Steppeler et al. 2002).

Forecast fields of the ALADIN-AUSTRIA operational model (Wang et al. 2006) are used as a first guess for the three-dimensional INCA fields temperature, humidity, and wind. The NWP output is 1-hourly, and has a horizontal resolution of 9.6 km. Two ALADIN forecast runs per day (00Z, 12Z) are used. The post-processed fields from these runs are available at about 04Z and 16Z, respectively.

The most important observational dataset for the INCA system is provided by surface stations. ZAMG operates a network of ~140 automated stations (TAWES) across the country. In the vertical, they span most of the topographic range in Austria, with highest stations Brunnenkogel (3440 m), and Sonnblick (3105 m). Although the distribution of stations is biased towards valley locations, there is a sufficient number of mountain stations to construct three-dimensional correction fields to the NWP model output on the basis of these observations. Radar data used in INCA is a two-dimensional composite of four radar stations. Due to the mountainous character of the country, radar data is of limited use in many areas in western Austria, especially during wintertime when precipitation may originate from shallow cloud systems. The Meteosat Second Generation (MSG) satellite provides data from which the derived quantity 'Cloud Type' is computed, which differentiates between cloud levels (low, medium, high) and varoius degrees of opaqueness.

The main conceptual difference between INCA and another Austrian analysis system VERA (Vienna Enhanced Resolution Analysis) is that INCA uses NWP model and remote sensing information for interpolation between observations, whereas VERA uses climatological information through a so-called fingerprint method (Steinacker et al. 2005). It would probably be best to combine these methods in order to make optimal use of both the climatological and the NWP data. A comparative evaluation of analysis skill of the two systems is planned.

15.3 Advection forecast

The operational INCA precipitation nowcast is an observation-based extrapolation that uses motion vectors determined from consecutive 15-min precipitation analyses. The analyses are generated by combining radar and surface station data. The station data basically determines the magnitude, and the radar data the spatial patterns of the precipitation field. Due to the problems associated with radar measurements of precipitation in mountainous terrain (e.g., Borga et al. 2000), the combination with surface observations in an alpine country like Austria is not straightforward. Before the radar field is used in the analysis, it is scaled 'climatologically' based on the ratio of monthly precipitation amounts observed at the stations and by the radar. It is then re-scaled on the basis of a comparison at analysis time of station observations and radar values at the stations. The re-scaling is done to ensure that, within grid resolution, surface observations are reproduced in the analysis. It is somewhat at odds with the conclusion of Germann and Joss (2002) that a short-term radar-gauge comparison is not suitable for correction of the radar field. However, in the INCA system the radar-gauge comparison allows for a shift of a few kilometers between the radar pixel and the station in order to allow for possible wind-drift, radar navigation, and finite fall speed effects. Furthermore, in alpine terrain the horizontal radius of influence of the correction is rather limited (on the order of 10 km).

The scaled radar field and the station interpolation are then combined. In this combination the relative weight of the fields in a given area depends on the visibility of that area by radar, as determined by the areal distribution of monthly precipitation ratios (Haiden et al. 2007). In areas where this ratio is high (radar visibility low), the analysis gives more weight to pure station interpolation. This method, which uses only two-dimensional radar data, gives a relatively smooth transition between radar precipitation patterns and those coming from station interpolation. However, in many alpine areas the effects of inhomogeneous radar coverage are still visible in the analyses. Therefore, a profile correction and downward extrapolation of the radar signal based on three-dimensional data, similar to what Germann and Joss (2002) have done for the Swiss Alps, is being developed.

The computation of motion vectors is based on a cross-correlation of consecutive precipitation analyses for different spatial shifts and determination of the shift which gives the highest correlation. The size of the correlation square is 100×100 km². The rather large size was chosen to capture the actual meso-scale movement of precipitation systems. In cases of orographic rainfall, small-scale (5–10 km) shower cells embedded in the flow may be advected towards a mountain barrier while at the same time the upslope precipitation area as a whole remains more or less stationary. The small-scale cells moving through the upslope area intensify, weaken and dissipate on the leeside. If the size of the correlation square is too small, the upslope precipitation in the advection nowcast will erroneously move downwind. A drawback of the large correlation square is that convective cells cannot be traced individually, and an average movement of the whole ensemble of cells is obtained instead. In practice the problem is alleviated somewhat by the fact that small-scale, non-severe, showery convection tends to move rather uniformly with the flow at some mid-tropospheric level. The horizontal scale of severe convection on the other hand is sufficiently large to locally dominate the motion vector computation. Largest motion errors will, therefore, occur for smaller cells close to severe ones.

An alternative method is to use an object-oriented approach where cells are identified by their centroid positions and advected individually. During the Sydney FDP it was found that extrapolation methods allowing for such differential motion performed slightly better, mostly because of high-impact storms which had motions different than surrounding storms (Wilson et al. 2004). However, the cell-tracking method cannot be applied in situations of widespread stratiform precipitation. In order to have in INCA the advantages of both approaches, we will continue to use the cross-correlation technique as standard method, but override its motion vectors by those derived from cell-tracking when appropriate.

Sometimes spurious correlations implying unrealistically large translation speeds are obtained due to dissipation of a radar echo at one location and initiation of another one in the vicinity. Such spurious motion vectors are meteorologically filtered by comparison with ALADIN wind fields at 500 and 700 hPa. The filtering is performed using the condition

$$\left|\vec{V}_{CORR}\right| + \left|\vec{V}_{CORR} - \vec{V}_{NWP}\right| \leq \left|\vec{V}_{NWP}\right| + 2\Delta, \tag{8}$$

where $\vec{V}_{CORR}$ is the motion vector derived by the correlation analysis, $\vec{V}_{NWP}$ is the NWP model 500 hPa or 700 hPa wind (whichever is closer to $\vec{V}_{CORR}$), and Δ is a prescribed wind speed scale which determines the amount of deviation permitted between $\vec{V}_{CORR}$ and $\vec{V}_{NWP}$. Operationally, the value $\Delta = 5$ m s^{-1} is used. The inequality expressed by Eq. (8) defines an elliptical area with its large semi-axis aligned with the vector $\vec{V}_{NWP}$.

Figure 3 shows a verification of areal precipitation of the operational INCA advection nowcast in comparison to a forecast obtained from a weighted combination of two NWP models for the Kamp catchment in the northeast parts of Austria.

Both forecasts were verified against INCA analyses, which means that the analysis error is not included in the verification. Shown is the mean absolute error (MAE) of both forecasts as a function of forecast time. The INCA nowcast has a significantly smaller MAE than the NWP model over the first 3 h. While this was a case of predominantly stratiform precipitation, similar results regarding the relative skill of INCA versus NWP forecasts are obtained in convective cases. In convective situations, both the NWP models and the INCA advection

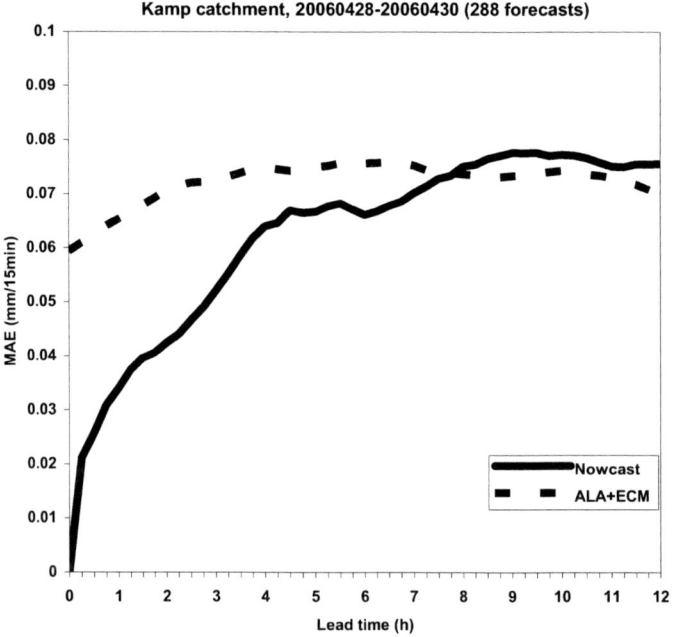

Fig. 3. Verification of areal precipitation of the operational INCA advection nowcast in comparison to a forecast obtained from a weighted combination of the two NWP models ECMWF and ALADIN for the Kamp catchment in the northeast parts of Austria (area #1 in Fig. 2). The verification period is 3 days, where every 15 min a new nowcast was made (288 nowcasts). The NWP forecast is updated four times a day

forecast have larger errors, and the advantage of nowcasting can again be seen for lead times of up to 2–3 h.

15.4 Convective analysis fields

From the three-dimensional INCA analyses of temperature, humidity, and wind, various convective analysis fields are derived. They are mostly standard diagnostic quantities commonly applied to observed radiosoundings or to pseudo-soundings generated from NWP model output. They are computed hourly, for every grid point in the horizontal, and the resulting field distributions provide additional convective guidance for the human forecaster. The convective diagnostics cover different aspects of moist convection. They include classical instability measures such as convective available potential energy (CAPE), Showalter index (SWI), and lifted index (LI). All three measure buoyancy and thus contain information about the intensity of deep convection. They do not, however, predict if, where, and when deep convection will actually be triggered. Therefore, there are also included the convective inhibition (CIN), the trigger temperature deficit (actual temperature minus trigger temperature), the lifted condensation level (LCL), the level of free convection (LFC), and the boundary-layer horizontal mass and moisture flux convergences CON and MFC. The equivalent potential temperature THETA_E is computed to indicate those areas which have the highest-energy near-surface air.

Lifted condensation level (LCL)
Level of free convection (LFC)
Convective available potential energy (CAPE)
Convective Inhibition (CIN)
Showalter index (SWI)
Lifted index (LI)
Trigger temperature deficit (DTRIG)
Equivalent potential temperature (THETA_E)
Boundary-layer mass convergence (CON)
Boundary-layer moisture flux convergence (MFC)

During the period of algorithm development, various combinations of the above fields were tested. The ones which proved most useful as predictors of cell evolution and were eventually used in the algorithm are indicated by bold letters.

The use of moisture convergence in deep convection forecasting can be traced back to early diagnostic studies of both tropical and mid-latitude systems, which resulted in the well-known Kuo (1965) parameterization of convective rainfall for numerical weather prediction models. The basic assumption underlying this parameterization is that the convective rainfall rate is proportional to the vertically integrated rate of moisture convergence due to the large-scale flow. Temporary storage of moisture, either in the form of increasing specific humidity, or through the build-up of cloud mass, is considered a small term in the overall moisture budget. As a result, the parameterization is successful in the case of quasi-equilibrium convection, where convective activity is directly related to synoptic-scale forcing. It is less suitable in cases of 'stored-energy' convection, when CAPE accumulated over several days may be released in a short interval even in the absence of synoptic-scale forcing. Nevertheless, the notion of moisture convergence, or more specifically, boundary-layer moisture convergence (here referred to as MFC), became increasingly important in convection forecasting also of the stored-energy type. From diagnostic studies (e.g., Hudson 1971) it became apparent that MFC could be related to convective rainfall intensity a few hours later and thus be used as a predictive tool (Waldstreicher 1989). It is now widely used in forecasting practice to identify those areas within a generally moist unstable air mass where initiation and development of deep convection is most likely to occur. Thus over the years the focus has shifted from the analysis of deep-tropospheric synoptic-scale MFC to near-surface meso-scale MFC. Its somewhat surprising usefulness in the prediction of convective initiation has been explained by Banacos and Schultz (2005). The MFC at any level can be written

$$\mathrm{MFC} = -\nabla \cdot (q\mathbf{V}_h) = -\mathbf{V}_h \cdot \nabla q - q\nabla \cdot \mathbf{V}_h, \tag{9}$$

which shows that it consists of an advection term and a convergence term. While the advection term dominates the MFC at the synoptic-scale, the convergence term becomes dominant at smaller scales relevant for the initiation of convection (Banacos and Schultz 2005). As a result, the spatial pattern of MFC at this scale becomes similar to that of mass convergence $-\nabla \cdot \mathbf{V}_h$. The latter quantity, integrated vertically across the boundary layer, is closely related to the vertical velocity near cloud base, which is a key factor in the triggering of new convective cells.

15.5 Cell evolution algorithm

Prior to the actual design of the cell evolution algorithm, the initiation, development, and weakening of convective cells in relation to the convective analysis fields listed above was studied subjectively on a day-to-day basis during the summer season of 2005. The objective of the study was to identify the most useful fields for the prediction of cell evolution. The selection of cases was done according to the following criteria. Within the period May–August 2005 all grid points in an INCA subdomain covering the northeastern province of Lower Austria were searched for daily precipitation amounts >30 mm/day and for 15-min amounts >15 mm/15 min. In this way 66 days with daily precipitation >30 mm, and 39 days (350 individual times) with 15-min amounts >15 mm were identified. The maximum analyzed daily total within the period was 237 mm (30 July 2005), the maximum 15-min amount was 32 mm.

By studying these cases it was found that *intensity changes* of already existing cells were hardest to predict. This agrees with the findings of Wilson and Mueller (1993) during their experimental thunderstorm nowcast study. Apparently such changes are influenced by meteorological characteristics smaller in scale than the ones resolved in the INCA system. Although the horizontal resolution of INCA is 1 km, the resolution of the ALADIN model's first guess is close to 10 km, and the mean distance between surface stations is about 20 km. Moreover, the evolution of the more intense thunderstorms is strongly governed by internal dynamics and nonlinear interaction between cells, the details of which cannot be predicted purely on the basis of convective analysis fields. Another potential problem is that the convective analysis fields are updated only once every hour, whereas significant convective evolution often occurs on shorter time scales.

A more tangible relationship was found between areas of cell initiation and some of the analysis fields, in particular when positive MFC coincided with a sufficient amount of CAPE. It was not possible, however, to predict exactly where and when on the <10 km, 15 min scale the formation of new cells would occur. It also became clear that visible satellite imagery would have to be used in addition to the convective analysis fields to avoid overprediction of cell initiation. The weakening and dissipation of cells showed some relationship with convective analysis fields as well. Here a combination of negative MFC, small CAPE, and significant trigger temperature deficit indicated general dissipation areas. How fast the weakening would proceed could not be predicted, probably because the negative effects of stability and divergence on a convective cell strongly depend on the properties of the cell itself and not just on environmental atmospheric conditions.

Summarizing the findings from the observation period it can be stated that MFC, CAPE and the VIS satellite image provided the most valuable information for convective evolution. In addition, CIN and DT_TRIG were found useful, in particular if they assumed large values. For the parameters LI, SWI, and THETA_E no additional prognostic value was found mostly because they are rather smooth fields. The LCL did not have any discernible relationship with convective evolution itself but was used to scale the MFC.

Based on the above findings, an algorithm was designed that diagnoses and predicts the evolution of convective cells in terms of the three categories of initiation, intensification, and weakening. In a first step, each grid point is classified as either 'convective' (using the condition that CAPE > 50 J kg^{-1} within a certain distance from the grid point) or 'non-convective'. This is similar to the classification convective versus non-convective in the GANDOLF system, but there a threshold of 100 J kg^{-1} is used (Pierce et al. 2000). For each convective grid point it is tested whether conditions for cell initiation, cell intensification, or cell weakening are fulfilled (Table 2).

In addition to the above fields, Table 2 contains criteria based on MSG satellite data, namely visible brightness and cloud type information. The visible brightness VIS is used to identify areas where non-precipitating cumulus convection is already present. It is scaled to its domain maximum value at a given time to allow use of a constant threshold value throughout the daytime diurnal cycle. It should be mentioned that the current algorithm is by design unable to predict initiation after sunset. Regarding MSG Cloud Type (CT) it was planned to allow initiation and intensification only in areas of cloud types classified as convective. However, the distinction between convective and stratiform cloudiness has not yet been implemented in the CT product. Therefore, the information is only used to exclude those areas from initiation and intensification that have one of the following types: cloud free land snow (CT=3), cloud free sea snow/ice (CT=4), high semi-transparent thin clouds (CT=15), high semi-transparent thick clouds (CT=16). The additional condition that the precipitation rate RR must be greater than a certain threshold value, had to be applied in order to avoid intensification of very small precipitation rates.

At each time-step the algorithm moves the precipitation field according to the motion vectors, tests whether the criteria for intensity changes are fulfilled, and performs them. The modified precipitation field is then moved further, and so forth. All three types of intensity changes (initiation, intensification, and weakening) are modeled as a Gaussian variation in time

$$I(t) = I_{MAX} \exp\left\{-\left[\underbrace{(t - t_{MAX})/\tau}_{t_{REL}}\right]^2\right\}, \qquad (10)$$

Table 2. Convective nowcast algorithm decision criteria and threshold values for cell initiation, intensification, and weakening

Criterion	Threshold value
Cell initiation	
CAPE > CAPEini	CAPEini = 100 J kg^{-1}
MFC > MFCini	MFCini = 2×10^{-6} s^{-1}
VIS > VISini	VISini = 0.5
CT ≠ CTexcl	CTexcl = [3, 4, 15, 16]
CIN < CINini	CINini = 200 J kg^{-1}
DTRIG > DTRIGini	DTRIGini = –2 °C
Cell intensification	
CAPE > CAPEint	CAPEint = 50 J kg^{-1}
MFC > MFCint	MFCint = 2×10^{-6} s^{-1}
CT ≠ CTexcl	CTexcl = [3, 4, 15, 16]
CIN < CINint	CINint = 200 J kg^{-1}
DTRIG > DTRIGint	DTRIGint = –2 °C
RR > RRint	RRint = 0.2 mm h^{-1}
Cell weakening	
RR > RRweak	RRweak = 0.0 mm h^{-1}
MFC < MFCweak	MFCweak = 0.0 s^{-1}
CAPE < CAPEweak	CAPEweak = min($\overline{CAPE}^{>0}$, 100 J kg^{-1})

where the three parameters I_{MAX} (maximum rainfall rate), t_{MAX} (time of maximum rainfall rate) or rather t_{REL} (time of maximum rainfall rate relative to time t), and τ (cell evolution time-scale) are determined as follows.

In the case of initiation and intensification the cell evolution timescale is set to the constant value $\tau_G = 30$ min. The time of maximum intensity of a newly initiated cell relative to time t is set to $t_{REL} = 2\tau_G$. Maximum cell intensity (mm/15 min) is parameterized as a function of specific humidity (at valley-floor level) and CAPE in the form

$$I_{MAX} = c_1 \rho q \sqrt{CAPE}, \qquad (11)$$

where $\rho = 1$ kg m^{-3}, and the non-dimensional coefficient $c_1 = 10$. The resulting precipitation intensity at time $t + \Delta t$ is given by

$$I(t + \Delta t) = I_{MAX} \exp\{-[(\Delta t + t_{REL})/\tau_G]^2\}. \qquad (12)$$

For cell intensification the time relative to the time of maximum intensity is derived from the ratio of maximum intensity to intensity at time t by inverting Eq. (12) and setting $\Delta t = 0$, which gives

$$t_{REL} = -\tau_G \sqrt{\ln[I_{MAX}/I(t)]}, \qquad (13)$$

where I_{MAX} is parameterized using Eq. (11). Precipitation intensity at time $t + \Delta t$ is again computed using Eq. (12). In the case of cell weakening it is assumed that $t_{REL} = 0$, $I_{MAX} = I(t)$. The timescale of weakening (in units of seconds) is parameterized based on moisture divergence

$$\tau_D = -\frac{c_2}{MFC}, \qquad (14)$$

where $c_2 = 9 \times 10^{-4}$. In addition, τ_D is limited by the constraint $30 \leq \tau_D \leq 60$ min. Precipitation intensity at time $t + \Delta t$ is again computed using Eq. (12).

15.6 Verification and parameter sensitivity

Verification of the convective nowcast algorithm has been performed for five small river catchments in different parts of the province of Lower Austria with terrain ranging from hilly (Kamp, Pulkau) to moderately alpine (Traisen, Triesting, Piesting). Catchment sizes range from 250 km^2 (Piesting) to 1500 km^2 (Kamp). It was decided to verify areal rather than point forecasts because of their greater relevance to hydrological applications and because we did not want to penalize small location errors of the order of a few km. The forecast range considered is +15 to +120 min, verification measures are root mean square error (RMSE) and bias or mean error (ME) of accumulated precipitation, computed against INCA analyses. Results of the convective nowcast algorithm are compared to those obtained by pure advection. The verification period consists of all forecasts made at full hours between 11:00 and 18:00 UTC for 10 objectively selected days of the year 2005 (80 forecasts). The selected days were the ones with the highest convective rainfall amounts in Lower Austria in 2005.

When interpreting the verification results of the convective nowcast it must be kept in mind that a systematic improvement above pure advection is generally hard to achieve. By predicting intensity changes, in particular in the case of initiation and intensification, the convective nowcast takes a greater 'risk' than the advective nowcast. Our attempt was to design an algorithm that gives significant improvements in individual cases without worsening the overall skill. On the other hand, the verification periods considered here cannot be considered entirely independent as they contain cases that have been used in the design of the algorithm. Thus we expected to see improvements at least in some of the areas.

Figure 4 shows the RMSE and ME for the reference and the convective nowcast. In the non-alpine areas the results are essentially neutral. Analysis of individual cases in these areas shows a number of days where there is in fact an improvement due to the convective nowcast but this is compensated by worsening on other days. Results for the alpine areas show a significant improvement in terms of RMSE for Piesting and Traisen, and rather neutral behavior for Triesting. In the former two areas there are also individual cases where the nowcast algorithm has a larger error than the reference, but overall the improvements dominate.

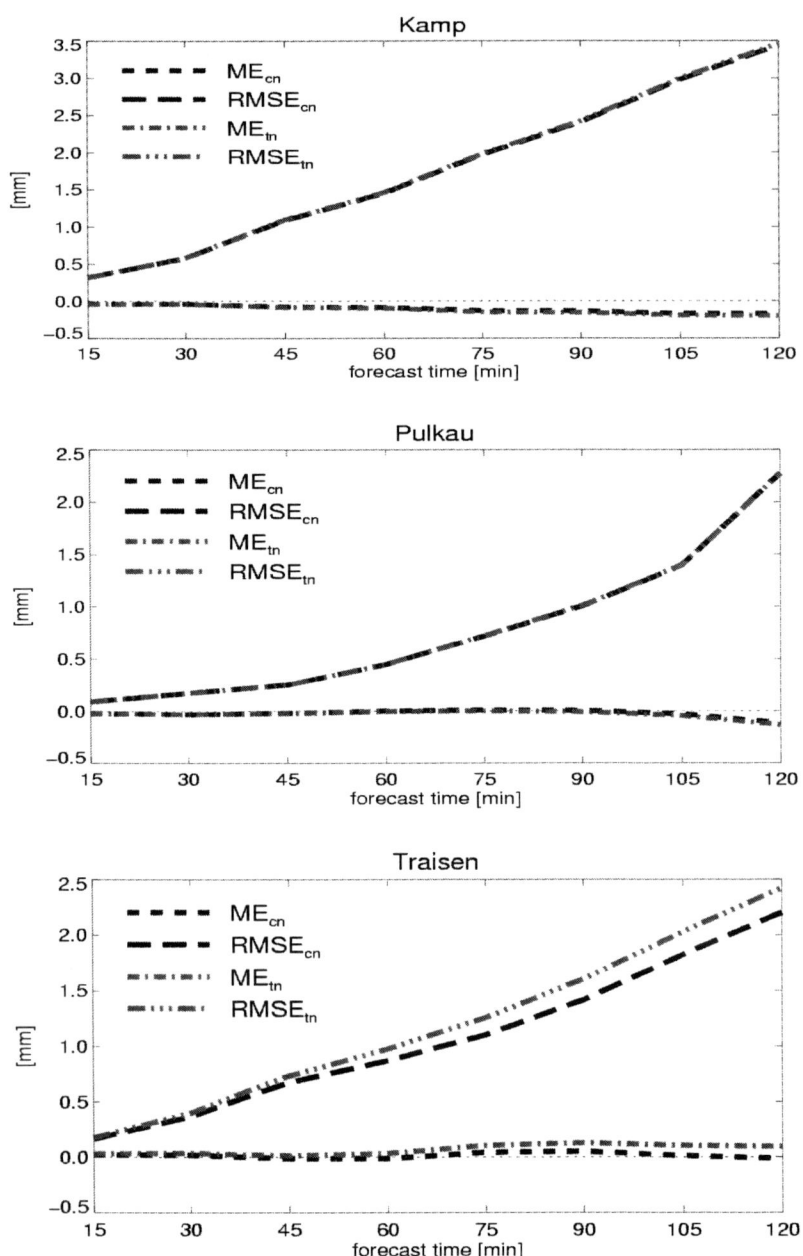

Fig. 4. Verification results for the non-alpine catchments Kamp and Pulkau, and the alpine catchments Traisen, Triesting, Piesting. Shown are root mean square error (RMSE) and bias (ME) of areal precipitation nowcasts as a function of forecast time for the convective nowcast (index cn) and the pure translational nowcast (index tn) (Steinheimer and Haiden 2007)

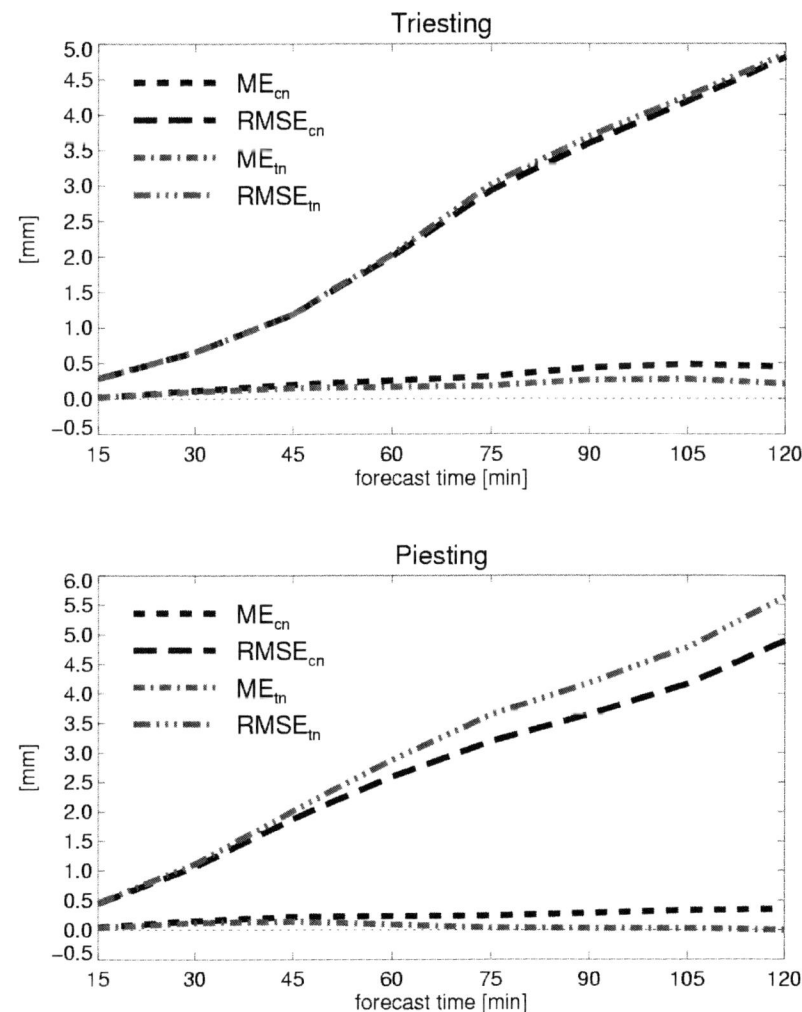

Fig. 4. (Continued)

One reason why the improvement is more pronounced in the alpine areas may be that, due to orographic effects, the moisture convergence is more predictable and thus better represented in the ALADIN wind field that is used as a first guess in the INCA wind analysis. The ME tends to increase slightly in areas where the RMSE is reduced. However, this is not always the case, as shown by the results for the Traisen catchment. The behavior of the ME in relation to the RMSE shows if the improvement in RMSE is primarily due to initiation and intensification, or weakening of cells. In the Traisen catchment, the correct prediction of cell weakening dominates, whereas in the Piesting and Triesting catchments the improvement is more strongly due to initiation and intensification. These two areas also contain well-known climatological 'hot spots' (Banta 1990) for orographically triggered convective developments. With regard to differences in error magnitude between catchments it is important to note that they do not necessarily indicate a difference in forecast skill but primarily reflect different precipitation amounts.

In order to confirm whether the verification results from 2005 can be generalized, an additional continuous verification was performed which covered all nowcasts made between 4 April 2006 and 16 May 2006 during a quasi-operational test run of the system. In this case no individual days were selected, thus the sample contains a mixture of non-convective and convective cases. Again, no significant change in RMSE was found for the non-alpine areas, whereas improvements similar in extent to the 2005 verification were found for two alpine areas (Triesting, Piesting) with the third (Traisen) giving neutral results.

Based on a number of individual cases, Wilson et al. (2004) provide a qualitative verification of the convective nowcast algorithm in the ANC system for up to +60 min. They demonstrate that it outperforms the purely translational forecast in most cases but also point out that further algorithm development is needed. In contrast to INCA, the ANC obtains wind fields derived from Doppler radar as input. It is likely that such input would increase the skill of the INCA convective nowcast especially in lowland areas, where the current wind analysis based on NWP results and station data appears to be unable to fully resolve the convergence characteristics relevant to convective intensity changes and CI. A quantitative evaluation and comparison with advection was also made for the GANDOLF system by Pierce et al. (2000). In contrast to INCA, GANDOLF uses an object-oriented algorithm to extrapolate cell movement and intensity. For all non-frontal convective events of 1995–1996, results similar to the ones presented here were obtained in the sense that both improvements and worsening compared to advection (NIMROD) were found for different catchments.

Some of the parameters of the cell evolution algorithm given in Table 2 were varied from their reference values to confirm that they were indeed set close to their optimum values and to determine the sensitivity of the results with regard to different parameters. In the categories of cell initiation and intensification, the non-dimensional coefficient c_1 in Eq. (11) was varied between 5, 7.5, 10, and 20. For small values intensification was too weak, for $c_1=20$ it was too strong, but differences were generally modest. An alternative scaling of the maximum rainfall rate I_{MAX} based on precipitation maxima already observed in the analysis (rather than using Eq. (11)) turned out to generate massive overpredictions of precipitation intensity, so it was deactivated. The time-scale τ_G of cell growth was not varied. In cases where intensification was correctly diagnosed by the algorithm and I_{MAX} had suitable values, the reference value of $\tau_G=30$ min gave quantitatively satisfactory values. When the moisture flux convergence threshold MFCini was changed from its reference value of 2×10^{-6} s^{-1} to 2.5×10^{-6} s^{-1} it led to an underestimation of cell initiation. A variable threshold that depends on the general magnitude of MFC present in the analysis field was also tested but did not show any clear advantage over the use of a fixed threshold. A variable threshold for CAPEini did not show a clear benefit either. It should be noted, however, that in the case studies there was always rather large CAPE present. A variable CAPE threshold may give some improvement for the prediction of convection in low CAPE environments. Raising the threshold VISini of satellite visible brightness from 0.5 to 0.75 gave some improvement in individual cases but not in the overall results. The condition CIN<CINini is useful for suppressing initiation in the presence of very large CIN values. Accordingly, the rather large value of 200 J kg^{-1} is used. If a smaller threshold is used, the suppression becomes too strong. Similarly, the condition that, in order for CI to occur, the temperature must not be lower than the trigger temperature by more than 2 K slightly improves the nowcast by reducing cases of false initiation.

In the category of cell weakening, using a negative value MFCweak$=-0.5\times 10^{-6}$ s^{-1} instead of 0.0 gives too little weakening. The variable CAPE threshold CAPEweak = min($\overline{CAPE}^{>0}$, 100 J kg^{-1}) is needed to avoid widespread weakening in cases of generally small CAPE. Using values higher than 100 J kg^{-1} gives too large weakening areas. It was also tested whether DT_TRIG is a useful predictor for weakening. In some cases, it led to significant improvements but overall it caused overestimation of cell weakening and was deactivated.

15.7 Orographic effects in convective initiation

In complex terrain, thermally driven flows strongly modify the diurnal evolution of the convective boundary layer (CBL), the formation of convective clouds, and the initiation of deep convection (Banta 1990). The problem of nowcasting CI before clouds have actually formed is also a problem of being able to predict the time-evolution of the CBL.

Over flat terrain, CBL evolution is well described by the concept of mixed-layer growth. Thermally-driven turbulent mixing keeps the vertical gradient of potential temperature small, and the CBL can be treated as a single layer, characterized by a mean potential temperature and, to a lesser degree, a mean specific humidity. Both the top of the CBL and the LCL usually rise during the day as a result of CBL warming and drying due to diurnal heating and entrainment of dryer air from above. Cumulus formation begins when the strongest thermals rise past their LCL. Typically, the atmospheric stratification in the morning exhibits a surface inversion, above which there is a less stable, often nearly dry-adiabatic 'residual' layer, which was formed by convection on the previous day. Before cumulus clouds can form, the surface inversion must be completely eroded so that thermals can penetrate to a sufficient height. The strength and depth of this inversion and the amount of sensible heat input determine to a large extent the time of onset of cumulus cloudiness (Haiden 1997).

In mountainous areas not all of the terrain is covered by cold air that has formed during the night. Katabatic flows lead to cold air pooling in valleys and basins while the upper parts of the slopes, and the peaks and ridges protrude out of the cold air. Over these parts of the terrain there is no surface inversion that must be eroded before cumulus initiation can start. Thus cumulus clouds begin to form earlier. Another difference compared to the flat terrain is that thermally-driven upslope flows provide a more organized and steady vertical motion field to trigger and support the clouds. The fact that onset of cumulus clouds is strongly favored over peaks and ridges is, therefore, basically due to horizontal inhomogeneities, which (1) allow the CBL to reach large heights even before the surface inversion at low elevations is completely eroded, and (2) lead to thermally-driven upslope flows.

Figure 5 shows a result of a mixed-layer model applied to the Sangre de Cristo mountain range in southern Colorado (Haiden 2004). The model was initialized with an observed early-morning sounding, and CBL evolution driven by surface heating, as well as the associated evolution of the LCL, was modeled. After 5 h of simulation, the top of the CBL over the foreland is still separated by a stable layer from the LCL. Over the mountain range the LCL is significantly higher, but the weaker initial stability there has allowed the CBL to grow past its LCL. The area where the CBL top has risen above the LCL coincides with the area of observed cumulus formation on that day. This suggests that a fine-scale real-time analysis of potential temperature (to estimate CBL top) and LCL should help in the nowcasting of the initiation of the first cumulus clouds.

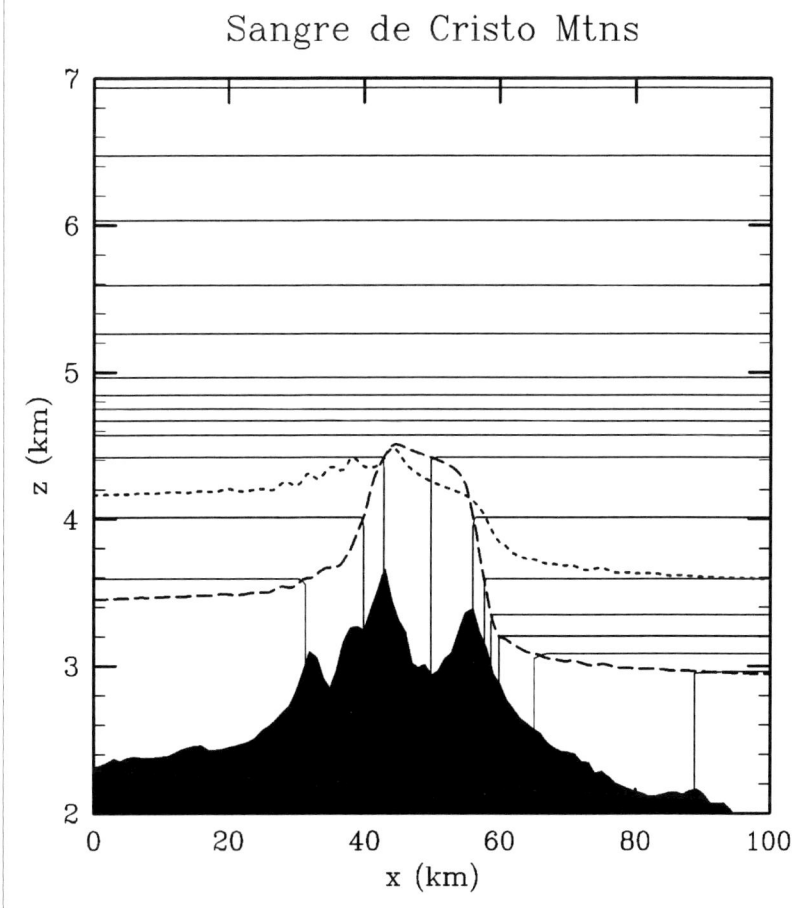

Fig. 5. Use of mixed-layer modeling to predict cumulus initiation over the Sangre de Cristo mountain range in southern Colorado. *Continuous lines* indicate potential temperature. Shown is the top of the CBL (*dashed*) and the LCL (*short dashes*) after 5 h of simulation. The area where the CBL top has risen above the LCL coincides with the area of observed cumulus formation (Haiden 2004)

If a topographic feature is favorable for cumulus initiation it does not necessarily mean it is also conducive to further cloud growth and development into a cumulonimbus cloud. It is well known, and can be shown theoretically, that for a given mountain height longer slopes can sustain a stronger upslope mass-flux than shorter, steeper ones (Schumann 1990; Haiden 2004). Thus, as a cloud grows both in vertical and horizontal extent, it is the larger-scale topography which becomes the most relevant for further convective developments. While slope flows may help to initiate the first cumulus clouds, meso-scale flows from the foreland towards the mountains, as well as up-valley flows (and their convergence) provide the necessary moisture supply and vertical motion for CI.

This upscale development explains why NWP models which have a horizontal resolution on the order 10 km and do not resolve individual valleys and slopes are nevertheless capable of predicting some of the preferred locations of orographically induced CI on that scale. It is probably one of the main reasons for the higher skill of the INCA cell evolution nowcast compared to the advection nowcast which was found for the alpine catchments. However, it is difficult for NWP models to accurately predict the timing of convective rainfall with regard to the diurnal cycle. Typically, the growth from non-precipitating shallow convection to precipitating, deep convection takes longer in the real atmosphere than in the models. Convective clouds are still mostly treated diagnostically in these models, and the build-up of cloud mass is not explicitly modeled.

What does this mean for the problem of convective precipitation nowcasting? The development of high resolution NWP models has a large potential for improved prediction of CI over complex terrain, where MFC and vertical motion are strongly tied to the orography and thus have a higher predictability than over flat terrain. This mainly applies to the initiation of primary convection. Once convective precipitation has formed, and thunderstorm outflows begin to modify the environment, meso-scale predictability is strongly reduced due to the complicated nonlinear interaction between existing cells and the mountain CBL, and their effect on subsequent cell growth. Thus it will be necessary to have either a rapid-update cycle for the NWP model, combined with a truly meso-scale data assimilation scheme, or a nowcasting method similar to the one proposed here, which takes into account the most recent observations of temperature, humidity, and wind within the boundary-layer. Either way, progress in methods of fine-scale meteorological analysis will be crucial for improvements in convection nowcasting.

15.8 Conclusions

An approach to convective nowcasting has been presented which is based on high resolution analyses of convective analysis fields. It is an attempt to extend existing nowcasting methods which use mass or moisture convergence to predict convective initiation to include temperature and humidity effects. It is found that the fields most relevant to the nowcasting problem are CAPE, CIN, moisture convergence, and trigger temperature deficit. Additionally, visible satellite imagery is needed in order to identify areas of incipient cell intensification and to avoid over-prediction of cell initiation. Verification of areal precipitation nowcasts for small catchments shows that on average the convective nowcast performs slightly better than the purely advective nowcast. While the improvement is small it is still encouraging because it was anticipated that it would be difficult to beat the 'conservative' advection nowcast with a more 'risky' cell evolution nowcast due to the double penalty effect. The advantage gained by cell evolution is more pronounced in the alpine catchments studied, whereas it is just marginal in the non-alpine areas. This appears to be due to the more predictable character of the orographically induced moisture convergence field in alpine terrain compared to that over the lowlands.

Further improvements could be achieved by improved analyses of the primary fields from which the convective diagnostics are derived. The typical length scale of CAPE, CIN, and MFC variations in the analysis is significantly larger than the scale of individual convective cells. The DT_TRIG field is a candidate for providing some of the missing small-scale information but is not yet sufficiently reliable. This leaves the visible satellite image as the main factor determining on the meso-γ scale where convective initiation or intensification takes place. It also means that the current nowcast algorithm is unable to predict cell initiation in clear air. In order to predict convective initiation *before* clouds are visible in the satellite data, a very good analysis and prediction of the boundary-layer wind field is needed. In the absence of clear-air Doppler radar information one must resort to fine-scale modeling. A high resolution wind-field analysis that takes into account physical processes like thermally driven upslope flows and mixed-layer growth could help to predict convective initiation in complex terrain.

References

Banacos PC, Schultz DM (2005) The use of moisture flux convergence in forecasting convective initiation: historical and operational perspectives. Weather Forecast 20:351–366

Banta RM (1990) The role of mountain flows in making clouds. In: Atmospheric Processes over Complex Terrain. Meteorol Monograph 23: 229–283

Borga M, Anagnostou EN, Frank E (2000) On the use of real-time radar rainfall estimates for flood prediction in mountainous terrain. J Geophys Res 105:2269–2280

Caracena F, Maddox RA, Hoxit LR, Chappell CF (1979) Mesoanalysis of the Big Thompson storm. Mon Weather Rev 107:1–17

Germann U, Joss J (2002) Mesobeta profiles to extrapolate radar precipitation measurements above the Alps to the ground level. J Appl Meteor 41: 542–557

Golding BW (1998) Nimrod: a system for generating automated very short range forecasts. Meteorol Appl 5:1–16

Haiden T (1997) An analytical study of cumulus onset. Q J Roy Meteor Soc 123:1945–1960

Haiden T (1998) Analytical aspects of mixed-layer growth in complex terrain. In: Preprints 8th Conference on Mountain Meteorology. American Meteorological Society, Flagstaff, Arizona, pp 368–372

Haiden T (2004) Extension of the mixed-layer concept to steep topography. In: Preprints 11th Conference on Mountain Meteorology. American Meteorological Society, New Hampshire

Haiden T, Kann A, Stadlbacher K, Steinheimer M, Wittmann C (2007) Integrated Nowcasting through Comprehensive Analysis (INCA)- System overview. ZAMG report, 49 pp. http://www.zamg.ac.at/fix/ INCA_system.doc

Hand WH (1996) An object-oriented technique for nowcasting heavy showers and thunderstorms. Meteorol Appl 3:31–41

Hudson HR (1971) On the relationship between horizontal moisture convergence and convective cloud formation. J Appl Meteorol 10:755–762

Kuo HL (1965) On formation and intensification of tropical cyclones through latent heat release by cumulus convection. J Atmos Sci 22:40–63

Mueller C, Saxen T, Roberts R, Wilson J, Betancourt T, Dettling S, Oien N, Yee J (2003) NCAR Auto-Nowcast System. Weather Forecast 18, 545–561

Pierce CE, Hardaker PJ, Collier CG, Haggett CM (2000) GANDOLF: a system for generating automated nowcasts of convective precipitation. Meteorol Appl 7:341–360

Pierce CE, Ebert E, Seed AW, Sleigh M, Collier CG, Fox NI, Donaldson N, Wilson JW, Roberts R, Mueller CK (2004) The nowcasting of precipitation during Sydney 2000: an appraisal of the QPF algorithms. Weather Forecast 19:7–21

Schumann U (1990) Large-eddy simulation of the up-slope boundary layer. Q J Roy Meteor Soc 116:637–670

Steinacker R, Ratheiser M, Bica B, Chimani B, Dorninger M, Gepp W, Lotteraner C, Schneider S, Tschannett S (2005) A mesoscale data analysis and downscaling method over complex terrain. Mon Weather Rev 134:2758–2771

Steinheimer M, Haiden T (2007) Improved nowcasting of precipitation based on convective analysis fields. Adv Geosci 10:125–131

Steppeler J, Bitzer H-W, Minotte M, Bonaventura L (2002) Nonhydrostatic atmospheric modeling using a z-coordinate representation. Mon Weather Rev 130:2143–2149

Waldstreicher JS (1989) A guide to utilizing moisture flux convergence as a predictor of convection. Nat Wea Digest 14:20–35

Wang Y, Haiden T, Kann A (2006) The operational limited area modelling system at ZAMG: ALADIN-AUSTRIA. Österr. Beiträge zu Meteorologie und Geophysik, Heft 37, 33 pp

Wilson JW, Mueller CK (1993) Nowcasts of thunderstorm initiation and evolution. Weather Forecast 8:113–131

Wilson JW, Schreiber WE (1986) Initiation of convective storms by radar-observed boundary layer convergence lines. Mon Weather Rev 114:2516–2536

Wilson JW, Ebert EE, Saxen TR, Roberts RD, Mueller CK, Sleigh M, Pierce CE, Seed A (2004) Sydney 2000 forecast demonstration project: convective storm nowcasting. Weather Forecast 19:131–150

16 Overview of methods for the verification of quantitative precipitation forecasts

Andrea Rossa[1], Pertti Nurmi[2], Elizabeth Ebert[3]

[1] Centro Meteorologico di Teolo, ARPA Veneto, Italy
[2] Meteorological Research, Finnish Meteorological Institute, Finland
[3] Centre for Australian Weather and Climate Research, Bureau of Meteorology, Australia

Table of contents

16.1 Introduction 419
16.2 Traditional verification of QPF and limitations for high
      resolution verification 423
      16.2.1 Common scores 424
      16.2.2 The double penalty issue 429
16.3 Scale-dependent techniques 433
      16.3.1 Neighborhood methods 433
      16.3.2 Spatial decomposition methods 437
16.4 Object and entity-based techniques 438
16.5 Stratification 440
      16.5.1 Seasonal, geographical and temporal stratification 441
      16.5.2 Weather-type dependent stratification 442
16.6 Which verification approach should I use? 448
References 449

16.1 Introduction

In the area of hydrological risk management, both Quantitative Precipitation Estimates (QPE) and Quantitative Precipitation Forecasts (QPF) are key in quantifying the potential for flooding, especially on the short time scales, i.e., for relatively small river and urban catchments. In such a context, forecasting can be viewed as the attempt to reduce the uncertainty of the future state of the hydrometeorological system and so anticipate mitigating actions. Authorities, however, often are still reluctant to devise and invest in such actions based on forecasts when their quality is unknown. In other words, for forecasts to be useful and effective the forecast quality and forecast uncertainty must be quantified.

Much effort has been and is being invested in the quest of working with imperfect precipitation observations and forecasts. A number of initiatives are underway, such as the Hydrologic Ensemble Prediction EXperiment (HEPEX, Schaake et al. 2007) which is an international project established by the hydrological and meteorological communities. The mission of HEPEX is to demonstrate how to produce reliable hydrological ensemble predictions that can be used with confidence by emergency management and water resources sectors to make decisions that have important consequences for economy, public health and safety. The COST 731 Action (Rossa et al. 2005) is a European initiative which deals with the quantification of forecast uncertainty in hydrometeorological forecast systems. It is linked to the MAP D-PHASE initiative (www.map.meteoswiss.ch), a WWRP Forecast Demonstration Project (FDP), which is to provide evidence of the progress meteorological and hydrological modeling has achieved over the last decade or so. A characteristic of an FDP is that strict evaluation protocols are established to demonstrate and document such progress. Indeed, many atmospheric and hydrological forecast systems participate in this effort. The atmospheric part includes nowcasting based on radar, very high resolution next-generation numerical weather prediction (NWP) models, operational models, as well as a number of limited area ensemble prediction systems.

In all of this verification, and verification of precipitation forecasts in particular, is fundamental! It is safe to say that the more detailed the forecasts the more complex the corresponding verification task. For example, verification of geostrophic flow can be viewed as relatively simple when compared to verification of turbulent flow. Precipitation is a stochastic quantity and exhibits fractal properties down to very small scales (e.g., Zawadzki 1973). It is difficult to observe, simulate and to verify. Furthermore, many more efforts have been invested in the development of forecasting techniques than in verification methodologies. This may be connected to the fact that the traditional approaches to verification of gridded forecasts were developed on relatively low resolution global NWP models to check the consistency of upper air fields against model analyses. Stanski et al. (1989) provide a thorough compilation of the statistics involved in NWP verification, while Wilks (2006) is an excellent text and reference book for statistical methods in the atmospheric sciences, covering forecast verification.

However, with increasing resolution of the limited area models, verification of weather elements against observations has become a more complex problem. For example, while for medium-range forecasting typically daily rainfall accumulations are verified, the higher resolution meso-scale models are expected to have skill also in shorter time scales. Their performance is tested for shorter accumulation periods where for instance the timing and location of a frontal passage is essential and the traditional verification methods are not necessarily sufficient. Small positioning errors in the forecasts may result in the so-called 'double penalty': the verification measure tends to penalize rather than reward the model's capability to provide some sort of information on small scale features (see Sect. 16.2.2).

These issues are accentuated when it comes to verifying high resolution QPFs. The necessity to evaluate and justify the advantages of the ever higher resolution over the computationally less expensive coarser resolution NWP in terms of QPF quality has stimulated radically different verification approaches for spatial forecast fields over the last decade or so. These methods go well beyond point-to-point pair verification and borrow ideas from fields such as image and signal processing. The main lines of extension to judge whether or not a precipitation forecast for a given time and location is correct is to ask the question whether the main *characteristics* of fields are captured in the simulation. In other words, conditions for right and wrong are relaxed from 'at a given point and time' in several ways. For example, in the class of neighborhood methods the condition of correct location is successively relaxed to yield an effective scale-dependent measure of forecast goodness (Ebert 2008). Harris et al. (2001) investigate whether the characteristic scales of rainfall fields are successfully reproduced, without necessarily requiring correspondence in location, while Ebert and McBride (2000) look for corresponding rain objects and decompose the measure for quality in components for matching location, amount and structure. Davis et al. (2006) take the description of precipitation objects one step further but still require object matching between the forecast and the observations.

For hydrological applications the localization of precipitation is important on the scale of the considered catchment, so that it is useful to perform QPF verification on river basin averages (e.g., Oberto et al. 2006). Wernli et al. (2008) combine the idea of verifying precipitation within a predefined area, say a medium to large river catchment, in which not just the average rainfall amount is evaluated but also the average capability of the model to predict location and structure of the rainfall field, measures that do not require object correspondence.

Datasets on which these methodologies are applied can span several years in order to try to document improvements in forecast quality. Improvements have been reported for parameters like the pressure or the temperature, but not for QPF (Hense et al. 2003). Performing verification over a full year will effectively mix a number of different flow regimes which, in theory, can present different challenges to a modeling system. Also, the verification results can be biased towards the most frequent regime, e.g. days with no intense weather. It is, therefore, quite common practice to differentiate verification for the four seasons, while it is far less common to perform a systematic separation of distinct flow regimes in which a forecast system may have different challenges to get realistic QPF.

The diversity of approach emerging from these examples, which are detailed further in Sects. 16.3 to 16.5, document the efforts of the scientific and operational community to find adequate measures to describe forecast quality of high resolution QPF. However, such a variety holds the risk that verification results become difficult to compare. There have been several efforts to harmonize verification activities in the recent past. ECMWF, for example, compiled a set of recommendations for their member states (Nurmi 2003), while the Joint Working Group on Verification (JWGV) provided a survey of verification methods of weather elements and severe weather events (Bougeault 2002) and recommendations for the verification and intercomparison of QPFs from operational NWP models (JWGV 2004). There is an ongoing exercise in which the more recent verification techniques are to be compared on a set of common cases (ICP 2007).

Probabilistic QPF is a promising avenue of improvement for high resolution rainfall prediction (e.g., Mittermaier 2007). The main ideas behind probabilistic forecasting are based on the imperfect knowledge of initial conditions and key parameters in parameterization schemes of mainly moist processes. Ensemble forecasting, i.e., forecasts starting from slightly differing initial conditions, is an established technique for estimating forecast uncertainty of the global models in the medium range. It has become increasingly popular also for high resolution limited area models in shorter time ranges, as well as in nowcasting. The radar community has started recently to produce probabilistic QPEs based on the error characteristics of radar measurements (Germann et al. 2006). Probabilistic forecasting is adding considerable complexity to the verification problem in that 'right and wrong' no longer have a strict sense when it comes to a single forecast observation pair. Verification needs to take the frequency of occurrence of events into account. These issues are, however, beyond the scope of this Chapter and will not be discussed.

This contribution aims at providing an overview on the standard techniques used in QPF verification and on recent, more sophisticated approaches, in order to provide a panorama of the tools available. The choice of technique for QPF verification may well be purpose-dependent, be it in hydrological applications for one or more catchments, in road weather forecasting for distinct stretches, or for model development where identifying specific model weaknesses is the necessary first step for improvement. It is, therefore, a specific goal of this writing to provide some sort of recommendations or guidelines to the collection of methods. For the sake of convenience, many of the illustrations are taken from the COSMO model (Steppeler et al. 2003), but the applied methods are by no means tied to this particular model. They are not even specific to NWP but can be applied to other comparisons of precipitation fields, e.g., QPE from different sensors (e.g., Ebert et al. 2007). An additional Chapter on QPF verification is presented by Tartaglione et al. (Chap. 17 in this book).

Section 16.2 reviews traditional verification scores and illustrates their limits for high resolution QPF verification. Section 16.3 deals with scale-dependent verification, while Sect. 16.4 with object-oriented approaches. Stratification of data sets to isolate model behavior in specific flow situations is dealt with in Sect. 16.5 before some recommendations are given in Sect. 16.6 as to the relative merits of the various techniques which have been discussed.

16.2 Traditional verification of QPF and limitations for high resolution verification

The strategy for any forecast verification application includes certain rational steps: choosing and matching a set of forecast/observation pairs, defining the technique to compare them, aggregating (pooling) and/or stratifying the forecast/observation pairs in appropriate data samples, applying the relevant verification statistics and, ultimately, interpreting the scores, not forgetting to analyze the statistical significance of the gained results. The latter is unfortunately quite often neglected both in verification studies as well as in operationally run forecasting systems.

Deterministic QPFs can be formulated and taken as either *categorical events* or *continuous variables* and verified correspondingly utilizing respective verification approaches and measures. Verifying QPFs as categorical events is clearly more common. The categorical approach involves issues like whether or not it rained during a given time period (rather than at a given instant) or, alternatively, whether the rainfall amount exceeded a given threshold. Verifying rainfall amount as a continuous variable brings about certain caveats because the rainfall amount is not a normally distributed quantity. Very large rainfall amounts may be produced by a forecasting system and, then again, in some cases very little or no rain. Many of the verification scores for continuous variables, especially those involving squared errors, are very sensitive to large errors. Consequently, categorical verification scores provide generally more meaningful information of the quality of the forecasting systems (or skill of the human forecasters) producing QPFs.

16.2.1 Common scores

There are a number of recent textbooks (Wilks 2006; Jolliffe and Stephenson 2003) and papers (Nurmi 2003; Bougeault 2002; Wilson 2001) as well as by JWGV (2004) which details the traditional precipitation verification methods and give an exhaustive account of their features. Reference is made to these publications rather than elaborating on these attributes here. However, a general definition and a short overview of the most common scores for the verification of categorical QPFs will follow, accompanied by brief comments of their pros and cons. Occasional references are made to current literature where one can embrace a deeper understanding of the behavior of these measures. Some additional, more recent scores are also introduced. Although these cannot be considered 'traditional' they are covered because they fit the framework properly.

Categorical events - Forecasts of the exceedance of precipitation thresholds

The joint distribution of binary (yes/no) forecasts and associated observed events and non-events is unambiguously defined by the four elements of a 2 × 2 contingency table: *hits, false alarms, misses* and *correct rejections*. Categorical statistics are applied to evaluate these binary events which in our case is the accumulated rainfall amount during a given time period exceeding a specified threshold. The most popular event is rainfall exceeding a threshold taken as rain versus no-rain. This threshold varies from country to country but is generally between 0.1 mm and 0.3 mm of accumulated precipitation during a 24-hour (or a 12-hour) period. These different definitions may have huge effects on the verification results (and their interpretation) since, as shown later, many of the categorical forecast verification measures are highly dependent on the observed frequency (or the *base rate*) of the event.

The seemingly simple definition of the binary event and the subsequent contingency distribution and its associated marginal distributions of forecasts and observations accommodate quite amazing complexity and there exist a large number of measures to tackle this ambiguity. Most of these scores have historical credentials as long as the history of forecast verification, dating back to the late 19th century. Consequently, they have been 're-invented' and renamed many times during later times.

The *Frequency Bias Index* (FBI) compares, as a ratio, the frequency of forecasts with the frequency of actual occurrences of the event. It ranges from zero to infinity and the optimal value for an unbiased forecasting system is one. The frequency bias is not a measure of accuracy as it does not provide information on the magnitude of forecast errors.

Probably the simplest and most intuitive performance measure providing some information on the accuracy of categorical forecasts is the *Proportion Correct* (PC) which gives the fraction of all correct forecasts (i.e., of the event and the non-event). This simplistic measure is easily very misleading since it rewards correct 'yes' and 'no' forecasts equally and is strongly influenced by the more common category which is normally the more uninteresting non-event. A prime educational example of the interpretation of PC and its consequences is the often cited, legendary Finley case (Finley 1884; Murphy 1996; see also e.g., Wilks 2006, pp. 267–268).

The *Probability Of Detection* (POD) measures the fraction of observed events that were correctly forecast, whereas the *False Alarm Ratio (FAR)* measures the fraction of forecast events that were observed to be non-events. In some literature, POD is called the *hit rate*, having as its complement the *miss rate* which gives the relative number of missed events. POD and FAR must always be examined together as neither of them is really adequate on its own. POD is sensitive to hits only and does not take into account false alarms, whereas FAR is sensitive to false alarms but takes no account of misses. Both of them can be artificially improved by producing excessive 'yes' forecasts (in the case of POD) or 'no' forecasts (to improve FAR). Such bogus human forecasting behavior is often called *hedging*. FAR is very sensitive to the climatological frequency of the precipitation event (Fig. 1, left panel), which is a property quite common to many of the traditional verification measures.

While FAR is a measure of false alarms given the forecasts, *False Alarm Rate (F)* is a kindred measure which measures the false alarms given the observed non-events. It is also called the *Probability Of False Detection* (POFD). F is almost exclusively associated with the verification of probabilistic QPFs by combining it with the hit rate to produce the so-called *Relative Operating Characteristic* (ROC) diagram or curve. The ROC measures the ability of the forecast to discriminate between observed events and non-events and is commonly used in the verification of probabilistic forecasts (for more on ROC diagrams see Jolliffe and Stephenson 2003).

A popular, historical measure for verifying categorical forecasts results from simply subtracting F from POD. This skill score has many 'inventors' and therefore many names, like *True Skill Statistics* (TSS), *Peirce Skill Score* (PSS) and *Hanssen-Kuipers Skill Score* (KSS). Idealistically, it measures the skill of a forecasting system to distinguish the 'yes' cases from the 'no' cases. It also measures the maximum possible relative economic value attainable by a forecast system, based on a Cost-Loss model (Richardson 2000). For rare events (e.g., heavy precipitation) the frequency of correct rejections is typically very high, leading to a very low F and, consequently, the score asymptotes to POD.

Another commonly used performance measure, especially for rare events is the *Threat Score* (TS), also known as the *Critical Success Index* (CSI). It is defined as hits divided by the sum of hits, false alarms and misses. Because TS takes into account both false alarms and misses it can be considered as a simple measure that tries to remove from consideration correct forecasts of the (simple) non-events. However, TS is known to be sensitive (again) to the local climatology of precipitation (Fig. 1, center panel). To overcome this feature the otherwise similar *Equitable Threat Score* (ETS) aims at removing the effects of hits that occur purely due to random chance. (Fig. 1, right panel).

One of the most widely used skill scores is the *Heidke Skill Score* (HSS). Its reference accuracy measure is Proportion Correct which is adjusted to eliminate forecasts which would be correct due to random chance. The HSS is related to ETS via a direct relationship and therefore does not provide any additional information.

The *Odds Ratio* (OR) measures the forecasting system's probability (odds) to score a hit (H) as opposed to the probability of

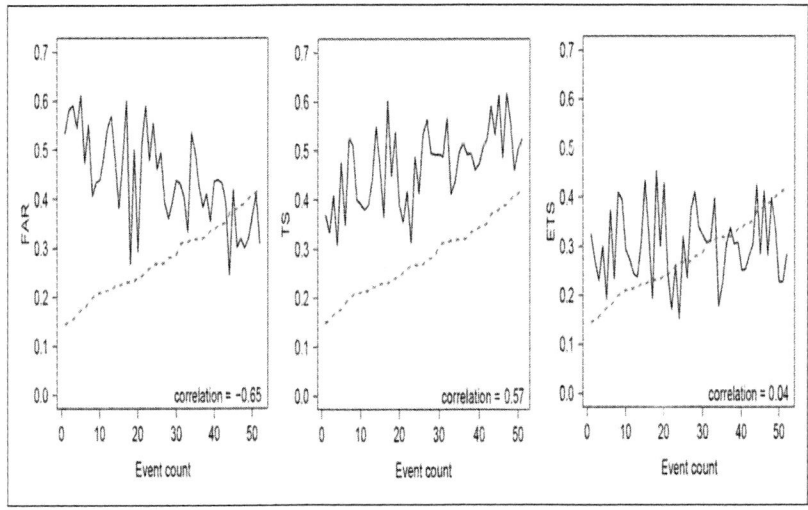

Fig. 1. Correspondence between a categorical verification measure (continuous line) and the frequency of observed rain events (dashed line) at a given observing station, for FAR (*left*), TS (*center*) and ETS (*right*). The rain event is defined as rainfall exceeding 0.3 mm during 24 hours and the vertical axis shows the relative frequency of such events arranged in ascending order

making a false alarm (F). It produces typically high numeric values because a no-skill system would equal one and a perfect system yields a score of infinity. A transformed *Odds Ratio Skill Score* (ORSS) is scaled to have values in the range [−1, +1] to be comparable with other verification scores of categorical events. The Odds Ratio cannot be considered a traditional verification measure and has been applied very scarcely in meteorological (QPF) verification. Nevertheless, it is advocated to possess several attractive properties (Göber et al. 2004; Stephenson 2000).

Stephenson et al. (2007) have proposed a new score, the *Extreme Dependency Score* (EDS), specifically for the verification of rare events like heavy QPF. The score is reported to be insensitive to the base rate, is not dependent on the potential frequency bias of the forecasts and will not encourage hedging.

Continuous variables - Forecasts of time-integrated accumulated precipitation

As already discussed in the previous Section, deterministic QPFs are often formulated as *continuous variables*, but perhaps more often verified as categorical events. Verification of continuous QPFs as such commonly involve statistics on how much the absolute forecast values depart from the corresponding observations, as well as the computation of relative (skill) measures against reference forecasts like climatology and persistence. What follows is a very brief description of the most common (traditional) verification methods applicable for the verification of continuous QPFs. In general, the intermittent and non-Gaussian distribution of precipitation strongly affects these measures which are generally sensitive to large errors.

The *mean value (arithmetic mean)* is always very useful to put forecast errors (see below) into their perspective. To define variability in rainfall the *sample variance* and the *sample standard deviation* are often used. The latter is conveniently in the same units as the original precipitation, being the square root of the previous.

The *Mean Error* (ME), or *bias*, is simply the arithmetic average of the difference between forecasts and observations. Like the frequency bias in the case of categorical QPF events, it is not an accuracy measure and does not produce information on the magnitude of forecast errors. The *Mean Absolute Error* (MAE) compensates for positive and negative forecast errors and is a scalar measure of forecast accuracy. The ME and the MAE viewed together provide useful information on the general behavior of forecast errors.

The *Mean Square Error* (MSE) is the average squared difference between forecasts and observations. Taking a square root of MSE produces the *Root Mean Square Error* (RMSE) which has the same units as the original entity. Due to the second power of these scores they are much more sensitive to large forecast errors than the MAE, which may be quite harmful in the presence of outliers in the dataset. The *correlation coefficient* (r) measures the degree of linear association between forecast and observed values, independent of absolute or conditional biases. This score is very sensitive to large errors and benefits from the square root transformation of precipitation amounts. The fear for high penalties when applying squared verification measures may easily lead a human forecaster to conservative forecasting (i.e., hedging).

Many of these accuracy measures, especially the MAE and the MSE, are commonly used to construct a *skill score* that measures the fractional (percentage) improvement of the forecast system over a reference forecast. The reference estimate is preferably persistence for forecasts with a lead time of c. 24 hours or less and climatology for longer range forecasts.

16.2.2 The double penalty issue

The traditional point-matching categorical and continuous verification measures are quite intuitive, easy to perceive and, above all, they have been used for many decades. There is no urge to cease applying them as long as their pros and cons and occasionally notorious behavior is known, understood and acknowledged. Whatever their pitfalls the traditional standard verification measures still do return optimum scores for, hypothetically, optimum forecasts, regardless of the underlying properties, like the resolution, of the models that produce these forecasts. A further aspect that favors preserving existing common verification methods is the lengthy time lag before new innovations in forecast verification research are mature enough to be accepted by the community and applicable for common use.

The verification endeavor has become more and more demanding and from a scientific perspective increasingly rewarding, during recent years with the continuously enhanced resolution of the NWP models, resulting effectively also in the detail in which a human forecaster depicts the weather. Today it is not uncommon to have detailed local, site-specific QPFs for several days ahead as compared to earlier times when forecasts were formulated rather as area and time-averaged entities. The most obvious and meaningful way to produce time/space focused precipitation forecasts would be using a probabilistic approach but the verification of probabilistic QPFs (PQPF) is not covered in this Chapter. Nevertheless, as long as NWP models do produce categorical QPFs, their quality needs to be evaluated from this perspective.

Let us consider, for example, a model forecast low pressure pattern having a phase error of half a wavelength and another model having not forecast the pattern at all. The former model would be punished twice, for not having the low where it is supposed to be and, secondly, for having the low where it is not supposed to be (*double penalty*). The latter model, however, would get penalized for only not having forecast the pattern.

It is quite common that high resolution, meso-scale, forecast models produce forecasts with seemingly realistic small scale (precipitation) patterns but with amplitude and gradients which may be somewhat misplaced. In the case of convective precipitation and/or narrow frontal rainbands such misplacements are hardly surprising but may show up as quite dramatic results when verified with common verification measures. The timing and space errors will result in a much larger RMSE than for the smoother lower resolution model forecast.

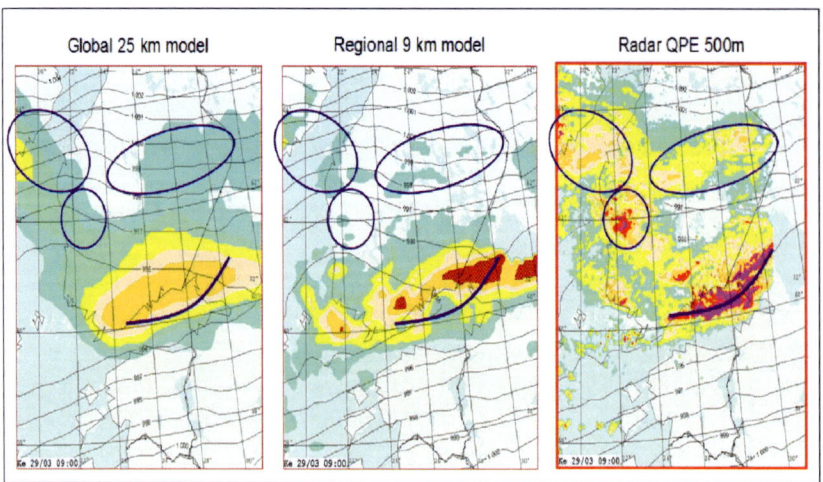

Fig. 2. A comparison of two NWP models operating at different spatial resolutions and the corresponding radar-based QPE. Some of the main features are indicated by the ovals and the arched curves

There are seldom, if ever, trivial cases in the real atmosphere. Figure 2 shows a comparison between a global and a regional NWP model in a case with well-defined precipitation patterns. The regional model shows some explicit small-scale structures with a reasonably realistic amplitude, albeit somewhat misplaced, when compared to a radar-based quantitative precipitation estimate (QPE), taken that the radar-based analysis is realistic. On the other hand, there are features in the global model (indicated by ovals) which are almost totally missing from the regional model. It would be quite hard to interpret intuitively or visually (applying *'eyeball verification'*) and even with objective verification measures which one of the forecasts is the better. As a matter of fact, one would need to first define the purpose of the forecast (end-user, application, etc) and of the verification.

Table 1 compares three NWP models operating at different resolutions. The QPF verification is done against three different 'observed truths', one based on rain gauge data, one on radar-based QPE and the third on merging these two data sources (the merging method is not relevant here). The results are somewhat mixed and incoherent. The highest scores (underlined numbers) are mostly gained for the coarse-scale model and the lowest ones (numbers with shaded backgrounds) for the fine-scale model, hinting at double penalty reminiscent behavior. However, the scores reflect also quite strongly the observation (or

Table 1. Verification statistics for three NWP models operating at different spatial resolutions and applying three different analysis types

	Global 25 km model			Local 22 km model			Local 9 km model		
	Gauge	Merge	Radar	Gauge	Merge	Radar	Gauge	Merge	Radar
Bias	1.44	1.45	.59	1.73	1.66	.73	.59	.59	.25
MAE	.33	.32	.47	.48	.48	.69	.31	.33	.67
RMSE	.57	.53	.93	.78	.77	1.14	.63	.66	1.25
r	.62	.70	.72	.40	.46	.42	.42	.45	.32
POD	.74	.74	.51	.65	.59	.39	.50	.42	.20
FAR	.49	.49	.14	.63	.64	.47	.15	.28	.19
KSS	.66	.66	.48	.52	.46	.28	.49	.41	.19
ETS	.38	.38	.39	.25	.22	.18	.43	.33	.15

analysis) type which has been applied as the 'truth' behind the verification. The important role of observations in verification is elaborated further in the Chap. 17, by Tartaglione et al., later in this book. The example here is presented merely to emphasize the complexity of verification. Nevertheless, it is advisable to use combined gauge-radar precipitation analyses whenever possible when verifying QPFs at high temporal and spatial resolutions.

The double penalty may be interpreted in terms of the categorical precipitation verification terminology: a forecast is penalized twice, for not getting the precipitation at the correct location (*miss*) and forecasting the precipitation at the wrong location (*false alarm*). This is illustrated schematically in Fig. 3 (a) where a high resolution forecast (left) would attain dramatically worse scores than its low resolution competitor (right) although the shape and amplitude appear perfect on the high resolution output.

The differences in scores are exclusively due to the misplacement of the entities. Applying a spatial translation and matching of the forecast pattern of the high resolution system with the observed field would result in the schematic shown as Fig. 3 (b). Such an exercise would result in perfect scores in this simplistically naive example. Object matching verification techniques are elaborated further in Sect. 16.4.

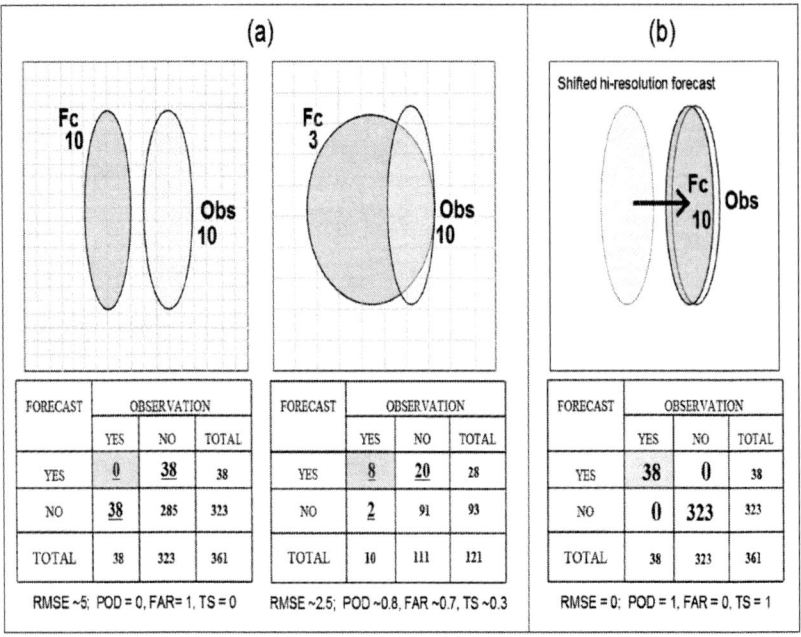

Fig. 3. A schematic of the double penalty effect on a high resolution forecast (**a**; *left*) compared to a low resolution forecast (**a**; *right*) and after applying the spatial matching technique such as that of Ebert and McBride (2000) (**b**)

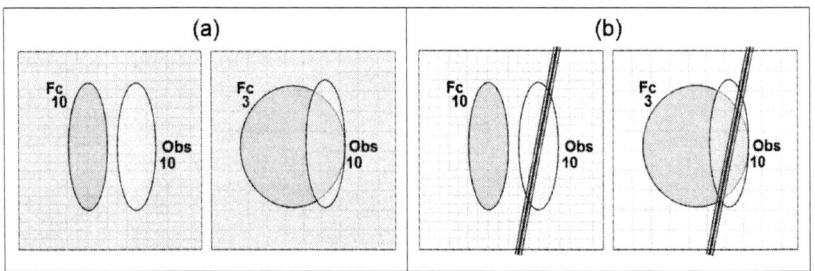

Fig. 4. A schematic of two different hypothetical forecast/verification applications, a hydrological catchment, indicated by the shading of the square domain (**a**) and a highway stretch, shown by the vertical line through the domain (**b**)

It is required in forecast verification, likewise in weather forecasting, that the target (end-) users and the purpose of verification/forecasting are known beforehand. This issue is briefly underlined using our previous example. The forecast/verification area of interest might be a distinct hydrological catchment area (indicated as the shaded rectangular area in Fig.4 (a)), or the focus of interest might be along a highway stretch (represented by the thick vertical line in Fig. 4 (b)). The quality of the forecasts would be evaluated quite differently for these two applications.

16.3 Scale-dependent techniques

As just seen, as precipitation forecasts from models and nowcasts are made at increasingly higher spatial and temporal resolution, the ability of the forecast to achieve an exact match with the observations becomes more difficult owing to the double penalty issue. The question becomes, if poor skill is shown at fine scales then at what scales does the forecast skill become acceptable? Scale-dependent verification methods address this question by measuring the correspondence between the forecast and the observations on a variety of space and time scales.

16.3.1 Neighborhood methods

Neighborhood (sometimes called 'fuzzy') verification approaches reward closeness by relaxing the requirement for exact matches between forecasts and observations. Ebert (2008) describes a framework for neighborhood verification using multiple methods. Some of these methods compute standard verification metrics for deterministic forecasts using a broader definition of what constitutes a 'hit'. Barnes et al. (2007) propose a conceptual framework to take into account close calls when evaluating U.S. National Weather Service weather warnings. Other methods treat the forecasts and/or observations as probability distributions and use verification metrics suitable for probability forecasts. Implicit in each neighborhood method is a particular decision model concerning what constitutes a good forecast. For example, one decision model could be that a good forecast must predict at least one event near an observed event.

The key to this approach is the use of a spatial window or neighborhood surrounding the forecast and/or observed points. The treatment of the points within the window may include averaging (upscaling), thresholding, or generation of a probability density function, depending on the metric used. Some methods compare neighborhoods of forecasts with neighborhoods of observations, while others compare the forecast neighborhood with the observation in the center of the neighborhood. Starting with the finest scale (neighborhood of one grid box) the size of the neighborhood is increased to provide verification results at multiple scales, thus allowing the user to determine at which scales the forecast has useful skill. Multi-dimensional windows can be used to represent closeness in space, time, intensity, and/or some other aspect.

Three of the most useful of the neighborhood techniques are described in this Section. They are demonstrated by verifying a high resolution (0.02°) forecast from the COSMO model against high-quality radar observations over Switzerland (Leuenberger 2005).

As seen in Fig. 5 the model predicted the rainfall structure quite well. However, the ETS (see Sect. 16.3.2) computed at grid scale for a 0.1 mm threshold was only 0.33. This illustrates the need for verification methods that give credit to 'close calls' and 'near misses'.

The most widely used neighborhood verification technique is upscaling, in which forecasts and observations are averaged to increasingly larger grid scales for comparison using a range of standard statistics (e.g., Zepeda-Arce et al. 2000; Cherubini et al. 2002; Yates et al. 2006). The implied decision model is that a good forecast has a similar mean rain amount as the observations. The upscaling verification of the COSMO forecast is shown in Fig. 6 in which the ETS is plotted as a function of spatial scale and rain intensity.

The verification scores generally improve with increasing scale and smaller rain thresholds, as expected. A relative peak in performance for the heavier rain rates is seen at a spatial scale of about 0.2 degrees, consistent with the good placement of the rain maximum.

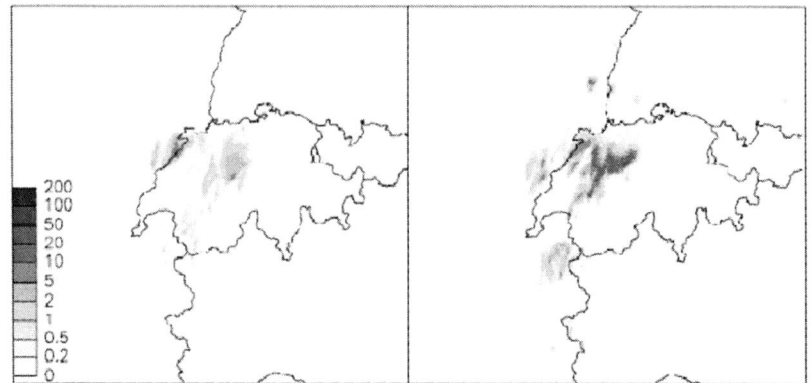

Fig. 5. Seventeen hour forecast from the COSMO model (*left*) and radar quantitative precipitation estimate (*right*) of hourly rainfall accumulation (mm) over Switzerland ending 17:00 UTC on 8 May 2003 (from Leuenberger 2005)

Chapter 16 - Verification of quantitative precipitation forecasts   435

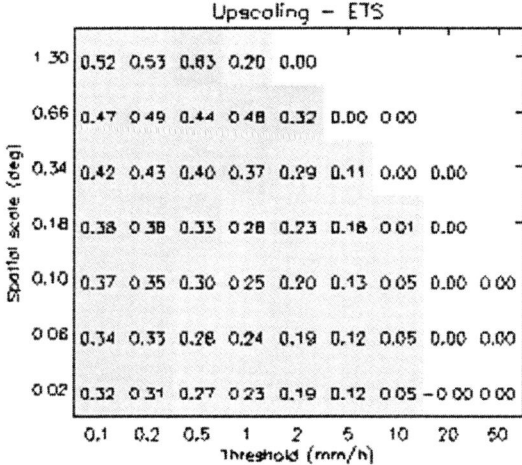

Fig. 6. Equitable threat score for the COSMO forecast shown in Fig. 5, as a function of spatial scale and rain threshold, when upscaling is used to average forecasts and observations to larger scales

Atger (2001) developed a multi-event contingency table method for comparing high resolution gridded rainfall forecasts to point observations. In this approach closeness is evaluated simultaneously in two or more 'dimensions' (spatial proximity, temporal proximity and similarity of rain intensity). A hit is counted whenever a forecast event is sufficiently close to an observed event. Multi-dimensional contingency tables are generated for varying thresholds, from which the hit rates can be plotted against the false alarm rates as points on a Relative Operating Characteristic (ROC) diagram.

The ROC in Fig. 7 suggests that the COSMO forecast successfully predicted rain close to where it was observed, both in terms of spatial location and intensity.

The fractions skill score (FSS) method of Roberts and Lean (2007) compares the forecast and observed fractional occurrences of rain exceeding a given threshold. The FSS is a version of the Brier Skill Score (see Jolliffe and Stephenson 2003) in which the observed occurrence is the event fraction within the neighborhood and the reference forecast is the no-overlap forecast. Roberts and Lean showed that the target value of FSS above which the estimates are considered to have useful skill is given by $0.5+f_{obs}/2$, where f_{obs} is the frequency of observed events over the full domain. The FSS values for the COSMO forecast (Fig. 8) are greater for light thresholds and larger scales, with useful skill displayed at spatial scales of 0.1 degree and larger for light rain, 0.2 to 0.7 degrees for moderate rain, and not at all for the heaviest rain rates.

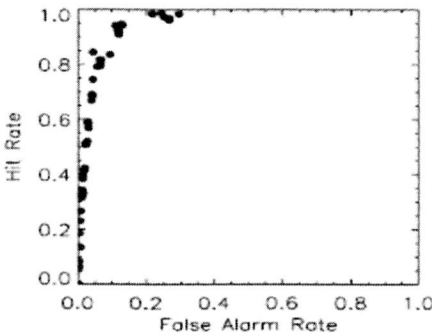

Fig. 7. Relative Operating Characteristic for the COSMO forecast shown in Fig. 5. Each point shows the hit rate and false alarm rate for a particular combination of spatial scale and rain intensity threshold

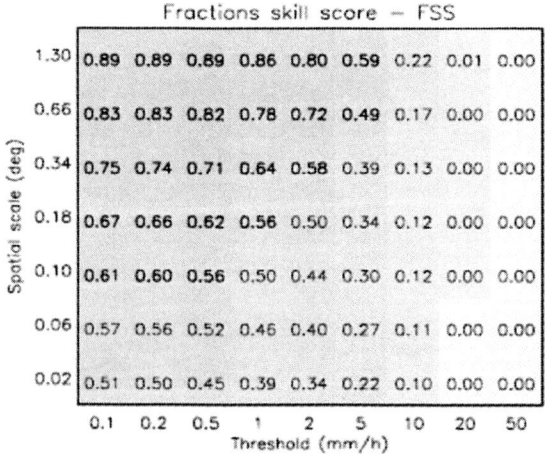

Fig. 8. Fractions skill score for the COSMO forecast shown in Fig. 5. FSS measures the similarity of the forecast and observed rain fractions for a variety of spatial scales and rain thresholds. The bold values indicate useful skill

16.3.2 Spatial decomposition methods

Another type of scale-dependent verification uses a spatial filter to decompose or separate the gridded forecasts and observations into different spatial scales and then computes the error separately for each scale. The scale-dependent errors sum to the total error. Scale decomposition allows errors associated with different phenomena to be isolated and identified. Several spatial filters have been proposed, including 2D Fourier transforms (Stamus et al. 1992), discrete cosine transforms (de Elia et al. 2002) and 2D discrete wavelet filters (Briggs and Levine 1997; Casati et al. 2004). Once the scale separation has been accomplished, then different continuous, categorical and probabilistic verification metrics may be applied.

In particular, the intensity-scale method of Casati et al. (2004) uses thresholding to convert the forecast and observations into binary images. Wavelet decomposition is applied to the binary error image and a skill score based on the mean squared error is computed for each scale. Figure 9 shows the application of the intensity-scale method to the COSMO forecast. The lowest skill is associated with small scales and high rainfall intensities while the greatest skill is found at large scales.

If the aim of the verification is to compare the multi-scale statistical properties of the forecast rainfall to those of the observed rainfall, that is, to evaluate whether the forecast rain looks realistic,

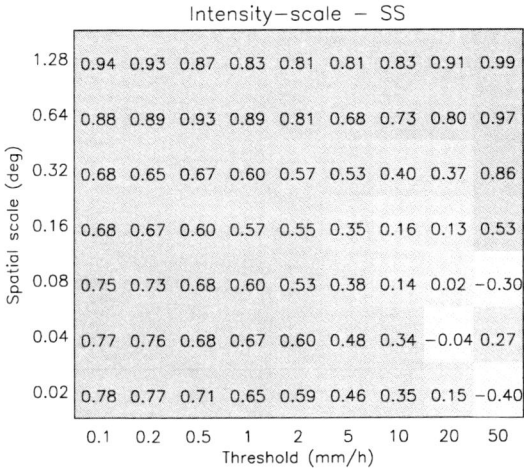

Fig. 9. Scale-dependent Heidke skill score for the COSMO forecast shown in Fig. 5, computed using the intensity-scale method of (Casati et al. 2004)

regardless of the actual placement of the rain relative to the observations, then the multi-scale approach described by Harris et al. (2001) may be used. They compare the power spectrum, structure function and moment scaling analysis of high resolution model output with radar data and to evaluate which scales are well represented by the model.

16.4 Object and entity-based techniques

The tendency of a human analyst, when presented with a rainfall map, is to focus on features of interest such as areas of heavy rain. Object- or entity-based verification techniques imitate this intuitive approach by identifying and comparing rain features in the forecast and observed fields, often using a pattern recognition methodology. By focusing on

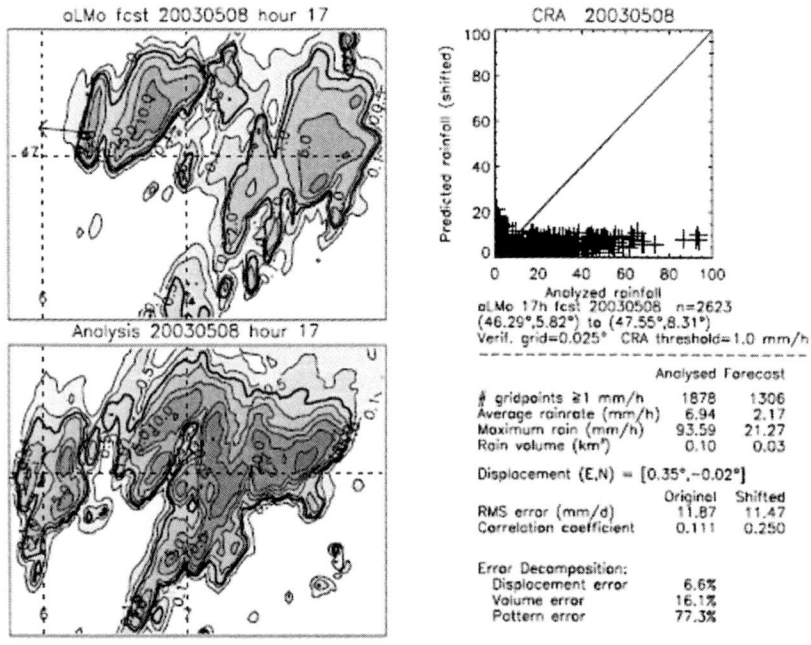

Fig.10. CRA verification of the COSMO forecast shown in Fig. 5. The red arrow in the upper left panel indicates that the best-fit of the forecast to the observations is made by translating the forecast approximately 30 km to the west

the properties of larger objects, the fine-scale errors take on lesser importance. Like most scale-dependent verification techniques, the object-based techniques require observations to be on the same grid as the forecast.

One of the early object-based approaches was the contiguous rain area (CRA) technique of Ebert and McBride (2000), in which a rain threshold is applied to identify overlapping or nearby entities in the forecast and observed fields. The entities are matched by spatially translating the forecast field over the observed field until a best-fit criterion is met and the properties of the matched entities are then compared. The total error can be decomposed into contributions from location, volume and pattern error.

Tartaglione et al. (Chap. 17 in this book) apply this methodology to precipitation verification over Cyprus. To see how the CRA verification compares to the neighborhood and scale decomposition approaches, Fig. 10 shows results for the COSMO forecast. According to the error decomposition, the majority of the error was due to differences in fine scale pattern, with only about 7% being due to incorrect location of the rain area.

A more sophisticated pattern recognition algorithm has recently been developed by Davis et al. (2006). Now called Method for Object-based Diagnostic Evaluation (MODE), it uses a convolution threshold approach to first identify objects in forecast and observed fields. The properties of the objects (e.g., location, area, shape, orientation, texture, etc) are then input to a fuzzy logic algorithm that both merges nearby objects in a scene and matches them between the forecast and observations. Verification consists of quantifying the differences in the properties of matched forecast and observed objects. The user can assign different weights to these properties in the fuzzy merging/ matching algorithm in order to emphasize certain important aspects of the forecast, for example, rain location or maximum intensity.

An image processing approach that has recently applied to spatial verification is morphing. Instead of trying to directly match objects in the forecast and observed fields, morphing distorts the forecast field until it optimally matches the observations. The 2D fields of distortion vectors and bias of the phase-corrected forecast give information about the forecast error. Application of morphing to precipitation verification is made difficult by the fact that rain features may exist in the forecast but not in the observations and visa versa. Recently, Keil and Craig (2007) proposed a forecast quality measure (FQM) that combines information about the displacement and amplitude errors. The distortion vectors are computed using a pyramidal matching algorithm where possible and, where no match can be found, an amplitude error is computed as the squared difference between the two fields. The FQM is the sum of the two normalized errors and reflects their subjective evaluation of forecast quality.

Cluster analysis also derives from the science of image processing and is a natural approach for associating pixels into objects in a high resolution rainfall grid, yet this strategy has only recently been used for verifying forecasts. In the verification method of Marzban and Sandgathe (2006, 2007) the forecast and observations are combined into a single field. K-means clustering is used to group pixels into k clusters based on their location and intensity and these clusters are further iteratively grouped using hierarchical agglomerative clustering. As the number of clusters is varied from k to 1, essentially increasing the spatial scale, the relative population of forecast and observed pixels in each cluster determines whether it is classified as a hit (forecast pixels between 20% and 80% of total), miss, or false alarm. These enable the calculation of categorical verification scores such as the threat score.

An object-based verification approach that assesses the structure of forecast rainfall in a pre-defined region such as a river basin is the Structure Amplitude Location (SAL) method of Wernli et al. (2008). As implied by the name, this approach compares the area mean structure, amplitude and location of threshold-defined precipitation objects in the forecast and observed fields, but does not attempt to match them. This approach is quite intuitive and computationally simple. Instead of a single number this method provides three: S, A and L (normalized structure, amplitude and location errors). An example of a SAL verification is given in Sect. 16.5.2.

16.5 Stratification

It is arguable whether QPF quality can be synthesized into one single number since one might be interested both in the forecast system's ability to predict the occurrence of rain, as well as how skillful it is in forecasting heavy rain. One might suspect that the performance in winter and in summer could be different, or that, for instance, model performance in anticyclonic conditions may differ from that in a vigorous northerly flow. These differences again may depend on the geographical location, especially with respect to the presence of a land-sea border or mountains. This kind of differentiated evaluation is achieved by appropriately stratifying the verification data set. If stratification into relatively homogeneous subsamples is not performed then the verification results may be artificially high (Hamill and Juras 2006). For example, a model may appear to perform well when it is, in fact, only differentiating winter and summer regimes. This is hardly useful. Various examples are presented for which differentiation with respect to event intensity, seasons, time of the day, geographical regions and weather types were applied, in order to illustrate the potential of stratification to unmask systematic model errors.

16.5.1 Seasonal, geographical and temporal stratification

Schubiger et al. (2006) present highlights of the comprehensive verification suite of COSMO, the operational NWP model of MeteoSwiss. Differentiation of rain intensity shows that occurrence of rain, or light rain, is generally overestimated, while COSMO tends to overestimate heavy rain over the mountains and underestimate it over the flatter Swiss Plateau. Averaging the diurnal cycle over a period of time effectively shows the forecast bias as a function of the hour of the day. Figure 11 shows such an averaged diurnal cycle for the months June and July 2006. In addition to singling out the hours of the day, model precipitation is verified separately for mountain stations (station height > 1500masl) and stations located over the Swiss Plateau (station height < 800masl). Given that a good part of the convective activity in the warm season consists of thermal convection in the mountains, this verification nicely isolates and reveals the problem that convection is triggered too early in the model, a misbehavior that was somewhat mitigated but not eliminated with a modified parameterization scheme for deep convection. More information on COSMO shortcomings were found looking at mountain and lowland stations separately (Schubiger et al. 2006).

Ebert et al. (2007) evaluate near-real-time satellite-derived QPE and NWP QPF on a global scale. They find that their performances are highly dependent on the rainfall regime and essentially opposed to each other, i.e., that satellite-derived QPE performs best in summer and at lower latitudes, whereas NWP has greatest skill in winter and at higher latitudes. Again, Ebert et al. (2003) report global NWP model QPF ETS values in the range of 0.4–0.5 in winter when synoptic weather is prevailing, while ETS values drop to 0.3 in summer when convective weather is predominant.

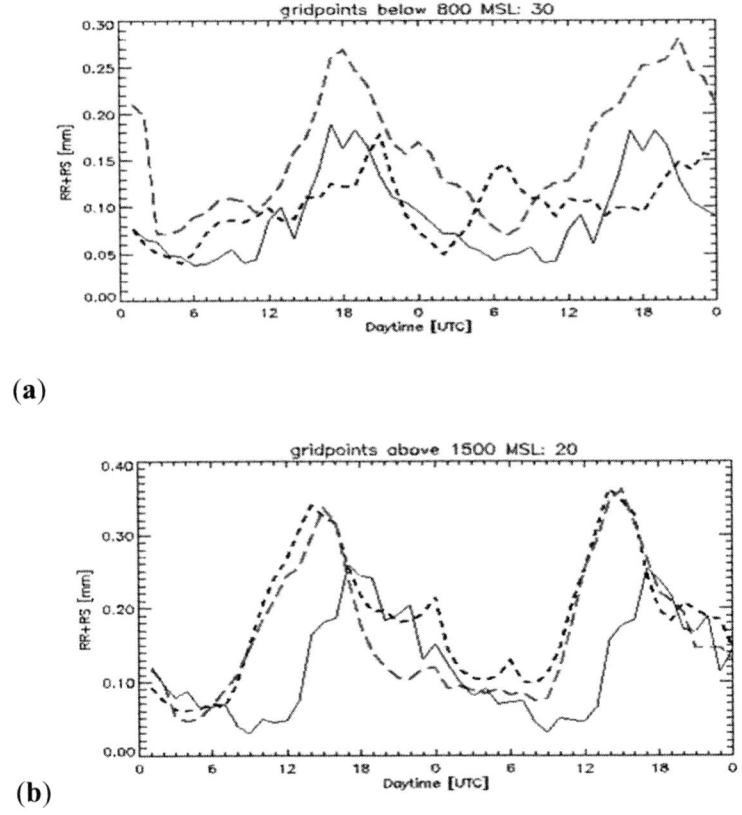

Fig. 11. Verification of the average diurnal cycle of precipitation (in mm) of COSMO forecasts for June and July 2006 for Swiss Plateau (<800masl, panel **a**) and Mountain stations (>1500masl, panel **b**). Continuous lines denote observations, short-dashed lines operational and long-dashed lines in grey a modified parameterization scheme for cumulus convection which is able to somewhat mitigate the early onset of convective precipitation in the mountains (courtesy F. Schubiger and S. Dierer, MeteoSwiss)

16.5.2 Weather-type dependent stratification

Monthly, seasonal and annual statistical verifications are limited in that their performance is judged over the whole spectrum of weather types the atmosphere can produce. The danger is that they can mask differences in forecast quality when the data, even in terms of flow regimes, are not homogeneous. Further, they can bias the results toward the most commonly sampled regime (for example days with no severe weather). A weather situation-dependent classification is another means by which stratification can be constructed. Rossa et al. (2003, 2004) have used the Schuepp Wetterlageneinteilung (Wanner et al. 1998) to perform a stratified COSMO QPF verification against QPE derived from the Swiss radar network (SRN) for years 2001 and 2002.

Zala and Leuenberger (2007) updated it for 2006 using a 'home made' classification into 11 classes comprising low flow configurations (cyclonic, anticyclonic and flat pressure distributions) and stronger flow configurations subdivided into the eight main wind directions.

Looking at the overall, unstratified, data set (Fig. 12) one is led to think that the COSMO 24 h QPF accumulations (forecast range +6 to +30 h) are quite decent. Bias values are smaller than 1 millimeter per day for large parts of the domain covered by the SRN, whereas the wet bias stays moderate even on some mountain peaks. There is a slight dry bias on the northern Swiss Plateau. However, looking at the various weather classes one can appreciate very significant differences in QPF quality in terms of precipitation bias. The most notable systematic behavior arises from the model's difficulty to partition orographic precipitation adequately between the up- and downwind side in that the upwind side generally receives too much and the lee side too little precipitation (Fig. 13a). This is especially true for the model version with instant fall-out of rain once this latter is formed (diagnostic precipitation scheme). This problem is somewhat mitigated, but not eliminated, with the introduction of the so-called prognostic precipitation scheme, which is capable of transporting formed raindrops with the wind. The most dramatic model error appears to occur in situations of southwesterly flow (Fig. 13b), when the model exhibits a widespread and quite marked dry bias over the Swiss Plateau and a portion of the northern foothills of the Alps, while retaining overestimation on the upwind side of part of the orography. In situations with northerly flow, including northwest and northeast, overestimation is substantial, while the dry bias over the Swiss Plateau is still there.

Jenkner et al. (2008a) construct a stratification based on the dynamic identification of distinct flow regimes. This is done by identifying upper-level streamers of potential vorticity (PV) and classifying them with respect to the orientation of their axis. As an example Fig. 14 shows the quantile-based Peirce skill score (PSS, 80% quantile, Jenkner et al. 2008b) of cases in which southwest-northeastward tilted PV streamers propagate past the Alps for two longitudinal ranges. Days with streamers upstream of the Alps (class

LC1_2, streamer axis between 10°W and 0°, panel a) are separated from days with streamers downstream of the Alps (class LC1_4, streamer axis between 10°E and 20°E, panel b). The former induce a southwesterly flow over Switzerland whereas the latter cause a northerly flow. The PSS identifies that the pixel-by-pixel matching of the COSMO re-forecast for LC1_2 (78 days) is low over the Swiss Plateau, while it is significantly higher for the class LC1_4 (51 days).

The SAL verification (Fig. 15) adds considerable information revealing COSMO's tendency to underforecast precipitation in cases of approaching troughs (panel a), both in quantity and areal extension. After the trough axis has passed the Alps (panel b), the model overforecasts precipitation with somewhat lesser error in terms of its structure.

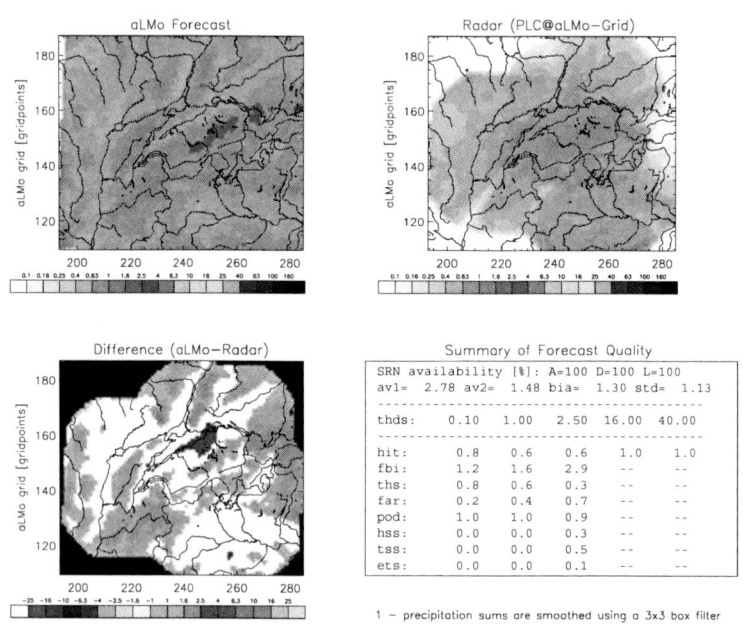

Fig. 12. Verification of operational COSMO precipitation forecasts (*upper left panel*, forecast range +06 h – +30 h) against the Swiss Radar Network (*upper right panel*) for the climatic year 2005. Shading denotes average daily precipitation in mm/24 h in a log scale (0.1, 0.16, 0.25, 0.4, 0.63, 1.0, 1.6, 2.5, 4.0, 6.3, 10, 16, 25 etc, darkest shading 10–16mm/24 h). Lower left panel denotes the average daily bias COSMO daily averaged QPF (white areas are within –1 and 1mm/24 h, darker areas denote a dry bias, lighter areas a wet bias, steps from 1mm/24 h up as in other plots), while the statistical scores are evaluated on all grid points

Chapter 16 - Verification of quantitative precipitation forecasts 445

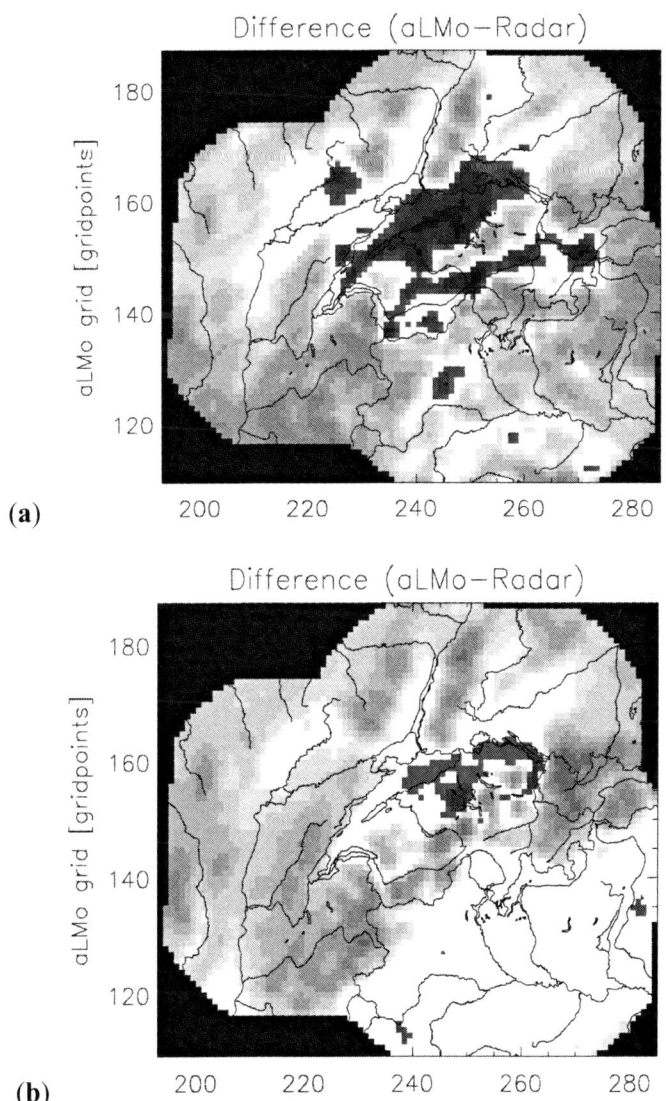

Fig. 13. As the bias field in Fig. 12, but weather classes 'southwest' (panel **a**, 50 days) and 'northwest' (panel **b**, 20 days)

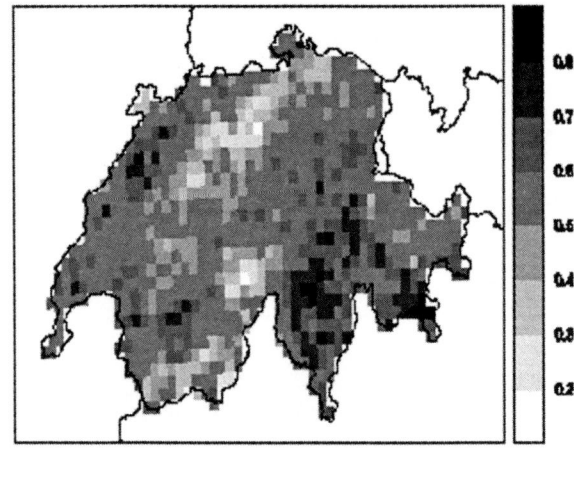

(a)

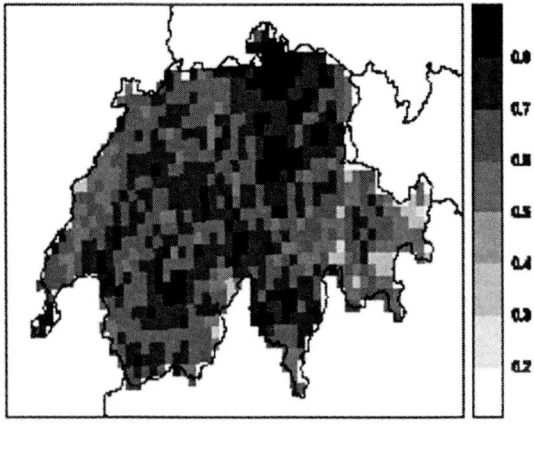

(b)

Fig. 14. Quantile-based Peirce Skill Score (PSS) for the 80% quantile in dependence of the flow classes LC1_2 (panel **a**, 78 days) and LC1_4 (panel **b**, 51 days, see text for explanation). The PSS measures how well the COSMO QPF match the observations for every individual pixel (courtesy J. Jenkner, ETH Zurich)

Chapter 16 - Verification of quantitative precipitation forecasts 447

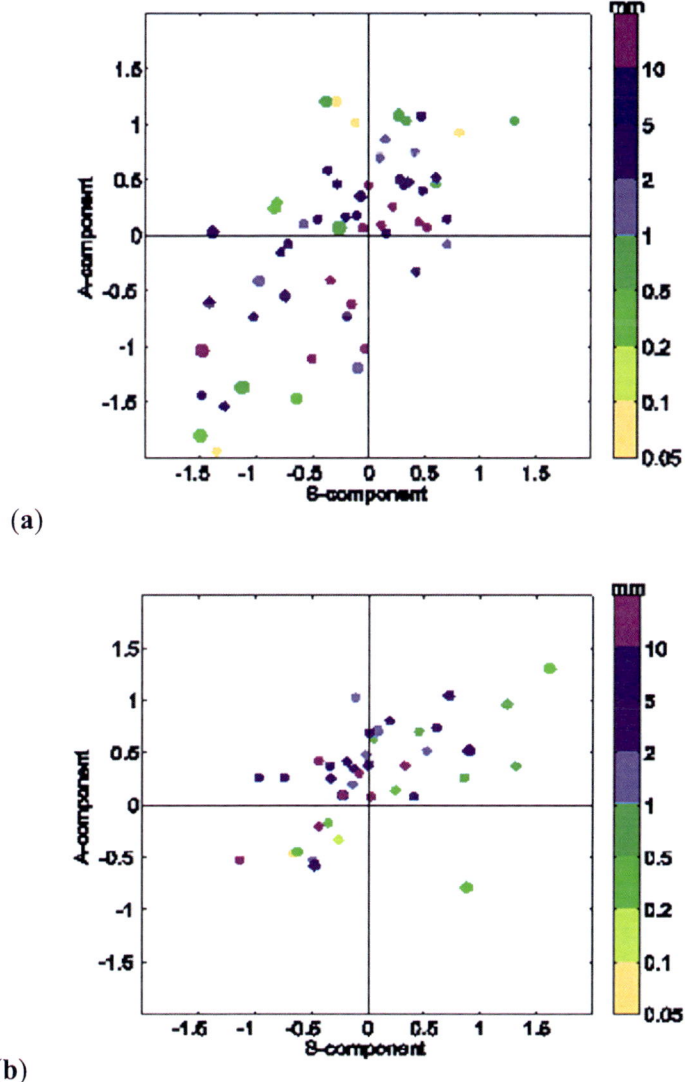

Fig. 15. SAL verification (Wernli et al. 2008) for the two flow regimes displayed in Fig. 14 for the area of Switzerland. The horizontal axis denotes how well the model matches the structure of the precipitation areas, the vertical axis how well it matches the rainfall amount, while the size of the dots denote average positioning errors. The color scale denotes the daily precipitation accumulations in mm/d for the days attributed to this flow regime (courtesy J. Jenkner, ETH Zurich)

16.6 Which verification approach should I use?

The verification approach that one should choose will depend in part upon the observations available for verifying the forecasts. As shown in Sect. 16.2 and discussed in greater detail by Tartaglione et al. (Chap. 17 in this book), the nature and accuracy of the 'truth' data have a profound effect on the verification results. If one has only rain gauge data available then the choices for verification approaches are limited to the traditional metrics and a few of the neighborhood techniques. We advocate using merged gauge-radar QPE where possible to take advantage of the additional spatial information available from these analyses and help to 'prove' the improvements in the new higher resolution models.

The standard continuous and categorical verification statistics computed from point match-ups are well understood and have been used for many years. In most cases it is advisable to continue computing such statistics, especially if a long time series of verification results is available and one wants to compare the accuracy of a new forecast system to that of an older system. However, if the resolution of the new forecast has been increased then the double penalty problem may lead to poorer verification results, even if one intuitively feels that the forecast is better. A more diagnostic evaluation using spatial verification methods may be desirable, especially if verifying data are available on a grid, say from radar or gauge analyses.

The neighborhood verification approach is useful when the forecasts are made at high resolution and it is unreasonable to expect a good match with the observations at the finest resolution. For verifying model forecasts at scales the model may be expected to resolve, the methods that compare against neighborhoods of observations may be more useful. If the aim is to evaluate the accuracy of the forecast for any given point of interest, it is better to use methods that compare forecasts to the observation in the center of the neighborhood. Among the neighborhood methods described in Sect. 16.3.1, the upscaling method is appropriate for users who wish to know if the rain amount is correct, for example, hydrologists using NWP forecasts to predict catchment rainfall and model developers evaluating the water balance of a model. The multi-event contingency table method is especially good for evaluating high resolution model output that may lead to advice and warnings for specific locations. NWP model developers and users can use the fractions skill score to determine at which scales the model has useful skill.

Scale decomposition methods separate the errors by scale, unlike neighborhood methods that filter out smaller scales. The scale decomposition methods are good for investigating the source of forecast errors when they are caused by processes occurring on different scales (for example, cloud-scale processes or large-scale advective processes). When the goal is to know whether a model's precipitation field resembles observed rainfall in a structural sense, then computing the multi-scale statistical properties is a sensible way to proceed. The SAL method is philosophically similar, but applied to objects rather than pixels.

Object-based verification approaches represent rain features as objects and are, therefore, quite intuitive. Many of these techniques give practical information about forecast quality such as location and amplitude errors. These methods tend to be more complex than other methods and also involve the choice of one or more parameters (the threshold used to define objects, for example) to which the method may be quite sensitive. Object-based approaches work well for well-defined rain areas appearing in both the forecast and observations (e.g., mesoscale convective systems, frontal systems and daily rainfall accumulations) but they do not handle noisy rain fields very well.

Independently of the chosen approach, appropriate stratification of verification data sets can help to isolate specific problems in the QPF systems. Hydrologists concerned with river catchments in mountainous terrain, for example, may well be interested in knowing in what regions of the forecast domain a QPF exhibits systematic errors.

References

Atger F (2001) Verification of intense precipitation forecasts from single models and ensemble prediction systems. Nonlinear Proc Geoph 8:401–417

Barnes LR, Gruntfest EC, Hayden MH, Schultz DM, Benight C (2007) False alarms and close calls: A conceptual model of warning accuracy. Weather Forecast 22:1140–1147

Bougeault P (2002) WGNE survey of verification methods for numerical prediction of weather elements and severe weather events. CAS/JSC WGNE Report No. 18, Appendix C. http://www.wmo.ch/web/wcrp/documents/wgne18rpt.pdf

Briggs WM, Levine RA (1997) Wavelets and field forecast verification. Mon Weather Rev 125:1329–1341

Casati B, Ross G, Stephenson DB (2004) A new intensity-scale approach for the verification of spatial precipitation forecasts. Meteorol Appl 11:141–154

Cherubini T, Ghelli A, Lalaurette F (2002) Verification of precipitation forecasts over the Alpine region using a high-density observing network. Weather Forecast 17:238–249

Davis C, Brown B, Bullock R (2006) Object-based verification of precipitation forecasts. Part I: Methods and application to mesoscale rain areas. Mon Weather Rev 134:1772–1784

Ebert EE (2008) Fuzzy verification of high resolution gridded forecasts: A review and proposed framework. Meteorol Appl (in press)

Ebert EE, McBride JL (2000) Verification of precipitation in weather systems: Determination of systematic errors. J Hydrol 239:179–202

Ebert EE, Damrath U, Wergen W, Baldwin ME (2003) The WGNE assessment of short-term quantitative precipitation forecasts. B Am Meteorol Soc 84:481–492

Ebert EE, Janowiak JE, Kidd C (2007) Comparison of near-real-time precipitation estimates from satellite observations and numerical models. B Am Meteorol Soc 88:47–64

de Elia R, Laprise R, Denis B (2002) Forecasting skill limits of nested, limited-area models: A perfect-model approach. Mon Weather Rev 130:2006–2023

Finley JP (1884) Tornado predictions. American Meteorological Journal 1:85–88

Germann U, Berenguer M, Sempere-Torres D, Salvadè G (2006) Ensemble radar precipitation estimation – a new topic on the radar horizon. In: Proceedings 4th European Conference on Radar in Meteorology and Hydrology. 18–22 September 2006, Barcelona, Spain. http://www.erad2006.org

Göber M, Wilson CA, Milton SF, Stephenson DB (2004) Fairplay in the verification of operational quantitative precipitation forecasts. J Hydrol 288:225–236

Hamill TM, Juras J (2006) Measuring Forecast Skill: Is it Real Skill or is it the Varying Climatology? Q J Roy Meteor Soc 132:2905–2923

Harris D, Foufoula-Georgiou E, Droegemeier KK, Levit JJ (2001) Multiscale statistical properties of a high-resolution precipitation forecast. J Hydrometeorol 2:406–418

Hense A, Adrian G, Kottmeier CH, Simmer C, Wulfmeyer V (2003) Priority Program of the German Research Foundation: Quantitative Precipitation Forecast. Research proposal available at http://www.meteo.uni-bonn.de/projekte/SPPMeteo/reports/SPPLeitAntrag_English.pdf

ICP (2007) Spatial Forecast Verification Methods Intercomparison Project (ICP). http://www.ral.ucar.edu/projects/icp/index.html

Jenkner J, Dierer S, Schwierz C, Leuenberger D (2008a) Conditional QPF verification using synoptic weather patterns – a 3-year hindcast climatology (in preparation)

Jenkner J, Dierer S, Schwierz C, Leuenberger D (2008b) Quantile–based short–range QPF evaluation over Switzerland (to be submitted)

Jolliffe IT, Stephenson DB (2003) Forecast verification. A practitioner's guide in atmospheric science. Wiley and Sons Ltd, 240 pp

JWGV (2004) Forecast verification – Issues, methods and FAQ. http://www.bom.gov.au/bmrc/wefor/staff/eee/verif/verif_web_page.html

Keil C, Craig GC (2007) A displacement-based error measure applied in a regional ensemble forecasting system. Mon Weather Rev 135:3248–3259

Leuenberger D (2005) High-resolution radar rainfall assimilation: Exploratory studies with latent heat nudging. Ph.D. thesis, ETH Zurich, Switzerland, Nr. 15884. http://e-collection.ethbib.ethz.ch/cgi-bin/show.pl?type=diss&nr=15884

Marzban C, Sandgathe S (2006) Cluster analysis for verification of precipitation fields. Weather Forecast 21:824–838

Marzban C, Sandgathe S (2007) Cluster analysis for object-oriented verification of fields: A variation. Mon Weather Rev (in press)

Mittermaier MP (2007) Improving short-range high-resolution model precipitation forecast skill using time-lagged ensembles. Q J Roy Meteor Soc 133:1–19

Murphy AH (1996) The Finley affair: A signal event in the history of forecast verification. Weather Forecast 11:3–20

Nurmi P (2003) Recommendations on the verification of local weather forecasts. ECMWF Tech. Memo 430:18.t http://www.ecmwf.int/publications/library/ecpublications/_pdf/tm430.pdf

Oberto E, Turco M, Bertolotto P (2006) Latest results in the precipitation verification over Northern Italy. COSMO Newsletter 6:180–184

Richardson DS (2000) Skill and relative economic value of the ECMWF ensemble prediction system. Q J Roy Meteor Soc 126:649–667

Roberts NM, Lean HW (2007) Scale-selective verification of rainfall accumulations from high-resolution forecasts of convective events. Mon Weather Rev (in press)

Rossa A, Arpagaus M, Zala E (2003) Weather situation-dependent stratification of precipitation and upper-air verification of the Alpine Model (aLMo). COSMO Newsletter 3:123–138

Rossa AM, Arpagaus M, Zala E (2004) Weather situation-dependent stratification of radar-based precipitation verification of the Alpine Model (aLMo). ERAD Publication Series 2:502–508

Rossa et al. (2005) The COST 731 Action MoU: Propagation of uncertainty in advanced meteo-hydrological forecast systems. http://www.cost.esf.org

Schaake JC, Hamill TM, Buizza R, Clarke M (2007) HEPEX, the Hydrological Ensemble Prediction Experiment. B Am Meteorol Soc 88:1541–1547

Schubiger F, Kaufmann P, Walser A, Zala E (2006) Verification of the COSMO model in the year 2006, WG5 contribution of Switzerland. COSMO Newsletter 6:9

Stamus PA, Carr FH, Baumhefner DP (1992) Application of a scale-separation verification technique to regional forecast models. Mon Weather Rev 120:149–163

Stanski HR, Wilson LJ, Burrows WR (1989) Survey of common verification methods in meteorology. World Weather Watch Tech. Rept. No.8, WMO/TD No.358, World Meteorological Organization, Geneva, Switzerland, 114 pp

Stephenson DB (2000) Use of the 'odds ratio' for diagnosing forecast skill. Weather Forecast 15:221–232

Stephenson DB, Casati B, Wilson CA (2007) The extreme dependency score: A new non-vanishing verification measure for the assessment of deterministic forecasts of rare binary events. Meteorol Appl (in press)

Steppeler J, Doms G, Schättler U, Bitzer H-W, Gassmann A, Damrath U, Gregoric G (2003) Meso-gamma scale forecasts using the non-hydrostatic model LM. Meteorol Atmos Phys 82:75–96

Wanner H, Salvisberg E, Rickli R, Schuepp M (1998) 50 years of Alpine Weather Statistics (AWS), Meteorol Z N.F. 7:99–111

Wernli H, Paulat M, Hagen M, Frei C (2008) SAL – a novel quality measure for the verification of quantitative precipitation forecasts, Mon Weather Rev (submitted)

Wilks DS (2006) Statistical methods in the atmospheric sciences. an introduction. 2nd edn. Academic Press, San Diego, 627 pp

Wilson C (2001) Review of current methods and tools for verification of numerical forecasts of precipitation. COST717 Working Group Report on Approaches to verification. http://www.smhi.se/cost717/

Yates E, Anquetin S, Ducrocq V, Creutin J-D, Ricard D, Chancibault K (2006) Point and areal validation of forecast precipitation fields. Meteorol Appl 13:1–20

Zala E, Leuenberger D (2007) Update on weather-situation dependent COSMO-7 verification against radar data. COSMO Newsletter 7:1–4

Zawadzki I (1973) Statistical properties of precipitation patterns. J Appl Meteorol 12:459–472

Zepeda-Arce J, Foufoula-Georgiou E, Droegemeier KK (2000) Space-time rainfall organization and its role in validating quantitative precipitation forecasts. J Geophys Res 105:10129–10146

17 Objective verification of spatial precipitation forecasts

Nazario Tartaglione[1], Stefano Mariani[2,3], Christophe Accadia[4], Silas Michaelides[5], Marco Casaioli[2]

[1] Department of Physics, University of Camerino, Italy
[2] Agency for Environmental Protection and Technical Services (APAT), Rome, Italy
[3] Department of Mathematics, University of Ferrara, Italy
[4] EUMETSAT, Darmstadt, Germany
[5] Meteorological Service, Nicosia, Cyprus

Table of contents

17.1 Introduction 453
17.2 The problem of observations in objective verification 456
17.3 Use of rainfall adjusted field for verifying precipitation 458
17.4 Statistical interpretation of position errors as derived by object-oriented methods 461
17.5 Assessing the difference between CMS indices from two different forecast systems 466
17.6 Conclusions 467
References 469

17.1 Introduction

Precipitation is surely one of the most important meteorological variables because of practical interest from the general public, hydrologists, power plant managers and other economic actors. Nevertheless, whoever works with numerical weather prediction (NWP) models knows how imperfect rainfall forecasts can be, especially at small scales. However, forecasters and NWP modelers, who care about the quality of their products, continuously strive to improve them.

Moreover, considering the importance of NWP models for a wide range of end-users, their accuracy in forecasting precipitation must be verified in order to determine their quality and value. As recalled by Doswell (1996), although it should be obvious, a forecast not verified is a worthless forecast. Brier and Allen (1951) proposed three main reasons for forecast verification: administrative, economic and scientific. Often these three reasons go together. Indeed, it is necessary to communicate verification results in an effective way to the end-users,

The first reason is the need for monitoring (e.g., see the ECMWF operational monitoring available online at http://www.ecmwf.int/products/forecasts/guide/Monitoring_the_ECMWF_forecast_system.html) an operational forecasting system in order to determine how well the system is performing (also considering changes in parameterization schemes, assimilation methods, configuration, etc.) and to guide possible future investments in the updating of weather forecast systems. The second reason is linked to the assessment of benefits of a correct forecast, from an economic point of view to decision-making activity or to particular end-user needs. Getting a good quality weather forecast is useful for civil protection, flooding risk management and agriculture. A last reason (but not the least!) to verify forecasts, involves examination of the forecast and the corresponding observations. Murphy et al. (1989) and Murphy and Winkler (1992) called this verification activity 'diagnostic'. It allows for the evaluation of model outputs with respect to observations. Verification activities provide, this way, valuable feedback to operational weather forecasters, giving indications on how to improve NWP models. Indeed, quantitative precipitation forecast (QPF) skill is considered as an indicator of the general capability of a NWP model to produce a good forecast (Mesinger 1996).

The standard verification techniques are based on the comparison of model outputs with observations (typically from rain gauges) valid at the same time and location. Detailed descriptions of such methods can be found in many books, such as Wilks (1995) and Jolliffe and Stephenson (2003). However, due to the difficulty of modeling the atmospheric processes related to rainfall (having, sometimes, short decorrelation lengths of about 5–20 km and high variability in space and time), it is not surprising that space-time distribution of a modeled precipitation field shows some differences from the real precipitation one. The resulting statistics can unjustly penalize high resolution models that make realistic forecasts of rainfall patterns but are shifted with respect to observations (Mass et al. 2002; Weygandt et al. 2004). In fact, high resolution models can actually reproduce precipitation patterns more accurately than coarse resolution ones, but they are often prone to displacement errors due to a variety of reasons (e.g. stochastic behavior of the atmosphere, lack of adequate initialization, difficulty to model microphysical processes), especially when convective precipitation is involved. Another important aspect of the verification activity is that verification results can depend upon the reliability of the observations. For instance, rain gauges give only point measurements, whereas areal estimates are needed to verify forecasts. Other ground-based or space-based sensors can give estimates of the actual precipitation field at different spatial scales, but they may also be affected by large errors. Consequently, both rain gauges and sensors suffer from some limitations. We shall shortly illustrate some of these limitations in the next section. Hence, we have to treat the verification of precipitation fields with much more care than the verification of other 'well behaved' meteorological variables, such as pressure and temperature.

Visual verification could provide a valid representation of model performance, but it is time-consuming and personal biases may affect the model evaluation. An objective technique verifying precipitation events, much in the way a human would in a subjective evaluation, would likely produce a more reliable assessment of model performance.

Differently from subjective verification, which is insufficient to verify many events, objective verification allows evaluating weather forecast systems and assessing variability on many time and space scales. The aim is to judge model performance taking into account the complexity of the problem, which is possible only with a lot of events. Objective verification is an on-going field of research and only some aspects can treated in a single chapter.

Several new verification techniques have been recently developed by the meteorological community. These new methods involve, for instance, the use of the Fourier spectra analysis (e.g., Harris et al. 2001; Zepeda-Arce et al. 2000) or an 'object-oriented' approach (e.g., Ebert and McBride 2000; Casati et al. 2004). We shall show a couple of applications of an object-oriented method, in particular the contiguous rain area (CRA) analysis (Ebert and McBride 2000).

The CRA technique searches for disagreement between forecast and observed patterns. The displacement disagreement is obtained shifting the forecast rainfall pattern over the observed pattern until a 'best-fit criterion' is satisfied. This criterion originally was the mean square error (MSE), especially when verification is about the forecast ability in matching the field maxima (e.g., Ebert and McBride 2000; Mariani et al. 2005), even though some authors (e.g., Tartaglione et al. 2005; Grams et al. 2006) have suggested also the correlation as best-fit criterion. Hereafter, with displacements we shall mean all those satisfying the 'best-fit criterion'.

The chapter is organized as follows. Section 17.2 is completely dedicated to discuss the issues of accuracy and representativeness of the rainfall observations. The authors feel that there is a need to care for observations in addition to the forecasts. This is particularly true for precipitation. In Sect. 17.3 an example is given of how observations can affect verification outcomes. Section 17.4 discusses the application of the CRA technique to a large number of precipitation events in order to define statistically robust and objective evaluation of location errors. Evaluation is performed in terms of two points of view: absolute (evaluation against observations) and comparative verification (model evaluation against observations and inter-model comparison). An example of such an analysis is shown in Sect. 17.5. Finally, conclusions are drawn in Sect. 17.6.

17.2 The problem of observations in objective verification

Evaluation of gridded precipitation forecasts by a NWP model is usually performed by using precipitation estimates. Such estimates are obtained by means of rain gauge networks or through measurements of radar reflectivity. Radars can be ground-based or spaceborne, such as the Precipitation Radar (PR) onboard the Tropical Rainfall Measuring Mission (TRMM), in orbit since 1997.

The accurate estimate of observed precipitation is very important in the verification process and searching for the 'true' precipitation is like a 'Holy Grail' search, especially when we wish to know the precipitation quantity fallen over an area. This problem becomes clearer, bearing in mind that gridded model outputs represent area averages. However, rain gauge measurements could be not fully representative of reality. For example, Nystuen (1999) showed that different types of gauges recorded different amounts of rainfall. Moreover, gauge estimates can also depend on type of precipitation, as well as on general meteorological conditions (wind, temperature, etc.). Errors in observations may affect scores and should be taken in account whenever possible (Bowler 2007).

Even working under a hypothesis of 'perfect' rain gauges, the point information has to be spatially interpolated on the model grid for comparison. The interpolation scheme often alters the precipitation values (e.g., Skok and Vhrovec 2006) and, consequently, the verification results.

The latter argument is also valid when we have to remap precipitation forecasts on a common verification grid, as needed for intercomparison of various models. In such a case, the interpolation schemes can modify the skill scores as well as the total amount of rainfall over the domain (Accadia et al. 2003). For instance, the bilinear interpolation scheme, which is usually employed by the European Centre for Medium-range Weather Forecast (ECMWF) does not conserve the total water amount when it is applied, differently from the remapping scheme (Accadia et al. 2003; Baldwin 2000). This latter method, which is operationally used at the National Centres for Environmental Prediction (NCEP) is indeed able to conserve, for a desired degree of accuracy, the total precipitation forecast of the native grid.

Even the climatology of areas where verification is performed affects statistical results (Hamill and Juras 2006) and should be taken into account as well.

Ground-based radar, although giving more detailed spatial information about rainfall, also suffers from several kinds of problems such as volumetric error, ground clutter, calibration errors, etc. Radar data upscaling, often performed before using radar data in verification, may introduce additional uncertainty. Joss et al. (2003) discussed the problem of upscaling of radar data, noting that the adopted upscaling strategies give different results.

Satellites, in theory, offer the perfect complement to ground observations in terms of spatial coverage (Ebert et al. 2007). However, the available precipitation estimates coming from passive microwave imagers and from infrared sensors on polar orbit are still prone to large errors and insufficient temporal sampling, providing valuable information only at seasonal scales and over large areas. Thus, such kind of observations is useful for climatological studies (e.g. Smith, 1988) but is outside of the scope of short-range forecast model verification. Achieving synergy between the spaceborne precipitation radar and microwave imagers offers the possibility to develop future generation operational monitoring capabilities (Bidwell et al. 2005). This concept is at the base of the TRMM mission and of the Global Precipitation Mission (GPM). The main limit of radar observations from space is generally is the limited swath (e.g., 200 km for TRMM PR).

Despite this, TRMM PR can be employed to range-adjust ground-based radar when the adjustment by means of rain gauges is not possible or is of limited use (e.g., over the sea). For example, the VOLTAIRE project (VOLTAIRE 2006 and Chap. 20 in this book) has shown that rain gauges are not suitable to range-adjust ground radar data over the island of Cyprus: the adjustment has been successfully performed using TRMM PR observations (Gabella et al. 2006). Tartaglione et al. (2006) have shown that observations adjusted by TRMM PR radar were more physically reliable than the original ground-radar ones, deducing that range-adjustment (even with TRMM precipitation radar) is mandatory before performing precipitation verification Thus, range-adjustment is desirable in a verification context and could provide additional scope for a satellite mission like GPM (see Chap. 6 in this book).

17.3 Use of rainfall adjusted field for verifying precipitation

The use of radar to verify precipitation is not new (e.g., Göber and Milton 2001; Kain et al. 2005; Baldwin and Elmore 2005) as well as the range-adjustment performed to reduce volumetric errors; for example, Harrison et al. (2000) describe the post-processing steps to correct errors and mitigate problems of radar data within the Nimrod system. Other than errors associated with radar, a major trouble is the obscuration of radar beam because of mountains. In such a case, other instruments could be used to estimate rainfall. In this section, we shall describe as some problems associated with observed precipitation estimates have been treated before performing forecast verification in an experiment over the island of Cyprus and how observations can affect verification finding. Observations discussed in this section were gathered from several types of instruments, rain gauges, a ground-based radar and the precipitation radar on board TRMM.

The verification technique applied for this study is the Ebert and McBride (2000) approach, known as CRA analysis. This approach gives, as a main result, a couple of numbers that identify the horizontal displacement (in longitude and latitude) for which the best match between the forecast and the observed fields is obtained. Regardless the criterion adopted in CRA analysis, the following result: (0,0) means that no location error is present and the forecast precipitation pattern is correctly positioned.

A problem that was encountered over the Cyprus area is a southeastern obscured area (see Chap. 19 in this book), and the explicit use of radar for performing a verification process is not correct. Fortunately, the rain gauge network of the island is dense (147 rain gauges over about 5900 km^2) and the region where the radar beam is obscured is satisfactorily covered by rain gauges. Thus, the observation to be used for the verification has to be a combination of rain gauge data analysis (interpolation is performed by means the Barnes algorithm) and ground-based radar data. The combination of such data can be performed by means of the following formula:

$$TP = \frac{RP \cdot RP + GP \cdot GP}{RP + GP} \tag{1}$$

This technique to combine precipitation estimates is very simple but it was found to be suitable for the site and the events considered here. It cannot be absolutely considered as a general method to combine precipitation data. More sophisticated merging methods are discussed in the literature (e.g., Stellman et al. 2001). From Eq. (1), the total precipitation TP on each grid point is a combination of radar precipitation estimates (RP) and gauge precipitation analysis (GP).

As a rule, radar data are usually range-adjusted to mitigate the volumetric errors and such a process is performed by mean of rain gauges. Over the island of Cyprus, such a kind of adjustment does not work, most likely because of the limited spatial area of Cyprus.

Thus, someone might think to use the radar data without any form of range-adjustment. The comparison of observations including original radar data led to a large displacement error of the precipitation forecast with respect observations available, which contradicts results obtained using only rain gauges (Tartaglione et al. 2005, 2006).

Since Cyprus is covered by TRMM passages, one of aim of a recently completed European Union project, namely, VOLTAIRE (see VOLTAIRE 2006 and Chap. 20 in this book) was to compare precipitation estimates with different instruments and sensors. During this project, Gabella et al. (2006) successfully tested range-adjustment procedures of the ground radar by means of the precipitation radar borne on TRMM satellite.

In this way, we can build three datasets of observations: the first one comprises rain gauge data (GP); the second one is obtained with a composition of rain gauge and radar data which includes the composite obtained only with original radar data (RPo); and the third one with range-adjusted radar data (RPa) using Eq. (1).

These datasets can be verified applying the CRA method, using as best-fit criteria the minimization of mean square error and the maximization of correlation. Results of such verification exercise can be seen in Table1.

Table 1. CRA verification for 24 h rainfall from 06:00 UTC 5 March 2003 to 06:00 UTC 6 March 2003. The observational type column indicates the type of precipitation field used in the comparison. A CRA rain rate contour of 0.5 mm $(24\ h)^{-1}$ has been selected, except for the previous analysis with only rain gauge data (GP; Tartaglione et al. 2005) for which a CRA rain rate contour equal to 0.0 mm $(24\ h)^{-1}$ was chosen

Observational type	No. of comparing grid points	MSE (mm^2)	Correlation	CRA pattern matching criterion	Displacement [E,N] (degree)	Shifted MSE (mm2)	Shifted Correlation
GP	73	84.28	0.51	MSE	[1.17, 0.54]	33.63	0.62
				CORR	[0.27, 0.09]	78.15	0.69
RPo	446	248.4	0.28	MSE	[0.54, 0.27]	182.14	0.34
				CORR	[0.45, 0.09]	215.42	0.42
RPa	454	273.9	0.36	MSE	[0.27, 0.09]	252.10	0.41
				CORR	[0.27, 0.00]	253.10	0.43

One difficulty that arises from the precipitation verification is the size of the observation domain. In our example, rain gauge network is obviously located over an island and the observation domain is confined to one containing only the land. When radar data are available together with rain gauge data, the observation domain can be extended up to the radar range. The extension of the observation domain has the natural consequence that the forecast has to correctly hit much more points. These arguments explain one of the reasons for which rain gauge data (GP) shows lower MSE and higher correlation than the other kinds of data. It is worth noting that the position error comes about with the minimization of MSE as best-fit criterion; this is different enough to the one obtained by maximizing correlation. The latter is equal to position errors obtained using RPa data.

The verification process is based on the assumption that we are able to correctly estimate precipitation, whereas the real precipitation, especially that falling over an area, can be only roughly estimated. A better estimation should make use of all available information and that information should correctly be merged. We have shown that radar adjustment is a fundamental process in verification of precipitation fields and when it cannot be performed, radar data cannot be absolutely used to verify modeled precipitation. For the use of ground-based radar data for verifying precipitation, the range-adjustment process is mandatory, even by means of other sensors other than rain gauges; otherwise, we risk to badly verifying a good forecast system.

17.4 Statistical interpretation of position errors as derived by object-oriented methods

It is well known that precipitation verification is a multidimensional issue and a single number cannot be exhaustive about the complexity and dimensionality of verification problems: it can lead to erroneous conclusions regarding the absolute and relative quantity of the forecast systems (Murphy 1991).

Let us consider a weather forecasting system that predicts the spatial distribution, on a gridded domain, of the precipitation over an area. For the discussion in this chapter, it is not relevant how the prediction comes about. It might have been computed with a NWP model or tossing dice. Once produced, the forecast pattern obtained is moved on the observations in order to satisfy a best-fit criterion, such as minimizing the MSE, in order to compute the location error. Such operation is performed, for example, by means of the CRA approach (Ebert and McBride 2000). Observations are placed on the same gridded verification domain of forecasts, by means of an interpolation algorithm such as Barnes' approach (Barnes 1964; Koch et al. 1983). As a result of this operation, we find out a couple of numbers: one is relative to the south-north direction and the other one to the west-east direction. Those numbers indicate how much grid points we have to displace the forecast precipitation pattern in order to satisfy the best-fit criterion chosen. When such a procedure is applied to many precipitation events it is possible to obtain a matrix whose columns and rows are the displacements needed to satisfy the best-fit criterion along the east-west and along the south-north direction, respectively. The values within the matrix are the numbers of events that needed to be displaced in the west-east and south-north direction. Figure 1 shows an example of a statistical outcome from CRA analysis.

Each number in Fig. 1 represents the number of events for which a displacement of n_lon and n_lat is needed in order to satisfy the best-fit criterion. The value eight in the central position of that matrix of Fig. 1 means that for eight events the best-fit criterion was obtained without shifting the precipitation pattern.

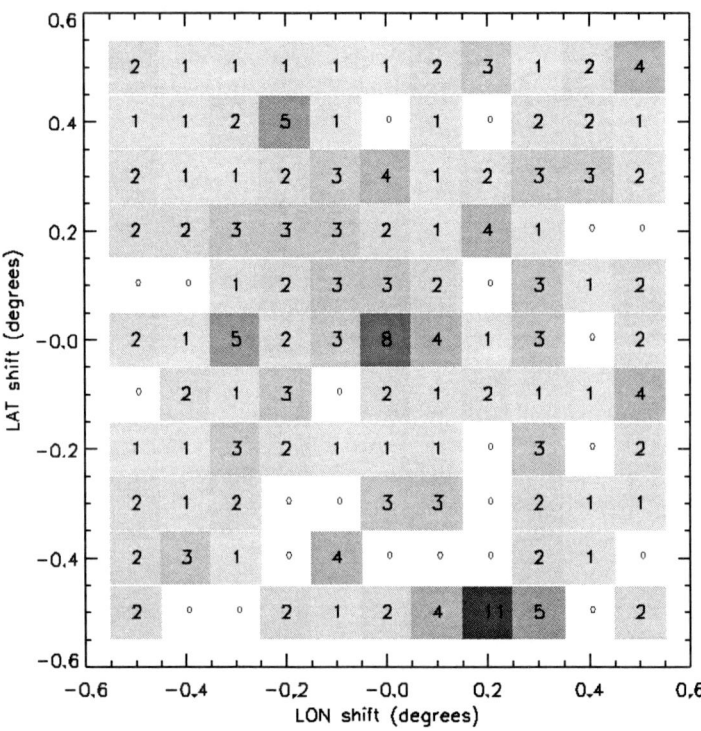

Fig. 1. This matrix shows the number of precipitation events, valid for a forecast system, in which the precipitation forecast pattern has to be zonally (columns) and meridionally (rows) shifted to satisfy a 'best-fit criterion' with respect to the observed pattern

Chapter 17 - Objective verification of spatial precipitation forecasts

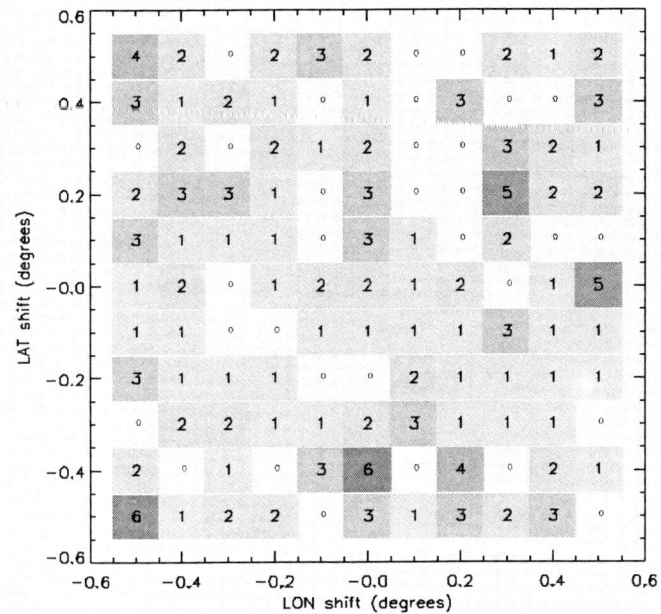

Fig. 2. As Fig. 1, but related to a second precipitation forecast system

Applying the same technique to another precipitation forecast system, we have the result shown in Fig. 2.[1]

On the one hand, it is true that the reduction of the vast amount of information from a set of forecasts and observations into a single measure can lead to misinterpretation of the verification outcomes; on the other hand, it is also true that from Figs. 1 and 2 we have a confusing framework because of too many numbers are displayed.

What information is given by numbers of the matrices shown above? Which forecast system is better? One may argue that those many numbers are not informative, at a first glance. Brier (1948) pointed out that 'the search for and insistence upon a single index' can lead to confusion; we do feel that it is a necessity to synthesize all the information, especially when we have to compare two forecast systems.

Taking into account only the events that have no shifting can be reductive, because many events can be wrong. Let us define an index,

---

[1] Figures 1 and 2 are the outcome of statistical verification of precipitation patterns forecast by two limited area models (Tartaglione et al. 2007).

which will be referred as to CRA Mean Shift index or CMS, in the following way:

$$\text{CMS} = \sum_{i=-N}^{N} \sum_{j=-N}^{N} f_{ij} \cdot w_{ij} \qquad (2)$$

so that the CMS index is proportional to the frequency f_{ij} of shifts with respect to the number of events. With this formulation, the frequency depends on the forecast system and criterion used; N is the maximum displacement (in grid point for instance) used during application of the CRA analysis; the term w_{ij} measures the distance from the center of the matrix (0, 0), which indicates no error shift; it is simply formulated in the following way:

$$w_{ij} = \sqrt{i^2 + j^2} . \qquad (3)$$

CMS goes from 0 to $N\sqrt{2}$, where N is the maximum displacement allowed; for Figs. 1 and 2, this is 5. In order to make CMS an absolute measure, it could be normalized dividing by N. Here, we shall consider the form $N\sqrt{2}$.

From a point of view of usefulness of data shown in Figs. 1 and 2, we can make two kinds of comparison: a relative one between two precipitation prediction systems and an absolute one.

The latter has to be performed allowing for a conceptual model of location errors. We start with the absolute evaluation of the numbers present on one of the matrixes of Figs. 1 and 2. What error model should we use? To give a statistical definition of what we are waiting as the 'ideal' result, let us consider our variable of interest, i.e., the number of events in which the best-fit criterion is obtained by shifting the precipitation patterns of n_lon and n_lat grid points. We will use the symbol Y to denote the observed value of that variable and X to denote the expected value. The corresponding lower cases y and x denote any possible value in the range of Y and X, respectively.

The probability density functions (pdf) over the space of all possible displacements are p(y) and p(x). Regardless the p(y) distribution is, we shall stress only on the differences between CMSs obtained from a p(y) distribution, whatever it is, and from the normal p(x) distribution. Under the hypothesis that location errors of precipitation patterns are independent, we expect a distribution of events that are normally distributed over the space of all allowed *displacements.* A perfect forecast system should have location errors distributed as a Dirac delta, i.e., the matrix of the location errors should be empty but in the position (0, 0), that is equivalent to no displacement. The CMS in such a case will be nil. However, we known that it is impossible to have a perfect forecast system, so we have to expect position errors normally distributed with a spread defined by a standard deviation; the smaller the standard deviation is, the better the forecast system should be.

From Fig. 3, it seems to be clear that a large departure from nil value of CMS is obtained from high spread bivariate normal distributions. The distribution becomes flatter by increasing the standard deviation, thus getting a uniform distribution considering the displacements' space only. In fact, if the position errors are uniformly distributed on the allowed displacements' space of Figs. 1 and 2, the correspondent CMS is about 4.2. A χ^2 test confirms the non-normality of the distributions of the events in the allowed displacements' space of Figs. 1 and 2.

In evaluating the two precipitation systems of Figs. 1 and 2, if the CMS has a value as high as 4 (the normalized value of CMS would be close to 0.8 that is equivalent to 4/N with N equal to five), this implies a poor precipitation forecast system. Position errors of precipitation

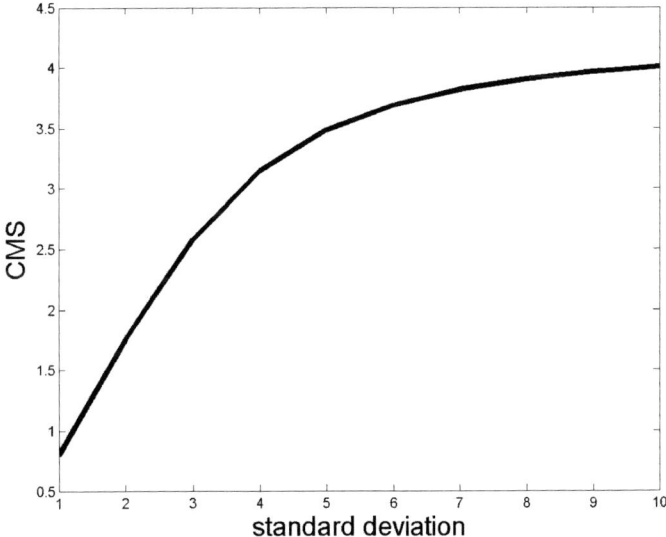

Fig. 3. The CRA Mean Shift Index, for location errors normally distributed on the space of allowed displacements like that of Figs. 1 and 2

patterns scatter almost uniformly on the allowed displacements' space. CMS of Figs. 1 and 2 are 4.11 and 4.48, respectively; consequently, it seems that position errors of the considered precipitation forecast systems are almost uniformly scattered over the domain.

17.5 Assessing the difference between CMS indices from two different forecast systems

In the previous section, we have shown that the CMS indices of the two forecast systems (hereinafter, CMS1 and CMS2) are very close to each other. A significance test is then required to assess whether the CMS difference is statistically different from zero. In other words, does the weather forecast system of Fig. 2 provide a forecast as good as the one of Fig. 1?

Since spatial errors of the considered systems are far from being normally distributed (as previously shown), it is better to use a bootstrap procedure to statistically assess the difference between the two CMS, that is, CMS2 – CMS1. For each forecast system, the bootstrap is performed by resampling the displacement errors along the west–east and south–north directions. The total number of resamples is 10000. Table 2 shows the summary statistics of the bootstrap procedure. The bootstrap standard error indicates the standard deviation of the bootstrap distribution. Figure 4 shows the distribution of differences between CMS2 and CMS1 indices.

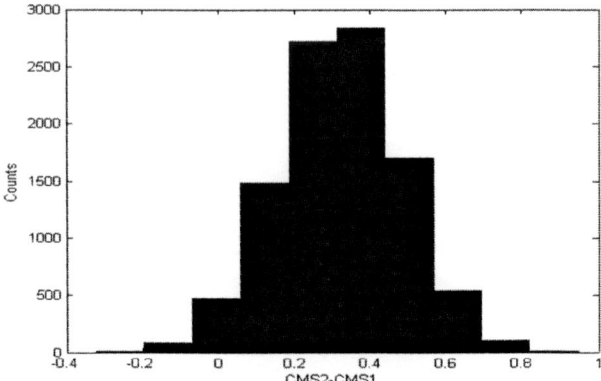

Fig. 4. Histogram of the 10000 resampling differences between CMS2 and CMS1 obtained with a bootstrap procedure

Table 2. Summary statistics of the bootstrap of 10000 resamples used to statistically assess the significance of the difference between the indices related to two precipitation forecast systems

	Observed	Bootstrap Mean	Standard Error
CMS2 – CMS1	0.370	0.320	0.165

The observed difference is 0.37. Is this number significantly different from 0? This question represents our null hypothesis. The answer obviously depends on the confidence interval chosen to test the significance of the hypothesis. For instance, the 90% bootstrap confidence interval is the interval between the 5% and 95% of the bootstrap distribution.

Significance can be deduced from Fig. 5a, whose zoom in is shown in Fig. 5b. Figure 5 shows that the location of zero difference is in the left tail of distribution, indicating that this value is unlikely to occur. The confidence interval does not cover the zero difference; the difference is significant at the 10% level. Hence, the precipitation prediction system 1 is significantly better than the precipitation prediction system 2. The choice of a larger confidence interval, for instance 95%, would reject the null hypothesis, even though it is close to the lower boundary of the confidence interval (Fig. 5b), and the two systems would not be considered as significantly different any more.

17.6 Conclusions

Many issues need to be considered in order to perform a robust rainfall forecast verification. One of the key issues is the accuracy of observations. Observations are often assumed as the 'truth' without further analysis. A quality check of observed data should be performed before using them to verify forecast precipitation, which has an areal meaning. This is not an easy task, due to the difficulty of rain gauges to accurately represent mean-areal rainfall and the uncertainty of radar-rainfall estimates. Other atmospheric observables, such as temperature, are much more correlated in space and time. Thus, estimating a spatial precipitation field and taking it as 'truth' is more complicated with respect to other meteorological variables.

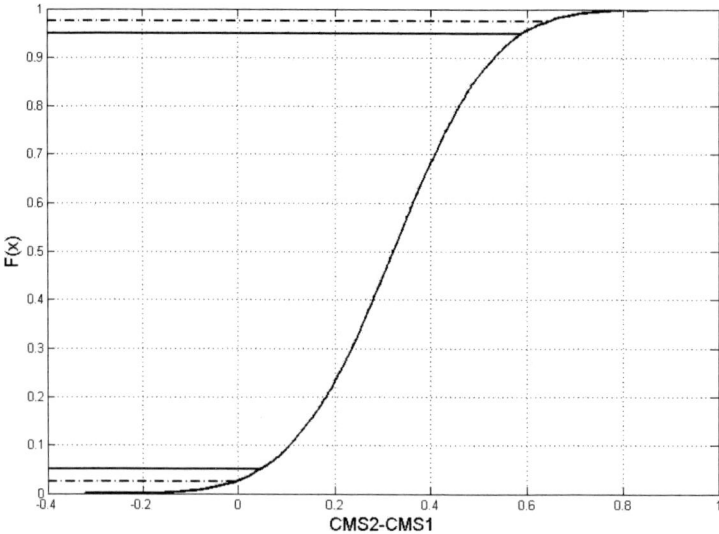

Fig. 5a. Cumulative distribution function F(x) of the resampling differences between CMS2 and CMS1. The solid and dash-dot lines delimit the 90% and 95% confidence intervals, respectively

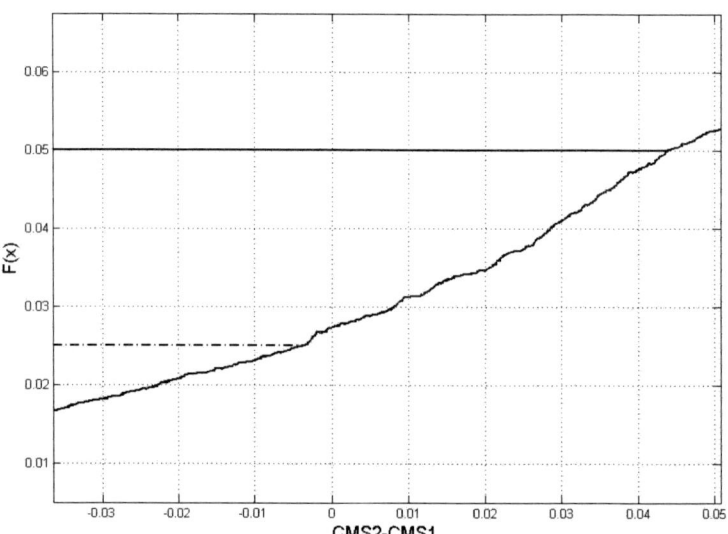

Fig. 5b. Zoom in of Fig. 5a

Several other problems affecting verification have been discussed and are briefly summarized as follows:

1. Comparison between observations and model forecasts can be problematic because of large differences at scale details. In other words, because of different variability of the two fields.
2. Radar may help to improve the observational analysis, provided that the in-built limitations are accounted for. In fact, it is widely recognized that radar-rainfall is estimated with a high degree of uncertainty due to a variety of causes (see Chap. 5 in this book). The volumetric error is one of the causes, and its correction is absolutely needed before verifying forecast precipitation.
3. Forecast precipitation patterns might be shifted with respect to the actual observations. This spatial displacement can be due to systematic model errors, for instance caused by numerical schemes, or due to the stochastic behavior of modeled atmosphere. When a set of matching rules to associate predicted objects with corresponding observed objects is established, one can estimate the spatial distance between observed and forecast precipitation patterns. However, the way of defining the measure of similarity to be used in comparisons is not unique. Some authors showed that verification results can be sensitive to measures used to compare forecasts with observations; for example Baldwin and Elmore (2005) discussed this problem.
4. The application of spatial verification leads to a variety of results whose physical interpretation cannot be easy, even in terms of forecast system quality. We have proposed the introduction of an index that allows characterizing spatial verification results both in absolute (of a single precipitation forecast system) and relative terms (with more forecast systems) and that allows handling the large amount of values obtained applying the CRA method to many cases.

References

Accadia C, Mariani S, Casaioli M, Lavagnini A, Speranza A (2003) Sensitivity of precipitation forecast skill scores to bilinear interpolation and a simple nearest-neighbor average method on high-resolution verification grids. Weather Forecast 18:918–932

Baldwin ME (2000) http://www.emc.ncep.noaa.gov/mmb/ylin/pcpverif/scores/docs/mbdoc/pptmethod.html

Baldwin ME, Elmore KL (2005) Objective verification of high-resolution WRF forecasts during 2005 NSSL/SPC spring program. In: Preprints 17th NWP Conference. American Meteorological Society, 1–5 August, Washington DC, paper 11B.4

Barnes SL (1964) A technique for maximizing details in numerical weather map analysis. J Appl Meteorol 3:396–409

Bidwell SW, Flaming GM, Durning JF, Smith EA (2005) The Global Precipitation Measurement (GPM) Microwave Imager (GMI) instrument: role, performance, and status. In: Proceedings Geoscience and Remote Sensing Symposium. IEEE International 1:25–29

Bowler N (2007) The effect of observation errors on verification. In: 3rd International Workshop on Verification Methods. 29 January – 2 February, Reading, UK

Brier GW (1948) Review of 'The verification of weather forecasts'. B Am Meteorol Soc 29:475

Brier GW, Allen RA (1951) Verification of weather forecasts. In: Malone TF (ed) Compendium of Meteorology. American Meteorological Society, pp 841–848

Casati B, Ross G, Stephenson DB (2004) A new intensity-scale approach for the verification of spatial precipitation forecasts. Meteorol Appl 11:141–154

Doswell CA III (1996) Verification of forecasts of convection: Uses, abuses, and requirements. In: Proceedings of the 5th Australian Severe Thunderstorm Conference. Avoca Beach, New South Wales, Australia

Ebert EE, Janowiack JE, Kidd CI (2007) Comparison of near-real-time precipitation estimates from satellite observations and numerical models. B Am Meteorol Soc 88:47–64

Ebert EE, McBride JL (2000) Verification of precipitation in weather systems: Determination of systematic errors. J Hydrol 239:179–202

Gabella M, Joss J, Perona G, Michaelides S (2006) Range adjustment for ground-based radar, derived with the spaceborne TRMM Precipitation Radar. IEEE T Geosci Remote 44:126–133

Göber M, Milton SF (2001) On the use of radar data to verify mesoscale model precipitation forecasts. In: Workshop report 1st SRNWP Mesoscale Verification Workshop. KNMI, 23–24 April, De Bilt, The Netherlands, p 11

Grams JS, Gallus WA, Koch SE, Wharton LS, Loughe A, Ebert EE (2006) The use of a modified Ebert-McBride technique to evaluate mesoscale model QPF as a function of convective system morphology during IHOP 2002. Weather Forecast 21:288–306

Hamill TM, Juras J (2006) Measuring forecast skill: is it real skill or is it the varying climatology? Q J Roy Meteor Soc 132:2905–2923

Harris D, Foufoula-Georgiou E, Droegemeier KK, Levit JJ (2001) Multiscale statistical properties of a high-resolution precipitation forecast. J Hydrometeorol 2:406–418

Harrison DL, Driscoll SJ, Kitchen M (2000) Improving precipitation estimates from weather radar using quality control and correction techniques. Meteorol Appl 7:135–144

Jolliffe IT, Stephenson DB (eds) (2003) Forecast Verification: A practioner's guide in atmospheric science, Wiley and Sons

Joss J, Gabella M, Perona G (2003) Needs and expectations after the 1st VOLTAIRE workshop, Upscaling and consequences of non-homogeneous beam filling. In: Proceedings 1st VOLTAIRE Workshop. ISBN 961-212-150-8. 6–8 October, Barcelona, Spain, pp 13–14

Kain JS, Weiss SJ, Baldwin ME, Bright D, Levit JJ (2005) Evaluating high-resolution configurations of the WRF model that are used to forecast severe convective weather: The 2005 SPC/NSSL Spring Program. In: Preprints 17th NWP Conference. American Meteorological Society, 1–5 August, Washington DC, paper 2A.5

Koch SE, desJardins M, Kocin PJ (1983) An interactive Barnes objective map analysis scheme for use with satellite and conventional data. J Clim Appl Meteorol 22:1487–1503

Mariani S, Casaioli M, Accadia C, Llasat MC, Pasi F, Davolio S, Elementi M, Ficca G, Romero R (2005) A limited area model intercomparison on the 'Montserrat-2000' flash-flood event using statistical and deterministic methods. Nat Hazard Earth Sys 5:565–581

Mass CF, Ovens D, Westrick K, Colle BA (2002) Does increasing horizontal resolution produce more skillful forecasts? B Am Meteorol Soc 83:407–430

Mesinger F (1996) Improvements in quantitative precipitation forecasting with the Eta regional Model at the National Centers for Environmental Prediction: The 48-km upgrade. B Am Meteorol Soc 77:2637–2649

Murphy AH (1991) Forecast verification: its complexity and dimensionality. Mon Weather Rev 119:1590–1601

Murphy AH, Brown BG, Chen Y.-S (1989) Diagnostic verification of temperature forecasts. Weather Forecast 4:485–501

Murphy AH, Winkler RL (1992) Diagnostic verification of probability forecasts. Int J Forecasting 7:435–455

Nystuen JA (1999) Relative performance of automatic rain gauges under different rainfall conditions. J Atmos Ocean Tech 16:1025–1043

Skok G, Vhrovec T (2006) Considerations for interpolating rain gauge precipitation onto a regular grid. Meteorol Z 15:545–550

Smith E A (1988) The Second Precipitation Intercomparison Project (PIP-2). J Atmos Sci 55:1481–1482

Stellman KM, Fuelberg HE, Garza R, Mullusky M (2001) An examination of radar and rain gauge-derived mean areal precipitation over Georgia watersheds. Weather Forecast 16:133–144

Tartaglione N, Mariani S, Accadia C, Casaioli M, Federico S (2007) Displacement errors of quantitative precipitation forecasts over the Calabria region. In: 3rd International Verification Methods Workshop. 29 January – 2 February, Reading, UK

Tartaglione N, Mariani S, Accadia C, Casaioli M, Gabella M, Michaelides S, Speranza A (2006) Sensitivity of forecast rainfall verification to a radar adjustment technique. Meteorol Z 15:537–543

Tartaglione N, Mariani S, Accadia C, Speranza A, Casaioli M (2005) Comparison of rain gauge observations with modeled precipitation over Cyprus using contiguous rain area analysis. Atmos Chem Phys 5:2147–2154

VOLTAIRE (2006) Validation of multisensor precipitation fields and numerical modeling in Mediterranean test sites. Final Reort, 135 pp

Weygandt SS, Loughe AF, Benjamin SG, Mahoney JL (2004) Scale sensitivities in model precipitation skill scores during IHOP. In: 22nd Conference on Severe Local Storms. American Meteorological Society, 4–8 October, Hyannis, MA

Wilks DS (1995) Statistical Methods in the atmospheric sciences. Academic Press, 467 pp

Zepeda-Arce J, Foufoula-Georgiou E, Droegemeier KK (2000) Space-time rainfall organization and its role in validating quantitative precipitation forecasts. J Geophys Res 105:10129–10146

Part IV. Integration of measurement, estimation and prediction of precipitation

18 Combined use of weather radar and limited area model for wintertime precipitation type discrimination

Roberto Cremonini, Renzo Bechini, Valentina Campana, Luca Tomassone

Arpa Piemonte, Area Previsione e Monitoraggio Ambientale, Torino, Italy

Table of contents

18.1 Introduction 475
18.2 Data source and precipitation type discriminating
      algorithms 478
      18.2.1 Data sources 478
      18.2.2 Precipitation type discriminating algorithms 480
18.3 Algorithm's validation 482
18.4 Results 485
      18.4.1 Ground network 2 m air temperature 485
      18.4.2 LAMI freezing level 486
      18.4.3 LAMI wet-bulb temperature 488
18.5 Summary and conclusions 489
References 490

18.1 Introduction

Snowstorms affect a variety of human activities such as transportation management in urban areas, highways and airports, commerce, energy and communications (Rasmussen et al. 2003). For these reasons, monitoring snowfalls in real-time with the maximum time-space resolution available is critical, as well as locating areas affected by these phenomena and nowcasting their evolution.

In wintertime, the problem is even more evident, especially in regions like the Piedmont, in Northwestern Italy, where the complex orography favors abundant but irregular snowfalls in densely populated areas. The knowledge of the exact location of the rain-snow boundary is also necessary to evaluate precipitation amounts for hydrological purposes. When radar reflectivity measurements are available, the precipitation estimation process implies different Z-R or Z-S relationships, depending on rain-snow precipitation type (Smith 1984). A better identification of dry snow, wet snow and rain brings to more accurate snow water equivalent accumulations that could be also obtained using snow density fields varying in time and space, according to the interpolated field of temperature derived from ground stations (La Chapelle 1961; Hedstrom and Pomeroy 1998).

The contour line of zero degree air temperature is one of the most simple and common indicator in order to classify the limit between snow and rain. Nevertheless, to identify operationally wintertime precipitation type at ground is not a simple task: several microphysical processes are involved in precipitation growth and in temperature profile evolution, making the snow-rain boundary strongly influenced by local scale processes. As a matter of fact, the precipitation type depends on lower-tropospheric air temperature and humidity profile, which are affected by horizontal and vertical advection, deep moist convection, vertical mixing/surface fluxes, atmospheric radiation and different latent heating (Olsen 2003). When solid precipitation passes through the freezing level, before reaching the ground, the latent heating generated by the melting, causes a negative tendency in temperature, thus creating, in conditions of weak advection, a zero degree isothermal layer and propagating downward the solid precipitation (Kain et al. 2000). Immediately when precipitation reaches sub-saturated air, evaporation begins. The heat required for transformation from water to vapor, proportional to both intensity of precipitation and relative humidity, is taken from the environmental air, hence causing cooling. Cooling by evaporation is one order greater than the melting one. Relative humidity must be considered if temperature at surface is several degrees above zero when wet-bulb temperature is near freezing (Matsuo and Sasyo 1981). The wet-bulb temperature profile is thus a key factor to find out the precipitation type at ground (Baumgardt 1999). Computation of Mitra et al. (1990) showed that inside clouds of 100% relative humidity and a lapse rate of 0.6 °C/100 m, 99% of ice mass of 10 mm snow flake melts within a fall distance of 450 m. This fall distance is about 100 m longer if the relative humidity is only 90%. These results are also consistent with radar observations which show that typically the bright band extends between 0 and 5 °C and encompasses several hundred meters (Pruppacher and Klett 1998).

Therefore, melting re-freezing and evaporation processes contribute to modify both in time and space the rain-snow boundary, making arduous to identify the precipitation type at ground.

A relatively new approach in precipitation type discrimination involves the use of dual-polarization radar data. Several studies have shown the utility of polarimetric radar observables for discriminating hydrometeor particle types (Straka and Zrnic 1993; Ryzhkov and Zrnic 1998). Due to the polarimetric signature overlap for different particle types, the fuzzy logic is the most widely used method to face the problem. Membership functions are usually defined for all available polarimetric observations (Z, Z_{dr}, K_{dp}, L_{dr}, ρ_{HV}) and for the vertical temperature profile which plays a key role in the classification process. A hydrometeor type is then assigned to each single radar cell.

Unfortunately, radar measurements always come from a certain altitude above ground, due to the Earth's curvature and to the complex orography, making dubious to assess precipitation type at ground without further assumptions or observations. Moreover, both the beam blockage and the rain attenuation at C-band or X-band can significantly affect differential reflectivity measurements, producing artifacts, whose consequence is a misleading classification of hydrometeor.

Several algorithms for discriminating precipitation type are currently available in literature: most of them use observed thermodynamic vertical profiles (Ramer 1993; Baldwin et al. 1994; Bourgouin 2000), others use the average virtual temperature calculated by geopotential heights of two pressure surfaces (Zerr 1997). A complete review of those algorithms can be found in Cortinas and Baldwin (1999).

In this study, we make a comparison between three algorithms, aimed at distinguishing between solid (snow, ice), mixed (wet snow, sleet) and liquid (rain) precipitation at ground over the Piedmont. These algorithms are based on reflectivity data, measured by operational C-band polarimetric radar, 2 m air temperature and wet-bulb temperature, derived from ground network observations and limited area numerical model (LAM) short-term forecasts.

The algorithm's verification is carried out by comparing each algorithm's output for several snow events occurred during 2005/2006 winter season with data collected by seven present weather Vaisala FD12P sensors, located in the Po valley.

Section 18.2 will provide an explanation of the data sources and a full description of the three algorithms tested for discriminating the precipitation type. In Sect. 18.3, the algorithm's verification method will be presented; some case studies and the results are in Sect. 18.4, followed by some concluding remarks in Sect. 18.5.

18.2 Data source and precipitation type discriminating algorithms

18.2.1 Data sources

Arpa Piemonte manages a ground network made by more than 350 automatic meteorological stations, collecting 2 m air temperature, humidity, pressure, wind and rainfall data.

The Piedmont is also monitored by two C-band polarimetric Doppler radars, located in Bric della Croce, close to Torino at 770 m ASL and in Monte Settepani, in Ligurian Apennines at 1,380 m ASL. The lowest beam reflectivity composite, derived from ground clutter free radar measurements, is used as input data in precipitation type algorithms.

Seven Vaisala FD12P automatic present weather sensors, installed in the Po plain at altitudes below 500 m ASL, have been used for algorithm's validation.

The FD12P sensor consists of a scatterometer that measures the amount of IR radiation scattered by 0.1 dm^3 air volume. The weather station is also equipped with precipitation and air temperature sensors: the combination of the measurements of each device (temperature and precipitation sensors and optical device) allows the determination of precipitation type.

Figure 1 shows Arpa Piemonte's ground network, the seven automatic Vaisala FD12P automatic present weather sensors used for results verification and C-band radar coverage for primary scan.

Both Torino-Caselle (m 287 ASL) and Cuneo-Levaldigi (m 386 ASL) FD12P sensors are installed within airports, co-located with manned stations, producing METAR messages. Hence it has been possible to compare automatic and manned present weather observations during the snow events presented in this study: the overall agreement was good, showing a Critical Success Index (CSI) of 90.2% with a little overestimation of snow cases by FD12P sensors.

For all 2005/2006 winter snow events, the precipitation type occurrence depending on wet-bulb temperature derived from FD12P weather sensors observations has been calculated. Results show that snow is identified for negative or slightly positive wet-bulb temperature. Mixed precipitation is identified for wet-bulb temperatures between −2 °C and +1.6 °C. When the wet-bulb temperature was greater than +2 °C, snow precipitation was never identified (Fig. 2).

Chapter 18 - Wintertime precipitation type discrimination   479

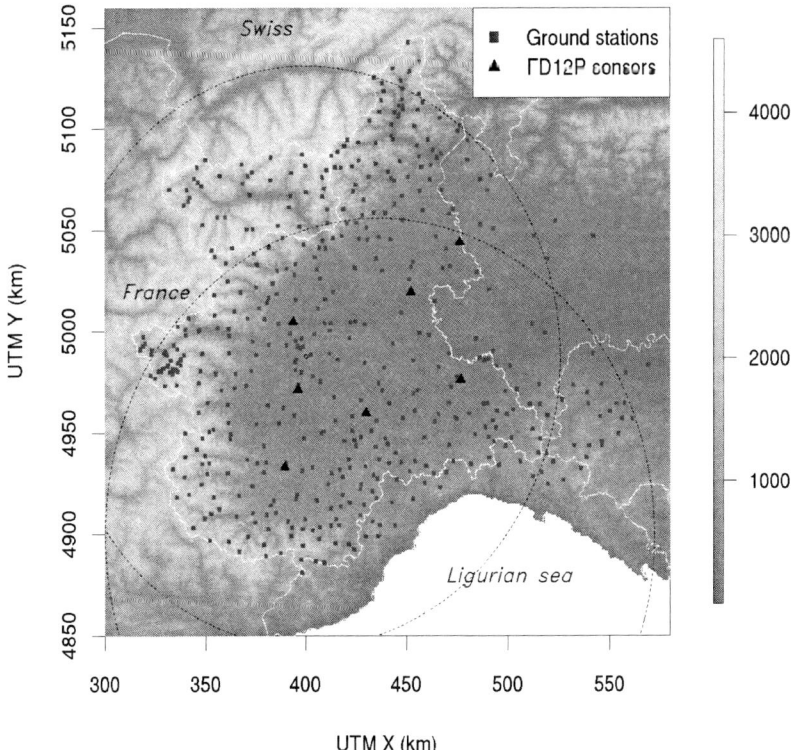

Fig. 1. Arpa Piemonte hydrometeorological ground network (*gray squares*), FD12P automatic present weather sensors (*black triangles*) and C-band primary scan coverage for Bric della Croce and Settepani (*dot circles*)

Short-term forecasts of air temperature and freezing level fields are derived from the Italian version of the non-hydrostatic limited-area model Lokal Model (LAMI), developed by COSMO Consortium. LAMI is a fully-compressible (non-hydrostatic) primitive equation model without any scale approximation. Due to the unfiltered set of equations, the vertical momentum equation is not approximate, allowing a better description of non-hydrostatic phenomena such as moist convection, breeze circulations and some kind of mountain-induced waves.

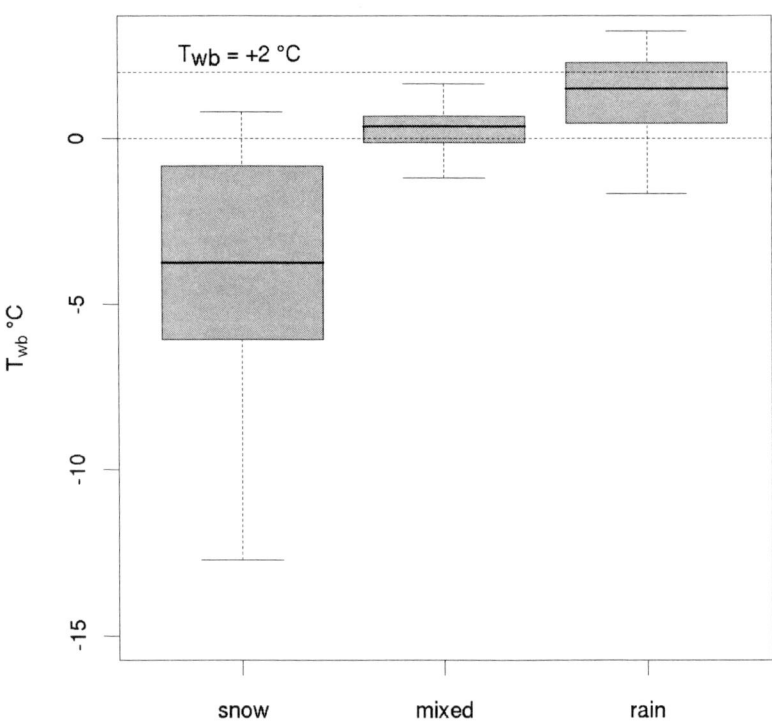

Fig. 2. FD12P automatic present weather precipitation classification for 2005/2006 winter snow events in Piedmont depending on 2 m wet-bulb temperature

LAMI implementation consists of two runs per day, at 00:00 and 12:00 UTC, with 7 km grid spacing and 35 levels in vertical (Paccagnella et al. 2005).

18.2.2 Precipitation type discriminating algorithms

All the algorithms classify precipitation in solid, mixed and rain. In this study, Cartesian lowest beam reflectivity data derived by radar composite are used to find out precipitating areas, meanwhile air temperature data observed or forecasted by LAM model are used as input data for precipitation type discriminating algorithms. The three algorithms used in this work are explained in the next Sections.

Ground network temperature

This algorithm makes use of only observed data. The basic idea is to assess the precipitation type by 2 m air temperature, measured by ground weather station of the regional meteorological gauge network. Every 30 minutes a regular 2 m air temperature field with 1 km² resolution is produced, interpolating sparse data with Kriging interpolator.

The precipitation type for each Cartesian reflectivity cell is then assigned depending on the interpolated 2 m air temperature:

$T \leq 0\,°C$	Solid precipitation
$0\,°C < T < 3\,°C$	Mixed precipitation
$T \geq 3\,°C$	Liquid precipitation

Freezing level

Considering the aforementioned importance of temperature profile in precipitation type classification methods, this algorithm is based on the simple *'rule of thumb'* of weather forecasters, which derives the rain-snow boundary by combined use of freezing level and precipitation intensity. The algorithm's inputs are reflectivity data for evaluating the rate of precipitation and the freezing level (FL) obtained by LAM short-term forecast.

The algorithm estimates snow level (SL) from LAM freezing level, applying a correction factor proportional to the precipitation intensity. This correction varies from 300 m under the FL for 20 dBZ radar echoes to 500 m under the FL for 40 dBZ.

Then, according to its ground altitude above sea level (H), each Cartesian reflectivity cell is classified in one of the three categories:

$H \geq FL$	Solid precipitation
$SL < H < FL$	Mixed precipitation
$H \leq SL$	Liquid precipitation

Wet-bulb temperature

In the third evaluated algorithm, wet-bulb temperature is calculated for each Cartesian reflectivity cell, interpolating the two standard pressure levels having the closest heights to the ground altitude, derived from LAM short-term forecast. Wet-bulb temperature is estimated with standard formulas starting from air temperature, relative humidity and pressure given by the LAM model analysis. The classification is carried out following the rule below, analogue to the first algorithm's one:

$T_{wb} \leq 0\ °C$ — Solid precipitation
$0\ °C < T_{wb} < 2\ °C$ — Mixed precipitation
$T_{wb} \geq 2\ °C$ — Liquid precipitation

For times between two standards model forecast output interval, wet-bulb temperature or freezing levels fields are estimated with a linear interpolation.

18.3 Algorithm's validation

The algorithm's performances are verified comparing the precipitation classification with 30 minutes present weather data, measured by seven Vaisala FD12P automatic present weather sensors installed in Po plains. Four snowfall events occurred during 2005–2006 late autumn – winter season are considered, summarized in Table 1, which shows event's duration, minimum and maximum 2 m air temperatures recorded by FD12P sensors in Torino-Caselle and Cuneo-Levaldigi.

In particular, during the 2–3 December 2005 snowstorm, the south Piedmont and Apennines recorded maxima in daily snowfalls with respect to the previous 6 years.

The 19–20 February 2006 snowstorms took place during XX Olympic Games and were characterized by a strong and abrupt intrusion of cold air in the Po plains in a warmer situation. On 19 February, precipitations covered the entire Piedmont with strong intensity: the snow limit suddenly decreased to 600 m ASL, with frequent and heavy snow showers, thunders and lightnings. On 20 February, precipitation was irregular and weaker with a higher freezing level (1100–1300 m ASL). Table 2 shows the freezing level behavior during the whole event from the Cuneo Levaldigi radiosounding. The sharp decrease of the freezing level during night between 18 and 19 February from 1700 m ASL to the ground is notable. The following day, the freezing level rose to about 900 m ASL and remained stable.

All events are characterized by a deep Atlantic depression approaching the Alps, anticipated by wet south-westerly flow.

In the first three events listed in Table 1, low temperatures and zero level near ground were recorded initially, maintaining snowfalls also on plains, due to the presence of a cold air layer at low levels. In the following hours, temperatures started to increase slowly, with snow turning into wet snow and rain at low heights.

Table 1. Algorithm's validation case studies

	FD12P Torino Caselle				
Case Study	Duration [hours]	Tmin [° C]	Hour (UTC)/Date	Tmax [° C]	Hour (UTC)/Date
30 Nov. 2005	24	−2,4	22:00 30 Nov. 2005	+7,4	12:00 30 Nov. 2005
2-3 Dec. 2005	20	−0,1	06:00 3 Dec. 2005	+3,1	12:00 2 Dec. 2005
26-27-28 Jan. 2006	68	−8,6	04:00 26 Jan. 2006	+2,7	18:00 28 Jan. 2006
19-20 Feb. 2006	25	+0,3	20:00 19 Feb. 2006	+3,1	11:00 19 Feb. 2006

	FD12P Cuneo Levaldigi				
Case Study	Duration [hours]	Tmin [° C]	Hour (UTC)/Date	Tmax [° C]	Hour (UTC)/Date
30 Nov. 2005	24	−1,4	22:00 30 Nov. 2005	+4,7	13:00 30 Nov. 2005
2-3 Dec. 2005	20	−5,5	07:00 3 Dec. 2005	+2,8	12:00 2 Dec. 2005
26-27-28 Jan. 2006	68	−7,4	04:00 26 Jan. 2006	+0,9	11:00 27 Jan. 2006
19-20 Feb. 2006	25	−2,0	23:00 19 Feb. 2006	+2,3	10:00 19 Feb. 2006

For algorithm's validation, a *contingency table* (Table 3) has been used, showing the relationship between known reference data by FD12P and the corresponding algorithm's output.

The accuracy of each class is calculated dividing the number of correctly classified observations in each category, by the total number in the corresponding column. The overall accuracy is computed by

Table 2. Freezing level (m ASL), recorded by Cuneo Levaldigi GTS sounding, during the snowstorms occurred on 19–20 February 2006

Date	Freezing level height (m ASL)
19 February 2006 00:00 UTC	1783
19 February 2006 06:00 UTC	1763
19 February 2006 12:00 UTC	682
19 February 2006 18:00 UTC	at ground
20 February 2006 00:00 UTC	–
20 February 2006 12:00 UTC	956
21 February 2006 00:00 UTC	963

Table 3. Contingency table for algorithm's validation

	Ground Reference (FD12P)		
Algorithm's output	Rain	Snow	Total
Rain	X	Z	X+Z
Snow	Y	W	Y+W
Total	X+Y	Z+W	N

dividing the total number of correctly classified observations by the total number of reference data.

The kappa coefficient (K) is a measure of the difference between the actual agreement between the reference data (present weather) and an automated classifier (snow/rain algorithm), on the one hand, and the chance agreement between the reference data and a random classifier, on the other hand. So, for example, with K equal to zero, it is suggested that the classification is only a random assignment of data; on the other side, a value of 0.7 for K is an indication that the observed classification is 70 percent better than one resulting by chance (Lillesand and Kiefer 2000).

The performance statistics, that can also be derived from contingency table (in Table 3), are Critical Success Index (CSI), Bias (BIAS), Probability Of Detection (POD) and False Alarm Ratio (FAR) (Wilks 1995).

These are defined as follows:
$CSI = W / (W + Y + Z)$
$BIAS = (W+Z) / (W + Y)$
$POD = W / (W + Y)$
$FAR = Y / (W + Y)$
where W is '*hit*', Y is '*false alarm*', Z is '*miss*' and X is '*correct null*'.

In the study, the *'no data'*, i.e. no precipitation observed by radar, were not considered, as they are not related with the algorithm's quality but they mainly depend on beam height respect to the ground and on radar sensibility.

18.4 Results

In the following three paragraphs, the performance of the three different algorithms is evaluated: in the first paragraph, results obtained with the algorithm taking as input the ground network temperature are shown; in the second paragraph, results using the LAMI freezing level; in the third paragraph, results using wet-bulb temperature deduced from LAMI data.

18.4.1 Ground network 2 m air temperature

For the validation of algorithm based on 2 m air temperature, 950 data collected by the FD12P sensors, located in the Po Valley, have been considered. In Table 4 there are the results obtained using the first algorithm described. For every event the 30 minutes precipitation type identified by the FD12P sensors is compared with the output of the algorithm.

Acceptable results are obtained for identification of snow, with an accuracy equal to 57.0%, with many cases of wrong mixed detection. The precipitation classified as mixed has been pretty well classified by

Table 4. Contingency table for algorithm's validation using ground network temperatures

		Present weather			
		no data/other	rain	mixed	snow
RADAR	no data	224	30	16	18
	Rain	6	25	3	0
	Mixed	44	143	69	123
	Snow	29	24	10	187
	Accuracy	73.93%	11.26%	70.41%	57.01%
	Overall accuracy			48.12%	
	Omission	26.07%	88.74%	29.59%	42.99%
	kappa coefficient	0.3857			

Table 5. As table 4, but considering only rain and snow precipitation type classes

		FD12P			
		No data/other	rain	snow	
RADAR	no data	224	30	18	CSI=88.6%
	rain	6	25	0	BIAS=112.8%
	snow	29	24	187	POD=100.0%
		Accuracy	51.0%	100.0%	FAR=11.4%
		Overall accuracy	89.8%		

the algorithm, with an accuracy of 70.4%. On the contrary, a heavy underestimation of rain is evident, the accuracy being only 11.2%, with many cases of wrongly mixed; the kappa coefficient is 0.386.

The algorithm demonstrates good capabilities in identifying snow, but on the other hand it has remarkable difficulties in distinguishing between mixed and rain, yielding a large overestimation of mixed and an underestimation of rain. This behavior suggests the necessity of a better tuning of the threshold chosen to distinguish between rain and wet-snow.

In order to evaluate whether the algorithm succeeds in the simpler but most important task, i.e. discriminating between rain and snow, the analysis has been carried on by considering rain or snow precipitation classes only. Limiting the contingency table to rain-snow classes (Table 5), it can be seen that CSI is equal to 88.6%, POD is equal to 100.0%, with overestimation of snow precipitation type (BIAS equal to 112.8%). Rain (compared only to snow) is very well estimated by this algorithm.

18.4.2 LAMI freezing level

Statistics on results obtained using LAMI freezing level short-term forecasts to discriminate between snow, mixed and rain, are shown in Table 6.

Quite good skills are obtained in rain and snow discrimination, with an accuracy of 52.7% and 59.4%, respectively. More difficult seems to be the identification of mixed precipitation, often classified by FD12P sensors as rain or snow. This behavior can be partially due to the way FD12P sensors classify mixed precipitation. The kappa coefficient is better than in previous case with a value of 0.446.

Table 6. Contingency table for algorithm's validation using LAMI freezing level

		Present weather			
		no data/other	rain	mixed	snow
RADAR	no data	224	29	17	20
	Rain	37	117	32	38
	Mixed	7	48	30	75
	Snow	34	28	19	195
	Accuracy	74.17%	52.70%	30.61%	59.45%
	Overall accuracy		58.76%		
	Omission	25.83%	47.30%	69.39%	40.55%
	kappa coefficient	0.4464			

The results in the rain-snow contingency table (Table 7) show an acceptable skill of the algorithm in recognizing the precipitation type in unambiguous cases: a CSI of 74.7% can be considered satisfying, in particular whether associated to a BIAS of 95.7% and to a False Alarm Rate of 12.6%.

Table 7. As table 6, but considering only rain and snow precipitation type classes

		FD12P			
		No data/other	rain	snow	
RADAR	no data	224	29	20	CSI=74.7%
	rain	37	117	38	BIAS=95.7%
	snow	34	28	195	POD=83.7%
	Accuracy		80.7%	83.7%	FAR=12.6%
	Overall accuracy		82.5%		

18.4.3 LAMI wet-bulb temperature

This Section describes verification results of the precipitation type diagnosis using the field of wet-bulb temperature calculated from temperature, relative humidity and geopotential height fields at standard pressure levels generated by LAMI model.

In Tables 8 and 9 overall statistics are shown. The kappa coefficient is 0.421.

Table 8. Contingency table for algorithm's validation using wet bulb temperature calculated by LAMI model short-term forecast

		Present weather			
		no data/other	rain	mixed	snow
RADAR	no data	231	34	20	20
	Rain	3	44	6	1
	Mixed	28	100	38	80
	Snow	40	29	32	221
	Accuracy	76.49%	21.26%	39.58%	68.63%
	Overall accuracy			54.99%	
	Omission	23.51%	78.74%	60.42%	31.37%
	kappa coefficient	0.4206			

Table 9. As Table 8, but considering only rain and snow precipitation type classes

		FD12P			
		No data/other	rain	snow	
RADAR	no data	231	34	20	CSI=88.0%
	rain	3	44	1	BIAS=112.6%
	snow	40	29	221	POD=99.5%
					FAR=11.6%
	Accuracy		60.3%	99.5%	
	Overall accuracy		89.8%		

With respect to the algorithms previously discussed, a further enlargement of the class identified by the algorithm as mixed can be seen. In 80 cases, according to FD12P, snow is classified as mixed and even in 100 cases rain is classified as mixed, where only 38 classifications are correct.

This result suggests that the wet-bulb temperature interval used in the algorithm (0 °C and +2 °C) seems unsuitable for mixed precipitation type classification, comparing the algorithm's outputs with FD12P classification.

In fact, Fig. 2 shows that mixed precipitation has been detected by FD12P sensors during all the events only for T_{wb} greater than -1.3 °C and less than $+1.1$ °C. It should be noticed that, while using freezing level, mixed was wrongly assigned to rain class, here, it is mostly assigned to snow one.

Again if only rain and snow precipitation types are considered (Table 9), it is evident that the choice of such thresholds leads to a satisfying result, with a snow overestimation. This behavior is well summarized by the calculated indexes: CSI = 88.0%, indicating very good performance of the algorithm, POD = 99.5% and BIAS = 112.6%, underlining that it overestimates snow versus rain.

18.5 Summary and conclusions

This study has analyzed the behavior of three different algorithms to estimate the precipitation type at ground, through the combined use of C-band radar reflectivity data and ground network observations or LAMI model short-term forecasts.

The algorithm's verification has been carried out comparing algorithm's output with seven Vaisala FD12P sensors, located in the Po plains, for four snowstorms that affected Piedmont, Italy, from November 2005 to February 2006.

The main problem in this work is mixed precipitation: in cases whereby this class is not considered in the validation, the results are better. It could be possible, of course, to enlarge the temperature range for mixed phase precipitation in the algorithms to increase its skill score, but with all the possible consequences of decreasing the ones associated with rain and snow.

This type of precipitation has been demonstrated difficult to deal with, due to ambiguous definition and poor performance of FD12P sensors in the detection of freezing drizzle or rain (Wauben 2002).

Limiting the analysis to rain and snow classification, all of the three algorithms show comparable skills, exhibiting slightly different behavior. The ground network and LAMI wet-bulb temperature have shown comparable skills (CSI > 88%), characterized by overestimation of the snow class. On the other side, the LAMI freezing level algorithm underestimates snow precipitating type at ground (BIAS equal to 95.7%).

On the one hand, a ground network based algorithm has good skills mainly due to Arpa Piemonte's meteorological stations dense coverage over regional territory (Fig. 1). In fact, results have been verified in the interior part of Piemonte plains, far from the borders where data become sparser. On the other hand, LAMI based algorithms have acceptable skills and especially the advantage in covering very well all the north west of Italy.

However, more work need to be done in tuning the algorithm's thresholds and in extending the verification to more snowstorms, in order to obtain more significant statistics. Improvements of the algorithm skill could be obtained by implementing precipitation type algorithms ensemble from standard forecast algorithms (Ramer 1993; Bourgouin 2000; Czys et al. 1996).

Acknowledgements

This work has been partially supported by ALCOTRA within project 113 FRAMEA 'Flood forecasting using Radar in Alpine and Mediterranean Areas' Interreg 3a Italia-Francia.

References

Baldwin M, Treadon R, Contorno S (1994) Precipitation type prediction using a decision tree approach with NMCs mesoscale eta model. In: Proceedings 10th Conference on Numerical Weather Prediction. American Meteorological Society, Portland, OR, pp 30–31

Baumgardt D (1999) Wintertime cloud microphysics review. SOO NWS La Crosse, WI. http://www.crh.noaa.gov/arx/micrope.html

Bourgouin P (2000) A method to determine precipitation types. Weather Forecast 15:583–592

Cortinas JV Jr., Baldwin ME (1999) A preliminary evaluation of six precipitation-type algorithms for use in operational forecasting. In: Postprints, 6th Workshop on Operational Meteorology. Halifax, Nova Scotia, Environment, Canada, pp 207–211

Czys RR, Scott RW, Tang KC, Przybylinski RW, Sabones ME (1996) A physically based, nondimensional parameter for discriminating between locations of freezing rain and ice pellets. Weather Forecast 11:591–598

Hedstrom NR, Pomeroy JW (1998) Measurements and modelling of snow interception in the boreal forest. Hydrol Processes 12(10–11):1611–1625

Kain JS, Goss SM, Baldwin ME (2000) The Melting Effect as a Factor in Precipitation-Type Forecasting. Weather Forecast 15:700–714

La Chapelle E (1961) Snow Layer Densification. Project F Progress Report No. 1. Alta Avalanche Study Center. US Department of Agriculture Forest Service, Wasatch National Forest

Lillesand TM, Kiefer RW (2000) Remote Sensing and Image Interpretation, 4th edn. John Wiley & Sons Inc., New York

Matsuo T, Sasyo Y (1981) Non-melting phenomena of snowflakes observed in sub saturated air below freezing level. J Meteorol Soc Jpn 59:26–32

Mitra SK, Vohl O, Ahr M, Pruppacher HR (1990) A Wind Tunnel and Theoretical Study of the Melting Behavior of Atmospheric Ice Particles. IV: Experiment and theory for snow flakes. J Atmos Sci 47:584–591

Olsen A (2003) Snow or rain? A matter of wet-bulb temperature. Examensarbete vid institutionen för geovetenskaper. pp 33, ISSN 1650-6553 Nr 48

Paccagnella T, Elementi M, Marsigli C (2005) High resolution forecast of heavy precipitation with Lokal Modell: analysis of two case studies in the alpine area. Nat Hazard Earth Sys 5:593–602

Pruppacher HR, Klett JD (1998) Microphysics of Clouds and Precipitation, 2nd Edn. Kluwer Academic Publishers, Boston, p 954

Ramer J (1993) An empirical technique for diagnosing precipitation type from model output. In: Preprint 5th International Conference on Aviation Weather Systems. American Meteorological Society, Vienna, VA, pp 227–230

Rasmussen R, Dixon M, Vasiloff S, Hage F, Knight S, Vivekanandan J, Xu M (2003) Snow nowcasting using a real-time correlation of radar reflectivity with snow gauge accumulation. J Appl Meteorol 42:20–36

Ryzhkov AV, Zrnic DS (1998) Discrimination between rain and snow with a polarimetric radar. J Appl Meteorol 37:1228–1240

Smith PL (1984) Equivalent radar reflectivity factors for snow and ice particles. J Appl Meteorol 23:1258–1260

Straka JM, Zrnic DS (1993) An algorithm to deduce hydrometeor types and contents from multiparameter radar data. In: Preprints 26th International Conference on Radar Meteorology. American Meteororological Society, Norman, OK, pp 513–516

Wauben WMF (2002) Automation of visual observations at KNMI: (I) Comparison of present weather. In: Proceedings 6th Symposium on Integrated Observing Systems. American Meteorological Society, 13–17 January 2002, Orlando, Florida

Wilks RS (1995) Statistical Methods in Atmospheric Science. Academic Press, London, pp 464

Zerr RJ (1997) Freezing rain: An observational and theoretical study. J Appl Meteorol 36:1647–1660

19 Adjusting ground radar using space TRMM Precipitation Radar

Marco Gabella[1], Silas Michaelides[2]

[1] Department of Electronics, Politecnico di Torino, Turin, Italy
[2] Meteorological Service, Nicosia, Cyprus

Table of contents

19.1	Introduction	494
	19.1.1 Monitoring hardware stability and measurements' reproducibility	494
	19.1.2 Calibration versus absolute calibration	494
	19.1.3 Adjustment	495
	19.1.4 Why to adjust Ground-based Radar (GR) data?	496
19.2	Radar/Gauge factor: range-dependence as seen by gauges	497
	19.2.1 Adjustment not directly related to physical variables	498
	19.2.2 Adjustment factor related to some physical variables	499
19.3	Comparing ground-based and spaceborne radar	500
	19.3.1 Range-dependence as seen by the TPR	501
19.4	Instrumentation and data description	503
	19.4.1 The TRMM Precipitation Radar (TPR)	503
	19.4.2 The Ground-based Radar (GR) in Cyprus	504
19.5	Results	505
	19.5.1 Bias and range-dependence derived from single overerpasses	505
	19.5.2 A robust range-adjustment equation: integrating more overpasses	507
	19.5.3 Comparing TPR and GR echoes	508
19.6	Summary and lessons learned	510
References		512

19.1 Introduction

Radar is an excellent tool to get a qualitative overview on the weather situation. Meteorological radars are the best sensors for providing precipitation observations over both land and sea areas. The surveillance area consists of several tens of thousands of square kilometers; the spatial resolution is of the order of several cubic kilometers; the temporal resolution can be as good as 5 min, or even less.

When aiming at applying radar data in a quantitative way, we see three important tasks to be carefully dealt with: (1) monitoring the stability of the hardware and reproducibility of the measurements; (2) taking care of the internal instrumental checking of the sensor (*calibration*); (3) transforming radar observations into the physical/meteorological variable of interest: for instance, transforming radar reflectivities aloft (somewhere high up in the sky) into rainfall intensity at ground level (*adjustment*).

19.1.1 Monitoring hardware stability and measurements' reproducibility

Relative accuracy defines our ability to reproduce in the future the values we measure today. Once relative accuracy and reproducibility are guaranteed, we may adjust our equipment with experience learnt in the domain of interest (e.g., precipitation measurements with other instruments).

19.1.2 Calibration versus absolute calibration

When we speak of radar 'calibration' we mean the internal instrumental checking of the sensor using e.g., microwave equipment. A reference power (noise source) is injected into the circulator (instead of the received power); then, the receiver should read exactly that value (± a given uncertainty). We are not yet measuring the power backscattered by a given object at a given distance. Calibration monitoring has to compensate for short-term variations of the radar equipment, i.e, to provide stable conditions for precipitation estimates. Calibration monitoring is the basis for long-term adjustments with, for example, rain gauges or spaceborne radar. Re-calibration may be needed after component replacement to monitor whether change in sensitivity has occurred: high reproducibility is the aim.

We, instead, refer to 'absolute calibration' to indicate the electrical/electromagnetic (including noise factor, waveguides, rotary joint, antenna gain and pattern, true pointing angle etc.) tuning of the radar system versus some known reference target (e.g., a metal sphere, a reflector with certified radar cross-sections) at various distances from the sensor itself. Absolute calibration, i.e., absolute accuracy, is not our present goal (nor may it be possible at good cost/benefit ratio). Instrumental checking of the calibration (monitoring transmitted power, etc.) should be carried out continuously.

19.1.3 Adjustment

We refer to 'adjustment' to indicate the physical/meteorological a posteriori tuning of the radar reflectivity estimates versus a (possibly large) set of in situ precipitation measurements, which, by integration in time and/or in space, should be made representative of the radar sampling volumes being several cubic kilometers at long ranges. Adjustment can also be performed versus other remotely sensed geophysical variables such as radar reflectivity seen from space. While calibration and instrumental checking should be carried out continuously, adjustment is a long-term task, which should be performed using several storms, preferably seasonal or yearly cumulative amounts.

Adjustment can be a tool to make our ground-based radar (GR) more quantitative. In the past, in situ point measurements were used as a reference. With the introduction of weather radar onboard the Tropical Rainfall Measuring Mission (TRMM) satellite, an additional reference is available since December 1997.

The 'adjustment concept' is based on *long-term reproducibility* of the hardware, therefore, permitting the use of an independent reference for long-term adjustment. The advantage of using gauges is the fact that they measure directly the quantity we are principally interested in. The disadvantage is their poor representativity associated with under-sampling of the spatial variability of the precipitation field. As a consequence, we do not foresee adjustment with gauge data using shorter time integration periods (or equivalently, smaller rainfall heights) because of poor representativity of both data sources. By frequent short-term adjustments, we may even make things worse, especially in complex-orography regions. Furthermore, it is also important that the products distributed over the network to the users are not adjusted from day to day. Long-term stability is needed: the resulting continuity creates a stable environment, which is known to all users and easier to relate on, rather than a short-term adjustment which is constantly being adapted to measurements of questionable representativity. Using spaceborne radar permits more robust results to be obtained because of: (a) the larger number of samples that are available and averaged at similar ranges. (b) The volumetric nature of these samples, instead of gauge 'point' measurements. (c) The possibility of covering land and sea, where no gauges are available.

19.1.4 Why to adjust Ground-based Radar (GR) data?

All ground-based radars have to measure rain from close to long distances from the radar. The radar sampling volume increases with the square of the distance. Since the variability of weather is high in the sampling volume at all ranges, radar echoes are blurred. The systematic component affected by the amount of blurring as well as overshooting with range can be investigated and compensated with a range-adjustment technique. We know that, on average, the weather signal significantly decreases with height. At longer ranges, the lower part of the sampling volume can be in rain, whereas, the upper part of the same pulse can be filled with snow or even be without an echo. This overshooting phenomenon at longer ranges is caused by the decrease in the vertical resolution, which amplifies the influence of the horizon and the Earth's curvature.

These facts cause an apparent decrease in sensitivity of the GR with range: images of cumulative radar-derived rainfall amounts, using large data sets spanning several months or years, clearly show unnatural circular features. The reader can refer, for instance, to Fig. 1 in Kracmar et al. (1999) (1 year's data in the Czech Republic); Fig. 3 of Gabella et al. (2005) (2 years' data in Switzerland), as well as the works by Vignal and Krajewski (2001) and Nelson et al. (2003). But how does one quantitatively assess the range-dependence? A suitable (i.e., sufficiently accurate) reference is probably still missing! Before the TRMM era, the reference was mainly limited to conventional, in situ, point measurements based on rain gauges on the ground. This is because remotely sensed, passive, measurements in the visible, infrared and microwave regions are associated to uncertainties that are even larger than active, radar measurements. This is not surprising, since passive measurements are not 'direct' raindrop measurements; they are instead measurements of (brightness) temperature (microwave and thermal infrared frequencies) or optical thickness (visible and near infrared). On the contrary, active radar measurements directly depend on the number of raindrops (although through a weighting function that depends on the diameter to the sixth power, in the case of Rayleigh scattering). It is well known that gauge observations represent local effects and not areal quantities. Because of the amazing variability of the precipitation field both in space and in time, the main problem of rain gauges is their poor spatial representativity. In those applications in which areal precipitation measurements are required, their main drawback is under-sampling, i.e., there are not enough observations to describe the variability of the precipitation field. Hence, it is obvious that point measurements would under-sample the precipitation fields, even though the measurements themselves were correct. Another problem of rain gauges is the small (too small) dimension of the sampled volume: even during a (continuous, no-intermittency) rainy day with, say, 1 mm raindrops, the sampling volume is less than 10^{-4} km^3, while in most applications we are probably interested in fractions of cubic kilometers. Consequently, it is not surprising that it is difficult to compare radar with rain gauges (e.g., Zawadzki 1975): in most cases, the former samples large volumes filled with snow and ice particles high (up) in the sky, whereas the latter sample small volume of raindrops close to the ground.

Sections 19.2 and 19.3 summarize procedures for compensating this variation of sensitivity using gauge and spaceborne radar observations, already published in the literature. The adjustment of GR observations, using the TRMM Precipitation Radar (TPR) as a reference has been originally performed on the island of Cyprus: a description of the instrumentation and experimental area is presented in Sect. 19.4. Section 19.5 shows the results of the six rainiest overpasses acquired in winter 2002, 2003 and 2004. Conclusions and outlook follow in Sect. 19.6.

19.2 Radar/Gauge factor: range-dependence as seen by gauges

Many adjustment techniques are based on the analysis of the Radar-to-Gauge (R/G) ratio, which we will call adjustment factor, F, and that can be seen as a measure of the 'multiplicative error' that affects radar estimates. Two components influence this multiplicative error: (1) a 'meteorological component', which is weather (and time) dependent and which is mainly caused by the fact that a reflectivity measurement taken aloft should be converted to a rainfall intensity at the ground; (2) a 'random component', which is caused by the great differences in sample size in space and time taken by the two types of instruments. Please note that in mountainous regions, the reflectivity measurement is also often affected by partial beam-occultation. Certainly, it is also affected by inhomogeneous beam-filling, maybe even at close ranges.

In Sect. 19.2.1, we give some examples of the successful application of the radar-gauge adjustment factor, applied a posteriori, to improve quantitative precipitation estimation using radar. In Sect. 19.2.2, we show how some authors have been able not only to derive but also to model the adjustment factor as a function of the distance from the radar site and eventually other physical variables involving radar observations.

19.2.1 Adjustment not directly related to physical variables

Quantitative Precipitation Estimation (QPE) in complex terrain is certainly a challenge. It is not surprising that, in complex orography regions, radar estimates derived aloft, using a single Z-R relationship, result in underestimation, which is the effect of a decreasing vertical profile of reflectivity with height, combined with beam occultation by relieves. One of the simplest remedies that can be used to compensate this bias is the so-called bulk adjustment correction. This 'global' adjustment technique consists in multiplying radar estimates by the ratio between the Gauges- and Radar-total (overall total, in time and space). In other words, the bulk adjustment consists of a single coefficient, which is applied to the whole area where QPE is attempted. The analysis of a large radar-gauge data set (from December 2000 to November 2002) in a very complex region like the Alps has shown that even a simple bulk-adjustment is able to reduce the root mean square of the 2-year cumulative radar-gauge differences from 1728 mm to 1425 mm (Gabella et al. 2005). A generalization of the bulk-adjustment is represented by an adjustment factor matrix (Wilson and Brandes 1979), which is variable in space (and constant in time). This idea has been recently successfully applied by, for example, Jessen et al. (2005) and Germann et al. (2006). The results of this last study are particularly significant and highly representative for complex orography regions, given the impressively large data set of Alpine radar-gauge data analyzed. In this work, the local bias adjustment factor, which is an adjustment factor variable in space, performs better than a single coefficient for the whole QPE region. This result is confirmed by several other experiments based on an adjustment factor, which is variable in space: the next Sect. 19.2.2 focuses on those experiments in which this variability is 'predicted' using some physical variables that characterize the radar detection environment such as the Height of Visibility and, most of all, the distance from the radar site. Among these techniques, the Weighted Multiple Regression has been applied to the same Alpine 2-year data set mentioned above: the root mean square of the radar-gauge differences is further reduced to 1182 mm. However, it is hard to realize that, for this long integration period, a 2-coefficent global adjustment based on a power-low between radar and gauge amounts, further reduces the root mean square of the radar-gauge differences to 667 mm!

19.2.2 Adjustment factor related to some physical variables

More than 2 decades ago, Koistinen and Puhakka (1981) proposed for Finnish radars a logarithmic adjustment factor, FdB, which depends on the range. This was the first attempt to relate the spatial variability of the adjustment factor to some physical variable affecting radar observations. The idea was then further developed and implemented: the adjustment technique operationally used in the Baltic Sea Experiment (BALTEX) is based on a parabolic regression between the adjustment factor, FdB, on a logarithmic, deciBel scale and the radar-gauge distance, both linear and squared (Michelson and Koistinen 2000).

In mountainous terrain, other factors, such as beam shielding by relief and orography, play also a negative role on radar estimates. Hence, Gabella et al. (2000) proposed an adjustment based on other two explanatory variables, in addition to the logarithm of the distance: the Height of Visibility, HV, plus the Height of the Ground, HG. HV is defined as the minimum Height that a weather target must reach to be Visible from the radar site: it reflects partial beam occultation by relieves as well as the vertical profile of reflectivity (water phase of the scattering hydrometeors). HG reflects the depth of the layer where precipitation growth related to orography can occur and can easily be obtained from a Digital Elevation Model. In the multiple regression proposed by Gabella et al. (2000), both the adjustment factor and the radar-gauge distance are transformed into the logarithmic dimension. This transformation aims at optimizing the performance of the regression. Log(AF) and Log(D) are regressed, instead of linear variables, because the Z-R relationship and the range-dependence of the radar signal follow power-laws. Since the vertical reflectivity profile decreases, on average, exponentially with height, Log(AF) is instead linearly related to HV and HG in the non-linear multiple regression. The radar-gauge distance often results to be the most important explanatory variable, since it reflects beam broadening, the altitude of the beam and, to some extent, attenuation. In an ordinary regression, the sum of the squares of the residuals is minimized to obtain the regression coefficients. However, Gabella et al. (2001) found that results can be further improved if the residuals are weighed according to the physical quantity of interest, that is, rainfall, in hydrological applications. The resulting adjustment technique is then called Weighted Multiple Regression (WMR). Mathematically, the relation between the adjustment factor in dB and the three explanatory variables is:

$$F(dB) = 10 \cdot Log(AF) = a_0 + a_D \cdot Log(D/D_0) + a_{HV} \cdot HV + a_{HG} \cdot HG \quad (1)$$

To reduce the range-dependence and height-dependence of a_0, we should divide the predictors by 'intermediate' values. While $HV_0 = HG_0 = 1$ km could be reasonable, $D_0 = 1$ km is certainly not. A value of 40 or 50 km is within the range used by the GR for quantitative precipitation estimation. In this way, a_0 will help us to modify the calibration of the GR so as to improve the average agreement between radar estimates aloft and gauge measurements at the ground.

19.3 Comparing ground-based and spaceborne radar

Since the introduction of the first spaceborne weather radar onboard the TRMM satellite in December 1997, it has been possible to monitor meteorological GR throughout the world (at latitudes covered by the satellite, namely within ±35°) using the TRMM Precipitation Radar (TPR), despite there are enormous differences between the TPR and GR. Mention can be made of the different sampling volumes, geometrical viewing angles, operation frequencies, attenuation, sensitivity and times of acquisition. Hence, a quantitative comparison between spaceborne and ground-based weather radar is a challenge, as can be seen in several references (e.g., Liao et al. 2001; Keenan et al. 2003; Bolen and Chandrasekar 2003; Amitai et al. 2004). In particular, there are also significant, non-correlated differences between the two radars, which are mainly caused by the different viewing angles. Indeed, ground-based and spaceborne sensors provide a complementary view: on the one hand, the GR measures rain from a lateral direction, while the spaceborne radar sees it from the top. The GR measures precipitation using a lateral view from close to long ranges. Because of the large variation, the scattering volume changes dramatically, increasing with the square of the distance. On the other hand, the TPR has the advantage of similar sized scattering volumes in all locations. This objectiveness stimulated the idea of using the TPR to estimate the influence of sampling volume of ground radars. There are two other important facts that would suggest using the spaceborne radar as a reference for the GR: (1) a great deal of effort has been made to provide the TPR with long-term, continuously monitored electronic stability; (2) the calibration factor is assumed to have an accuracy of within 1 dB (Kumagai et al. 1995). Section 19.3.1 shows how TPR data could be used to adjust ground-based radar echoes as a function of the range from the radar site itself.

19.3.1 Range-dependence as seen by the TPR

The concept proposed by Gabella et al. (2006) uses the TPR as a reference to range-adjust the GR radar reflectivity estimates. The TPR, on the one hand, allows this assessment to be made, because its measurements originate from similar 400–420 km distances. The GR, on the other hand, has to measure rain from 'close' to the radar (10 km in the present analysis) to long distances from it (110 km in the present analysis). Consequently, the GR scattering volume, which increases with the square of the distance, changes significantly. This beam broadening with distance effect, in combination with the average decrease of the vertical reflectivity profile with height and inhomogeneous beam filling causes, on average, underestimation with range of the GR. As an example, at longer ranges, the lower part of the sampling volume could be in rain, whereas the upper part of the same pulse could be filled with snow, or even be without an echo. On the contrary, the change in distance from the TPR is small (between 400 and 420 km range) and can be only slightly or not even correlated to the range, D, from the GR site. Therefore, an eventually larger underestimation with increasing range from the GR site can be estimated, by using the TPR as a reference. To check the existence of the above mentioned range-dependence, the average, linear radar reflectivity, in circular rings around the GR site is computed. This average reflectivity, $<Z(D)>_{2\pi}$, which is a function of the distance, D, from the GR site, is computed in the same circular ring for both radars. We use 7 rings 'centered' at 25, 50, 65, 75, 85, 95 and 105 km: the last rings are 10 km wide; the 1st and 2nd ones are 30 and 20 km wide.

Hence, the regions used to determine the average radar reflectivity values are large, considerably larger than the rather coarse TPR horizontal resolution. The selected large sampling rings reduce mismatches caused by different beamwidths (mismatches in space) and GR-TPR time lag (mismatches in time, up to 25 min! for the Cyprus radar). Let <GR(D)>$_{2\pi}$ and <TPR(D)>$_{2\pi}$ be average reflectivity values (averaged in azimuth) at a distance D from the GR site for both the GR and the TPR. These two variables show similar behavior, although they are hidden by the underestimating trend (with increasing distance) of the GR. Deviations caused by mismatches in space and time are reduced by averaging over the large area of the rings. While <TPR(D)>$_{2\pi}$, does not correlate with the distance from the GR site, <GR(D)>$_{2\pi}$ tends to decrease with the distance. The factor F(D) = (<GR(D)>$_{2\pi}$)/(<TPR(D)>$_{2\pi}$), which is the ratio between the radar under investigation and the reference, is introduced as the dependent variable. The distance from the GR site, D, is used as the explanatory variable. The regression between the dependent and the independent variable is performed in the logarithmic domain, where power laws lead to linear relationships. Obviously, the relationship between Log(F) and Log(D) is much more complex than a linear dependence. In purely stratiform rain, for instance, as shown by Fabry et al. (1992) and Rosenfeld et al. (1993), we can first expect an almost constant value of F(D), then an increase (bright band contamination), followed by a rapid decrease (more parabolic than linear). However, we prefer to deal with an easy to understand, essential, 'first-order' correction model, which is simply based on two coefficients: the first coefficient represents the bias at the 'reference' distance; the second one represents the tendency of GR to underestimate with range. In mathematical form:

$$10 \cdot Log_{10} \left(\frac{\langle GR(D) \rangle_{2\pi}}{\langle TPR(D) \rangle_{2\pi}} \right) = F_{dB}(D) = a_0 + a_D \cdot Log_{10} \left(\frac{D}{D_0} \right). \quad (2)$$

To reduce the range-dependence of a_0, we divide the predictor D by an intermediate, 'reference' value, which is within the used 10–110 km range (D_0 = 40 km in this paper). In this way, a_0 can help us to modify the calibration of the GR so as to improve the average agreement between the radar estimates aloft and the measurements on the ground. The slope a_D in Eq. (2) reflects the deviation of the radar sensitivity from the common $1/r^2$ law (i.e., it reflects the rate of change of the calibration with distance). Negative values can be expected and are in fact found (see next Sections), since the sampling volume of the

GR increases with the distance, as already mentioned. This is a consequence of inhomogeneous beam filling, caused mainly by the overshooting of precipitation.

19.4 Instrumentation and data description

19.4.1 The TRMM Precipitation Radar (TPR)

Kummerow et al. (1998) offered a comprehensive description of the TRMM sensor packages. A complete description of the Ku-band TRMM Precipitation Radar can be found in Kozu et al. (2001). The TPR data used in this study are attenuation-corrected radar reflectivities obtained at 13.8 GHz with the TRMM 2A25 algorithm described in Iguchi et al. (2000); this algorithm produces the best estimate of radar reflectivity close to the ground level. The TPR vertical resolution, V, at the nadir is dominated by the 'equivalent' (chirp radar signal) pulse length: V is ~250 m. With increasing distance from the Nadir, the resolution of the TPR samples becomes poorer in the vertical. This is caused by the inclination of the TPR pulse volume, as can be seen in Fig. 1 in Joss et al. (2006). The TPR beam is scanned electronically from the Nadir using 24 adjacent beam positions on both sides. The antenna phase shifters are programmed to increment in constant 0.75 steps from angle bin to angle bin. This scanning program leads to a swath scene made up of 49 footprints (24 for each side plus one at the Nadir). Let α be the off-Nadir angle, Ψ the beamwidth of the TPR pencil-beam antenna, H the altitude of the satellite (above the Earth). At increasing off-Nadir angles, the vertical layer depth, Δh, increases according to the following law:

$$\Delta h = H \cdot \Psi \cdot \tan(\alpha) + P \cdot \cos(\alpha) \tag{3}$$

In other words, the vertical resolution is no longer dominated by the equivalent pulse length of the TPR: the 'pulse limited case' is replaced by the 'beamwidth limited case' (see e.g., Nathanson et al. 1991, p. 73, for definitions and explicative figures). It is worth noting that the '3 dB one-way' angular resolution is often assumed for Ψ (0.71° for the TPR). However, this assumption is too optimistic. Because of the variability of precipitation, we should consider the contribution of radar reflectivities in a larger angle, e.g., twice the '3 dB angle' (in this case, using a Gaussian approximation, we obtain 24 dB rejection, two-way, at the edge of the radar beamwidth respect to the beam center).

Of the many output variables that are available from the TRMM 2A25 product, we deal here with the attenuation-corrected radar reflectivity calculated for the lowest TPR pulse volume, the so-called *NearSurfZ*. The echo heights range between 2 and 3 km above-sea-level. Below this altitude, TPR echoes are influenced by ground clutter: this limitation depends on the backscattering coefficient of the surface, the height of the topography over the land and the rain intensity.

19.4.2 The Ground-based Radar (GR) in Cyprus

In 1995, the Meteorological Service of Cyprus purchased a C-band Doppler radar, designed for nowcasting use. Since its installation on the Kykkos site, the radar has been used by weather forecasters to issue hazardous weather warnings. The interpretation of the radar products is purely qualitative for this application.

The radar was installed on the northwestern, mountainous region of the island, near Kykkos, the site of a medieval monastery. The radar site (Latitude: 34.98°; Longitude: 32.73°), named Kykkos, is at 1310 m above-sea-level; the antenna tower is ~15 m. Figure 1 in Gabella et al. (2006) shows a digital elevation map of the island, the radar site and, above all, the two sectors with considerable beam occultation caused by the Troodos massif in the Southeast direction and the Tripylos hill in the Northwest direction. The nearby (10–15 km range) high Olympus peak (1951 m above sea level) of the Troodos massif causes considerable ground clutter and, consequently, beam shielding behind it in a 'large' sector (approximately between 100° and 140° azimuth). A much narrower sector (between 190° and 200° azimuth) is shielded by the closer Tripylos hill (1450 m above sea level). Obviously, regions affected by ground clutter and beam shielded have been masked out and not used for deriving the correction coefficients of Eq. (2): these unreliable data are labeled 'Not used' and shown as dark grey in Figs. 1–2.

With the antenna focus at 1325 m above sea level and in standard refractivity conditions, the beam axis at the lowest elevation (0° elevation) reaches a maximum altitude of ~2000 m at a 110 km range, which is the maximum distance referred to in this Chapter. The main features of the GR are listed in Table 1 in Gabella et al. (2006): in this study, (2 µs) echoes were transmitted with a pulse repetition frequency of 250 Hz. The raw echoes were sampled using a 1° interval in azimuth and 500 m radial resolution range-bins.

19.5 Results

19.5.1 Bias and range-dependence derived from single overpasses

Six overpasses with 'remarkable' rain have been selected to check the presence of a residual range-dependent effect affecting GR echoes ('remarkable' both in terms of average rainfall intensity and areal extension, given the semi-arid rainfall regime of the island of Cyprus). The selected overpasses occurred on 11 and 12 February 2002; on 3 and 4 February 2003; on 5 March 2003 and on 5 December 2003. The corresponding regression coefficients, derived from Eq. (2), are shown in column 4 and 5 of Table 1 (first six rows). The negative values of a_D imply that, for increasing ranges from the GR site, the GR tends to underestimate with respect to the TPR in all the six rainy situations. The derived slope ranges from −7.7 dB/decade to −14.1 dB/decade. The negative values of a_0 imply that the GR underestimates with respect to the TPR at the intermediate range of 40 km. This underestimation ranges from −1.4 dB to −6.6 dB. The degree of uncertainty affecting the derived values of a_0 and a_D is smaller for the last two overpasses which are characterized by larger values of the square of the correlation coefficient (last column of Table 1). It is easy to see that the last two overpasses are characterized by larger values of the slope in addition to the explained variance. Note that on 5 March, the time leg between GR and TPR was just 2 min. Suppose that we would like to use the derived coefficients to correct GR images: what couple of a_0 and a_D should we use? The values derived on 3 February 2003 are certainly the most unreliable, not only because a much smaller variance is explained (39%), but also because weather echoes are weaker and limited to smaller areas. Hence, to overcome the problem and to find an answer to the question we propose to integrate the echoes of all the six TPR and GR acquisition into two cumulative images that will be used to derive more robust and representative coefficients. These more robust values of a_0 and a_D are presented and discussed in the next section.

Table 1. Coefficients in dB used to explain the GR/TPR ratio, F_{dB}, as a function of the logarithm of the distance from the GR (Eq. 2). The 'offset' coefficient a_0 reflects the radar 'calibration' at the 'intermediate' distance of 40 km. The 'slope' coefficient a_D shows the decrease in sensitivity of the GR with range. The first four lines give the results of single overpasses while the last line refers to an integration of all the four overpasses. The increased sample size allows more robust results. From the increased explained variance (last line, last column), we conclude that the summation of the four overpasses helps to reduce the uncertainties associated to the regression coefficients

Date of the TRMM overpass	UTC	Nearest-in-time GR scan	a_0 (dB)	a_D (dB/decade)	r^2
11 February 2002	22:50	23:15	−3.4	−6.7	0.47
12 February 2002	00:28	00:15	−4.8	−11.1	0.53
3 February 2003	11:43	11:45	−1.4	−15.9	0.39
4 February 2003	10:47	10:30	−5.6	−12.3	0.49
5 March 2003	20:02	20:00	−2.0	−20.8	0.87
5 December 2003	01:17	01:15	−6.6	−20.6	0.84
Two overpasses together, added reflectivity			−4.5	−9.4	0.59
Three overpasses together, added reflectivity			−4.5	−10.3	0.72
Four overpasses together, added reflectivity			−4.1	−10.6	0.75
Five overpasses together, added reflectivity			−3.9	−11.5	0.82
Six overpasses together, added reflectivity			−4.0	−12.1	0.82

19.5.2 A robust range-adjustment equation: integrating more overpasses

As previously described, small samples can lead to errors, e.g., of single overpasses. In this Section, the available overpasses are added step-by-step to increase the sample size.

The resulting parameters are shown in the lower part of Table 1. By adding together more overpasses, we tend to reach the desired result of obtaining a surveillance area that is filled by weather echoes. Hence, our hope is that the representativity and robustness of the derived coefficients tend to increase with the number of integrated overpasses. We are also happy to see that by increasing the number of overpasses the explained variance seems to increase. Being confident that the summation of the data of six overpasses helps to reduce the uncertainties associated to the regression coefficients, in the next Sect. 19.5.3, we will correct the GR echoes using the following range-adjustment factor in dB:

$$F_{dB}(D) = -4.0 - 12.1 \cdot Log_{10}\left(\frac{D}{D_0}\right). \tag{4}$$

From the derived values of the coefficients in Eq. (4), it can be seen that the GR/TPR ratio decreases with distance and is smaller than one for ranges larger than 21 km. A first order correction of the GR, based on these two derived coefficients, can be attempted. The TPR calibration is assumed to have an accuracy of ~1 dB (Kozu et al. 2001), which is smaller than the absolute value of a_0. As far as the range-dependence is concerned, ~4 dB of overestimation could be expected at 10 km, while ~10 dB should be added to the GR reflectivity that are 110 km far from the radar site. Equation (4) gives large underestimation at far ranges. This fact is considered to be mainly caused by an overshooting problem (increasing sampling volume of the GR with range combined with inhomogeneous beam filling and, on average, a decreasing vertical profile of radar reflectivity). The overestimation at short ranges should not be emphasized, as many difficult guesses had to be made to solve the radar equation. The regression coefficients in Eq. (4) are derived from 10 GR/TPR reflectivity values, that is eight degrees of freedom; the explained variance is 82%. The remaining part of the variability of the GR/TPR ratio, which is not explained by the regression coefficients, is caused by mismatches in space and time in addition to the different operation frequencies (and residual attenuation). Note that the disagreement between the GR and the TPR caused by mismatches in

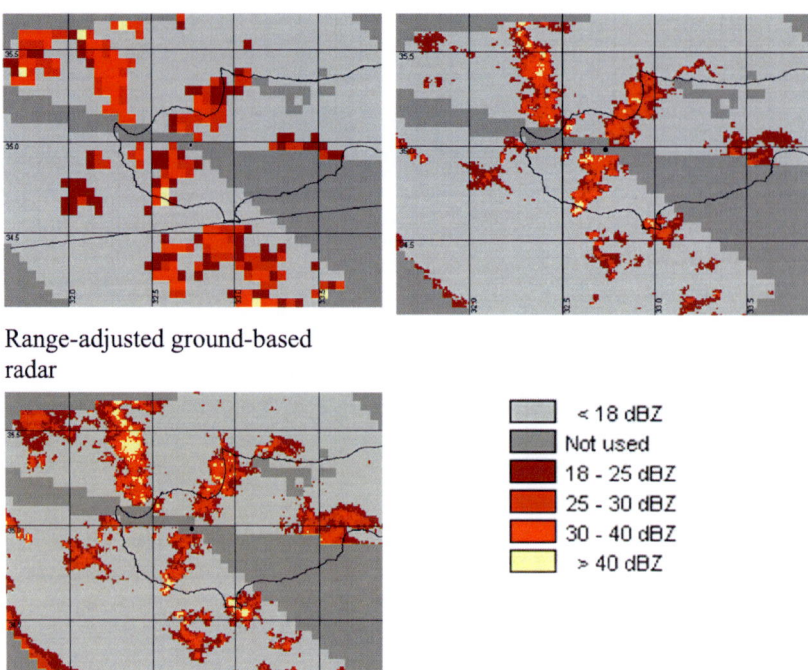

Fig. 1. Lat-Lon representation of the spaceborne and ground-based radar reflectivity echoes acquired on 11 February 2002. The observations of the TRMM Ku-band precipitation radar (*top left* picture) are calculated with native resolution of 0.05°×0.05°, the highest possible one with the satellite at 402 km altitude. The ground track (nadir) is shown with a black line. The TRMM satellite overpass (orbit 24205) was at 22:50 UTC. The GR data were acquired 25 min after the overpass with 1°×500 m resolution then resampled on a 0.01°×0.01° Lat-Lon grid, corresponding to approximately 1100×900 m. The top right picture shows the original GR data while the bottom left one shows the range-adjusted GR data

time and space of the corresponding sampling volumes are exacerbated the well-known extreme spatio-temporal variability of the precipitation field, including the vertical reflectivity profile (and the presence of the bright band).

19.5.3 Comparing TPR and GR echoes

In this Section, we show some visual comparison between TPR *NearSurfZ* and GR *0° elevation scan* images. The top left picture in

Fig. 1 shows TPR echoes acquired at 22:50 UTC on 11 February 2002 (TRMM orbit 24205). In comparing this attenuation-corrected spaceborne radar image with the ground-based radar image (central picture), we should bear in mind the enormous difference between the two sensors (different operating frequency, attenuation, sampling volume sizes, geometrical viewing angles, sensitivity etc.).

A large part of the disagreement is also caused by mismatches in space, since *NearSurfZ* echoes are geometrically coincident with the lowest scan of the GR (0° elevation) only in some sub-domains of the surveillance area. For a more accurate comparison, it is necessary to merge full 3D data of both the TPR and GR by using geo-referencing algorithms like the ones described by Liao et al. (2001) and Bolen and Chandrasekar (2003). However, for this first overpass, also mismatch in time is relevant (GR image was acquired 25 min later). Finally, as explained in Sect. 19.3.1 and shown in Sect. 19.5.1, part of the disagreement at far (and very close) ranges of the GR can also be statistically explained in terms of distance from the GR site: as shown in Table 1 (first row, last column), the explained variance as a function of the logarithm of the distance is 47%, with a negative slope of ~7 dB/decade. In case of a perfect agreement, the slope would tend to 0 and the explained variance to 100%. The bottom left picture in Fig. 1 shows the range-adjusted GR echoes according to Eq. (4), which makes use of all overpasses. Since the absolute value of the slope in Eq. (4) is considerably larger than 7 dB/decade (while the offset, a_0, in the first and last row of Table 1, is similar), it is likely that the GR reflectivity values at far ranges are over-compensated in the bottom left picture of Fig. 1. Figures 4 and 5 in Gabella et al. (2006) show GR and TPR images for the overpasses on 12 February 2002 and 4 February 2003.

As a further example, we show in Fig. 2 the comparison for 5 March 2003. Figure 1 shows the overpass with the worst time lag (25 min). Figure 2 is characterized by the best time lag (2 min). Let us focus our attention on the two (yellow) small cells in the southeastern corner with TPR reflectivity values larger than 40 dBZ (top-picture). Apparently, only one cell is detected by the GR (top right picture). After the range-adjustment, two cells are clearly visible (bottom left picture). On the contrary of Fig. 1 (bottom left picture), here it is likely that GR echoes at far ranges are still slightly underestimated: the compensation factor used is, in fact, 14.1 dB/decade, while for this particular overpass the derived compensation factor resulted to be 20.8 dB/decade (5th row and 5th column in Table 1).

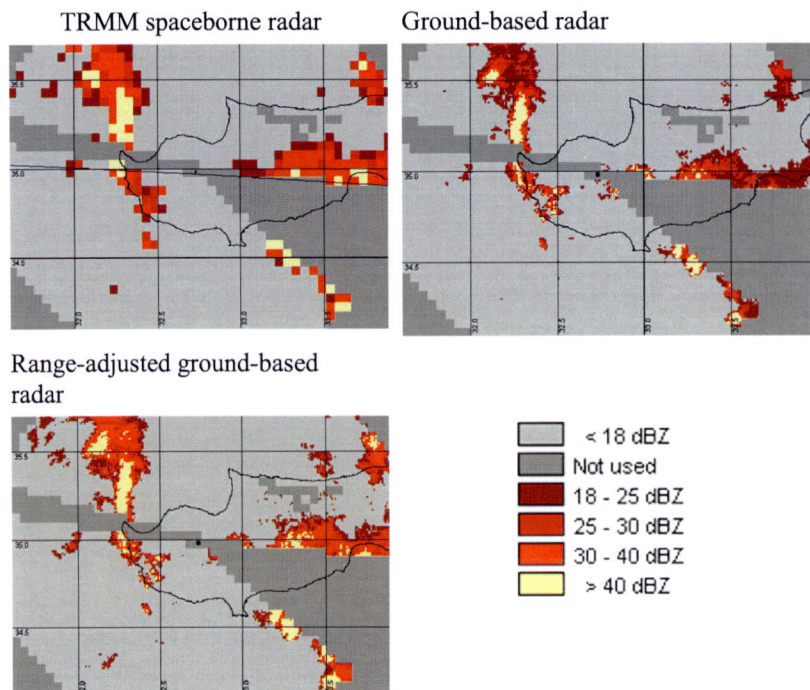

Fig. 2. As Fig. 1, but for the TRMM overpass of 5 March 2003 at 20:02 UTC, orbit #30235 (*top left* picture); the ground-based radar scan was recorded at 20:00 UTC (*top right* picture: original data; *bottom left*: range-adjusted data)

19.6 Summary and lessons learned

The radar sampling volume increases with the square of the range, which is often referred to as beam broadening. If the hydrometeors were homogeneously distributed over the volume itself, beam broadening had no effect on the radar measurements. However, this is rarely the case.

On average, the vertical radar reflectivity profile tends to decrease with height. Because of the Earth's curvature, the larger the range the higher the radar sampling volume. Consequently, a possible cause of the systematic range-dependence of the ground-based radars is the old, well-known problem of overshooting, which, combined with the vertical decrease in the radar echo, can lead to serious underestimation. The effect becomes severe at distant ranges or in the case of partial beam occultation by relief. Furthermore, at distant ranges, the upper part of the radar beam can be in the region of weak snow echoes or even above the echo top. Overshooting of precipitation can even affect the lowest ground-based radars scans. This Chapter not only illustrates the possible causes of the apparent decrease in sensitivity of the GR with range, but also presents a procedure that can be used to assess and eventually compensate this range-dependence, using the radar in space as a reference. The correction is obtained by calculating the 'azimuth-integral' of the GR and TPR radar reflectivity at a constant range (hence comparing the range-dependence of both radars along the radial direction). Through a regression, the ratio GR/TPR (on a logarithmic scale) is related to the distance (also on a logarithmic scale) from the GR site. As a result, the adjustment factor versus distance is found.

When and where available, spaceborne radars can be used to monitor and adjust meteorological Ground-based Radars (GR). The GR scattering volume increases with the square of the distance, while it is almost constant for spaceborne radars. This paper shows how spaceborne radars can be used to adjust GR estimates. The correction is obtained by calculating the 'azimuth-integral' of the radar reflectivity at a constant range. In this way, the range-dependence of both radars along the radial direction can be compared. Their ratio, on a logarithmic scale versus distance (which is also on a logarithmic scale), is statistically analyzed. As a result, the adjustment factor versus distance is found. The underestimation of ground-based radars at far ranges has often been verified in literature using rain gauges (Koistinen and Puhakka 1981; Gabella et al. 2000; Michelson and Koistinen 2000). It has been verified using time-cumulated rain gauge amounts. Here, we use a spaceborne radar that permits more robust results to be obtained because of: (a) the larger number of samples that are available and averaged at similar ranges; (b) the volumetric nature of these samples, instead of gauge 'point' measurements; (c) the possibility of covering land and sea, where no gauges are available.

The possibility of extending the useful range by correcting GR observations, from the island of Cyprus to over maritime areas is highly desirable in the south-eastern Mediterranean. Using radar and gauges, we are forced to extrapolate the results derived over the island, to the area over the sea. The combination of the TPR with the GR is valid in the whole range of the two radars.

From this novel comparison between average GR and TPR reflectivity values we have learned that:

1. It is worth using a correction to adjust the range-dependence of ground-based radars. This subject has often been discussed in literature. Here, a procedure is described, to correct data of ground radars using the TPR.

2. At the intermediate radar-gauge distance of 40 km, the GR reflectivity is found to be approximately −4.0 dB lower than TPR reflectivity. This is equivalent to −2.7 dB in the rain rate (1 dB in rainfall intensity [dBR] corresponds to ~1.5 dB in radar reflectivity [dBZ]), thus modify the constant in the radar equation by adding 4 dB.
3. Using the TPR-reflectivity Z as a reference, we find an apparent decrease in the sensitivity of the GR equal to ~14 dB per decade of the distance between the GR and the rain, thus use a different exponent in the range correction of the radar equation: −3.4 instead of −2.0.
4. The values of a_0 and a_D in Eq. (4), derived through the integration of all the six available overpasses, are not too dissimilar from the average value on a logarithmic scale of the values derived from single overpasses (Table 1).

It will be useful to investigate to what extent long-term, climatological data can be used to substitute TPR-data in regions where the TRMM satellite is not available. The usefulness of climatological data would be limited by the assumption about the homogeneity of average precipitation amounts in circular rings around the GR site of interest.

Acknowledgments

The Meteorological Service of Cyprus provided the ground-based radar data; the TRMM 2A25 data were provided by NASA DAAC at the Goddard Space Flight Center. Cooperation between Politecnico di Torino and Meteorological Service of Cyprus started within the EU COST717 Action, continued and was strengthened thanks to the contract EVK2-CT-2002-00155 (VOLTAIRE) funded by the EC and the 'Joint Italian-Cypriot cooperation program on research and development'.

References

Amitai E, Nystuen JA, Liao L, Meneghini R, Morin E (2004) Uniting space, ground, and underwater measurements for improved estimates of rain rate. IEEE Geosci Remote S 1:35–38

Bolen SM, Chandrasekar V (2003) Methodology for aligning and comparing spaceborne radar and ground-based radar observations. J Atmos Ocean Tech 20:647–659

Fabry F, Austin GL, Tees D (1992) The accuracy of rainfall estimates by radar as a function of range. Q J Roy Meteor Soc 118:435–453
Gabella M, Bolliger M, Germann U, Perona G (2005) Large sample evaluation of cumulative rainfall amounts in the Alps using a network of three radars. Atmos Res 77:256–268
Gabella M, Joss J, Michaelides S, Perona G (2006) Range adjustment for Ground-based Radar, derived with the spaceborne TRMM Precipitation Radar. IEEE T Geosci Remote 44:126–133
Gabella M, Joss J, Perona G (2000) Optimizing quantitative precipitation estimates using a non-coherent and a coherent radar operating on the same area. J Geophys Res 105:2237–2245
Gabella M, Joss J, Perona G, Galli G (2001) Accuracy of rainfall estimates by two radars in the same Alpine environment using gauge adjustment. J Geophys Res 106:5139–5150
Germann U, Galli G, Boscacci M, Bolliger M (2006) Radar precipitation measurements in a mountainous region. Q J Roy Meteor Soc 132:1669–1692
Iguchi T, Kozu T, Meneghini R, Awaka J, Okamoto K (2000) Rain-profiling algorithm for the TRMM Precipitation Radar. J Appl Meteorol 39:2038–2052
Jessen M, Einfalt T, Stoffer A, Mehlig B (2005) Analysis of heavy rainfall events in North Rhine-Westphalia with radar and rain gauge data. Atmos Res 77:337–346
Joss J, Gabella M, Michaelides S, Perona G (2006) Variation of weather radar sensitivity at ground level and from space: case studies and possible causes. Meteorol Z 15:485–496
Keenan T, Ebert E, Chandrasekar V, Bringi V, Whimpey M (2003) Comparison of TRMM satellite-based rainfall with surface radar and gauge information. In: Preprints 31st International Conference on Radar Meteorology. Seattle, WA, pp 383–386
Koistinen J, Puhakka T (1981) An improved spatial gauge-radar adjustment technique. In: Preprints 20th Conference on Radar Meteorology. Boston MA, USA, pp 179–186
Kozu T, Kawanishi T, Kuroiwa H, Kojima M, Oikawa K, Kumagai H, Okamoto K, Okumura M, Nakatsuka H, Nishikawa K (2001) Development of precipitation radar onboard the Tropical Rainfall Measuring Mission (TRMM) satellite. IEEE T Geosci Remote 39:102–115
Kracmar J, Joss J, Novak P, Havranek P, Salek M (1999) First steps towards quantitative usage of data from Czech weather radar network. In: Collier GC (ed) COST-75 Final International Seminar on Advanced Weather Radar Systems. Commission of the European Communities, Brussels, Belgium, ISBN 92-828-4907-4, pp 91–101
Kumagai H, Kozu T, Satake M, Hanado H, Okamoto K (1995) Development of an active radar calibrator for the TRMM Precipitation Radar. IEEE T Geosci Remote 33:1316–1318
Kummerow C, Barnes W, Kozu T, Shiue J, Simpson J (1998) The Tropical Rainfall Measuring Mission (TRMM) sensor package. J Atmos Ocean Tech 15:809–817

Liao L, Meneghini R, Iguchi T (2001) Comparisons of rain rate and reflectivity factor derived from the TRMM Precipitation Radar and the WSR-88D over the Melbourne, Florida, site. J Atmos Ocean Tech 18:1959–1974

Michelson DB, Koistinen J (2000) Gauge-radar network adjustment for the Baltic Sea experiment. Phys Chem Earth (B) 25:915–920

Nathanson FE, Reilly JP, Cohen MN (1991) Radar Design Principles, 2nd edn. McGraw-Hill, New York

Nelson BR, Krajewski WF, Kruger A, Smith JA, Baeck ML (2003) Archival precipitation data set for the Mississippi River Basin: algorithm development. J Geophys Res 108(D22): 8857, doi:10.1029/2002JD003158, 2003

Rosenfeld D, Atlas D, Wolff D (1993) General probability matching relations between radar reflectivity and rain rate. J Appl Meteorol 32:50–72

Vignal B, Krajewski WF (2001) Large sample evaluation of two methods to correct range-dependent error for WSR-88D rainfall estimates. J Hydrometeorol 2:490–504

Wilson JW, Brandes EA (1979) Radar measurement of rainfall – a summary. B Am Meteorol Soc 60:1048–1058

Zawadzki I (1975) On radar-raingage comparison. J Appl Meteorol 14:1430–1436

20 Implementing a multiplatform precipitation experiment

Giovanni Perona[1], Marco Gabella[1], Riccardo Casale[2]

[1]Politecnico di Torino, Electronics Department, Torino, Italy
[2]Research Directorate General, European Commission, Brussels
 Belgium

Table of contents

20.1   Introduction 515
          20.1.1   Scientific/technological objectives of the
                       VOLTAIRE project 517
          20.1.2   Project organization 518
20.2   VOLTAIRE project summary and recommendations 520
          20.2.1   Summary 520
          20.2.2   Main lessons learned 523
          20.2.3   Recommendations 524
20.3   VOLTAIRE technical conclusions 526
20.4   Outlook for QPE using radar 527
          20.4.1   Where we stand today 527
          20.4.2   Proposed solution: use of many inexpensive,
                       redundant, short-range radars 528
20.5   General conclusions 529
20.6   Appendix 530
References 530

20.1 Introduction

Knowledge of the quantitatively accurate amount and spatial distribution of precipitation fields is highly sought after in climate research, civil protection and weather forecasting. Unfortunately, precipitation is one of the most difficult atmospheric phenomena to measure and model, especially in mountainous terrain. Given the complexity of the Mediterranean area, an improvement in observational techniques is a prerequisite for precipitation estimates related, on the one hand, to the frequency of extreme rainfall events and, on the other hand, to droughts. Better estimates of precipitation may lead to a better understanding and forecasting of floods, which are one of the major natural hazards in Europe, especially in Mediterranean regions.

The above issues were addressed within the VOLTAIRE project the scope of which was to build a European methodology for a more accurate precipitation monitoring in Mediterranean areas, taking into account the specific technological and dynamical problems related to such a composite region. The project was funded by the European Commission under its VI Framework Programme.

VOLTAIRE stands for 'Validation Of muLTisensor precipitAtion fields and numerical modeling In mediterRanEan test sites'. Why an acronym coincident with the nickname of the famous philosopher François Marie Arouet? *Our motivations* can be found in the paragraph below:

When King Louis XVI, shut away in prison, saw the works by Voltaire that covered the walls of his cell, he exclaimed 'This man has destroyed France'. He had destroyed much more: that way of being, of thinking and acting, that concept of life, that culture, that system that still today is known as 'ancient regime'. In no period has there ever been a more 'modern' intellectual in Europe. And he continues to be so, more than three centuries after his birth. It is impossible to think in a freer way than he did; it is not possible to write or communicate in a more penetrating way. He was, and remains, a master. Voltaire, though the result of a particular society and environment, interprets the requirements of eternal order. His battles against fanaticism and intolerance are not out of date, as each and every era suffers from fanaticism and intolerance. At his school, one becomes a free, independent and anticonformist spirit. Whatsoever society or community will always need someone like Voltaire who, with his accusations and reproaches, will protect them from their own abuses.

Dynamical Meteorology, in fact, traditionally considered two extreme space-scales of cloud-cover and precipitation at extratropical latitudes: the cumulonimbus scale (a few kilometers) and the frontal scale (hundreds of kilometers). However, it has become clear, since the first application of remote sensing, that other scales of aggregation are dynamically relevant. The so-called *wide precipitation bands* - which develop in extratropical cyclones on a scale of approximately 50–100 km across and a few hundred km along the band itself and are, surprisingly, stable, on average, with respect to ordinary vertical convection – are, for example, of particular interest. These 'intermediate-scale' phenomena are particularly significant in the Mediterranean area where the interaction of synoptic-scale perturbations with the complex orography and land-sea structures generates a whole series of quite complex 'meso-scale' features.

As far as precipitation fields are concerned, their complexity and high variability in time and space represents a challenge for both observations and numerical models: indeed, in current operational forecast models, precipitation represents a diagnostic and not a prognostic variable. Consequently, even the quality of modeled precipitation fields in the Mediterranean area needs to be verified; however, the possibility of their validation using 'ground truths' is difficult due to measuring problems that originate from the same surface complexity which, in turn, is responsible for the numerical forecasting problems themselves.

It is well known that the best way to remotely sense precipitation fields is by means of radar and, in fact, many ground-based Doppler radars have been installed in Europe over the last decade. Monitoring the meteorological phenomena in the tropical areas that are not covered by ground-based radars is now possible through the precipitation radar onboard the recent Tropical Rainfall Measuring Mission (TRMM) satellite. The highly successful TRMM program motivated the designation of a future mission, the so-called Global Precipitation Measurement (GPM, see Chapter 6 in this book), which will be able to extend TRMM observations to higher latitudes than the present ones, which are limited at 35°. As part of the plans to prepare for the future use of GPM, it is important to promote comparisons of TRMM radar with ground-radar data. In Europe, this is only possible in the southern part of Cyprus.

20.1.1 Scientific/technological objectives of the VOLTAIRE project

In more detail, the scientific/technological objectives of the VOLTAIRE project are listed below:

1. to improve the accuracy of ground-based radar precipitation fields in Mediterranean test sites using adjustment techniques based on:
   A) in situ rain gauge measurements (tailored to mountainous and hilly regions);

B) spaceborne weather radar measurements (where available, e.g., in Cyprus);
2. to compare data quality schemes for ground-based radar;
3. to focus on the 'variability' of precipitation fields when trying to improve quantitative precipitation estimates obtained from ground-based radars. The improvement in accuracy has been continuously sought for by addressing the various sources of error in mountainous terrain through a painstaking systematic approach based on the following milestones, which are strictly related to objectives B) and A): clutter elimination, correction for visibility and/or vertical profile, gauge-adjustment;
4. to compare ground-based radars, rain gauges and spaceborne radars; to use range-adjusted and gauge-adjusted ground-based radar observations as ground validation for the Tropical Rainfall Measuring Mission (TRMM) radar. We aimed at assuring TRMM radar data validity in Mediterranean Sea areas not covered by either ground-based radar or by rain gauges;
5. to gain experience with the TRMM mission, its Ground Validation program and to prepare European participation in the future Global Precipitation Measuring mission;
6. to quantitatively compare precipitation fields represented by numerical models, rain gauges, adjusted ground-based radar and the spaceborne radar.

20.1.2 Project organization

The composition of the VOLTAIRE consortium is given in the Appendix. The project was organized in ten *Work Packages* (*WPs*). *WP10* comprises the management, at both a scientific/technical and financial/administrative level and was led by the Project Leader, Prof. Giovanni Perona, who was responsible for programming the activities, supervision, evaluation and interfacing with the European Commission. A Steering Committee assisted the Project Leader in the general organization, planning and development of the project. The Steering Committee, chaired by the Project Leader, was composed of the Scientific Board, the Exploitation Board and the Project Office. The Scientific and Exploitation Board composition and details concerning the management can be found at http://www.voltaireproject.net/ proj_manag.htm of the project website.

An important accompanying activity was the external reviewing by the Quality Manager. This role has been covered by an independent, external, highly skilled scientist, with more than 40-years experience in monitoring precipitation in mountainous terrain using radar, for research and operational use, namely Jürg Joss, former Director of the Swiss-Italian branch of MeteoSwiss.

WP9, led by the University of Ljubljana, dealt with the dissemination of the results, at all levels and in all possible forms. *WP1*: 'Database building, standardization and management', led by Politecnico di Torino, was the common reference and link for all the research and technological development *WPs*, namely from *WP2* to *WP8*. The main objective of *WP1* was to organize the data bank to be used by the Partners to develop and test their algorithms. *WP2*: 'Data quality control of time variable data', led by Einfalt & Hydrotec, was devoted to objective 2, as described in the Sect. 20.1.1, 'Scientific objectives of the VOLTAIRE project'. The Meteorological Service of Cyprus was leading *WP3,* 'Radar adjusted rain fields in Cyprus and comparison with TRMM data'. The objective of WP3 is coincident with VOLTAIRE objective D). *WP4*: 'Improved radar-gauge adjusted rain fields based on the TRMM validation program', led by NASA/GMU, together with *WP8*: 'Preparation of a Validation Supersite for GPM in the western Mediterranean site', led by the Universitat Politecnica de Catalunya, are mainly devoted to objective 4. The important objective 3 was developed within *WP5,* 'Structural characterization of precipitation fields', led by MeteoSwiss. Politecnico di Torino led *WP6,* 'Optimized/adjusted rain fields in complex-orography regions', which focused on objective 1 A). Objective 6 is carried out within *WP7*, 'Numerically modeled rain fields and comparison with observations', which was led by Università di Camerino. Several *WPs* have been concerned with objective 1 B), in particular *WP7*, *WP8*, *WP4*, and most of all, *WP6* and *WP3*. The first comparison in Europe of ground-based and spaceborne weather radar was indeed very stimulating and has involved, with various degrees of involvement, all the scientists, resulting in many useful suggestions and original ideas.

To sum up, the VOLTAIRE partnership was highly complementary and each Partner, according to its scientific background, was an expert in his field. Individual tasks have been performed in a way that was consistent with the overall project objectives.

20.2 VOLTAIRE project summary and recommendations

20.2.1 Summary

Precipitation is much more variable in time and space than other meteorological variables, even from a climatological point of view. Reliable long-term records and field distributions with high spatio-temporal resolution only exist over land and even there, the coverage is far from being complete. Within VOLTAIRE, precipitation was studied not only through conventional in situ point measurements (i.e., rain gauges) but most of all, by using meteorological radar at both ground level and from space.

Radar is a unique tool to obtain an excellent overview of the weather situation both in time and space. However, radar is also a delicate tool that needs maintenance, monitoring and quality checks. As far as these problems are concerned, within VOLTAIRE it was possible to capitalize on the many years of experience and know-how of MeteoSwiss (which has been working with radar in mountainous terrain since the Sixties) while developing routines and algorithms with the purpose of obtaining the quality check and quality control of radar-derived and gauge-derived precipitation amounts (*WP2* and *WP4*). Indeed, not only echoes from the ground have to be eliminated whenever possible, but also less reliable measurements should receive smaller weights than observations with small uncertainties; remotely sensed and in situ observations should be combined and mutually checked to reach better results (Amitai et al. 2005; Golz et al. 2005). The correction of radar data based on image processing methods and physical understanding can improve radar data. More than 15 methods from literature and from our own development were implemented in the course of this project. They can be applied to PPI Cartesian, PPI polar and volume polar data. Even when there are 'only' PPI data available, problematic areas with bright band or beam shielding can be corrected under certain conditions (Golz et al. 2005). These algorithms are now available as a C++ software library.

A fundamental problem in quantitative precipitation estimation is the asymmetry and the large dispersion of the distributions of particles and with it of the precipitation rate. Indeed, the distributions are wide and skewed-to-the-high-end at the same time. This fact, combined with the variability of particle type and number density, causes complex behavior even in derived integral quantities such as the total amount of water per unit volume, rain rate and snow rate, radar reflectivity. As a direct consequence, heavy rain, when present, significantly contributes to the volume of precipitation during an event, although it is concentrated in both time and space (e.g., Joss et al. 2006). In turn, the chance of detecting weak rain is much greater than strong rain and we may be tempted to extrapolate the properties of weak rain to strong rain. Such an extrapolation would involve large errors, since the causing mechanisms are different. Then, the spatio-temporal variability of the precipitation field needs to be analyzed in its full complexity: *WP5* proposes the variogram as a tool

Significant information from radar data can usually only be extracted at short ranges from the radar itself, simply because the radar detects the rain echo not at ground level, as would be desirable, but aloft, at variable heights due to obstacles as well as the Earth's curvature. All these difficulties increase rapidly with range from the radar location. *WP3* and *WP6* show that at least a partial remedy is sometimes possible, even in mountainous terrain. A gauge-based algorithm, to be used operationally to adjust radar estimates, was developed and tested (Gabella and Notarpietro 2004; Gabella 2004; Gabella et al. 2005). One important objective of VOLTAIRE, namely to use the electronic stability of spaceborne radar to monitor the status of ground-based radar (see the DoW, end of page 2), was achieved. For the first time, the electronic stability and reproducibility of TRMM Precipitation Radar (TPR) has been used not only to assess the overall bias but also to range-adjust single scans of the Ground-based Radar (GR) using the TPR as a reference. It has been shown that approximately 10dB have to be added to the measured radar reflectivity Z (in mm^6/m^3), when the range increases from 10 km to 100 km. It appears that the radar echo decreases with range as r^{-3} instead of as r^{-2} (Gabella et al. 2006). More recently, a similar procedure has been used to analyze the effects of the decreasing vertical resolution of TPR when the radar beam is steered away from the nadir (Joss et al. 2006). Using range-adjusted data of the GR for reference, a few dB correction has to be added to the measured values, when increasing the distance from close to Nadir to the edge of the swath (~120 km far away from nadir).

Within VOLTAIRE, an attempt was made to meet the challenge of the new Global Precipitation Measuring (GPM). In *WP4*, a framework was developed for global verification of spaceborne radar estimates of precipitation, presenting new opportunities and challenges (Amitai et al. 2005; Amitai et al. 2006). The framework is based on comparing ground and satellite probability distribution functions (pdf) of rain rate after rain type classification. The comparisons reveal large discrepancies which vary with rain type. A discussion on opportunities and challenges to determine and reduce the uncertainties in space-based and ground-based radar estimates of rain rates distribution is included in Amitai et al. (2006).

In *WP8*, a simulation-based framework was developed (Llort et al. 2004, Deliverable 8.2). This framework generates a 3D high resolution precipitation field and simulates the measurements derived from different spatial sampling strategy of ground and spaceborne radars. The interaction between ground and spaceborne radar data and the effects of various factors (errors with distance, attenuation effects, etc.) were studied using this tool as a base. A proposal was developed to complement the equipment which is presently available in Catalonia with new equipment (Llort et al. 2006).

The fusion of conventional in situ observations of precipitation amounts with non-conventional, remotely sensed estimates (derived using both spaceborne and ground-based weather radars) was certainly a problematic and challenging task. Even more challenging was the comparison between observations and numerically predicted precipitation values, performed in *WP7*. Some problems had to be solved before comparing observations and forecasts. Rain gauge networks, ground and space radar and numerical models have significantly different resolutions from each other and the necessary upscaling or downscaling processes are inherently difficult. In particular, the upscaling/downscaling processes should be performed conserving total water (Accadia et al. 2003). Even simple range-adjustment techniques applied to ground-based radar data affect the comparison of observed and forecast precipitation. In this context, the influence of a TRMM-based range-adjustment on ground radar data in Cyprus was investigated (Tartaglione et al. 2006). Radar data provide a much better space-time resolution than a network of rain gauges: this fact dramatically changes the model verification processes. Finally, it has to be noted that different optimization criteria (e.g., maximizing the correlation or rather minimizing the differences) give different results in the comparison between observations and forecast (Tartaglione et al. 2005; Tartaglione et al. 2006).

The importance of disseminating research results not only among scientists but especially among meteorological services and agencies is well known. Within VOLTAIRE, the presence of two meteorological services and one national agency has driven scientific activities towards operational and technological development results: these results were highlighted in a specific part of this final report, namely the Technological Implementation Plan (see www.voltaireproject.net). The diffusion of the Technological Implementation Plan (TIP) to operational services, the VOLTAIRE website and data bank, the innovative results published in international journals and presented at international meetings, the special issue of Meteorologische Zeitschrift entirely devoted to VOLTAIRE results, all represent managerially effective means of dissemination (*WP9* and *WP10*).

20.2.2 Main lessons learned

1. In mountainous terrain, precipitation is even more variable both in space and time because of orographic effects and the interaction of the wind fields with the mountains. This variability within the scattering volume is in contradiction with the usual assumption of homogeneous beam filling in the radar equation. In turn, this assumption is the basis for estimating reflectivity, attenuation and phase shift along the beam.
2. The vertical variability of the radar reflectivity profile, which is often complicated by the bright band, together with orographic effects on visibility, make the extrapolation from the estimated reflectivity aloft to rainfall intensity at the ground quite a difficult task. Furthermore, approximate procedures validated in moderate rain may lead to large deviations in the rare, but important, events of heavy rain. Extrapolating experience from moderate to heavy precipitation may lead to significant errors, which could be even more significant for hydrological applications.
3. The radar data provided by weather services are often not sufficiently reliable for hydrological use. However, the automatic check and correction procedures are able to significantly improve the quality of these data – or to detect the regions and intervals which are no good for further use. From this point of view, the management of the network of radars conducted over recent years by the 'Radar and Satellite' group of MeteoSwiss in Locarno Monti is certainly paradigmatic. Some important aspects that should be considered are the following:

Task A: Effective ground clutter rejection (Germann and Joss 2004);
Task B: Automatic calibration of hardware;
Task C: Periodical maintenance of hardware (Joss et al. 1998);
Task D: Sutomatic monitoring of hardware (Joss et al. 1998);
Task E: Use of quality descriptors (Galli's Chapter in Joss et al. 1998);

Task F: Continuous improvements in data processing and objective large-sample validation (Germann et al. 2006);

Task G: Comprehensive monitoring and adjustment of the system through comparison/fusion with data from other sources (network of gauges and, where available, TRMM precipitation radar).

4. Tasks A to E were developed in Switzerland before the VOLTAIRE kick-off meeting (November 2002). However, thanks to the VOLTAIRE technological development and research activities, quantitative precipitation estimation in mountainous terrain based on a large weather radar data set has been performed (Gabella et al. 2005): these results would not have been possible without the high quality data set made available by MeteoSwiss. Task F is continuously under way. Important milestones have been reached in recent years and the main results were presented at the European radar conference (Germann et al. 2004) and in literature (Germann et al. 2006) also related to task G.

5. A comparison between the output of NWP models, on one hand, and the fusion of rain gauge, ground-based-radar and TRMM radar observations, on the other hand, shows that NWP precipitation forecasts are more similar to range-adjusted radar observations than to original radar values. This fact confirms the importance of adjusting radar echoes at far ranges.

20.2.3 Recommendations

The main recommendation is to invest in quality of ground-based RADAR (see e.g., Sect. 20.2.2, Tasks A-G). A second step of investing in quality of ground-based radar is to check and correct the provided data. This should be done by any data user before applying the data. A first initiative to simplify these tasks resulted in the creation of the VOLTAIRE radar data control library, a C++ implementation (Golz and Einfalt 2006).

It is well known that ground-based radars suffer from an apparent decrease in sensitivity with distance. Within VOLTAIRE, this effect has been verified for the first time using TRMM spaceborne radar observations. Indeed, the TRMM radar can be used as a reference, since it looks at the clouds in its field of view approximately from the same distance (400–420 km). The resulting adjustment technique has been called range-adjustment (Gabella et al. 2006) and is proposed in the VOLTAIRE Technological Implementation Plan (Sect. 6.2, Part II of the VOLTAIRE Final Report, Perona et al. 2006).

Ground-based Radar data need *adjustment*. Adjustment coefficients between rainfall amounts at the ground (as measured by gauges) and radar estimates aloft for quantitative precipitation estimation can be derived on a purely statistical basis or trying to add up some physical insights. In the first case, within VOLTAIRE, procedures adopting sets of local, 'geographical' coefficients (e.g., Germann et al. 2006; Jessen et al. 2005) proved to be useful. However, even just a simple power-law regression between gauges and radar estimates significantly improves the agreement. For instance, using a network of 3 radars in Switzerland (427 gauges in ~40 000 km^2) the root mean square difference (RMSD) between 2-year cumulated radar and gauges rainfall amounts (average precipitation 3031 mm) is significantly reduced (from 1728 mm to 667 mm). In complex terrain, though, where overshooting effects are important, even just a single adjustment coefficient for the whole surveillance area (bulk-adjustment) can improve the original data (from 1728 mm to 1425 mm). These statistically-based procedures often failed when applied in real-time to specific events.

Alternatively, the so-called Weighted Multiple Regression (WMR) relates the spatial variability of the Adjustment Factor to some physical variables that certainly influence it, namely the minimum radar visibility level, the ground level and the distance from the radar site. For instance, during intense events in the Western Alps and Florida, the use of WMR-derived coefficients computed on previous days proved to be satisfactory; in Switzerland, WMR-coefficients computed in May 2001 clearly improved the estimates in September 2001 (and vice versa) Since the WMR is simple to use, fast and able to correct several effects (overshooting, partial beam occultation, non-uniform beam filling, beam broadening), it could be recommended for operational services. Consequently, it has been inserted in the Technological Implementation Plan. Recently, the WMR technique has been successfully applied for Quantitative Precipitation Estimation in arid and semi-arid areas in the southwestern Mediterranean region (Morin and Gabella 2007).

Adjustment of observed data turns out to be essential also when comparing those data with precipitation forecast by NWP models, generally leading to a better match. In addition, some more suggestions have come out for the comparison between observed and forecast precipitation (see *WP7*), such as: (1) applying up/down-scaling techniques that conserve the total amount of water; (2) using object-oriented methods to compare observations and forecasts in order to evaluate the shift in precipitation patterns: models may forecast a precipitation pattern reasonably well, but the pattern can be shifted with respect to the real event.

20.3 VOLTAIRE technical conclusions

In this Section, an attempt is made to draw up some general conclusions from the external Quality Manager's point of view, as presented during the Final Meeting in Torino (30 March 2006) and discussed with the European Commission (EC) Scientific Officer, the Project Leader and the Partners.

1. The resolution of our observations is coarse – too coarse for the high spatio-temporal variability of the precipitation fields.
2. The variability of 'instantaneous' precipitation fields is so large that a few percentages of time and/or space can be responsible for half of the amount of rain, or more.
3. We should be careful when interpreting overall statistics on precipitation fields – extremely careful when extrapolating from the frequently-occurring, weak precipitations to rare, harmful storms.
4. Resolution, data quality and variability require special care!
5. Parameterizing sub-grid variability is not an easy job – because of the shape of the distributions and non-linearities of relations that are involved.
6. In many applications, we may have to explain to customers that we cannot obtain the desired accuracy, not even with radar or – at present - with any other concept.

Particularly, when dealing with radars, it is important to remember that:

- The relevant width of the beam is at least twice the 3 dB beamwidth.
- The sample volume is not homogeneously filled with one type of particles.
- The sample volume size varies by a factor of 30 dB or more within the radar range of interest.
- Adjusting with data from a few gauges, over a long period of time, is useful – but many gauges over a long period of time are even better.
- A comparison between ground-based and spaceborne radar is helpful – but the task is complicated by the variability of precipitation fields, combined with the changes of sample volume size.
- Data provided by radar systems cannot be expected to be free from measurement errors and
  uncertainties. These errors and uncertainties may be variable in space and time, and only some of them can be detected and corrected.

Regarding the verification of forecast precipitation:

- Range-adjustment is necessary before using ground-based radar data for model verification.
- Any transformation of forecast and observed precipitation from a grid to another one requires special care since it may affect verification results.

20.4 Outlook for QPE using radar

20.4.1 Where we stand today

Radar is a unique tool to get an overview on the weather situation. It gives an excellent overview in both time and space. Over 40 years, researchers have been investigating ways for obtaining the best use of radar. As a result, we often find qualitative assurances on how much radar is a useful tool, and it is! After this initial statement, however, regularly comes a long list on how to increase the accuracy of radar or in what direction to move for improving it. Perhaps we should rather ask: is the resulting data good enough for our application?

Perhaps more of what we would like to do can only be achieved at short ranges from the radar. This is not because we miss careful investigations, but simply, because radar is seeing the rain aloft, while we need to know what is arriving at ground level. Echoes from the ground have to be eliminated. Obstacles as well as the Earth's curvature lead to a variable horizon, allowing us to see precipitation at variable height, often too far from the ground. All these difficulties increase rapidly with range from the radar location.

Furthermore, precipitation is too variable for the coarse resolution of our instruments. The variability of natural precipitation is so large that the radar beam often does not resolve it. As a result, we find different types of particles and inhomogeneous reflectivity in the pulse volume, aloft and compared to the ground level.

We may wonder: why is it so difficult to grasp a realistic precision out of 'long-range' (say up to 100/150 km) weather radar? Perhaps, the main reason is found in the difficulty of reproducing the results verified with large effort at close ranges. We cannot extrapolate them to the full range displayed by our operational, meteorological radars. At short ranges, problems caused by shielding, inhomogeneous beam filling, attenuation, clutter and vertical profile may be dealt with. They are not at longer ranges. This statement does not exclude the use for weather forecasting in full range of our radars. The radar tells us, where and when something is coming. Radar data will help us to validate the forecasts of the NWP. Here, combining the information from many radars may help a lot. The combination of data from more than one radar may also mitigate the effects caused by the range-dependence of a single radar.

20.4.2 Proposed solution: use of many inexpensive, redundant, short-range radars

The meteorological radar is a useful tool in weather monitoring. Traditionally, such radar is designed to investigate large areas, as to achieve a wide view of weather phenomena over contiguous regions and last but not least, cushion the costs of quite an expensive device. Then, it is usually installed on a high place and works with relevant peak power, performing a full volumetric scan by increasing elevation in subsequent azimuthal rotations.

Nevertheless, new needs have recently appeared in *nowcasting*: local authorities and private weather services require now accurate knowledge of *short-term* weather dynamics in a *short-range* context, such as a single region or valley, as to plan daily activities and get ready with countermeasures. Clearly, they look for cost-effective solutions, easy installation and long-lasting reliability.

For that purpose, an innovative project of cheap X-band radar has been developed by the Remote Sensing Group at Politecnico di Torino, Department of Electronics: in this approach, the radar is intended to be placed down a valley, launching low-power pulses to analyze the above sky through an elevation scanning strategy. Moreover, a potential added value can be reached linking radars of that kind into a 'radar network', also for the cross-validation of rain gauge measurements. The project has been funded by the EC within the InterReg IIc FORALPS project.

Many 'vertically-scanning' low-cost short-range X-band radars for rain estimates can be a valid solution and alternative to a long-range C-band radar. Long range radars have proved to be useful for weather forecasting and qualitative surveillance. However, the results, verified with large effort at close ranges, *cannot* be generalized. It seems impossible to reproduce the results easily obtained close to the radar for *quantitative applications* at far ranges. This is *especially true in mountainous terrain*. Therefore, an interesting solution could be to combine the data of many, small, low-cost and short-range X-band radar for QPE.

20.5 General conclusions

Precipitation is certainly one of the meteorological quantities that most directly affect human life, much more than the other key quantities that characterize the weather from a physical point of view, e.g., pressure, humidity, temperature. On the one hand, rain is important for every day life, in agriculture, in causing disasters, but on the other hand, it is not well known, as it is so difficult to measure and to forecast quantitatively. The rain field is much more complex and more variable in time and space than, for instance, pressure or temperature, which, instead, can be measured with a much higher accuracy. The multi-fractal nature of the rain field is certainly very poorly characterized by conventional instruments (rain gauges), by long-range weather radar or by new spaceborne radar like TRMM Precipitation Radar. These are the reasons why the EC is promoting projects related to rain aspects in its various forms, from the lack of it (droughts), to the excess of it (floods, erosions).

In this context the VOLTAIRE project is certainly a comprehensive up-to-date research and technological development project that has specifically been related to a systematic study of various representations of rain fields and rain-related topics, namely:

- rain field characterization in complex environments such as mountainous terrain and the Mediterranean area;
- quantitative precipitation estimation using various measuring systems: rain gauge networks, ground-based meteorological radars and spaceborne weather radar;
- comparisons between rain forecasting and rain observations;
- design of a ground validation site for the future Global Precipitation Measuring mission.

The results obtained within VOLTAIRE, which have been published in international journals (e.g., in vol 15 – Special Issue of Meteorologische Zeitschrift) and presented in Conferences, will certainly have to be taken into account by researchers operating in other rain-related EC projects of the VII Framework Programme. However, the complex multi-fractal structure of rain fields still needs to be thoroughly characterized with new instruments for a better understanding of rain related phenomena of interest to a wider public (droughts, floods, erosion).

20.6 Appendix

The VOLTAIRE consortium was made up of nine Partners from six different countries:

1. polito — Politecnico di Torino, Italy.
2. e&h — Einfalt & Hydrotec, Germany.
3. cymet — Meteorological Service of Cyprus.
4. gmu — George Mason University – NASA, USA
5. mch — MeteoSwiss, Switzerland.
6. apat — Agenzia per la Protezione dell'Ambiente e per i servizi Tecnici, Italy.
7. unicam — Università di Camerino, Italy.
8. upc — Universitat Politecnica de Catalunya, Spain.
9. unilj — University of Ljubljana, Slovenia.

References

Accadia C, Mariani S, Casaioli M, Lavagnini A, Speranza A (2003) Sensitivity of precipitation forecast skill scores to bilinear and a simple nearest neighbor average method on high resolution verification grids. Weather Forecast 18:918–932

Amitai E, Liao L, Llort X, Meneghini R (2005) Accuracy verification of spaceborne radar estimates of rain rate. Royal Meteorol Soc Atmos Sci Lett 6:2–6

Amitai E, Llort X, Sempere-Torres D (2006) Opportunities and challenges for evaluating precipitation estimates during GPM mission. Meteorol Z 15:551–557, DOI: 10.1127/0941-2948/2006/0157

Gabella M (2004) Improving Operational measurement of precipitation using radar in mountainous terrain – Part II: Verification and Applications. IEEE Geosci Remote S 1:84–89

Gabella M, Notarpietro R (2004) Improving operational measurement of precipitation using radar in mountainous terrain – Part I: Methods. IEEE Geosci Remote S 1:78–83

Gabella M, Michaelides S, Perona G (2005) Preliminary comparison of TRMM and ground-based precipitation radars for a European test site. Int J Remote Sens 26:997–1006

Gabella M, Joss J, Michaelides S, Perona G (2006) Range adjustment for Ground-based Radar, derived with the spaceborne TRMM Precipitation Radar. IEEE T Geosci Remote 44:126–133

Germann U, Joss J (2004) Operational Measurement of Precipitation in Mountainous Terrain. In: Meischner P (ed) Weather Radar: Principles and Advanced Applications – Physics of Earth and Space Environment, vol XVII Chap. 2, Springer-Verlag, pp 52–77

Germann U, Galli G, Boscacci M, Bolliger M (2006) Radar precipitation measurement in a mountainous region. 132:1669–1692, doi: 10.1256/qj.05.190

Golz C, Einfalt T (2006) Radar Data Quality Control Methods in VOLTAIRE. Meteorol Z 15:497–504

Golz C, Einfalt T, Gabella M, Germann U (2005) Quality control algorithms for rainfall measurements. Atmos Res 77:247–255

Jessen M, Einfalt T, Stoffer A, Mehlig B (2005) Analysis of heavy rainfall events in North Rhine-Westphalia with radar and raingauge data. Atmos Res 77:1-4, 337–346

Joss J, Schädler B, Galli G, Cavalli R, Boscacci M, Held E, Della Bruna G, Kappenberger G, Nespor V, Spiess R (1998) Final Report: Operational Use of Radar for Precipitation Measurements in Switzerland. vdf Hochschulverlag AG an der ETH Zürich, 108 pp, ISBN 3 7281 2501 6

Joss J, Gabella M, Michaelides S, Perona G (2006) Variation of weather radar sensitivity sensitivity at ground level and from space: case studies and possible causes. Meteorol Z 15:485-496, doi: 10.1127/0941-2948/2006/0150

Llort X, Sánchez-Diezma R, Amitai E, Sempere-Torres D (2004) A simulation framework to better understand comparison between TRMM and ground radar rainfall measurements. In: Proceedings of the 1st VOLTAIRE Workshop, (VOLTAIRE Deliverable 9.2) ISBN 961-212-150-8, p 18

Llort X, Berenguer M, Franco M, Sánchez-Diezma R, Sempere-Torres D (2006) 3D downscaling model for radar-based precipitation fields. Meteorol Z 15:505–512

Morin E, Gabella M (2007) Radar-based quantitative precipitation estimation over Mediterranean and dry climate regimes. J Geophys Res 112, D20108, doi: 10.1029/2006JD008206

Perona G, Joss J, Gabella M, Michaelides S, Amitai E, Galli G, Speranza A, Sempere Torres D, Vrhovec T, Rakovec J, Germann U, Monacelli G, Sanchez-Diezma R, Bolliger M, Golz C, Skok G (2006) VOLTAIRE (EVK2-CT-2002-00155) Final Report, ISBN 961-212-187-7 [http://www.voltaireproject.net/]

Tartaglione N, Mariani S, Accadia C, Speranza A, Casaioli M (2005) Comparison of raingauge observations with modeled precipitation over cyprus using contiguous rain area analysis. Atmos Chem Phys 5:2147–2154

Tartaglione N, Mariani S, Accadia C, Casaioli M, Gabella M, Michaelides S, Speranza A (2006) Sensitivity of forecast rainfall verification to a radar adjustment technique. Meteorol Z 15:537–544

Author index

Ahrens, B., 367
Amitai, E., 343
Anagnostou, E.N., 219
Anagnostou, M.N., 313
Anselm, M., 83
Bechini, R., 475
Bendix, J., 171
Bringi, V.N., 233–279
Campana, V., 475
Casaioli, M., 453
Casale, R., 515
Chandrasekhar, V., 285
Cremonini, R., 475
Duchon, C., 33
Dufournet, Y., 285
Ebert, E.E., 419
Einfalt, T., 101
Gabella, M., 493, 515
Glasl, S., 83
Grecu, M., 219
Gultepe, I., 59
Haiden, T., 389
Hou, A., 132
Jaun, S., 368
Kokhanovsky, A., 172

Kummerow, C., 134
Lammer, G., 3
Lensky, I.M., 172
Levizzani, V., 135
Mariani, S., 455
Michaelides, S., 101, 493
Moisseev, D.N., 287
Nauss, T., 172
Nurmi, P., 422
Nystuen, J.A., 344
Perona, G., 518
Randeu, W.L., 235
Rossa, A.M., 420
Russchenberg, H., 285
Schönhuber, M., 3
Shepherd, J.M., 131
Skofronick-Jackson, G., 131
Spek, L., 285
Steinheimer, M., 389
Tartaglione, N., 453
Thies, B., 171
Thurai, M., 233
Tomassone, L., 475
Turek, A., 171
Unal, C., 283

Subject index

2D-Video-Distrometer (2DVD): 3–29, 37–39, 50–57, 61, 235, 238, 239, 247, 248, 250, 253–255, 257, 259, 261, 263–265, 267, 269–271, 273, 275, 276, 278, 325, 326, 334, 336, 351, 352, 356
Acoustic Rain Gauge (ARG): 344, 351–360
acoustic rainfall measurement: 343–361
active (retrieval, sensors): 134, 135, 139, 140, 144, 145, 154, 155, 196, 197, 200–202, 314, 496
Advanced Microwave Sounding Unit (AMSU): 139, 147, 153, 200, 212
aerosol: 211, 212
ALADIN model: 397–409
anomalous propagation: 109, 119, 121
attenuation (correction): 117, 140, 222, 225, 226, 236, 237, 247, 248, 252, 269, 313
attenuation (differential): 19, 150, 156, 247, 248, 316–318, 337, 503, 504, 509
attenuation (gaseous): 136, 140, 155
attenuation (path): 142, 156, 157, 314, 315, 318, 319, 337, 352
attenuation (rain): 3, 24, 88, 109, 110, 116, 117, 121, 122, 157, 221, 222, 236, 251, 477
attenuation (specific differential): 247, 316
attenuation (specific): 156, 221, 237, 247, 316
Auto Estimator (AE): 197
Autonowcaster system (ANC): 392, 410

beamblock (*or* beam blockage): 117, 118, 122, 477
Bergeron-Findeisen process: 174–176

Bernoulli effect: 42
best fit criterion: 439, 455, 456, 461–464
bootstrap procedure: 466, 467
Brier Skill Score (BSS): 370, 376–385, 435
bright band: 110, 119, 140, 142, 243, 258, 476, 502, 508, 520, 523
brightness temperature *see* temperature (brightness)

calibration (of radar): 144–147, 149, 150, 154, 161, 163, 203, 204, 251, 257, 268, 269, 304, 314, 352, 457, 494, 500–502, 506, 507, 523
calibration (of rain gauge): 9–11, 14–17, 21, 22, 26, 35, 44, 46, 57, 64, 86, 87, 96
canting angle: 19, 24, 27, 234, 239, 241, 269, 271–273, 278, 290, 322
catchment-mean precipitation: 369
classic tall 2DVD: 9, 20, 21, 29
Climate Prediction Center Merged Analysis of Precipitation (CMAP): 204
Climate Prediction Center morphing method (CMORPH): 204
cloud condensation nuclei (CCN): 210, 211
cloud imaging probe: 66, 68, 74, 75, 79
cloud microphysics: 205–207, 212, 285–310
cloud optical thickness: 173, 174, 177–180, 205, 496
Cloud Resolving Model (CRM): 146, 151, 152, 159, 199, 200, 223, 224
cloud top temperature: 172, 174, 188, 190, 197, 198, 205–208, 314
cloud water path (cwp): 175–181, 188, 189

Subject index

CloudSat: 152, 153, 159, 202, 213
clutter: *see* ground clutter
combined retrieval: 157, 220–228
Contiguous Rain Area (CRA): 439, 455
Contiguous Rain Area Mean Shift (CMS): 464–468
Continental cloud: 151, 209
Convective Available Potential Energy (CAPE): 395, 401–406, 411–415
Convective Boundary Layer (CBL): 412–414
Convective Initiation (CI): 391, 392, 395, 402, 410–415
co-polar attenuation: 318, 319
co-polar correlation coefficient: 242, 253, 278
co-polar reflectivity: 238, 241, 242
COSMO: 423, 434–446, 479
COSMO-LEPS: 367–386
COST 717: 107, 512
COST 731: 420
Critical Success Index (CSI): 426
cross-polar reflectivity: 19
crosstalk: 19, 24, 26

depolarization (linear ratio): 238–241
depolarization (rain-induced): 237
Deterministic Observational Reference (DOR): 374–382
differential attenuation: 19, 150, 247, 248, 316
differential back scatter phase: 242, 254
differential propagation phase: 236–252, 314–319
differential reflectivity: 5, 24, 26, 238, 239, 287, 288, 294, 296, 297, 302–306, 310, 314, 318–323, 357, 477
differential tension: 36
disdrometer: 4, 35, 38, 50, 51, 57, 60, 61, 68, 75, 84, 96, 117, 234, 235, 243, 248, 251, 254, 257, 258, 260, 268, 271, 276, 316, 319, 321, 325–337, 348, 349, 352–357
distrometer: 3–29, 351, 352
double penalty: 415, 429–433, 448
drizzle: 59–63, 68–73, 75–79, 245, 346, 350, 353, 359, 489

drop axis of symmetry: 20, 234
drop axis ratio: 26, 27, 238–242, 248, 253, 263–265, 269, 321–323
drop horizontal velocity: 11, 19, 20, 271
drop size distribution (DSD): 4–6, 11, 16, 18, 23–25, 110, 155, 158, 234–278, 293–300, 304, 308–310, 315, 316, 320–325, 329–332, 337, 343–348, 358
drop terminal velocity: 17, 68, 173, 247, 254, 263, 275, 279
droplet spectrometer: 84–96
Dual-frequency Precipitation Radar (DPR): 145–163, 201–203

effective droplet radius: 172–174, 179, 180, 206, 208, 211
effective measuring area: 8, 16–19
effective radius: 180, 206–208, 211
electromechanical distrometer: 4
Enhanced Convective Stratiform Technique (ECST): 173, 183–189
ensemble prediction (*or* ensemble forecasting): 367–386, 420–422
Ensemble Prediction System (EPS): *see* ensemble prediction.
entity-based verification: 438
Equitable Threat Score (ETS): 426
equivolumetric sphere diameter: 11, 15, 16, 18, 19, 26
error (in measurement): 18, 20, 38, 39, 45–47, 61, 64, 69, 78, 87, 94, 106
error (in prediction): 374, 380, 386, 400, 401, 407, 321, 424, 425, 428, 429, 437–444, 447–449, 455–459, 461, 464–466
error (in radar): 107–114, 236, 248, 257, 264, 300–303, 310, 318, 328, 331–335, 457–459, 469, 497, 521–523, 526
error (in satellite retrieval): 143, 144, 152, 156, 220, 227, 457, 458, 507
extinction: 5, 75, 110, 135, 136, 155
Extreme Dependency Score (EDS): 427
eyeball verification: 430

fall velocity (*or* fall speed): 4–7, 13–19, 22, 24–26, 30, 50, 61, 64, 84–86, 94,

Subject index

97, 207, 234, 235, 272–278, 287, 291, 292, 294–296, 398
False Alarm Ratio (FAR): 183–186, 425–427, 435, 436, 484–487
FD12P: 61 64, 66, 68–73, 75–79, 477–480, 482–489
fog measuring device (FMD): 66, 68, 75
fog settling rate: 75
fog: 60, 62, 66, 68–71, 75, 76, 78, 79, 93
Forecast Demonstration Project (FDP) 392, 399, 420
Forecast Quality Measure (FQM): 439
Fractions Skill Score (FSS): 435, 436, 448
FRAM project: 59–80
freezing level: *see* zero degree isotherm
Frequency Bias Index (FBI): 425
fuzzy logic: 252, 433, 439
fuzzy verification: 433

gamma distribution: 155, 221, 237–250, 260, 293, 308, 309, 320–322
GANDOLF system: 391, 392, 404, 410
Global Precipitation Climatology Project (GPCP): 204, 205
Global Precipitation Measurement (GPM) mission: 131–164, 199, 201, 213, 457, 458, 517–519, 521
Goddard Profiling (GPROF) technique: 199
GOES: 133, 197, 206, 213
GOES Multispectral Rainfall Algorithm (GMSRA): 206
GOES Precipitation Index (GPI): 197
GPM Microwave Image (GMI): 145–150, 157, 158, 161, 163
ground clutter: 107, 109, 111, 114–118, 122, 251, 457, 478, 504, 518, 523, 527
Ground Validation (GV): 142, 143, 147, 151–153, 163, 278, 314, 518, 529

hail (*or* hailstone): 4, 6, 8, 9, 15, 16, 26, 27, 95, 110, 136, 150, 203, 211, 236, 251, 289–292, 295
Hanssen-Kuipers Skill Score (KSS): 426
Heidke Skill Score (HSS): 426, 437
Hydro Estimator (HE): 197

Hydrologic Ensemble Prediction Experiment (HEPEX): 420
hydrometeor classification: 251
hydrometeor: 6, 11, 13, 16, 17, 19, 50, 84, 110, 132–136, 139, 142, 150, 154, 158–160, 196, 199, 235, 251, 252, 286, 287, 289–291, 294, 296, 315, 321, 477, 499, 510
hydrophone: 344–352

ice crystal orientation: 288–292
ice crystal shape: 289–295
ice crystal velocity: 291, 292
ice water content (IWC): 286, 293, 303, 304, 307–310
imaging distrometer: 4
INCA system: 392–415
Indoor User Terminal (IUT): 10, 11, 12
inhomogeneous beam filling: 498–501, 503, 507, 523, 527

Joss-Waldvogel disdrometer (JW, JWD): 61, 84, 96, 234, 243, 248, 254, 257, 326, 327, 334, 336, 348

kinetic energy (of drops): 91, 95

LAMI model: 479–490
lifted condensation level (LCD): 394, 401
light extinction distrometer: 5
light precipitation: 60, 62, 70, 73, 75, 78, 79, 288
Lightning Imaging Sensor (LIS): 142
lightning: 198, 202
line scan camera: 5, 6, 8, 10–13, 19, 20, 28, 235, 265, 271
liquid water content (LWC): 240, 245, 293, 304, 321, 323
liquid water path (lwp): 173–175
low-profile 2DVD: 9, 20–22, 29

MAP D-PHASE: 420
maritime cloud: 208–211
mass density (of ice crystals): 290
mass-weighted mean diameter: 240, 245, 246, 249, 260, 276

matrix camera distrometer: 5
Meteorological Operational satellite (MetOP): 139, 147
Meteosat Second Generation (MSG): 172, 175, 183, 209, 213, 397
Method for Object-based Diagnostic Evaluation (MODE): 439
microphysical retrieval: 211, 319–324, 328
Mie scattering: 110, 155, 160, 242
mixed phase precipitation (*or* mixed precipitation): 6, 25–27, 150, 153, 157, 160, 208, 209, 477–490
morphing: 161, 439
multiple deterministic realization reference: 374–379, 380–384
multispectral analysis/data: 171–190, 196
Mutual Information Skill Score: 370, 378, 385

neighborhood verification: 433–435, 448, 449
NGI Geonor T-200B: *see* vibrating-wire gauge
NRL blended technique: 203, 204

object oriented verification method: 391, 399, 410, 438–440, 455, 461–466, 525
Odds Ratio (OR): 426, 427
Odds Ratio Skill Score (ORSS): 427
optical array probe distrometer: 5
optical gauge: 61
optical path: 9, 16, 22, 28
orographic effect (on precipitation): 121, 124, 152, 393, 398, 409, 410, 412–415, 443, 523
Outdoor Electronics Unit (OEU): 10–12
overshooting (of precipitation): 110, 496, 503, 507, 510, 525

particle size distribution (PSD): 136, 145, 146, 150–160, 173, 219–223, 228, 286, 296, 310
passive (acoustic detection): 345, 350
passive (algorithm, retrieval; radiometer, sensor, sounder, technique): 134–145, 150–154, 159–163, 196, 198, 200, 202, 221, 228, 314, 496
path integrated attenuation (PIA): 117, 221–228
Peirce Skill Score (PSS): 426, 443–446
piezoelectricity: 87, 88
Plan Position Indicator (PPI): 117, 121, 252, 255–258, 268, 269, 274, 351, 520
polarimetric covariance matrix: 238
polarization: 19, 24, 110, 112, 123, 137, 138, 141, 150, 212, 224–227, 234–241, 247, 254, 260, 263, 271, 272, 276, 286, 287, 294, 303, 313–337, 351, 352, 477
pollution: 209, 211
precipitation discrimination: 171–190, 202, 475–490
precipitation nowcasting: 389–415
precipitation occurrence sensor system (POSS): 61, 65, 66, 70–73, 235, 254, 255
precipitation rate: *see* rain rate
precipitation suppression: 210, 211
precipitation type classification: *see* rain type classification
probabilistic forecasting (of precipitation): 376, 422, 426
probabilistic observational reference (POR): 374–376, 380–385
Probability Of Detection (POD): 183–188, 425, 426, 484–489
Probability Of False Detection (POFD): 183–188, 426
Proportion Correct (PC): 425
Pyro-cloud: 211, 212

quality control: 8, 101–123, 142, 519, 520
quantitative precipitation estimate (QPE): 419–448, 498, 527, 528
quantitative precipitation forecast (QPF): 419–448
quantization: 14, 16, 265, 271

radar (adjustment): 461, 494–512
radar (C band): 110, 234, 236, 247–249, 251–253, 259, 268, 269, 276, 278, 315, 316, 477–479, 489, 504, 528

radar (calibration): 155, 157, 251, 257,
    268, 269, 314, 352, 457, 494, 495,
    500–502, 506, 507, 523
radar (Doppler): 65, 109, 110, 114, 115,
    123, 151, 254, 286, 287, 289, 296,
    302–307, 314, 319, 325, 410, 415,
    478, 504, 517
radar (dual polarized, polarimetric):
    5, 19, 23, 24, 27, 29, 123, 138, 141,
    153, 234–238, 243, 247, 250–252,
    257, 259, 260, 263, 268, 271, 273,
    276, 278, 286, 287, 314–337, 351,
    352, 360, 477, 478
radar (Ka band): 133, 145, 146, 148,
    151, 163
radar (Ku band): 133, 134, 145, 146,
    148, 151, 163, 503, 508
radar (micro rain radar): 235, 254
radar (S band): 116, 234, 236, 249, 251,
    293, 315, 322, 323
radar (W band): 133, 134
radar (X band): 65, 110, 116, 117, 234,
    236, 237, 247–249, 251, 254, 276,
    314–330, 337, 351–354, 360, 477, 528
Rain Area Delineation Scheme (RADS):
    183–189
rain propagation effects: 236
rain rate (rainfall rate, precipitation
    rate): 5, 6, 11, 16, 18, 22–24, 26, 35,
    39–42, 50–53, 57, 60–68, 76, 77, 94,
    114, 116, 121, 133, 134, 138,
    140–146, 148, 153–159, 161, 163,
    197, 200, 203, 204, 206, 222, 234,
    237, 238, 241–244, 246, 247, 250,
    252, 253, 255–259, 269, 271, 276,
    278, 286, 315, 316, 320, 322, 325,
    333–335, 345–350, 354–360, 402,
    404, 406, 411, 434–436, 460, 512,
    520–522
rain type classification: 276, 358, 481,
    489, 521
raindrop oscillation: 232, 235, 236, 261,
    262, 265, 266, 276
raindrop oblateless: 11, 15, 17–20, 27,
    50, 94, 238, 247, 248, 265–269, 278,
    287, 288, 297, 314, 321
raindrop orientation: 19, 20, 26,
    234–242, 269–271, 274

raindrop shape (or drop shape): 6, 16, 19,
    24, 27, 28, 50, 94, 153, 154, 205,
    234–248, 263–269, 278, 287, 320–323
raindrop vertical velocity: 11, 22, 25, 27
rainfall rate: *see* rain rate
rainfall retrieval: 183, 197, 199, 200,
    204, 313–337
Range-Height Indicator (RHI):
    255, 257
Rayleigh scattering: 156, 238–250
Relative Operating Characteristic
    (ROC): 184–188, 426, 435, 436
resonant frequency: 35, 345
Rosenfeld-Lensky Technique (RLT):
    207–212

scattering amplitudes: 28, 238, 240, 320
scattering: 25, 28, 70–72, 110, 135–139,
    142, 152, 155, 157, 174, 179,
    197–200, 238, 240–242, 246, 248,
    250, 254, 286, 290, 293–295, 319,
    320, 323, 351, 497, 499, 501, 504,
    511, 523
Semi-Analytical Cloud Retrieval
    Algorithm (SACURA): 174, 175
Sensor Unit (SU): 7–10, 20–22, 28, 29
shape parameter: 222, 240, 243,
    246–248, 258, 308, 309, 321
snow, snowflake: 4–6, 13, 16, 23–29,
    44, 50, 53, 56, 57, 61–67, 76, 109,
    110, 132, 136–138, 146, 149, 150,
    152, 153, 157, 160, 163, 200, 202,
    251, 258, 286, 289, 404, 475–490,
    496, 497, 501, 510, 521
solid precipitation: 27, 46, 56, 202, 476,
    481, 482
spectral polarimetry: 285–310
Spinning Enhanced Visible and Infrared
    Imager (SEVIRI): 172, 175, 176,
    178, 180, 183, 188, 190, 209, 213
stability (of radar): 494
Structure Amplitude Location (SAL)
    method: 440

temperature (brightness): 135, 137, 138,
    140, 149, 158, 159, 161, 175, 178,
    179, 185, 187, 197, 206, 212, 223,
    224, 228, 496

temperature (effect on precipitation measurement): 9, 35, 38, 47–50, 54–57, 67, 72
temperature (effect on prediction): 391–397, 401, 403, 411–415, 422, 455, 456, 467, 476–478, 480–482, 488
temperature (effect on radar): 108, 109, 118–120
Terra-MODIS: 174
Threat Score (TS): 426
tipping bucket rain gauge (tipping-bucket): 9, 18, 23, 49, 55, 60, 61, 67, 75, 79
total precipitation forecast: 457, 459
total precipitation sensor (TPS): 62, 67, 80
TRMM Microwave Imager (TMI): 138, 141–143, 201, 213, 214, 220, 225–228
TRMM Multisatellite Precipitation Analysis (TMPA): 204
TRMM Precipitation Radar (TRMM PR, TPR): 135, 139, 140, 142, 145, 155, 200, 201, 213, 395, 456–459, 497, 500, 503, 508, 517, 521, 524, 529
Tropical Rainfall Measuring Mission (TRMM): 134, 138, 140–143, 147, 151, 153, 155, 157–159, 162, 163, 196, 199–201, 203, 204, 206, 207, 211, 220, 222–225, 228, 457–459, 495–512, 517–519, 522–524, 529
True Skill Statistics (TSS): 426

uncertainty (in measurement): 60, 61, 64–66, 69, 73, 79, 106–111, 113, 114, 119, 368, 369
uncertainty (in prediction): 368–370, 374, 375, 379–386, 420, 422
uncertainty (in radar retrieval): 107, 248, 314, 318, 332, 337, 457, 467, 469, 494, 505–507, 522, 526
uncertainty (in satellite retrieval): 140, 152–159, 219, 223, 224, 228, 496, 522

variability (in drop size distribution): 219, 242, 257, 276, 321

variability (in precipitation): 115, 131, 132, 143–145, 164, 350, 369, 384, 428, 454, 495, 497, 503, 508, 517, 518, 521, 526, 527
variability (in radar reflectivity): 523
verification (of precipitation forecast): 400, 407–410, 415, 419–448, 453–469, 475–490, 522, 527
vibrating-wire gauge (or NGI Geonor T-200B): 33–39, 41, 42, 50–53, 55–57
visibility (versus precipitation): 62, 63, 73, 75–77, 79
visibility (versus radar): 117, 118, 122, 398, 499, 518, 523, 525
visibility map: 117, 118, 122
visual verification: 455
VOLTAIRE project: 515–530
VRG101: 62–65, 69–73, 77–80

water content (liquid or ice): 16, 135, 150, 156, 196, 222, 226, 227, 240, 245, 246, 286, 293, 303, 304, 307–310, 321–323
water equivalent of snow: 26, 64, 476
Weighted Multiple Regression (WMR): 500
wide precipitation band: 516
wind (effect on acoustic measurement): 344, 345, 348, 350, 353–355, 359
wind (effect on prediction): 391–399, 401, 409, 410, 414, 415, 433, 443, 456
wind (effect on polarimetry): 287, 239, 263, 264, 271, 278, 287, 290, 296, 297, 300–304, 310
wind (effect on precipitation measurement): 9, 11, 13, 20, 29, 38, 42, 43, 46–50, 52, 53, 57, 60, 65, 67, 71–73, 94, 270, 271, 386
wind (effect on satellite estimations): 173, 202, 224, 227
wind (retrieval from satellite): 136, 138, 139

zero degree isotherm (or freezing level): 23, 120, 160, 286, 476, 479, 481–490